The Nuclear Envelope

D1744465

EXPERIMENTAL BIOLOGY REVIEWS

Series Advisors:

D.W. Lawlor
AFRC Institute of Arable Crops Research, Rothamsted Experimental Station, Harpenden, Hertfordshire AL5 2JQ, UK

M. Thorndyke
School of Biological Sciences, Royal Holloway, University of London, Egham, Surrey TW20 0EX, UK

Environmental Stress and Gene Regulation
Sex Determination in Plants
Plant Carbohydrate Biochemistry
Programmed Cell Death in Animals and Plants
Biomechanics in Animal Behaviour
Cell and Molecular Biology of Wood Formation
Molecular Mechanisms of Metabolic Arrest
Environment and Animal Development: genes, life histories and plasticity
Brain Stem Cells – SEB Symposium Series Vol. 53
Endocrine Interactions of Insect Parasites and Pathogens
Vertebrate Biomechanics and Evolution
Osmoregulation and Drinking in Vertebrates
Host-Parasite Interactions

The Nuclear Envelope

Edited by

D. E. EVANS
Research School of Biological and Molecular Sciences, Oxford Brookes University, Oxford, UK

C. J. HUTCHISON
Department of Biological Sciences, University of Durham, Durham, UK

J. A. BRYANT
Department of Biological Sciences, University of Exeter, Exeter, UK

BIOS Scientific Publishers
Taylor & Francis Group

© Garland Science/BIOS Scientific Publishers, 2004

First published 2004

All rights reserved. No part of this book may be reprinted or reproduced or utilised in any form or by any electronic, mechanical, or other means, now known or hereafter invented, including photocopying and recording, or in any information storage or retrieval system without permission in writing from the publishers.

A CIP catalogue record for this book is available from the British Library.

ISBN 0 41534 645 2

Garland Science/BIOS Scientific Publishers
4 Park Square, Milton Park, Abingdon, Oxon, OX14 4RN, UK and
270 Madison Avenue, New York, NY 10016, USA
World Wide Web home page: www.bios.co.uk

Garland Science/BIOS Scientific Publishers is a member of the Taylor & Francis Group.

Distributed in the USA by
Fulfilment Center
Taylor & Francis
10650 Toebben Drive
Independence, KY 41051, USA
Toll Free Tel.: +1 800 634 7064; E-mail: taylorandfrancis@thomsonlearning.com

Distributed in Canada by
Taylor & Francis
74 Rolark Drive
Scarborough, Ontario M1R 4G2, Canada
Toll Free Tel: +1877 226 2237; E-mail: tal_fran@istar.ca

Distributed in the rest of the world by
Thomson Publishing Services
Cheriton House
North Way
Andover, Hampshire SP10 5BE, UK
Tel: +44(0)1264 332424; E-mail: salesorder.tandf@thomsonpublishingservices.co.uk

Production Editor: Andrew Watts
Typeset by Saxon Graphics Ltd, Derby, UK
Printed by Cromwell Press, Trowbridge, UK

Cover image courtesy of Martin Goldberg, University of Durham, UK

Contents

Contributors

Ali, G.S., Department of Biology and Program in Cell and Molecular Biology, Colorado State University, Fort Collins, Colorado 80523, USA

Allen, S., Department of Structural Cell Biology, Paterson Institute for Cancer Research, Christie Hospital NHS Trust, Manchester, M20 9BX

Barron, Y.D., Department of Plant Pathology, Cornell University, 334 Plant Science Building, Ithaca, NY 14850, USA

Brandizzi, F., Department of Biology, 112 Science Place, University of Saskatchewan, Saskatoon SK S7N 5E2 Canada

Broers, J.L.V., Dept. of Molecular Cell Biology, Research Institutes CARIM and GROW, University of Maastricht, 6200 MD Maastricht, The Netherlands

Burke, B., University of Florida, Department of Anatomy and Cell Biology, 1600 Archer Road Gainesville, Florida L32610, USA

Carvalho, M., Department of Plant Pathology, Cornell University, 334 Plant Science Building, Ithaca, NY 14850, USA

Clarke, P.R., 1 Research Centre, Level 5, Ninewells Hospital and Medical School, University of Dundee, Dundee DD1 9SY, Scotland

Collas, P., Institute of Medical Biochemistry, University of Oslo, PO Box 1112 Blindern, Oslo 0317 Norway

Crisp, M., University of Florida, Department of Anatomy and Cell Biology, 1600 Archer Road Gainesville, Florida L32610, USA

Debela, M.H., Research School of Biological and Molecular Sciences, Oxford Brookes University, Headington Campus, Oxford OX3 0BP, UK

Dreger, M., Freie Universität Berlin, Inst. f. Chemie – Biochemie, AG Neurochemie, Thielallee 63, 14195 Berlin, Germany

Drummond, S., Department of Structural Cell Biology, Paterson Institute for Cancer Research, Christie Hospital NHS Trust, Manchester, M20 9BX

Evans, D.E., Research School of Biological and Molecular Sciences, Oxford Brookes University, Headington Campus, Oxford OX3 0BP, UK

Fromm, H., School of Biology, University of Leeds, Leeds LS2 9JT, UK

Galbraith, D.W., Department of Plant Sciences, University of Arizona, Tucson AZ 85721 USA

Goldberg, M., University of Durham, School of Biological and Biomedical Sciences, University of Durham, Science Laboratories, South Road, Durham, DH1 3LE, UK

Golovkin, M., Department of Biology and Program in Cell and Molecular Biology, Colorado State University, Fort Collins, Colorado 80523, USA

Hutchison, C.J., School of Biological and Biomedical Sciences, University of Durham, Durham DH1 3LE UK

Irons, S.L., Research School of Biological and Molecular Sciences, Oxford Brookes University, Headington Campus, Oxford OX3 0BP, UK

Jackson, D.A., Department of Biomolecular Sciences, UMIST, PO Box 88, Manchester, M60 1QD, UK

Landsverk, H.B., Institute of Medical Biochemistry, University of Oslo, PO Box 1112 Blindern, Oslo 0317 Norway

Lazarowitz, S.G., Department of Plant Pathology, Cornell University, 334 Plant Science Building, Ithaca, NY 14850, USA

Lee, K.K., Stowers Institute for Medical Research, 1000 East 50th St., Kansas City MO 64110, USA

Määttä, A., School of Biological and Biomedical Sciences, University of Durham, Durham DH1 3LE UK

Martins, S.B., Institute of Medical Biochemistry, University of Oslo, PO Box 1112 Blindern, Oslo 0317 Norway

Maske, C.P., Sir William Dunn School of Pathology, South Parks Road, Oxford OX1 3RE, UK

McGarry, R.C., Department of Plant Pathology, Cornell University, 334 Plant Science Building, Ithaca, NY 14850, USA

Meier, I., Plant Biotechnology Center and Dept. of Plant Biology The Ohio State University 244 Rightmire Hall, 1060 Carmack Road, Columbus, OH 43210, USA

Morris, G.E., Biochemistry Group, North East Wales Institute, Mold Road, Wrexham, LL11 2AW, UK

Otto, H., Freie Universität Berlin, Inst. f. Chemie – Biochemie, AG Neurochemie, Thielallee 63, 14195 Berlin, Germany

Patel, S., Plant Biotechnology Center and Dept. of Plant Biology The Ohio State University 244 Rightmire Hall, 1060 Carmack Road, Columbus, OH 43210, USA

Ramaekers, F.C.S., Dept. of Molecular Cell Biology, Research Institutes CARIM and GROW, University of Maastricht, 6200 MD Maastricht, The Netherlands

Reddy, S.N., Department of Biology and Program in Cell and Molecular Biology, Colorado State University, Fort Collins, Colorado 80523, USA

Rose, A., Plant Biotechnology Center and Dept. of Plant Biology The Ohio State University 244 Rightmire Hall, 1060 Carmack Road, Columbus, OH 43210, USA

Salina, D., University of Florida, Department of Anatomy and Cell Biology, 1600 Archer Road Gainesville, Florida L32610, USA

Taleb, F., Department of Plant Sciences, Tel Aviv University, Tel Aviv, Israel

Vaux, D.J., Sir William Dunn School of Pathology, South Parks Road, Oxford OX1 3RE, UK

Watson, M.D., School of Biological and Biomedical Sciences, University of Durham, Durham DH1 3LE UK

Wilson, K.L., Department of Cell Biology, Johns Hopkins University School of Medicine, 725 N. Wolfe Street, Baltimore MD 21205 USA

Worman H.J., Departments of Medicine and of Anatomy and Cell Biology, College of Physicians and Surgeons, Columbia University, New York, NY 10032, USA

Zhang, C., Department of Cell Biology and Genetics, Peking University, Beijing 100871, China

Abbreviations

ABD	actin-binding domain
AhR	aryl hydrocarbon receptor
AML	acute myelogenous leukaemia
AN	annulate lamellae
CC	commitment complex
CH	calponin homology
CID	collision-induced dissociation
CLSM	confocal laser scanning microscopy
CP	coat protein
EDMD	Emery–Dreifuss muscular dystrophy
EM	electron microscopy
ER	endoplasmic reticulum
ESI	electrospray ionization
FESEM	field effect scanning electron microscopy
FISH	fluorescence *in situ* hybridization
FO	fibrous organelle
FRAP	fluorescence recovery after photobleaching
GCL	germ cell less
GFP	green fluorescent protein
HAT	histone acetyl transferase
HDAC	histone deacetylase complex
HMT	histone methyltransferase
IBB	importin β binding
IEF	isoelectric focusing
IF	intermediate filament
INM	inner nuclear membrane
LAP	lamina-associated polypeptide
LBR	lamin B receptor
LCR	locus control region
MA	mitotic apparatus
MALDI	matrix-assisted laser desorption ionization
MAP	microtubule-associated protein
MAR	matrix attachment region
MCM	mini-chromosome maintenance
MDS	myelodysplastic syndrome
MEF	mouse embryonic fibroblast
MP	movement protein
MT	microtubule
MTOC	microtubule organizing centre
NE	nuclear envelope
NES	nuclear export signal/sequence
NIS	nuclear import signal/sequence
NLS	nuclear localization signal

NMD	nonsense mediated decay
NSP	nuclear shuttle protein
NPC	nuclear pore complex
ONM	outer nuclear membrane
ORC	origin recognition complex
PKA	protein kinase A
POM	pore membrane
PSD	post source decay
rER	rough endoplasmic reticulum
SAR	scaffold attachment region
SEM	scanning electron microscopy
SIR	Silent Information Regulatory
SR	spectrin repeat
ss	single strand
SSDB	sequence-specific DNA binding
TMD	transmembrane spanning domain
WGA	wheat germ agglutinin

Preface

The nuclear envelope is one of the defining architectural elements in eukaryotic cells. Its components form a selective barrier that separate transcriptional and translational events and which allows eukaryotes to maintain additional levels of gene control (chapter 1). Recent evidence has shown that a key element of metazoan nuclear envelopes, that nuclear lamina, is also critical for longevity. The nuclear lamina is an intermediate filament scaffold that supports the nuclear membrane and which determines the size and shape of the nucleus. It is also a dynamic structure, which is at the heart of the breakdown and reassembly of the nuclear envelope during mitosis (chapters 10 and 12). When the proteins that build this structure, the lamins, are mutated the result is that cells in certain tissues age prematurely leading to diseases ranging from white fat disorders and diabetes to heart disease (chapters 3 and 4). Given the pivotal role of this structure it is curious that not all eukaryotes appear to possess a lamina. This book sets out to investigate the similarities and differences in the structure and function of the nuclear envelope in plants, animals and fungi. The book arose from a conference sponsored by the Society for Experimental Biology and held in Durham in 2003 in which researchers from very different backgrounds gathered to discuss whether, principally in plants and animals the nuclear envelopes are different. Through the proceeding chapters the reader will find that some structures within the nuclear envelope, principally the nuclear pore complexes, are remarkably similar between the different eukaryotic kingdoms. In keeping with this the mechanisms of transport of proteins and RNA across nuclear pores are also very similar (chapters 6, 7, 9 and 17). In contrast, the nuclear envelope itself displays fascinating differences when plants and animals are compared. Plants apparently do not have lamins although they are fully capable of targeting mammalian nuclear membrane proteins, which bind to lamins, to the inner nuclear membrane (chapter 14). Perhaps they have evolved a completely different solution to the same problem of providing a scaffold to support the nuclear membrane and one highlight of the meeting was the discovery of coiled-coil proteins in arabidopsis that could easily fulfil the function of lamins (chapter 5). This raises the question as to whether dynamic behaviours in plants and animals operate through different or convergent mechanisms (chapter 13). Finally, the nuclear envelope cannot be treated as an entity that is isolated from the rest of the cell. This leads back to its key function which is communication and regulation of gene expression. This is most commonly thought about in terms of transport through nuclear pores. However, rapidly emerging evidence now shows that the lamina is directly linked to the cytoskeleton allowing for the transmission of force between these two elements (chapters 10 and 16). In addition the nuclear envelope is an important relay station for signalling molecules such as calcium or GTPases (chapters 11 and 18). We hope that this book will both inform and stimulate its readers and will promote closer collaboration between researchers working on plant and animal models.

The nuclear envelope: a comparative overview

David E. Evans, John A. Bryant and Chris Hutchison

The eukaryota, including animals, plants and fungi all share a separation of nucloplasm and genetic material from the cytoplasm effected by a membrane structure, the nuclear envelope. The nuclear envelope has great significance in the maintenance of the structure of the nucleus and in the regulation of its function. The structure of the nuclear envelope and transport across it are highly organized and many of its functions and properties appear to be conserved across the kingdoms. However, there are also key differences; for instance, during cell division, the nuclear envelope of most higher organisms breaks down resulting in what is known as an 'open' mitosis. However, in contrast, yeast and other fungi undergo a closed mitosis where the nuclear envelope does not break down and chromosome segregation and separation occurs within the nucleoplasm. Similarly, while animals and fungi possess discrete centrosomes (microtubule organizing centres, known as spindle pole bodies in yeast) to which microtubules are attached, such structures are absent in plants and the entire NE appears able to act to bind and organize microtubules.

The properties of various aspects of the nuclear envelope have been most intensively studied in mammals, for which the most complete information on transport and protein composition exists. Other metazoans that have been studied, and for which comparative data on the nuclear envelope exist, include amphibians, birds, fish, flies and the model nematode *Caenorhabditis elegans*. Yeast has been used extensively as a model organism for studies on nuclear pores, transport and targeting. While knowledge of the nuclear envelope in plants remains limited, sufficient information is available to show conservation of structure and function with important differences in protein composition.

The purpose of this chapter is to introduce the nuclear envelope and to highlight similarities and differences in structure, function and protein composition between the kingdoms as an introduction to the detailed understanding of the nuclear envelopes of a variety of species presented subsequently in this volume.

The Nuclear Envelope, edited by D.E. Evans, C. Hutchison & J.A. Bryant.
© 2004 Garland Science/BIOS Scientific Publishers

1. A general description of the nuclear envelope

The nuclear envelope is made up of three membrane regions (*Figure 1*). The first is an outer membrane, connected to and closely associated with rough endoplasmic reticulum and often coated with ribosomes. This outer nuclear envelope is associated with intermediate filaments of the cytoskeleton. The second region is the inner membrane, separated from the outer membrane by a lumenal space of 20–40 nm. It is intimately associated with the filamentous proteins of the nuclear lamina. The third membrane region, termed the pore membrane (Wozniak and Blobel, 1992), is a domain linking the inner and outer membranes through multi-protein pores, the nuclear pore complexes (NPCs). While the outer NE appears to have a protein composition very similar to that of the rough endoplasmic reticulum to which it is connected, the inner nuclear envelope (NE) and the pore membrane have unique protein compositions. The pore membrane contains membrane integral protein components of the NPC. Any membrane proteins entering the inner NE have to pass through this membrane to reach their destination and are retained in the inner NE by interaction with proteins of the nucleoplasm (Chapters 6–8).

While nuclear envelopes are frequently depicted as a spherical, or near-spherical structure, the NE is frequently invaginated with deep folds extending into the nucleoplasm in both plants and animals (Collings *et al.*, 2000; Fricker *et al.*, 1997). Micrographs of plant cells show the nuclei to be frequently confined to a narrow band of cytoplasm between the vacuole and cell wall, though plant cells capable of dividing generally (but by no means always) lack such large vacuoles.

2. Nuclear envelope formation

Lower eukaryotes, including yeast and other fungi, undergo a 'closed' mitosis in which the nuclear envelope does not break down and chromosome replication, segregation

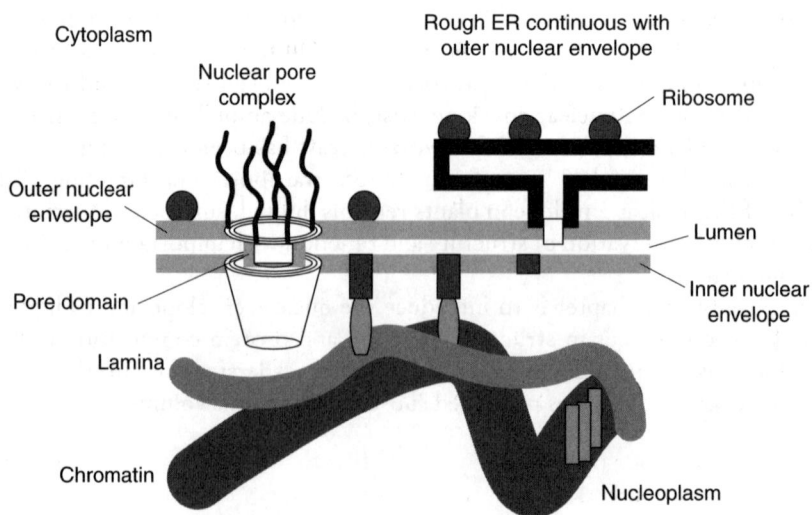

Figure 1. A summary structure of the nuclear envelope, including the major membrane domains.

and separation occur within the confines of the nucleoplasm. Here, new membrane and pores are added to an existing structure. Most higher eukaryotes, including animals and higher plants, undergo an open mitosis as the mitotic spindle is essentially a cytoplasmic structure. In these species NE breakdown permits the kinetochore microtubules to contact the chromosomes and initiate their segregation.

In those species that undergo an open mitosis, this structure breaks down in its entirety in prophase to become part of the membranes of the mitotic apparatus and then reforms in telophase. Questions of the mechanism of nuclear envelope breakdown, the destination of its components during division and of the mechanism of NE reformation are all addressed subsequently in this volume (Chapters 10–14).

3. Nuclear pore complexes and transport across the nuclear envelope

NPCs in all three kingdoms display eight-fold symmetry and penetrate the lumen of the NE, forming a pathway for the diffusion of small molecules and the active transport of large ones (Cronshaw *et al.*, 2002). The cytoplasmic face of NPCs is decorated with rod-like filaments extending into the cytoplasm, while the nucleoplasmic face has a highly organized filamentous basket-like structure. Proteomic analysis suggests a roughly equal number of around 30 nucleoporins (the proteins of the nuclear pore) in both animals and yeast (Cronshaw *et al.*, 2002; Chapters 6–8). However, the mammalian NPC is considerably larger than that in yeast (125 MDa, 1450 x 800 A, compared with 50–80 MDa, 960 x 350 A in yeast). While some of the proteins of animal and plant NPCs are glycosylated, those from yeast are not. Plant nucleoporins bear five *N*-acetylglucosamine (GlcNAc) residues, while those from vertebrates, although glycosylated with the same sugar, bear only one (Heese-Peck *et al.*, 1995). In plants and animals undergoing an open mitosis, NE reformation commences with the recruitment of nucleoporins to chromatin stimulated by RanGTP (Zhang *et al.*, 2002; Chapter 11). The mechanism for the generation of new nuclear pores in the closed mitosis of yeast is less clear as they are inserted into existing membrane but RanGTP is once again involved, as it is in the recruitment of additional pores to existing envelope (Ryan *et al.*, 2003; Chapter 11).

The transport function of the nuclear pores is conserved in all kingdoms. Transport occurs at a remarkable rate – ten import and ten export events per second per pore, or 600 000 ribosomes transported per minute per nucleus in HeLa cells (Gorlich and Kutay, 1999; Macara, 2001; Mattaj and Englmeier, 1998). Transport through the pore appears to be regulated not by a physical barrier, but by the activation energy required to enter the gate (Chapters 6–8). Transport in plants, animals and yeast requires the activity of specific transport proteins and is directed by RanGTP activity (Chapters 6–9). Most proteins transported across the nuclear envelope possess a nuclear localization signal (NLS). Three major types of NLS identified in animals also function in plants (Smith and Raikhel, 1999). Import to the nucleus in animal cells and in yeast involves the activity of importins α and β. Plants also have homologues of both importins; plant importin α for instance interacts with both animal and yeast NLS to facilitate transport in chimeric assays. In addition to the standard pathways for protein import it is also becoming clear that proteins without any obvious nuclear localization can gain entry to the nucleus. It is probable that some of these are piggy-backed into the nucleus because the protein in question forms a complex or association with a protein that does have a conventional NLS. In contrast to these proteins that enter the nucleus without an NLS it is also clear that proteins possessing an NLS may not in fact enter the nucleus. Thus, some proteins that possess NLSs are partitioned between nucleus and cytosol rather than being completely taken up into the nucleus. It is likely that some form of post-translational modification marks a protein, albeit a protein that is already 'addressed' to the nucleus, for actual nuclear uptake. Thus we need to understand the role of the cargo in the nuclear uptake process.

4. Nuclear lamina

Underlying the NE of many (but not all) organisms, and directly associated with it via lamin-binding proteins embedded in the INE, is a well-ordered lamina of intermediate-type filament proteins (Hutchison, 2002). This lamina, made up of proteins of the lamin family (lamins A, B and C; see Chapters 10–14) forms a two-dimensional (and in some instances a three-dimensional) structure with both structural and regulatory properties. A nuclear lamin structure is found in all metazoans. The most primitive metazoans have a single lamin polypeptide, whereas vertebrates have seven lamins in two families, A and B (see *Table 1*). No lamin genes have been identified in the genomes of yeast or *Arabidopsis* (Hutchison, 2002; Meier, 2001) and they appear to be absent from plants and fungi. The lamins are members of the type V intermediate filament family of proteins. Electron microscope studies show that they form a regular 2D orthogonal arrangement interconnecting adjacent NPCs. The position of the NPCs is governed by interaction with the lamina as deletion of the lamina results in clustering of the pores in the membrane (LenzBohme *et al.*, 1997; Liu *et al.*, 2000). The importance of lamins in mammals is illustrated by the variety of disease states resulting from their absence or mutation (Chapters 3 and 4). These laminopathies result in the failure of cells of certain tissue types. As well as determining nuclear shape and size, the lamins also are involved in organization of integral membrane proteins in the NE and their association with centres of DNA replication and splicing sites suggests that they are regulators of gene expression (Chapters 15–20). Thus laminopathies may be the result of altered gene expression, or due to lack of mechanical strength in the nucleus or both.

The absence of lamin homologues from plant and fungal nuclei suggests that their functions are either redundant in these species or are undertaken by other proteins. Masuda and co-workers (Masuda *et al.*, 1997, 1999) identified a novel 134 M_r protein containing a central helical coiled-coil domain and with sequence similarities to myosin, tropomyosin and intermediate filaments that is associated with the NE and may fulfil some lamin-like functions. The presence of large filamentous proteins has also been suggested in plant nuclei by electron microscopy, but these do not appear as the ordered meshwork seen in animals (Beven *et al.*, 1991; Minguez and Delaespina, 1993; McNulty and Saunders, 1992). It is possible that some of the mechanical/structural functions of the lamins are not required in organisms with a cell wall because their nuclei are sufficiently protected from mechanical stress to survive without a lamina. In some instances, plant nuclei are highly mobile within the cell and observation of nuclei migrating along root hair cells reveals a remarkable degree of flexibility of the structure. Plant nuclei are also frequently polyploid as a result of DNA endoreduplication and vary greatly in size as a consequence. Microscopy often reveals them to be flattened against the plasma

Table 1. *Lamin polypeptides in different phylogenies*

Organism	Gene	Lamin peptides produced
Mammal	*LMNA* gene	A-type lamins[1] A, AD10, C, C2
	LMNB gene	B-type lamins[1] B1, B2, B3
Avians	Lamin A gene	A-type lamin A
	Lamin B gene	B-type lamin B1, B2[1]
Amphibians	Lamin A gene	A,C
	L gene family (Li – Liv)	Li, Lii, Liii, Liv
Arthropods	Lamin C gene	A type lamin C
	DmO	B-type lamin DmO
Nematodes	*LMN-1*	B-type lamin LMN-1

[1]Multiple forms produced by alternative splicing.
(After Hutchison, 2002 and references therein.)

membrane/cell wall in a very narrow strip of cytoplasm compressed by the vacuole. The gene regulatory function of the lamins may also be taken over by other proteins; the Silent Information Regulatory (SIR) proteins of yeast, for instance, have a structural organization similar in part to the lamins and do not form filaments but are involved in gene silencing (Diffley and Stillman, 1989).

5. Proteins of the nuclear envelope

The proteins of the outer NE are considered to be equivalent to those of the rough endoplasmic reticulum to which the outer NE is directly connected. Those of the inner NE, however, are a specific subset of proteins that have passed through the nuclear pore membrane and are retained in the INE by their association with constituents of the nucleoplasm. *Table 2* summarizes currently described INE proteins. Most have been described in metazoans, including mammals, amphibia and fish. A family of three (LAP2, Emerin and Man1) contain a conserved domain, the LEM domain, at or near their N-termini. Apart from one INE-associated protein, MFP1 (see *Table 2*), much less is known about proteins resident in the INE of plants or fungi; however, interestingly, chimeric expression of the N-terminal 138 residues of the human lamin B receptor is targeted to the NE in plants (Irons and Brandizzi, 2003; Chapter 14) and yeast (Ellenberg *et al.*, 1997; Smith and Blobel, 1993) suggesting similar targeting and retention mechanisms are present. *Caenorhabditis elegans*, a metazoan having just one B-type lamin, has homologues of the three LEM domain

Table 2. Proteins of the inner nuclear envelope

Protein name	Characteristics	Function
Emerin	Membrane protein, Single pass membrane integral, INE. Contains a LEM[1] domain	Binds lamins A and C Altered phosphorylation associated with the cell cycle
Lamin B receptor	Membrane protein with homology to ERG4/ ERG24 family of sterol reductases. 8 transmembrane domains, located in INE with N-terminus in nucleoplasm	Anchors the lamina and heterochromatin to the INE. Also has sterol reductases activity. Phosphorylation by cdc2 kinase results in dissociation from lamina in cell division
LAP1 (lamina-associated polypeptide 1)	Type 2 (single pass) membrane protein located in the INE	Binds A and B type lamins; role in attachment of nuclear lamina
LAP2 (lamina-associated polypeptide 2)	Type 2 (single pass) membrane protein located in the INE. Contains a LEM[1] domain	Binds lamin B1 and chromatin. Binding regulated by phosphorylation during mitosis
MAN1	Three transmembrane domains, LEM[1] domain, located at INE	Binds chromatin; functions in chromosome segregation and cell division
Nurim	Multipass protein, 5 transmembrane domains	Large hydrophilic domain
Unc-84 (Matefin)	Single pass transmembrane protein described in *C. elegans*. Contains a SUN[2] domain	Structural protein bridging cytoskeleton and nucleus. Binding with nuclear material broken during NE disassembly
MFP1 (matrix filament attachment protein 1)	Described in plants; interacts with chromatin and the INE	Structural protein linking INE and nuclear matrix

[1]LEM domain: a 50-residue domain containing two parallel alpha-helices. Found in the INE proteins LAP2, Emerin and MAN1.
[2]SUN domain (Unc-84 domain). Found in Unc84; involved in nuclear anchoring.

proteins suggesting their conserved importance in nuclear envelope function in metazoans (Lee *et al.*, 2000).

6. Conclusion

The major structures and functions of the nuclear envelope are conserved in eukaryotes; however, the protein composition and interactions of the membrane with nuclear and cytoplasmic structures as well as its behaviour during cell division all show divergence. Subsequent chapters of this volume not only present current understanding of the nuclear envelope and its function across the kingdoms, but also facilitate the enhanced understanding of structure and function that results from detailed comparison.

References

Beven, A., Guan, Y.H., Peart, J., Cooper, C. and Shaw, P. (1991) Monoclonal-antibodies to plant nuclear matrix reveal intermediate filament-related components within the nucleus. *J. Cell Sci.* **98**: 293–302.

Collings, D.A., Carter, C.N., Rink, J.C., Scott, A.C., Wyatt, S.E. and Allen, N.S. (2000) Plant nuclei can contain extensive grooves and invaginations. *Plant Cell* **12**: 2425–2439.

Cronshaw, J.A., Krutchinsky, A.N., Zhang, W.Z., Chait, B.T. and Matunis, M.J. (2002) Proteomic analysis of the mammalian nuclear pore complex. *J. Cell Biol.* **158**: 915–927.

Diffley, J.F.X. and Stillman, B. (1989) Transcriptional silencing and lamins. *Nature* **342**: 24.

Ellenberg, J., Siggia, E.D., Moreira, J.E., Smith, C.L., Presley, J.F., Worman, H.J. and Lippincott-Schwartz, J. (1997) Nuclear membrane dynamics and reassembly in living cells: Targeting of an inner nuclear membrane protein in interphase and mitosis. *J. Cell Biol.* **138**: 1193–1206.

Fricker, M., Hollinshead, M., White, N. and Vaux, D. (1997) Interphase nuclei of many mammalian cell types contain deep, dynamic, tubular membrane-bound invaginations of the nuclear envelope. *J. Cell Biol.* **136**: 531–544.

Gorlich, D. and Kutay, U. (1999) Transport between the cell nucleus and the cytoplasm, *Ann. Rev. Cell Develop. Biol.* **15**: 607–660.

Heese-Peck, A., Cole, R.N., Borkhsenious, O.N., Hart, G.W. and Raikhel, N.V. (1995) Plant nuclear–pore complex proteins are modified by novel oligosaccharides with terminal N-acetylglucosamine. *Plant Cell* **7**: 1459–1471.

Hutchison, C.J. (2002) Lamins: Building blocks or regulators of gene expression? *Nature Rev. Mol. Cell Biol.* **3**: 848–858.

Irons, S.E. and De Brandizzi, F. (2003) The first 238 amino acids of the human lamin B receptor are targeted to the nuclear envelope in plants. *J. Exper. Bot.* **54**: 1–8.

Lee, K.K., Gruenbaum, Y., Spann, P., Liu, J. and Wilson, K.L. (2000) *C. elegans* nuclear envelope proteins emerin, MAN1, lamin, and nucleoporins reveal unique timing of nuclear envelope breakdown during mitosis. *Mol. Biol. Cell* **11**: 3089–3099.

LenzBohme, B., Wismar, J., Fuchs, S., Reifegerste, R., Buchner, E., Betz, H. and Schmitt, B. (1997) Insertional mutation of the Drosophila nuclear lamin Dm(0) gene results in defective nuclear envelopes, clustering of nuclear pore complexes, and accumulation of annulate lamellae. *J. Cell Biol.* **137**: 1001–1016.

Liu, J., Ben-Shahar, T. R., Riemer, D., Treinin, M., Spann, P., Weber, K., Fire, A. and Gruenbaum, Y. (2000) Essential roles for *Caenorhabditis elegans* lamin gene in nuclear organization, cell cycle progression, and spatial organization of nuclear pore complexes. *Mol. Biol. Cell* **11**: 3937–3947.

Macara, I.G. (2001) Transport into and out of the nucleus. *Microbiol. Mol. Biol. Rev.* **65**: 570.

Masuda, K., Xu, Z.J., Takahashi, S., Ito, A., Ono, M., Nomura, K. and Inoue, M. (1997) Peripheral framework of carrot cell nucleus contains a novel protein predicted to exhibit a long alpha-helical domain. *Experi. Cell Res.* **232**: 173–181.

Masuda, K., Haruyama, S. and Fujino, K. (1999) Assembly and disassembly of the peripheral architecture of the plant cell nucleus during mitosis. *Planta* **210**: 165–167.

Mattaj, I.W. and Englmeier, L. (1998) Nucleocytoplasmic transport: The soluble phase. *Ann. Rev. Biochem.* **67**: 265–306.

McNulty, A.K. and Saunders, M.J. (1992) Purification and immunological detection of pea nuclear intermediate filaments – evidence for plant nuclear lamins. *J. Cell Sci.* **103**: 407–414.

Meier, I. (2001) The plant nuclear envelope. *Cell. Mol. Life Sci.* **58**: 1774–1780.

Minguez, A. and Delaespina, S.M.D. (1993) Immunological characterization of lamins in the nuclear matrix of onion cells. *J. Cell Sci.* **106**: 431–439.

Ryan, K.J., McCaffery, J.M. and Wente, S.R. (2003) The Ran GTPase cycle is required for yeast nuclear pore complex assembly. *J. Cell Biol.* **160**: 1041–1053.

Smith, H.M. and Raikhel, N.V. (1999) Protein targeting to the nuclear pore. What can we learn from plants? *Plant Physiol.* **119**: 1157–1164.

Smith, S. and Blobel, G. (1993) The 1st membrane spanning region of the lamin-B receptor is sufficient for sorting to the inner nuclear-membrane. *J. Cell Biol.* **120**: 631–637.

Wozniak, R.W. and Blobel, G. (1992) The single transmembrane segment of Gp210 is sufficient for sorting to the pore membrane domain of the nuclear-envelope. *J. Cell Biol.* **119**: 1441–1449.

Zhang, C.M., Goldberg, M.W., Moore, W.J., Allen, T.D. and Clarke, P.R. (2002) Concentration of Ran on chromatin induces decondensation, nuclear envelope formation and nuclear pore complex assembly. *Euro. J. Cell Biol.* **81**: 623–633.

2

The nuclear envelope proteome

Mathias Dreger and Henning Otto

1. Introduction

The proteome is by definition the complete set of proteins of a cell at any defined moment. This definition includes in its widest sense all possible modifications of cellular proteins at any given time (Kahn, 1995; Wilkins *et al.*, 1996). Due to the lack of technical means to display and analyse the whole proteome at once, the term 'proteomics' which describes the analytical approach to investigate the proteome of a given system refers to a general concept rather than to any particular experiment. Among the experimental approaches currently used in proteomics, the mapping of the proteins of particular subcellular structures proves to be an increasingly successful concept which is applied to explore more and more defined subcellular structures that can be isolated and sufficiently purified (Dreger, 2003). Most of the recent subcellular proteomics studies have been performed at this mapping level. The results of such experiments are mainly qualitative, and at best semi-quantitative.

Therefore, more refined studies are required, which exceed this basic level of protein mapping and include a quantification of the results as well as the analysis of molecular interactions and post-translational modifications to appropriately describe a subcellular proteome. Optimally, advanced proteomics studies should relate such refined data on a subcellular proteome with a defined physiological state of the investigated system. For the nuclear envelope, this could for example mean identifying changes in protein composition and post-translational protein modifications related to the phases of the cell cycle. Although still rare at the moment (Dreger, 2003), such studies should clearly be the focus of future projects in subcellular proteomics.

In this chapter, we first briefly describe some structural aspects of the nuclear envelope, followed by a summary of the experimental approaches to map nuclear envelope proteins. In addition to approaches that apply modern protein chemical techniques including mass spectrometry, we will discuss strategies such as the screening of expression libraries designed to generate fusion proteins from cDNAs in combination with a fluorescent protein. Then, we will provide a short introduction to the proteins identified from nuclear envelope and lamina nuclear-pore-complex preparations. While substructures of the

nuclear envelope such as the nuclear pore complexes have been analysed in great detail
(Cronshaw *et al.*, 2002; Rout *et al.*, 2000), our knowledge of the proteins building the
remainder of the nuclear envelope, let alone their interactions still is, despite recent
advances, rather limited. Therefore, we will finally outline possible next steps and
necessary developments in nuclear envelope proteomics that might improve our understanding of the molecular composition and organization of the nuclear envelope in
different life stages of the cell.

2. Structure of the nuclear envelope

2.1 *Subcompartments of the nuclear envelope*

Eukaryotic cells are characterized by the sequestration of their genetic material and of
gene expression-related processes in the cell nucleus. The nuclear space is separated
from the remainder of the cell by the nuclear envelope, which can be considered, at
least in part, as a specialized domain of the endoplasmic reticulum (ER) (*Figure 1*). Its
two membranes, the outer nuclear membrane (ONM) and the inner nuclear
membrane (INM), join at the nuclear pores and enclose the perinuclear space that is
part of the ER lumen (Gant and Wilson, 1997). The outer nuclear membrane is part of
the rough endoplasmic reticulum (rER), which is indicated by ribosomes attached to
it. It appears to be different from other parts of the rER since some of its protein
constituents seem to be exclusively located at this membrane where they are involved
in positioning the nucleus within the cell by mediating contacts to the cytoplasmic
cytoskeleton (Zhen *et al.*, 2002). Such contacts are essential for developing organisms
since a disturbed nuclear migration can lead to severe developmental defects (Morris,
2000). The pore membrane (POM), joining outer and inner nuclear membranes at the
nuclear pores, is required by the cell for anchoring the nuclear pore complexes which
are large protein assemblies that control the exchange of macromolecules between
cytoplasm and nucleoplasm (Davis, 1995). Integral membrane proteins, which are
constituents of the INM have to pass both nuclear pore complex (NPC) and POM, to
reach their cellular location. The INM faces the nucleoplasm and, in higher
eukaryotes, is supported by the filaments of the nuclear lamina. INM proteins are
involved in structuring the chromatin. In particular, the heterochromatin found juxtaposed to the INM, is apparently positioned there by interactions with some of the
INM proteins (Furukawa, 1999; Ye *et al.*, 1997). The nuclear lamina of higher
eukaryotes, finally, is a very stable network of intermediary filaments formed from
A- and B-type lamins. It serves as a mechanical support of the nuclear membranes, but
apparently also acts as an organizing grid for the different molecular interactions at the
inner nuclear membrane that are required for the structural and functional organization of the whole nucleus (Hutchison, 2002). In its framework, INM proteins
interact with lamins as well as with chromatin components. An exact regulation of
such interactions is apparently strictly required for the cellular survival. Many
examples show that a disturbed balance of such interactions results in disturbed
nuclear organization and eventually apoptosis. Various processes including transcription, splicing or replication are affected by misregulated interactions around the
lamins (Bell *et al.*, 2001; Jenkins *et al.*, 1993; Spann *et al.*, 2002). A loss of lamin A or of
some of its interactions has to date been linked with at least seven different genetic
diseases ranging from muscular dystrophies to lipid dystrophies, the most recent

example being Hutchinson–Gilford progeria syndrome. Together, they are called laminopathies due to their common pathogenic culprit (Bonne *et al.*, 1999; Burke and Stewart, 2002; Eriksson *et al.*, 2003; Hutchison *et al.*, 2001).

2.2 *Cell cycle dependence of the nuclear envelope structure*

In higher eukaryotes that undergo an open mitosis, the nuclear envelope breaks down at the prophase of mitosis. This requires force both to rupture the nuclear envelope and, probably subsequently due to mixing of nucleoplasmic and cytoplasmic proteins, to disassemble the different molecular contacts between lamins, chromatin and inner nuclear membrane proteins (Beaudouin *et al.*, 2002). The nuclear membranes at that point lose their identity and disperse their proteins throughout the endoplasmic reticulum. After mitosis, the interphase interactions are re-established and the nuclear envelope forms in an ordered fashion around the decondensing chromatin. Such a drastic reconfiguration of macromolecular interactions seems to be mainly achieved by phosphorylation events that specifically occur during mitosis (Burke and Ellenberg, 2002; Gant and Wilson, 1997).

In contrast to higher eukaryotes, simpler eukaryotes like baker's yeast (*Saccharomyces cerevisiae*) lack a lamina. Also, their nuclear envelope retains its integrity throughout the whole cell cycle. This requires a controlled mechanism for separating both daughter nuclei without opening the nuclear envelope. While the molecular mechanisms involved in mitotic events at the nuclear envelope are rather well characterized, much less is known about changes throughout the interphase which are much more subtle and may in many cases just consist of changes in post-translational modifications of proteins. A few reports demonstrate, for example, the phosphorylation of nuclear envelope proteins during interphase (Collas *et al.*, 1997; Dreger *et al.*, 1999; Hennekes *et al.*, 1993; Stuurman *et al.*, 1995).

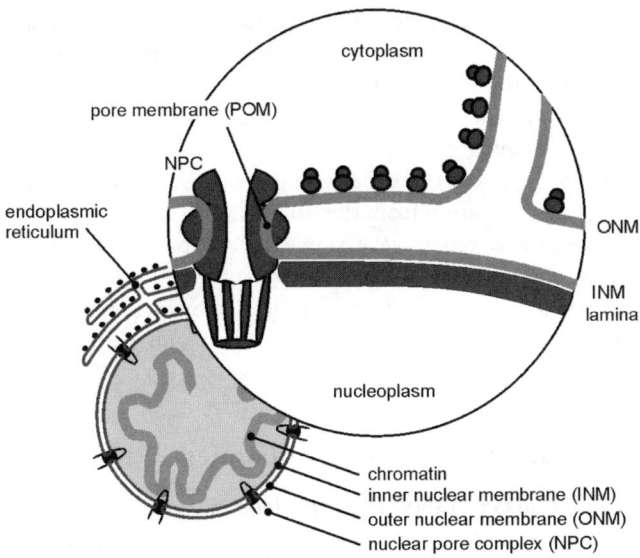

Figure 1. Schematic representation of the nuclear envelope structure.

2.3 *Analytical challenges in nuclear envelope proteomics*

A structure like the nuclear envelope that is an ER domain and connected to the total ER, consists of more or less specialized membranes with partly distinct protein compositions and has high-affinity binding interactions between its different substructures poses several problems that have to be solved, experimentally and analytically, to give reliable and biologically relevant information on the protein composition of the structure.

Any set of nuclear envelope proteins identified by broad screening approaches of isolated nuclear envelopes has to be analysed in terms of protein localization to the substructures of the nuclear envelope. Membrane proteins, for example, could be common to all ER membranes or could be specifically located to ONM, POM or INM. ONM proteins, for example, may have biologically significant contacts to the cytoplasmic cytoskeleton. Proteins of the INM, on the other hand, could either bind to the lamina, to chromatin, or both. Chromatin components or other nucleoplasmic proteins that are present throughout the interior of the nucleus finally may show highly significant binding to the lamina or to inner nuclear membrane proteins. Having identified large numbers of proteins in a nuclear envelope preparation, it is therefore a major task to validate the identified proteins. For INM proteins, this may be simply achieved by demonstrating that they are enriched at the nuclear rim and bind directly to lamins. Therefore, their properties allow for a clear assignment to the INM. A validation is, however, much more difficult for proteins of the ONM or for intranuclear proteins, since their assignment to the nuclear envelope will require the detection of specific interactions at the proposed location which are associated with a biological function at the nuclear envelope.

The description of the full set of nuclear envelope proteins common to all cells, which is at the moment far from being complete, still falls short of a comprehensive proteomic characterization of the structure. Some proteins may be present at the nuclear envelope only in selected tissues or at a particular cellular status. Also, some proteins may associate with the nuclear envelope in a highly dynamic fashion.

Therefore, a proteome analysis of isolated nuclear envelopes requires an exact definition and maintenance of the status of the cells from which such nuclear envelopes are obtained. This should include differences due to differentiation, to the phase of the cell cycle, or to signalling events.

Such differentiation-dependent or otherwise transient nuclear envelope constituents are only accessible to identification from the isolated nuclear envelope when their interaction with the structure is either very stable or can be stabilized by suitable, for example chemical, means.

Post-translational protein modifications should also be included in a comprehensive description of the nuclear envelope proteome, because frequently they induce changes in molecular interactions.

3. Strategies in nuclear envelope proteomics

3.1 *Experimental approaches to subcellular proteomes*

To identify previously unknown protein constituents of a subcellular structure, three different principal strategies have been successfully applied: (i) a visual protein-trap approach based on the expression of fluorescent fusion proteins (Gonzalez and Bejarano, 2000); (ii) a visual gene-trap approach using the expression of fusion proteins generated by inserting a reporter gene at random into the genome (Sutherland *et al.*, 2001); and (iii) subcellular fractionation followed by a protein chemical analysis of resulting protein mixtures (*Table 1*) (Dreger, 2003).

The visual protein-trap approach (Gonzalez and Bejarano, 2000) is based on the insertion of cDNAs into a suitable expression vector whose expression leads to fusion proteins carrying a fluorescent protein that allows a direct visual screening for the localization of the expression product. As a result, the cloning of a particular cDNA linked to a protein with a defined subcellular localization already holds information on the subcellular structure to which such a protein is assigned. This has been successfully applied, for example, in a screen which returned nurim as a constituent of the nuclear envelope (Rolls *et al.*, 1999). The appeal of this method is a fast screening of many different cDNAs simultaneously for their capability to code for proteins of a particular subcellular location. In the process of isolating clones of such cDNAs, the localization of the encoded protein is inherently validated. A successful isolation of a protein of interest simultaneously provides the tool to study the intracellular dynamics of the protein *in vivo*.

Disadvantages are a possible influence of the added sequence on protein interactions which could result in a mislocalization of the fusion protein compared to the endogenous protein, toxic or other unwanted side effects of fusion protein overex-pression, and a bias for abundant cDNAs which may mask low copy proteins. Also, screening and cloning in a eukaryotic context requires a rather difficult and 'expensive' isolation strategy. Finally, this approach reveals the subcellular localization of a protein but cannot provide any information on its molecular interactions.

A second screening approach based on molecular biology is the visual gene-trap approach, which has already been used successfully to identify proteins of different nuclear locations (Sutherland *et al.*, 2001). Here, an exon encoding a reporter gene is randomly inserted into the genome. Whenever this reporter inserts into an intron such that it is spliced into the transcript, it will be expressed as a carboxy-terminal domain of a fusion protein containing amino-terminal parts of the endogenous gene product. Provided the targeting signals of the endogenous protein remain intact, the fusion protein will have the same subcellular distribution as the protein encoded by the altered gene. Therefore, this approach also allows the identification of proteins of a subcellular organelle by visual screening. In contrast to the protein-trap approach, expression is controlled by the endogenous promoter, which prevents problems with overexpression. The generation of a carboxy-terminal truncation of the endogenous protein, on the other hand, inherits the risk of a non-functional protein that may be aberrantly distributed. In addition, all targeting signals from sequences on the carboxy-terminal side of the insertion are lost. This also implies that targeting infor-mation encoded by the exon that generates the carboxy-terminus of a protein is not accessible to his gene trap method.

Subcellular proteomics, a third approach to proteins of subcellular structures, will be the focus of this chapter (Dreger, 2003). Here, the aim is to map the protein compo-sition of isolated subcellular structures at the level of endogenous proteins with modern protein chemical and mass spectrometric methods (*Figure 2*). The approach relies on established and well-characterized subcellular fractionation techniques. Optimally, all

Table 1. Approaches to subcellular proteomes

	Visual protein-trap screen	Visual gene-trap screen	Subcellular proteomics
Method	Screening of cDNA library by expression as fusion proteins with fluorescent protein	Random insertion of an exon encoding a fluorescent protein as carboxy-terminus	Subcellular fractionation, isolation of proteins and identification with protein chemical and mass spectrometric methods
Reference	Rolls *et al.* (1999)	Sutherland *et al.* (2001)	Dreger (2003)
Visual screen	Yes	Yes	No
Expression level	Overexpression	Endogenous	Endogenous
Advantages	Fast screening of large cDNA-pools	Screening of the whole genome	Proteins identified while still in the structure of interest
			Protein interactions (partially) preserved
			Analysis of endogenous complexes possible
	Inherent validation of positives	Inherent validation of positives	Post-translational modifications accessible to analysis
	Direct access to protein dynamics *in vivo*	Direct access to protein dynamics *in vivo*	
Disadvantages/ risks	Fluorescent protein may influence localization	Carboxy-terminal truncation and fluorescent protein may influence localization	Artefacts due to preparation by the loss of proteins or an artificial association / aggregation (contaminants)
	Overexpression may change subcellular distribution	Truncation may result in loss of function and may compromise cell viability	Difficult access to small (<20 kDa), large (> 200 kDa) or hydrophobic (e.g. membrane) proteins
	Biased by mRNA abundance Toxic effects of over expression	Low copy proteins may not be detectable	Abundant proteins may obscure identification of less abundant proteins
	No access to protein interactions	Carboxy-terminal exon always missed	Identification depends on quality of sequence databases
	No access to post-translational modifications	No access to protein interactions No access to post-translational modifications	Validation of positively identified proteins by independent methods required

protein constituents are still retained within the investigated structure and their relevant interactions are still maintained throughout the purification procedure. By further subfractionating such a structure, substructures and finally isolated protein complexes can be analysed resulting in a refined picture of the structure and the molecular interactions established therein. Opening the cell, on the other hand, can cause well-known problems, particularly through the loss of proteins during the purification procedure or through the artificial association or aggregation of proteins with the purified structure. Such contaminants have to be identified and excluded which makes it necessary to validate the localization of each identified protein by independent, for instance, immunochemical methods. The isolation of complex protein mixtures within the context of their subcellular structure poses other technical problems. In particular, it is difficult to find small proteins (< 20 kDa), large proteins (> 200 kDa) or highly hydrophobic proteins, for example membrane proteins.

3.2 Protein identification techniques

The successful identification of proteins from the prepared protein mixtures requires the separation of individual proteins, proteolytic digestion and a processing of the resulting peptide mixtures for mass spectrometric analysis. In most cases, a separation of the complex protein mixtures is necessary before digestion, because only peptide

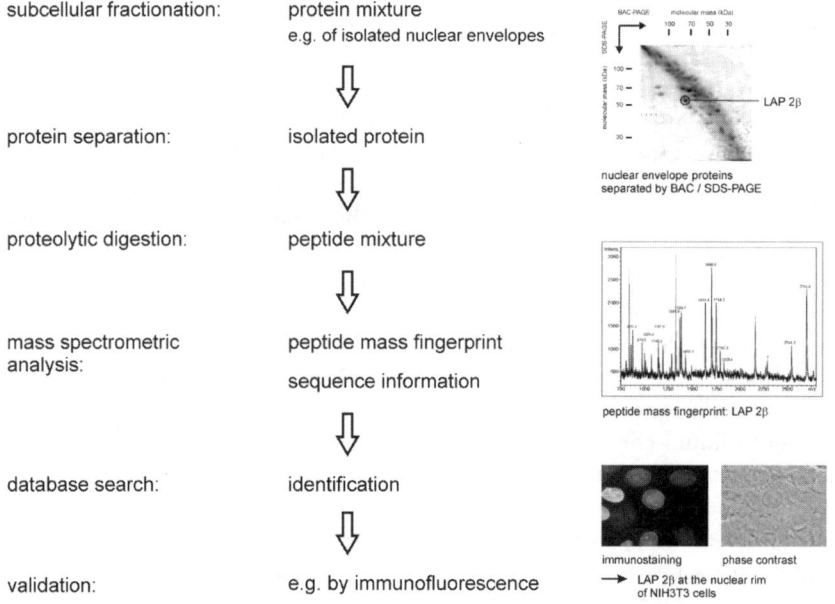

Figure 2. *Subcellular proteomics: steps of analysis. As an example, experimental data on the identification of LAP 2β from isolated nuclear envelopes is shown in the left column: (top) separation of nuclear envelope proteins by BAC/SDS-PAGE, Coomassie-stained gel. The LAP 2β protein spot is encircled; (middle) MALDI-mass spectrometry: the peptide mass fingerprint of LAP 2β is shown; (bottom) validation by immunofluorescence with an antibody specific for the LAP 2 membrane isoforms (mainly LAP 2β). As expected, a nuclear rim staining is obtained.*

mixtures from a small number of proteins can be analysed simultaneously. This is about to change with the automated analysis of peptide mixtures separated multidimensionally by chromatographic techniques and coupled directly to electro-spray-ionization (ESI) mass spectrometry (Washburn et al., 2001).

In general, a separation of proteins for mass spectrometric analysis is done at least two-dimensionally. Combinations of different chromatographic separations, of reversed-phase chromatography and SDS-polyacrylamide gel electrophoresis (SDS-PAGE), isoelectric focusing (IEF) and SDS-PAGE or two-dimensional BAC/SDS-PAGE have been used. Each separation technique gives different answers in efficiently separating different sets of proteins (Galeva and Altermann, 2002). The identification of integral membrane proteins, in particular, is the main challenge in the initial characterization of nuclear envelope proteins. Here, the only methods that are useful are those that allow hydrophobic proteins to stay soluble. This could be either reversed phase chromatography and SDS-PAGE (Erdmann and Blobel, 1996; Lee et al., 1996; Schrotz-King et al., 1999) or the two-dimensional BAC/SDS-PAGE which uses the cationic detergent 16-BAC under acidic conditions in the first dimension and a regular SDS-PAGE as a second dimension (Hartinger et al., 1996). Other techniques, in particular the very common combination of IEF and SDS-PAGE in all its variations, are highly biased against the isolation of membrane proteins in that such proteins either do not focus or transfer poorly into the second dimension (Santoni et al., 2000).

Isolated proteins or small protein complexes are next proteolytically, commonly tryptically, cleaved into peptides and the peptide mixtures supplied to a mass spec-trometer for analysis. More complex peptide mixtures, for example obtained when a higher number of proteins are present in the digested sample, then require another dimension of separation (such as liquid chromatographic peptide separation) before mass spectrometry in order to give interpretable results in the mass spectrometric analysis. Different instrument settings are suited for particular separation and identifi-cation strategies. Because the sample is introduced as a liquid solution in ESI mass spectrometry, analytes can be eluted from a chromatography column directly into the mass spectrometer in an on-line fashion. This strategy is especially efficient in the identification of proteins subsequent to one-dimensional, low-resolution separation by gel electrophoresis. Partial mass spectrometric sequencing is essential for protein identification using this strategy, but can be performed routinely in a highly automated manner (Aebersold and Mann, 2003).

Matrix-assisted laser desorption ionization (MALDI) mass spectrometry requires the co-crystallization of analyte with the matrix which renders it less suitable for coupling to liquid chromatography. However, due to the occurrence of the peptide ions in a comparably uniform charge state ($[M+H^+]^+$), MALDI mass spectra allow a more straightforward identification of proteins based solely on the set of observed peptide masses after proteolytic digestion (peptide-mass fingerprinting). Often, obtaining partial sequence information is only required, when the simultaneous digestion of several proteins results in a very complex peptide mixture. In combination with high-resolution electrophoretic separation of protein mixtures and in-gel digestion, MALDI-mass spectrometry is an ideal choice.

The primary data of a successful mass spectrometric analysis, finally, are lists of peptide masses and partial sequence information linked to proteins of a sample of a defined cellular structure and species origin. These data have to be analysed, commonly by publicly accessible search interfaces matching masses and sequences to database entries in large sequence databases. A successful identification relies on accurate sequence entries, preferably for the species background from which the sample was taken. For a high sequence coverage, as many peptide masses as possible must find a database counterpart within a small window of mass accuracy. This should be supported by one or more sequences from peptides obtained by peptide fragmentation (post source decay (PSD) in MALDI-MS-MS or collision-induced dissociation (CID) in ESI-MS-MS), fitting to the sequence in question. Ideally, high sequence coverage in combination with peptide sequences that are part of the proposed sequence hit, which comes from the same species from which the cell organelle has been prepared, should guarantee a high confidence in protein identification. In reality, such ideal results are often prevented by a lack of sequence entries for the species background under analysis or by erroneous sequence entries in the databases. Fortunately, such problems with database entries are often rescued by a remarkably high degree of sequence similarity between organisms that are supposedly closely related. There are, however, examples like the mouse lamin B receptor, whose sequence shows sufficient sequence dissimilarity to the rat lamin B receptor, that its unequivocal identification could only be achieved after the mouse sequence had been entered into the sequence database. A comprehensive characterization of the proteome of any cellular substructure results in a list of putative protein constituents of that structure. Since any subcellular fractionation or isolation of cellular structures is prone to a preparation-induced change in its composition, the candidate constituents of such a structure must be shown to reside in the structure *in vivo*. Therefore, the validation of the mass spectrometrically identified proteins by independent methods is the biggest part of the proteomic characterization of an organelle or another cellular structure. The validation is commonly done by either immunofluorescence staining of identified proteins or by the transient expression of the identified proteins as fusion proteins with either a fluorescent protein or a tag for immunodetection (Cronshaw *et al.*, 2002; Dreger *et al.*, 2001; Rout *et al.*, 2000). Demonstrating the right localization of the endogenous protein requires detectable amounts of that protein and suitable reagents, mostly antibodies. Since it is not yet possible to provide sufficient high-quality antibodies for many proteins, the transient expression of tagged proteins or fluorescent proteins is very popular. Such an intervention in a living cell may, however, lead to ambiguous or even erroneous results due to the overexpression of an otherwise moderately or little expressed protein or due to an influence of the additional sequences of the fusion protein on its localization.

3.3 *Nuclear envelope proteomics*

Nuclear envelope fractions
For a first characterization of the nuclear envelope proteome, a preparation of highly purified nuclear envelopes has to be established. This is best done from cultured cells since this avoids the heterogeneity of different cells forming a tissue. The procedure starts with the purification of nuclei, from which nuclear envelopes can be obtained, for example by treating the nuclei in a hypo-osmotic solution with nucleases. The successful isolation of nuclei from Neuro2a mouse neuroblastoma cells achieved in our laboratory may partially depend on the size of the cells compared to the size of the

nucleus (Dreger *et al.*, 2001). The preparation from other cell types has to be optimised individually.

Depending on the purification conditions, the isolated nuclear envelopes contain variable amounts of chromatin. Essentially, nuclear membranes and lamina are obtained from such a preparation, which in addition contains variable amounts of nuclear pore complexes, additional endoplasmic reticulum membranes and chromatin components that bind directly or indirectly, but with high affinity, to lamins or inner nuclear membrane proteins. Thus, proteins identified have to be assigned to one of the different substructures present in the preparation.

One option to distinguish between nuclear envelope proteins and proteins of the ER could, for example, be a subtractive approach that relies on the comparison of the protein composition of isolated nuclear envelopes and a rER preparation. Such a comparison, however, would not exclude proteins found in both preparations from being present in the nuclear membranes as there is currently no way to identify proteins that are common to all domains of the endoplasmic reticulum.

Another approach that has been successfully applied to gain insight into substructures of the nuclear envelope is the differential extraction of nuclear envelopes, which has been used, for example, for the biochemical characterization of LAP-2β (Foisner and Gerace, 1993; Furukawa *et al.*, 1995). A considerable contribution to the analysis of NE substructures has been achieved by the analysis of Triton X-100-extracted nuclear envelopes. This fraction represents the lamina–nuclear pore complex fraction of the nuclear envelope. Additionally, it contains many proteins of chromatin and the INM that bind with high affinity to lamins. Another NE subfraction, which supports such an analysis, is obtained by treating nuclear envelopes with chaotropes. This fraction contains predominantly integral proteins of the nuclear membranes and, to a variable extent, of other parts of the ER. The comparison between such subfractions allows a preliminary assignment of proteins to one of the nuclear envelope substructures (Dreger *et al.*, 2001).

Similarly, preparations for the isolation of intact, pure nuclear pore complexes have been described, that were successfully used to identify the proteins of both yeast and mammalian nuclear pore complexes (Cronshaw *et al.*, 2002; Rout *et al.*, 2000).

Validation of putative nuclear envelope constituents

As discussed above, subcellular fractionation and subfractionation of the initial nuclear envelope preparation may be prone to the loss of relevant protein constituents of the analysed structure or the non-specific association of other proteins. Therefore, results of the proteomic analysis of nuclear envelopes and NE substructures must be validated. This is commonly achieved by showing the intracellular distribution of identified proteins. Endogenous proteins or transiently expressed, tagged proteins are visualized by immunofluorescence.

Alternatively, fusion proteins with a fluorescent protein are used. Taking into account caveats regarding the overexpression of modified proteins as discussed above, the validation of proteins that are exclusively associated with the nuclear envelope is easily achieved by showing a distribution of such a protein to the nuclear rim, either in a dotted fashion as for nuclear pore complex proteins or in a continuous fashion as for proteins of the nuclear lamina or the inner or outer nuclear membrane. Demonstrating its membrane association further increases the confidence of positive identification for an integral membrane protein of the nuclear envelope; additionally showing an inter-action with lamins or chromatin compounds indicates that it is an INM protein. Most of the characterized INM proteins have in fact been shown to bind to lamins, chro-matin proteins or both.

Recently, evidence for the localization of a spectrin repeat protein has been presented, which relies on the immunodetection of the protein at the nuclear rim in conventionally as well as in digitonin-permeabilized cells (Zhen *et al.*, 2002). Since digitonin permeabilizes the plasma membrane but leaves the nuclear membranes intact, this result hints at a cytoplasmic orientation of the protein (see also Section 4.1).

To a certain extent, all membrane proteins of the NE should be present throughout the ER, since they are generated in the rER and have to accumulate at the nuclear envelope. INM proteins are thought to accumulate at the nuclear rim by anchorage through high affinity binding to nuclear envelope or intranuclear binding sites (Gant and Wilson, 1997). A similar anchorage should exist for proteins of the ONM, which has to depend on interactions reaching the inner face of the nucleus.

The validation of other intranuclear proteins (for example chromatin constituents) or ER proteins is much more complicated. For the intranuclear proteins, this has to be done individually by demonstrating a specific interaction with one of the *bona fide* nuclear envelope proteins. Such a validation for common proteins of the ER, which may also be present in the nuclear membranes, may reasonably not be achieved at all, since it probably would require individual demonstration of the presence at the nuclear envelope by immuno-electron microscopy.

Like any subcellular proteome analysis, nuclear envelope proteomics result in large lists of proteins that are part of the different nuclear envelope substructures and are identified as such with different degrees of certainty. An appropriate way of presenting such data is a database linking analytical, sequence and other information on the iden-tified protein and making such structured data collections accessible to the public. Attempts to establish such comprehensive data collections on the protein compo-sition of organelles or of cellular complexes are being made in many places, although with very different architecture and organization. Good examples are the nuclear protein database (http://npd.hgu.mrc.ac.uk (Dellaire *et al.*, 2003)) and the nucleolar protein database (http://www.dundee.ac.uk./lifesciences/lamonddatabase (Andersen *et al.*, 2002)).

4. Proteins of the nuclear envelope

In this section, we will give a brief overview of proteins found at the nuclear envelope according to their putative localization at one of the substructures (see also *Table 2*).

Four different categories of nuclear envelope proteins can be distinguished according to their localization, their biochemical properties, in particular their membrane integration, and their molecular interaction (*Figure 3*). From the cyto-plasmic to nucleoplasmic face of the NE, these are: (i) proteins of the outer nuclear membrane and their cytoplasmic interaction partners; (ii) pore membrane proteins and the constituents of the nuclear pore complex; (iii) inner nuclear membrane proteins; and (iv) proteins of the lamina with their intranuclear interaction partners. Most of the known inner nuclear membrane proteins bind with a high affinity to

Table 2. Proteins at the nuclear envelope

Protein[a]	Isoforms	m.w.[b] (kDa)	No. of TMS[c]	Species	Motifs[d]	Similarity[d]	Interactions[d]	Functions	Identifier[e]
Outer nuclear membrane									
ANC-1		27	1	C. *elegans*	actin binding domain, KASH-domain	Calponin, α-actinin	Actin	possibly INM nuclear anchoring, nuclear migration?	Uni: Cel.21656
klarsicht		245	1	Drosophila	KASH-domain		Kinesin, dynein?	microtubule-based vesicle transport	Uni: Dm.2787
Msp-300 (*CG18251-PA, -PB*)	PB PA	888 134	1	Drosophila	actin binding domain KASH domain	Calponin, α-actinin Myosin heavy chain non-msucle type B	Actin	muscle specific	Uni: Dm.20325
nesprin 1 (*SYNE-1, MYNE-1*)	β myne-1 α	380 131 112	1	human	actin-binding domain (*Isoform* 1β) KASH domain bipartite NLS, spectrin repeats	calponin, α-actinin	actin, MUSK ?	muscle specific, anchoring of subsynaptic nuclei; nuclear migration?, nuclear differentiation	Uni: Hs. 19102
nesprin 2 (*SYNE-2, NUANCE*)	NUANCE γ β β2 α α2 NUANCE-N33	796 377 87 76 62 48 30	1	human	actin binding domain (*NUANCE*) KASH-domain, bipartite NLS, leucine zipper, spectrin repeats	calponin, α-actinin (*NUANCE*)	actin (*NUANCE*)	nuclear anchoring? nuclear migration?	PID: 29839588 Uni: Hs.57749
Pomfil	1 2 3	205 >256 203	1 1 —	human	actin-binding domain, leucine zipper	unc-53, Pom121	NPC	neuron-specific, ONM only	no sequence entries

Table 2. continued

Protein[a]	Isoforms	m.w.[b] (kDa)	No. of TMS[c]	Species	Motifs[d]	Similarity[d]	Interactions[d]	Functions	Identifier[e]
Pore membrane and nuclear pore complex									
ALADIN (*adracalin*)		60		human	WD repeat		NPC	linked to triple A (achalasia, adrenocortical insufficiency, alacrimia)	Uni: Hs.125262
Brr6 (*bad response to refrigeration 6*)		23	2	yeast			NPC	partially INM	PID: 6321190
ndc1		74	6	yeast			NPC, spindle pole body		PID: 6323610
nem1		51	1–2	yeast			Nup84 (indirectly), Spo7p	spherical nucleus (nuclear envelope morphology)	PID: 6321791
Pom34p		34	1?	yeast			NPC	NPC anchoring	PID: 6323046
Pom121p		121	1	rat			NPC	glycoprotein, NPC anchoring	Uni: Rn.10474
Pom152p		152	3	yeast			NPC	NPC anchoring, NPC formation in interphase	PID: 6323777
Pom210p (*gp210, NUP210*)		204	1	rat			NPC	glycoprotein, NPC anchoring	Uni: Rn.3189
spo7p		30	1–2	yeast			Nup84 (indirectly), nem1	spherical nucleus (nuclear envelope morphology), sporulation	PID: 6319310

Table 2. continued

Protein[a]	Isoforms	m.w.[b] (kDa)	No. of TMS[c]	Species	Motifs[d]	Similarity[d]	Interactions[d]	Functions	Identifier[e]
Inner nuclear membrane									
emerin		29	1	human	LEM		BAF, lamin A	linked to Emergy-Dreifuss muscle dystrophy	Uni: Hs.2985
LAP-1	C C short B A	57 52	1	rat			lamin B, lamin A/C (LAP-1B)	sequences of LAP-1A and LAP-1B not determined. Splice variants from a single gene	Uni: Rn.11373
LAP-2 (*old: thymopoietin*)	β γ δ ε	50 38 43 46	1	mouse	LEM domain, LEM-like domain		BAF, DNA, GCL, lamin B, HA95	regulation of gene expression, nuclear architecture (functional and structural)	Uni: Mm.124
LBR		71	8	human	RS-domain TUDOR domain	ERG4/ERG24 sterol reductases	HP1, histone H3, lamin B, LBR-kinase, p34/32	heterochromatin anchoring, sterol reductase?, linked to Pelger-Huet-anomaly	Uni: Hs.152931
LUMA (*hypothetical protein BAB23556*)		45	3–4	mouse					PID: 12836214
MAN1		100	2	human	LEM domain		BAF, DNA, lamins		Uni: Hs.7256
nurim		29	5	human					Uni: Hs.57222
otefin (*CG5581-PA*)		47		drosophila			lamin Dm, YA	vesicle attachment to chromatin	Uni: Dm.3658
RFPB		>152	9	rabbit		type IV, P-type ATPase	RUSH transcription factors	partial sequence, appr. 18 residues missing at N-terminus	PID: 7715417
SUN-1 (*KIAA 0810*)		92	2(-4)	human	SUN domain				PID: 3882341
SUN2 (*KIAA 0668*)		83	2(-3)	human	SUN domain				PID: 3327150
UNC-84	A B	126 99	1	c. elegans	SUN domain			nuclear anchoring, nuclear migration	Uni: Cel.17328
UNC-83	A	118	1	c. elegans	coiled coil, LL- (?)MS			nuclear migration	Uni: Cel.17239

Table 2. continued

Protein[a]	Isoforms	m.w.[b] (kDa)	No. of TMS[c]	Species	Motifs[d]	Similarity[d]	Interactions[d]	Functions	Identifier[e]
lamina									
lamin	A AΔ10 C	74 71 65		human	Intermediate-filament (IF) head domain, IF rod domain, IF tail domain		emerin, LBR, LAP2, MAN1	linked to several laminopathies, precursor, C-terminal peptide removed.	Uni: Hs.377973
lamin	B1 B2	66 68		human			emerin, LBR, LAP1, LAP2,	isoprenylated	Uni: Hs.89497 Uni: Hs.76084
nucleoplasmic face									
BAF (*barrier-to-autointegration factor, Breakpoint cluster region protein 1*)		10		human			DNA, LEM domain	cross-bridges DNA without sequence specificity	Uni: Hs.433759
GCL		60		mouse	BTP/POZ-domain		DNA, LAP-2β	transcription factor	Uni: Mm.74594
HA95 (*HAP95, NAKAP95*)		72		human	NLS, Zn-finger	AKAP95	chromatin, LBR, LAP-2β, PKA catalytic subunit	replication initiation	Uni: Hs.96200
histone H3	F3A F3B	15		human			DNA, histones, LBR?		Uni: Hs.181307 Uni: Hs.180877
HP-1	α β γ	22 21 21		human	chromo domain, chromo-shadow domain		BGR1 (HP-1α), histone H4, LBR, TAF II 130	stabilizes heterochromatin	Uni: Hs.89232 Uni: Hs.77254 Uni: Hs.406384
Ya (*fs(1)Ya*)		78		drosophila			otefin		Uni: Dm.2083

a) alternate names in parentheses
b) calculated molecular weight
c) in most cases, transmembrane sequences were predicted by using TMHMM Server v 2.0 (http://www.cbs.dtu.dk/services/TMHMM-2.0/)
d) in parentheses; isoform, to which a property applies
e) Uni: Unigene cluster number; PID: general protein identifier. Access via the Entrez-browser at http://www.ncbi.nlm.nih.gov/entrez/query.fcgi

lamins. Both groups of proteins therefore share their interaction partners, being part of the same complexes with chromatin proteins and DNA. Also, some interactions between outer and inner nuclear membrane probably exist, which couple nuclei to the cytoplasmic cytoskeleton. Such interactions, however, have not yet been convincingly demonstrated.

4.1 *Outer nuclear membrane proteins*

Until recently, the ONM was considered to be exclusively part of the rER. This property is well documented by electron micrographs showing ribosomes attached to the ONM. In a proteomic screen of the nuclear envelope, one can therefore expect to find proteins generally present in the endoplasmic reticulum.

Very recently, however, a member of a group of spectrin repeat proteins has been shown to be specifically located at the ONM (Zhen *et al.*, 2002). These spectrin repeat proteins are very large integral type II membrane proteins, which share a common carboxy-terminal KASH-domain containing a single transmembrane domain, which mediates nuclear envelope localization. It is named from its presence in the protein klarsicht (drosophila), which is required for nuclear migration and normal development of the photoreceptors of the drosophila composite eye, ANC-1 (*Caenorhabditis elegans*) and nesprins (nuclear envelope spectrin repreat proteins; also called 'syne' proteins) (Starr and Han, 2002; Zhang *et al.*, 2002; Zhen *et al.*, 2002). The nesprins and the drosophila protein Msp300 of this group have a central rod domain consisting of multiple spectrin repeats, while the rod domain of ANC-1 shows different repeats (Starr and Han, 2003). The full-length variants of these proteins all bind actin with an amino-terminal α-actinin-like actin-binding domain (Korenbaum and Rivero, 2002).

Several isoforms or truncated forms of nesprins, found independently by different groups, are derived from two genes encoding the very large full-length forms. Nesprin-1 (Zhang *et al.*, 2002), also called syne-1 (Apel *et al.*, 2000) or myne-1

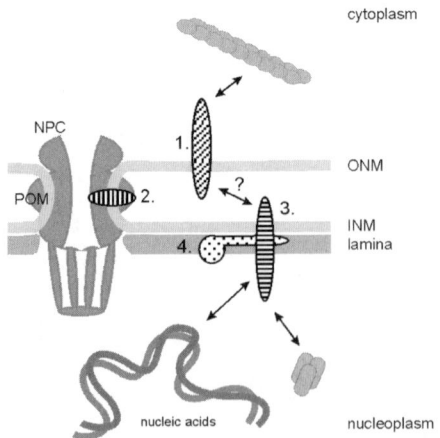

Figure 3. Classification of nuclear envelope proteins.

(Mislow *et al.*, 2002b), and its variants are muscle-specific proteins. The drosophila-orthologue Msp300 is required for normal muscle development (Zhang *et al.*, 2002), and all forms of nesprin-1 are developmentally expressed in smooth and skeletal muscle (Mislow *et al.*, 2002b). In developing myotubes, nesprin-1 is involved in nuclear migration and anchoring (Apel *et al.*, 2000). The localization of nesprin-1 is ambiguous. Binding of lamin A and emerin by nesprin-1a points towards an intranuclear localization (Mislow *et al.*, 2002a). The interaction of the nesprin-1α2 form with the muscle-specific protein kinase MUSK, however, hints at a cytoplasmic orientation, since MUSK is incorporated into the postsynaptic membranes of muscle fibres (Zhang *et al.*, 2001).

In contrast to nesprin-1, nesprin-2 (Zhang *et al.*, 2002), also called syne-2 (Apel *et al.*, 2000) or NUANCE (Zhen *et al.*, 2002), and its shorter isoforms are ubiquitously expressed with the lowest expression levels being in muscle. They are targeted to the ONM by their KASH-domain in combination with the last two spectrin repeats. By immunofluorescence, the actin-binding domain of full-length nesprin-2 has been detected in the cytoplasm of digitonin-permeabilized cells, which retain a nuclear envelope impermeable to antibodies. So far, this is the strongest indication of a localization of nesprin-2 to the ONM. Other variants of nesprin-2, however, may well be present at the INM. In addition, some nesprin-2-related immunoreactivity apparently shows a co-localization with nucleoli of Triton X-100-permeabilized cells (Zhen *et al.*, 2002).

Finally, three neuronal proteins have been identified, which may link nuclear pores and actin filaments. Two of these Pomfil proteins are integral membrane proteins with an α-actinin-like actin-binding domain, which have been localized by immuno-electron microscopy to the outer face of the nuclear envelope. The expression of these proteins may be linked to neuron growth or neural tumour formation (Coy *et al.*, 2002).

4.2 *Proteins of the nuclear pore complex*

The nuclear pore complexes are biochemically integrated into the lamina of the nuclear envelope. Therefore, nucleoporins are to a variable extent present in proteomic screens of the nuclear envelope, depending on preparation and subfractionation conditions. The proteomic analysis of the nuclear pore complex from yeast as well as from higher eukaryotes, including a thorough validation, has been achieved recently (Cronshaw *et al.*, 2002; Rout *et al.*, 2000) and is described in detail in Chapter 6. Briefly, nuclear pore complexes from yeast as well as from mammalian cells are built from approximately 30 nucleoporins. Copies of these proteins are present in multiples of eight, reflecting the eightfold-symmetry of the nuclear pore complex. Most nucleoporins are arranged symmetrically on both the cytoplasmic and the nucleoplasmic side of the NPC, with a few exceptions, which show a preference for one side. The sequences of a subset of nucleoporins are enriched in phenylalanine and glycine, which are arranged in FG repeats. These FG nucleoporins presumably provide a binding surface for transport complexes of the nucleocytoplasmic transport.

The NPCs are anchored at the nuclear pore by integral membrane proteins of the pore membrane. In yeast, three such proteins, POM152p, Ndc1p and POM34p have been identified (Rout *et al.*, 2000). POM152p, a type II membrane protein that is similar to the vertebrate Pom210 exposes its very large carboxy-terminal domain to the lumen of the endoplasmic reticulum (Tcheperegine *et al.*, 1999; Wozniak *et al.*, 1994). It presumably stabilizes the pore by adopting a grommet-like structure, interacts with other nucleoporins, and may also be involved in the *de novo*-formation of new pore complexes (Strambio-de-Castillia *et al.*, 1995; Tcheperegine *et al.*, 1999). In this aspect, it might act synergistically with Ndc1p, a second integral pore membrane protein (Marelli *et al.*, 2001). This protein is apparently also part of the spindle pole body of yeast (Chial *et al.*, 1998).

Besides these three nucleoporins, four other proteins interact at least functionally with nuclear pore complexes. Without being part of the NPC, Brr6p locates to the nuclear pores, where it may be involved in the spatial distribution of nuclear pore complexes (de Bruyn Kops and Guthrie, 2001). Spo7p and Nem1p interact genetically with the nucleoporin Nup84p (Siniossoglou *et al.*, 1998), and Snl1p that is structurally related to POM152p may be involved in nuclear pore formation and stabilization (Ho *et al.*, 1998; Sondermann *et al.*, 2002).

In higher eukaryotes, two integral proteins of the pore membrane, Pom210 (gp210) and Pom121, organize the anchoring and stabilization of the NPC. Pom121 has a single transmembrane domain near its amino-terminus and a large carboxy-terminal domain oriented towards the centre of the nuclear pore, which contains FG repeats (Hallberg *et al.*, 1993; Soderqvist and Hallberg, 1994). Together with Pom210, it may form a functional subcomplex involved in nuclear pore complex formation after mitosis (Cronshaw *et al.*, 2002). Pom210 features a short carboxy-terminal tail inside the nuclear pore, a single membrane-spanning sequence, which targets it to the NPC, and a glycosylated amino-terminal domain of 200 kDa inside the ER lumen (Greber *et al.*, 1990; Wozniak and Blobel, 1992; Wozniak *et al.*, 1989). Its integrity and its interactions, in particular with Pom121, are essential for NPC formation and for a functional NPC (Drummond and Wilson, 2002).

Interestingly, at least one nucleoporin, ALADIN (also named adracalin), is linked to a congenital disease called triple A-syndrome (Tullio-Pelet *et al.*, 2000). Nothing is, however, known about pathogenic mechanisms arising from ALADIN mutations.

4.3 *Inner nuclear membrane proteins*

The first integral proteins of the INM, the lamin B receptor (LBR) and the lamina-associated polypeptides 1 and 2 (LAP-1 and LAP-2), were detected by their ability to bind directly to lamins (Foisner and Gerace, 1993; Worman *et al.*, 1988). Since then, several additional proteins, emerin, MAN-1, nurim, LUMA, SUN-1, SUN-2, unc-83 and the RING-finger domain-binding protein RFBP have been described as INM proteins. These proteins also bind either directly or indirectly to lamins since they are present in Triton X-100-extracted nuclear envelope fractions, which contain nuclear pore complexes, the lamina and tightly bound proteins (Dreger *et al.*, 2001). As well as these integral membrane proteins, the drosophila protein otefin is a peripheral membrane protein, which binds directly to the INM (Ashery-Padan *et al.*, 1997a, b).

Among these proteins, two groups related to each other by shared domains are emerging, the LEM-domain proteins LAP2, emerin, MAN-1 and Otefin and the SUN-domain proteins SUN-1 and SUN-2. A third group, which seems to be present at but not exclusively located to the INM, probably consists of some of the KASH-domain proteins that have been briefly discussed in Section 4.1 on ONM proteins.

The lamin B receptor is a polytopic integral membrane protein with eight trans-membrane sequences (Worman *et al.*, 1990). Its amino terminus is exposed to the nucleoplasm and anchors the LBR to the INM by its ability to bind DNA, lamins and the heterochromatin protein HP-1 (Soullam and Worman, 1993; Ye and Worman, 1994; Ye *et al.*, 1997). The interaction with HP-1, however, may not be a direct one, but mediated by the core histones H3/H4 in an acetylation-dependent manner (Polioudaki *et al.*, 2001). Besides these interactions, a complex of LBR with a p58/LBR-kinase, and the LBR-associated protein p34/p32 has been described (Simos and Georgatos, 1992, 1994) The kinase constitutively phosphorylates Ser-Arg dipeptide motifs in the amino-terminal domain of the LBR and of splicing factors of the SR-family (Nikolakaki *et al.*, 1997, 1996). The carboxy-terminal part of the lamin B receptor comprising the eight membrane-spanning sequences is homologous to sterol reductases and exhibits such activity *in vitro*. The relevance of this activity being present at the inner nuclear membrane is still unknown (Holmer *et al.*, 1998; Schuler *et al.*, 1994; Silve *et al.*, 1998).

The presence of the LBR at the INM is essential for normal development. A loss or strong reduction of LBR in humans is linked to the rare Pelger–Huet anomaly, which is characterized by a reduced lobulation of granulocyte nuclei and, in homozygous human patients, causes a developmental delay, skeletal abnormalities and epilepsy (Hoffmann *et al.*, 2002).

A second type II integral INM protein, LAP-1, probably exists in three isoforms (A, B and C), which are produced from a single gene by alternative splicing. Isoforms LAP-1A and LAP-1B are preferentially expressed in differentiated cells. All isoforms bind with their amino terminal domain to A- and B-type lamins. It contains a single transmembrane sequence followed by a well-conserved carboxy-terminal domain exposed to the ER lumen (Foisner and Gerace, 1993; Kondo *et al.*, 2002; Martin *et al.*, 1995). For LAP-1C, a complex with lamin B and a yet unknown protein kinase has been found. From its distribution in mitotic vesicles and its asso-ciation with spindle microtubules, it was suggested that LAP-1 and LAP-2 proteins (see below) are located in discrete inner nuclear membrane territories (Maison *et al.*, 1997).

Members of the membrane isoforms of LAP-2, another important group of INM proteins, have already been characterized in considerable detail. LAP-2β is the largest of four isoforms, which are generated by alternative splicing from a single gene by omitting an increasing number of exons, which encode small sequences inserted between shared amino-terminal and carboxy-terminal sequences (Harris *et al.*, 1995). All of them are integral type II membrane proteins with a single transmembrane sequence at the carboxy-terminus. They share a common amino-terminal domain with LAP-2α, a fifth member of the LAP-2 family which is not a membrane protein, has a completely different carboxy-terminal part and is distributed throughout the nucleus (Dechat *et al.*, 1998). The common domain of LAP-2 proteins consists of two domains that confer DNA-binding, an amino-terminal LEM-like domain that binds directly to DNA and a more central LEM domain, which is named after LAP-2, emerin and MAN-1, all of which contain at least one LEM domain (Lin *et al.*, 2000). The LEM domain is a motif of 40 amino acid residues that binds to the DNA-bridging hete-rochromatin protein BAF (barrier-to-autointegration factor), which, by binding to such INM proteins is probably involved in heterochromatin attachment to the nuclear envelope, and in the regulation of chromatin condensation (Furukawa, 1999;

Shumaker *et al.*, 2001; Zheng *et al.*, 2000). LAP-2β interacts furthermore with the A kinase anchoring protein AKAP95-like protein HA-95 that probably accumulates kinases at the lamina. This interaction seems critically involved in regulation of chromatin decondensation (Eide *et al.*, 1998; Martins *et al.*, 2000; Örstavik *et al.*, 2000). The latest addition of interaction partners for LAP-2β is the transcription factor GCL (germ cell less), whose translocation and binding to the nuclear envelope reduces its transcriptional activity (de la Luna *et al.*, 1999; Nili *et al.*, 2001). Functionally, LAP-2β seems to be regulated via phosphorylation. The phosphorylation during mitosis at a not yet identified site is important for its release from lamina and chromatin (Foisner and Gerace, 1993). Four other identified phosphorylation sites, either between the LEM-like and the LEM domain or adjacent to the LEM domain, are occupied during interphase of the cell cycle (Dreger *et al.*, 1999). Their functional relevance is not yet clear. Balanced amounts of LAP-2β and its interaction partners are necessary for the functional integrity of processes like nuclear envelope assembly, nuclear envelope growth, transcription and DNA replication (Gant *et al.*, 1999; Martins *et al.*, 2003; Nili *et al.*, 2001; Yang *et al.*, 1997).

Emerin, the second member of the LEM-domain proteins, contains a single amino-terminal LEM domain and a single transmembrane sequence at its carboxy-terminus. It binds BAF:DNA complexes via its LEM domain and lamin A via a binding site within the central region of the molecule (Bione *et al.*, 1994; Manilal *et al.*, 1996; Nagano *et al.*, 1996).

Emerin is linked to a congenital disease, the X-linked recessive Emery–Dreifuss muscular dystrophy. Mutations of emerin that cause a mislocalization of the protein lead to a selective, slow progressive degeneration and loss of muscle fibres with characteristic contractures of joints and muscle wasting. The disease-associated mutations cause either an aberrantly phosphorylated protein variant or a carboxy-terminally truncated variant that lacks the transmembrane domain (Ellis *et al.*, 1998; Nigro *et al.*, 1995). *Vice versa*, a lack of lamin A, one of the emerin binding partners, leads also to a loss of emerin at the INM in the autosomal dominant form of Emery–Dreifuss muscle dystrophy (Bonne *et al.*, 1999; Sullivan *et al.*, 1999).

MAN-1 is the third LEM domain protein at the INM, which presumably binds BAF:DNA complexes with its amino-terminal domain. It probably binds lamins as it co-fractionates with them. MAN-1 is an integral membrane protein containing two membrane-spanning sequences (Lin and al, 1998; Paulin-Levasseur *et al.*, 1996). Its function is still unclear.

The SUN proteins SUN-1 and SUN-2 and their non-human orthologues are likewise integral proteins of the inner nuclear membrane with unknown function in mammals (Dreger *et al.*, 2001). Their topology with respect to transmembrane sequences is not clear. Both proteins share a short carboxy-terminal stretch of about 60 amino acid residues which show a remarkable similarity with the *C. elegans* protein UNC-84 and the yeast protein Sad1p which is present at the inner nuclear membrane and associated with the spindle pole bodies (Hagan and Yanagida, 1995). UNC-84 recruits UNC-83 to the nuclear envelope with its carboxy-terminal SUN domain. Both proteins are essentially required for a correct nuclear migration in developing *C. elegans*. It has been suggested that UNC-84 is involved in the attachment of centrosomes to nuclei, while UNC-83 might be important for microtubule attachment at the nuclear envelope (Lee *et al.*, 2002; Malone *et al.*, 1999; Starr *et al.*, 2001).

Nurim, a protein with five membrane-spanning sequences and short linker sequences, has been identified in a visual protein-trap screen, as mentioned above. Similarly, LUMA has been found in a subcellular proteomic analysis of Triton X-100-extracted nuclear envelopes. LUMA has three to four transmembrane sequences and a large hydrophilic loop between the first and second membrane-spanning sequence, which is probably oriented towards the nucleoplasm (Dreger *et al.*, 2001). Functions for both proteins are still lacking. Also, the sequences show no similarity with characterized proteins and no sign of characterized functional domains or sequence signatures. Both proteins behave biochemically like lamin-interacting proteins although no direct binding has yet been described.

Finally, the RING finger-binding protein RFBP is an integral INM protein, which binds the RING finger motif of the SWI/SNF-related RUSH transcription factors (Hayward-Lester *et al.*, 1996; Mansharamani *et al.*, 2001). It is a newly discovered type IV P-type ATPase that is ubiquitously expressed, with its expression level being regulated hormonally. RFBP spans the membrane nine times and is capable of binding to euchromatin via its flexible loop, which is a characteristic structural feature of all P-type ATPases (Mansharamani *et al.*, 2001).

Otefin is the only peripheral membrane protein of the INM known so far. It shows no homology to other proteins. A large hydrophilic domain rich in serine and threonine residues is followed by a stretch of 17 hydrophobic amino acids near its carboxy-terminus, which mediates binding to membranes and is essential for nuclear envelope targeting. The nuclear envelope localization is stabilized by other interactions of otefin (Ashery-Padan *et al.*, 1997b; Harel *et al.*, 1989; Padan *et al.*, 1990; Ulitzur *et al.*, 1997). Otefin is expressed in the drosophila egg and in somatic cells but not in sperm. It binds to the drosophila lamin Dm0 and stays at the nuclear envelope together with the lamin at early cell divisions in a so-called spindle envelope. Although otefin is mainly localized at the nuclear envelope, it is present in the nucleoplasm and the cytoplasm at certain developmental stages of the drosophila embryo where it may be bound to nuclear membrane vesicles (Ashery-Padan *et al.*, 1997a; Goldberg *et al.*, 1998).

4.4 *Lamins*

Lamins, which are intermediate filament proteins, build the cortical cytoskeletal element of the lamina beneath the inner nuclear membrane (Hutchison, 2002; Hutchison *et al.*, 1994; Strelkov *et al.*, 2003). Two major classes of lamins, A-type and B-type lamins are derived from two genes. By alternative splicing, lamina A and C and several B-type lamin isoforms are generated. B-type lamins are ubiquitously expressed and present at all developmental stages. A-type lamins, in contrast, are not present in germ cells and in embryonic cells (Georgatos *et al.*, 1994; Moir and Spann, 2001). Lamins consist of an amino-terminal head domain, a central coiled coil, α-helical rod domain, and a carboxy-terminal tail domain followed by a CAAX motif that post-translationally mediates a farnesylation in combination with a carboxy-methylation. The farnesylated carboxy-terminus is, however, cleaved from A-type lamins, which renders them soluble during mitosis while B-type lamins stay at the membrane of the endoplasmic reticulum (Aebi *et al.*, 1986).

The lamins form dimers via coiled coil–helix interactions and filaments of variable length by assembling these dimers in a head-to-tail fashion. The lamin filaments form to a variable extent a very stable meshwork beneath the inner nuclear membrane, which can in its extreme be a very dense and regular structure in some cells as for example the *Xenopus* oocyte (Aebi *et al.*, 1986; Zhang *et al.*, 1996). Lamins are also present throughout the nucleus (Moir and Spann, 2001; Stuurman *et al.*, 1998). The lamina, although extremely stable during interphase, is solubilized by PKC- and cdc2-kinase-mediated phosphorylation at key phosphorylation sites, which results in a mobilization of the lamins. A-type lamins then adopt a cytosolic distribution, while B-type lamins are still attached to the membrane, and are dispersed throughout the endoplasmic reticulum (Collas, 1999; Enoch *et al.*, 1991; Peter *et al.*, 1990). Besides providing a mechanically stable meshwork supporting the nuclear envelope, lamins also provide binding sites for the regulated attachment of chromatin and inner membrane proteins. Such interactions have proven to be of utmost importance for the structural and functional integrity of the nucleus. A whole group of genetic diseases, called laminopathies, arises from the lack of either lamin A or the lamin B receptor or from some of its interactions. As mentioned above, lamin A mutations in particular are involved in the pathogenic mechanisms leading to severe congenital diseases (Burke and Stewart, 2002; Östlund and Worman, 2003; Worman and Courvalin, 2002).

4.5 *Proteins associated with inner nuclear membrane and lamina*

Into this category fall such proteins that are confirmed to be functionally present at the lamina or at the INM. Proteins of this category may conditionally accumulate near the membranes, but do not have to. Instead, their functional involvement has been shown by demonstrating binding to lamins or INM proteins and, ideally, by showing an associated function.

BAF has been identified as a ligand for LEM domains. It is a small protein that had been described before as a factor that prevents auto-integration of virus DNA (Furukawa, 1999; Zheng *et al.*, 2000). It bridges DNA without any sequence preference and binds in its DNA-bound state with an enhanced affinity to the LEM domains of LAP2, Emerin or MAN-1 (Shumaker *et al.*, 2001). This interaction seems to be critical for the regulation of the chromatin condensation and for anchoring heterochromatin to the nuclear envelope. A loss of the INM proteins, for example tested in knockdown experiments for Ce-MAN1 and Ce-emerin in *C. elegans*, resulting in a loss of the BAF interaction leads to pathogenic changes in chromatin structure, aberrant nuclear morphology, and defects in cell division timing, which is often developmentally lethal (Liu *et al.*, 2003).

HP-1 is a heterochromatin protein involved in silencing gene expression. It binds specifically to the amino-terminal domain of the LBR. The protein exists in three isoforms in humans. It consists of a chromo-domain, which mediates binding to histone H3 methylated at the lysine residue at position 9, and a chromo-shadow-domain, which is necessary for self-association, for binding to the transcriptional co-activators, TIF1α and TIF1β, and for binding to the lamin B receptor (Bannister *et al.*, 2001; Lachner *et al.*, 2001; Ye and Worman, 1996; Ye *et al.*, 1997). A direct interaction between HP-1 and the lamin B receptor has been questioned. Both proteins might come together by binding to histone H3/H4 heterodimers (Polioudaki *et al.*, 2001). Interaction of HP-1 with modified histones is apparently a highly dynamic process. HP-1 interacts with both euchromatin and heterochromatin. An increased methylation of histone H3 and a higher grade of chromatin condensation, however, significantly reduce HP-1 mobility (Cheutin et al., 2003).

HA95 (HAP95, NAKAP95) is present throughout the nucleus. Its sequence and its functionality are similar to that of the nuclear A-kinase anchoring protein, AKAP-95. Like AKAP-95, it possesses nuclear localization signals and two putative zinc fingers. In contrast to AKAP-95, it does not have a binding site for the A kinase-regulatory subunit, isoform 2. Instead, HA95 seems to directly bind the catalytic a-subunit of protein kinase A (PKA), which might be important for regulatory events (Martins et al., 2000; Örstavik et al., 2000; Seki et al., 2000). In viral infection, HA95 associates, for example, with the repeat region of the Epstein–Barr virus protein, EBNA-LP. This leads to a relocation of HA95 and associated catalytic subunits of PKA to PML bodies and affects transcription (Han et al., 2001, 2002). HA95 could therefore serve as scaffold for transcriptional regulation. HA95 is tightly bound to chromatin and remains bound to chromatin throughout mitosis. It also binds to the INM proteins LBR and LAP-2β. The latter interaction seems to be responsible for replication initiation but is not required for the elongation phase of replication (Martins et al., 2003). By interacting with chromatin and with INM proteins, HA95 is involved in attaching nuclear membranes to chromatin. Functionally, HA95 seems not to interact directly with the lamina, but is required for the detachment of the nuclear membranes from chromatin and for subsequent chromatin condensation, and for the disassembly of nuclear membranes in mitosis. It does, however, not affect nuclear envelope reassembly after mitosis (Martins et al., 2000).

Finally, YA is a cell cycle-dependent nuclear envelope constituent that is required for embryonic mitosis, particularly for the initiation of the first mitosis in the drosophila embryo. It is provided by the oocyte. Its expression starts during post-oogenic maturation and persists throughout embryogenesis (Lin and Wolfner, 1991). YA associates with the nuclear envelope from interphase to metaphase. During the remainder of mitosis, it is dispersed within the cytoplasm. It localizes to the nuclear envelope by binding to lamin Dm0 with its carboxy-terminal lamin interaction region. Besides lamin, it also binds DNA, with a preference for double-stranded DNA, and histone H2B (Goldberg et al., 1998; Mani et al., 2003; Yu and Wolfner, 2002).

4.6 Proteins putatively associated with the nuclear envelope

In proteomic screens of the nuclear envelope, as in any other screen on a subcellular structure, many more proteins can be identified, whose participation in building the structure is unclear or questionable. For example, mitochondrial proteins are regularly found in such screens despite an effective removal of mitochondria. Such proteins probably have to be judged as contaminants, whose presence in the analysed fraction results from artificial interactions generated during the purification procedure. It is, however, much more difficult to validate all the other possible proteins. For the nuclear envelope, these would include cytoplasmic and endoplasmic components such as transport factors. For the nuclear envelope, the identification of a large number of such proteins could be expected (for an example, see Dreger et al., 2001). The validation of these proteins in terms of functionally relevant interactions at the nuclear envelope, however, is in most cases far from clear. The most interesting interactions at the inner face of the nuclear envelope, for example, have to be proven individually. All possible interactions could also arise from proteins being kept together via macromolecular frameworks like lamina, intranuclear lamin filaments, chromatin structures, and, possibly, a kind of nuclear matrix.

4.7 Post-translational modifications at the nuclear envelope

Finally, the description of the proteome would not be complete if post-translational modifications were not taken into account. Little is known about the occurrence and importance of such modifications. Best characterized are phosphorylation events in the context of mitosis, which are used for solubilizing the structures around the lamina. Furthermore, there is a body of data on the O-glycosylation of nucleoporins, which can be used to block nucleocytoplasmic transport by applying wheat germ agglutinin. During mitosis, phosphorylation of lamins leads to the disassembly of the lamina (Collas *et al.*, 1997; Fields and Thompson, 1995). Likewise, the phosphorylation of LAP2 and nuclear envelope proteins disrupts interactions with lamins and other proteins enabling the detachment of the nuclear membranes and the retraction of the mitotic endoplasmic reticulum from the chromatin (Favreau *et al.*, 1996; Foisner and Gerace, 1993). Much less is known about interphase phosphorylation of nuclear envelope constituents. The phosphorylation of lamin B during S-phase and the phosphorylation at different sites of the common domain of LAP2 have been reported (Dreger *et al.*, 1999; Kill and Hutchison, 1995; Stuurman *et al.*, 1995). Also, phosphorylation at different sites of the LBR by its LBR kinase (RS-kinase) and other kinases has been well documented (Nikolakaki *et al.*, 1996, Nikolakaki *et al.*, 1997, Simos and Georgatos, 1992). The functional relevance of these phosphorylation sites is still unknown. Finally, tyrosine phosphorylation at the nuclear envelope has been reported, with a few candidate substrates of tyrosine kinases at the nuclear envelope (Otto *et al.*, 2000). Up to now, it has however not been possible to identify phosphorylation sites or clarify the context of such tyrosine phosphorylations.

5. Conclusions

More than 20 years after the identification of the lamin B receptor as the first example of an inner nuclear membrane protein that binds to the lamina, our knowledge of the nuclear envelope and its proteins and molecular interactions still is very limited. A major breakthrough has of course been the thorough identification of presumably all nucleoporins of the nuclear pore complexes of yeast and of higher eukaryotes. But despite first attempts to systematically characterize the total nuclear envelope and its substructures by protein chemical means, only a handful of new proteins of the inner nuclear membrane and one putative constituent of the outer nuclear membrane have been added to the list of validated nuclear envelope proteins. Today, we are still far away from knowing even the basic constituents of the nuclear envelope common to most eukaryotic cells let alone tissue- or differentiation-dependent nuclear envelope proteins. To complicate the situation, only a minimal set of functional assays exists to investigate the biological functions of even the known NE constituents.

Therefore, the analysis of the protein composition of the nuclear envelope will have to continue and to be refined. Proteins at a concentration below that of cytoskeletal elements are still not sufficiently accessible and small (below 25 kDa) as well as very large proteins (above 200 kDa) are difficult to analyse. Furthermore, differences linked to differentiation or to phases of the cell cycle should be analysed. Finally, mass spectrometric techniques like precursor ion scans make post-translational modifications of proteins accessible for analysis such that differences due to the cellular status could be analysed.

One major task for such proteomic studies is the validation of candidate constituents of the structure under scrutiny. At the moment, this has to be done for each protein individually. A strategy that could ease this step somewhat would be to isolate complexes formed with *bona fide* constituents of the nuclear envelope. To achieve this, methods should be developed that allow the isolation of such endogenous complexes despite high affinity binding to components of the macromolecular networks of lamina and chromatin. Such methods should focus on avoiding harsh treatments that would first dissolve native interactions, making it necessary to reconstitute such interactions subsequently. Likewise, the analysis of endogenous complexes would be preferable to the analysis of complexes formed around recombinant proteins or protein fragments. The benefit of such methods would be the observation of identified nuclear envelope proteins in their native environment, which would optimally include the preservation of molecular interactions established in the intact cell.

References

Aebersold, R. and Mann, M. (2003) Mass spectrometry-based proteomics. *Nature* **422**: 198–207.

Aebi, U., Cohn, J., Buhle, L. and Gerace, L. (1986) The nuclear lamina is a meshwork of inter-mediate-type filaments. *Nature* **323**: 560–564.

Andersen, J.S., Lyon, C.E., Fox, A.H., Leung, A.K., Lam, Y.W., Steen, H., Mann, M. and Lamond, A.I. (2002) Directed proteomic analysis of the human nucleolus. *Curr. Biol.* **12**: 1–11.

Apel, E.D., Lewis, R.M., Grady, R.M. and Sanes, J.R. (2000) Syne-1, a dystrophin- and Klarsicht-related protein associated with synaptic nuclei at the neuromuscular junction. *J. Biol. Chem.* **275**: 31986–31995.

Ashery-Padan, R., Ulitzur, N., Arbel, A., Goldberg, M., Weiss, A.M., Maus, N., Fisher, P.A. and Gruenbaum, Y. (1997a) Localization and posttranslational modifications of otefin, a protein required for vesicle attachment to chromatin, during Drosophila melanogaster development. *Mol. Cell Biol.* **17**: 4114–4123.

Ashery-Padan, R., Weiss, A.M., Feinstein, N. and Gruenbaum, Y. (1997b) Distinct regions specify the targeting of otefin to the nucleoplasmic side of the nuclear envelope. *J. Biol. Chem.* **272**: 2493–2499.

Bannister, A., Zegerman, P., Partridge, J.F., Miska, E.A., Thomas, J.O., Allshire, R.C. and Kouzarides, T. (2001) Selective recognition of methylated lysine 9 on histone H3 by the HP1 chromo domain. *Nature* **410**: 120–124.

Beaudouin, J., Gerlich, D., Daigle, N., Eils, R. and Ellenberg, J. (2002) Nuclear envelope breakdown proceeds by microtubule-induced tearing of the lamina. *Cell* **108**: 83–96.

Bell, A.C., West, A.G. and Felsenfeld, G. (2001) Insulators and boundaries: versatile regulatory elements in the eukaryotic. *Science* **291**: 447–450.

Bione, S., Maestrini, E., Rivella, S., Mancini, M., Regis, S., Romeo, G. and Toniolo, D. (1994) Identification of a novel X-linked gene responsible for Emery-Dreifuss muscular dystrophy. *Nat. Genet.* **8**: 323–327.

Bonne, G., Di Barletta, M.R., Varnous, S., *et al.* (1999) Mutations in the gene encoding lamin A/C cause autosomal dominant Emery-Dreifuss muscular dystrophy. *Nat. Genet.* **21**: 285–288.

Burke, B. and Ellenberg, J. (2002) Remodelling the walls of the nucleus. *Nature Rev. Mol. Cell Biol.* **3**: 487–497.

Burke, B. and Stewart, C.L. (2002) Life at the edge: the nuclear envelope and human disease. *Nature Rev. Mol. Cell Biol.* **3**: 575–585.

Cheutin, T., McNairn, A.J., Jenuwein, T., Gilbert, D.M., Singh, P.B. and Misteli, T. (2003) Maintenance of stable heterochromatin domains by dynamic HP1 binding. *Science* **299**: 721–725.

Chial, H.J., Rout, M.P., Giddings, T.H. and Winey, M. (1998) *Saccharomyces cerevisiae* Ndc1p is a shared component of nuclear pore complexes and spindle pole bodies. *J. Cell Biol.* **143**: 1789–1800.

Collas, P. (1999) Sequential PKC- and Cdc2-mediated phosphorylation events elicit zebrafish nuclear envelope disassembly. *J. Cell Sci.* **112**: 977–987.

Collas, P., Thompson, L., Fields, A.P., Poccia, D.L. and Courvalin, J.C. (1997) Protein kinase C-mediated interphase lamin B phosphorylation and solubilization. *J. Biol. Chem.* **272**: 21274–21280.

Coy, J.F., Wiemann, S., Bechmann, I., Bachner, D., Nitsch, R., Kretz, O., Christiansen, H. and Poustka, A. (2002) Pore membrane and/or filament interacting like protein 1 (POMFIL1) is predominantly expressed in the nervous system and encodes different protein isoforms. *Gene* **290**: 73–94.

Cronshaw, J.M., Krutchinsky, A.N., Zhang, W., Chait, B.T. and Matunis, M.J. (2002) Proteomic analysis of the mammalian nuclear pore complex. *J. Cell Biol.* **158**: 915–927.

Davis, L.I. (1995) The nuclear pore complex. *Annu. Rev. Biochem.* **64**: 865–896.

de Bruyn Kops, A. and Guthrie, C. (2001) An essential nuclear envelope integral membrane protein, Brr6p, required for nuclear transport. *EMBO J.* **20**: 4183–4193.

de la Luna, S., Allen, K.E., Mason, S.L. and La Thangue, N.B. (1999) Integration of a growth-suppressing BTB/POZ domain protein with the DP component of the E2F transcription factor. *EMBO J.* **18**: 212–228.

Dechat, T., Gotzmann, J., Stockinger, A., Harris, C.A., Talle, M.A., Siekierka, J.J. and Foisner, R. (1998) Detergent-salt resistance of LAP2alpha in interphase nuclei and phosphorylation-dependent association with chromosomes early in nuclear assembly implies functions in nuclear structure dynamics. *EMBO J.* **17**: 4887–4902.

Dellaire, G., Farrall, R. and Bickmore, W.A. (2003) The nuclear protein database (NPD): sub-nuclear localisation and functional annotation of the nuclear proteome. *Nucl. Acids Res.* **31**: 328–330.

Dreger, M. (2003) Subcellular proteomics. *Mass Spectrom. Rev.* **22**: 27–56.

Dreger, M., Otto, H., Neubauer, G., Mann, M. and Hucho, F. (1999) Identification of phosphorylation sites in native lamina-associated polypeptide 2 beta. *Biochemistry* **38**: 9426–9434.

Dreger, M., Bengtsson, L., Schoneberg, T., Otto, H. and Hucho, F. (2001) Nuclear envelope proteomics: novel integral membrane proteins of the inner nuclear membrane. *Proc. Natl Acad. Sci. USA* **98**: 11943–11948.

Drummond, S.P. and Wilson, K.L. (2002) Interference with the cytoplasmic tail of gp210 disrupts "close apposition" of nuclear membranes and blocks nuclear pore dilation. *J. Cell Biol.* **158**: 53–62.

Eide, T., Coghlan, V., Örstavik, S., *et al.*, (1998) Molecular cloning, chromosomal localization, and cell cycle-dependent subcellular distribution of the A-kinase anchoring protein, AKAP95. *Exp. Cell Res.* **238**: 305–316.

Ellis, J.A., Craxton, M., Yates, J.R. and Kendrick-Jones, J. (1998) Aberrant intracellular targeting and cell cycle-dependent phosphorylation of emerin contribute to the Emery-Dreifuss muscular dystrophy phenotype. *J. Cell Sci.* **111**: 781–792.

Enoch, T., Peter, M., Nurse, P. and Nigg, E.A. (1991) p34cdc2 acts as a lamin kinase in fission yeast. *J. Cell Biol.* **112**: 797–807.

Erdmann, R. and Blobel, G. (1996) Identification of Pex13p a peroxisomal membrane receptor for the PTS1 recognition factor. *J. Cell Biol.* **135**: 111–121.

Eriksson, M., Brown, W.T., Gordon, L.B., *et al.*, (2003) Recurrent de novo point mutations in lamin A cause Hutchinson-Gilford progeria syndrome. *Nature* **423**: 293–298.

Favreau, C., Worman, H.J., Wozniak, R.W., Frappier, T. and Courvalin, J.C. (1996) Cell cycle-dependent phosphorylation of nucleoporins and nuclear pore membrane protein Gp210. *Biochemistry* **35**: 8035–8044.

Fields, A.P. and Thompson, L.J. (1995) The regulation of mitotic nuclear envelope breakdown: a role for multiple lamin kinases. *Prog. Cell Cycle Res.* **1**: 271–286.

Foisner, R. and Gerace, L. (1993) Integral membrane proteins of the nuclear envelope interact with lamins and chromosomes, and binding is modulated by mitotic phosphorylation. *Cell* **73**: 1267–1279.

Furukawa, K. (1999) LAP2 binding protein 1 (L2BP1/BAF) is a candidate mediator of LAP2–chromatin interaction. *J. Cell Sci.* **112**: 2485–2492.

Furukawa, K., Pante, N., Aebi, U. and Gerace, L. (1995) Cloning of a cDNA for lamina-associated polypeptide 2 (LAP2) and identification of regions that specify targeting to the nuclear envelope. *EMBO J.* **14**: 1626–1636.

Galeva, N. and Altermann, M. (2002) Comparison of one-dimensional and two-dimensional gel electrophoresis as a separation tool for proteomic analysis of rat liver microsomes: cytochromes P450 and other membrane proteins. *Proteomics* **2**: 713–722.

Gant, T.M., Harris, C.A. and Wilson, K.L. (1999) Roles of LAP2 proteins in nuclear assembly and DNA replication: truncated LAP2beta proteins alter lamina assembly, envelope formation, nuclear size, and DNA replication efficiency in Xenopus laevis extracts. *J. Cell Biol.* **144**: 1083–1096.

Gant, T.M. and Wilson, K.L. (1997) Nuclear assembly. *Ann. Rev. Cell Dev. Biol.* **13**: 669–695.

Georgatos, S.D., Meier, J. and Simos, G. (1994) Lamins and lamin-associated proteins. *Curr. Opin. Cell Biol.* **6**: 347–353.

Goldberg, M., Lu, H., Stuurman, N., Ashery-Padan, R., Weiss, A.M., Yu, J., Bhattacharyya, D., Fisher, P.A., Gruenbaum, Y. and Wolfner, M.F. (1998) Interactions among Drosophila nuclear envelope proteins lamin, otefin, and YA. *Mol. Cell Biol.* **18**: 4315–4323.

Gonzalez, C. and Bejarano, L.A. (2000) Protein traps: using intracellular localization for cloning. *Trends Cell. Biol.* **10**: 162–165.

Greber, U.F., Senior, A. and Gerace, L. (1990) A major glycoprotein of the nuclear pore complex is a membrane-spanning polypeptide with a large lumenal domain and a small cytoplasmic tail. *EMBO J.* **9**: 1495–1502.

Hagan, I. and Yanagida, M. (1995) The product of the spindle formation gene sad1+ associates with the fission yeast spindle pole body and is essential for viability. *J. Cell Biol.* **129**: 1033–1047.

Hallberg, E., Wozniak, R.W. and Blobel, G. (1993) An integral membrane protein of the pore membrane domain of the nuclear envelope contains a nucleoporin-like region. *J. Cell Biol.* **122**: 513–521.

Han, I., Harada, S., Weaver, D., Xue, Y., Lane, W., Örstavik, S., Skalhegg, B. and Kieff, E. (2001) EBNA-LP associates with cellular proteins including DNA-PK and HA95. *J. Virol.* **75**: 2475–2481.

Han, I., Xue, Y., Harada, S., Örstavik, S., Skalhegg, B. and Kieff, E. (2002) Protein kinase A associates with HA95 and affects transcriptional coactivation by Epstein-Barr virus nuclear proteins. *Mol. Cell. Biol.* **22**: 2136–2146.

Harel, A., Zlotkin, E., Nainudel-Epszteyn, S., Feinstein, N., Fisher, P.A. and Gruenbaum, Y. (1989) Persistence of major nuclear envelope antigens in an envelope-like structure during mitosis in Drosophila melanogaster embryos. *J. Cell Sci.* **94**: 463–470.

Harris, C.A., Andryuk, P.J., Cline, S.W., Mathew, S., Siekierka, J.J. and Goldstein, G. (1995) Structure and mapping of the human thymopoietin (TMPO) gene and relationship of human TMPO beta to rat lamin-associated polypeptide 2. *Genomics* **28**: 198–205.

Hartinger, J., Stenius, K., Högemann, D. and Jahn, R. (1996) 16-BAC/ SDS-PAGE: a two-dimensional gel electrophoresis system suitable for the separation of integral membrane proteins. *Anal. Biochem.* **240**: 126–133.

Hayward-Lester, A., Hewetson, A., Beale, E.G., Oefner, P.J., Doris, P.A. and Chilton, B.S. (1996) Cloning, characterization, and steroid-dependent posttranscriptional processing of RUSH-1 alpha and beta, two uteroglobin promoter-binding proteins. *Mol. Endocrinol.* **10**: 1335–1349.

Hennekes, H., Peter, M., Weber, K. and Nigg, E.A. (1993) Phosphorylation on protein kinase C sites inhibits nuclear import of lamin B2. *J. Cell Biol.* **120**: 1293–1304.

Ho, A.K., Raczniak, G.A., Ives, E.B. and Wente, S.R. (1998) The integral membrane protein snl1p is genetically linked to yeast nuclear pore complex function. *Mol. Biol. Cell* **9**: 355–373.

Hoffmann, K., Dreger, C.K., Olins, A.L., *et al.*, (2002) Mutations in the gene encoding the lamin B receptor produce an altered nuclear morphology in granulocytes (Pelger Huet anomaly). *Nat. Genet.* **31**: 410–414.

Holmer, L., Pezhman, A. and Worman, H.J. (1998) The human lamin B receptor/sterol reductase multigene family. *Genomics* **54**: 469–476.

Hutchison, C.J. (2002) Lamins: building blocks or regulators of gene expression? *Nature Rev. Mol. Cell Biol.* **3**: 848–858.

Hutchison, C.J., Alvarez-Reyes, M. and Vaughan, O.A. (2001) Lamins in disease: why do ubiquitously expressed nuclear envelope proteins give rise to tissue-specific disease phenotypes? *J. Cell Sci.* **114**: 9–19.

Hutchison, C.J., Bridger, J.M., Cox, L.S. and Kill, I.R. (1994) Weaving a pattern from disparate threads: lamin function in nuclear assembly and DNA replication. *J. Cell Sci.* **107**: 3259–3269.

Jenkins, H., Holman, T., Lyon, C., Lane, B., Stick, R. and Hutchison, C. (1993) Nuclei that lack a lamina accumulate karyophilic proteins and assemble a nuclear matrix. *J. Cell Sci.* **106**: 275–285.

Kahn, P. (1995) From genome to proteome: looking at a cell's proteins. *Science* **270**: 369–370.

Kill, I.R. and Hutchison, C.J. (1995) S-phase phosphorylation of lamin B2. *FEBS Lett.* **377**: 26–30.

Kondo, Y., Kondoh, J., Hayashi, D., Ban, T., Takagi, M., Kamei, Y., Tsuji, L., Kim, J. and Yoneda, Y. (2002) Molecular cloning of one isotype of human lamina-associated polypeptide 1s and a topological analysis using its deletion mutants. *Biochem. Biophys. Res. Commun.* **294**: 770–778.

Korenbaum, E. and Rivero, F. (2002) Calponin homology domains at a glance. *J. Cell Sci.* **115**: 3543–3545.

Lachner, M., O'Carroll, D., Rea, S., Mechtler, K. and Jenuwein, T. (2001) Methylation of histone H3 lysine 9 creates a binding site for HP1 proteins. *Nature* **410**: 116–120.

Lee, K.K., Starr, D., Cohen, M., Liu, J., Han, M., Wilson, K.L. and Gruenbaum, Y. (2002) Lamin-dependent localization of UNC-84, a protein required for nuclear migration in *Caenorhabditis elegans. Mol. Biol. Cell* **13**: 892–901.

Lee, R.P., Doughty, S.W., Ashman, K. and Walker, J. (1996) Purification of hydrophobic integral membrane proteins from *Mycoplasma hyopneumoniae* by reversed-phase high-performance liquid chromatography. *J. Chromatogr. A* **737**: 273–279.

Lin, F., Blake, D., Callebaut, I., McBurney, M., Paulin-Levasseur, M. and Worman, H.J. (1998) MAN1 is an integral protein of the nuclear envelope inner membrane that shares a conserved sequence motif with LAP2/thymopoietin and emerin. *Mol. Biol. Cell* **(Suppl.) 9**: 447a.

Lin, F., Blake, D.L., Callebaut, I., Skerjanc, I.S., Holmer, L., McBurney, M.W., Paulin-Levasseur, M. and Worman, H.J. (2000) MAN1, an inner nuclear membrane protein that shares the LEM domain with lamina-associated polypeptide 2 and emerin. *J. Biol. Chem.* **275**: 4840–4847.

Lin, H.F. and Wolfner, M.F. (1991) The Drosophila maternal-effect gene fs(1)Ya encodes a cell cycle-dependent nuclear envelope component required for embryonic mitosis. *Cell* **64**: 49–62.

Liu, J., Lee, K.K., Segura-Totten, M., Neufeld, E., Wilson, K.L. and Gruenbaum, Y. (2003) MAN1 and emerin have overlapping function(s) essential for chromosome segregation and cell division in *Caenorhabditis elegans*. *Proc. Natl Acad. Sci. USA* **100**: 4598–4603.

Maison, C., Pyrpasopoulou, A., Theodoropoulos, P.A. and Georgatos, S.D. (1997) The inner nuclear membrane protein LAP1 forms a native complex with B-type lamins and partitions with spindle-associated mitotic vesicles. *EMBO J.* **16**: 4839–4850.

Malone, C.J., Fixsen, W.D., Horvitz, H.R. and Han, M. (1999) UNC-84 localizes to the nuclear envelope and is required for nuclear migration and anchoring during *C. elegans* development. *Development* **126**: 3171–3181.

Mani, S.S., Rajagopal, R., Garfinkel, A.B., Fan, X. and Wolfner, M.F. (2003) A hydrophilic lamin-binding domain from the Drosophila YA protein can target proteins to the nuclear envelope. *J. Cell Sci.* **116**: 2067–2072.

Manilal, S., Nguyen, T.M., Sewry, C.A. and Morris, G.E. (1996) The Emery-Dreifuss muscular dystrophy protein, emerin, is a nuclear membrane protein. *Hum. Mol. Genet.* **5**: 801–808.

Mansharamani, M., Hewetson, A. and Chilton, B.S. (2001) Cloning and characterization of an atypical type IV P-type ATPase that binds to the RING motif of RUSH transcription factors. *J. Biol. Chem.* **276**: 3641–3649.

Marelli, M., Lusk, C.P., Chan, H., Aitchison, J.D. and Wozniak, R.W. (2001) A link between the synthesis of nucleoporins and the biogenesis of the nuclear envelope. *J. Cell Biol.* **153**: 709–724.

Martin, L., Crimaudo, C. and Gerace, L. (1995) cDNA cloning and characterization of lamina-associated polypeptide 1C (LAP1C), an integral protein of the inner nuclear membrane. *J. Biol. Chem.* **270**: 8822–8828.

Martins, S., Eikvar, S., Furukawa, K. and Collas, P. (2003) HA95 and LAP2beta mediate a novel chromatin-nuclear envelope interaction implicated in initiation of DNA replication. *J. Cell Biol.* **160**: 177–188.

Martins, S.B., Eide, T., Steen, R.L., Jahnsen, T., Skalhegg, B.S. and Collas, P. (2000) HA95 is a protein of the chromatin and nuclear matrix regulating nuclear envelope dynamics. *J. Cell Sci.* **113**: 3703–3713.

Mislow, J.M., Holaska, J.M., Kim, M.S., Lee, K.K., Segura-Totten, M., Wilson, K.L. and McNally, E.M. (2002a) Nesprin-1alpha self-associates and binds directly to emerin and lamin A in vitro. *FEBS Lett.* **525**: 135–140.

Mislow, J.M., Kim, M.S., Davis, D.B. and McNally, E.M. (2002b) Myne-1, a spectrin repeat transmembrane protein of the myocyte inner nuclear membrane, interacts with lamin A/C. *J. Cell Sci.* **115**: 61–70.

Moir, R.D. and Spann, T.P. (2001) The structure and function of nuclear lamins: implications for disease. *Cell. Mol. Life Sci.* **58**: 1748–1757.

Morris, N.R. (2000) Nuclear migration. From fungi to the mammalian brain. *J. Cell Biol.* **148**: 1097–1101.

Nagano, A., Koga, R., Ogawa, M., Kurano, Y., Kawada, J., Okada, R., Hayashi, Y.K., Tsukahara, T. and Arahata, K. (1996) Emerin deficiency at the nuclear membrane in patients with Emery-Dreifuss muscular dystrophy. *Nat. Genet.* **12**: 254–259.

Nigro, V., Bruni, P., Ciccodicola, A., Politano, L., Nigro, G., Piluso, G., Cappa, V., Covone, A.E., Romeo, G. and D'Urso, M. (1995) SSCP detection of novel mutations in patients with Emery-Dreifuss muscular dystrophy: definition of a small C-terminal region required for emerin function. *Hum. Mol. Genet.* **4**: 2003–2004.

Nikolakaki, E., Meier, J., Simos, G., Georgatos, S.D. and Giannakouros, T. (1997) Mitotic phosphorylation of the lamin B receptor by a serine/arginine kinase and p34(cdc2). *J. Biol. Chem.* **272**: 6208–6213.

Nikolakaki, E., Simos, G., Georgatos, S.D. and Giannakouros, T. (1996) A nuclear envelope-associated kinase phosphorylates arginine-serine motifs and modulates interactions between the lamin B receptor and other nuclear proteins. *J. Biol. Chem.* **271**: 8365–8372.

Nili, E., Cojocaru, G.S., Kalma, Y., *et al.*, (2001) Nuclear membrane protein LAP2beta mediates transcriptional repression alone and together with its binding partner GCL (germ-cell-less). *J. Cell Sci.* **114**: 3297–3307.

Örstavik, S., Eide, T., Collas, P., Han, I.O., Tasken, K., Kieff, E., Jahnsen, T. and Skalhegg, B.S. (2000) Identification, cloning and characterization of a novel nuclear protein, HA95, homologous to A-kinase anchoring protein 95. *Biol. Cell* **92**: 27–37.

Östlund, C. and Worman, H.J. (2003) Nuclear envelope proteins and neuromuscular diseases. *Muscle Nerve* **27**: 393–406.

Otto, H., Dreger, M., Bengtsson, L. and Hucho, F. (2000) Tyrosine phosphorylation at the nuclear envelope. In: *Dynamic Organization of Nuclear Function*, p. 17. Cold Spring Harbor Laboratory, Cold Spring Harbor, New York.

Padan, R., Nainudel-Epszteyn, S., Goitein, R., Fainsod, A. and Gruenbaum, Y. (1990) Isolation and characterization of the Drosophila nuclear envelope otefin cDNA. *J. Biol. Chem.* **265**: 7808–7813.

Paulin-Levasseur, M., Blake, D.L., Julien, M. and Rouleau, L. (1996) The MAN antigens are non-lamin constituents of the nuclear lamina in vertebrate cells. *Chromosoma* **104**: 367–379.

Peter, M., Nakagawa, J., Doree, M., Labbe, J.C. and Nigg, E.A. (1990) In vitro disassembly of the nuclear lamina and M phase-specific phosphorylation of lamins by cdc2 kinase. *Cell* **61**: 591–602.

Polioudaki, H., Kourmouli, N., Drosou, V., Bakou, A., Theodoropoulos, P.A., Singh, P.B., Giannakouros, T. and Georgatos, S.D. (2001) Histones H3/H4 form a tight complex with the inner nuclear membrane protein LBR and heterochromatin protein 1. *EMBO Rep.* **2**: 920–925.

Rolls, M.M., Stein, P.A., Taylor, S.S., Ha, E., McKeon, F. and Rapoport, T.A. (1999) A visual screen of a GFP-fusion library identifies a new type of nuclear envelope membrane protein. *J. Cell Biol.* **146**: 29–43.

Rout, M.P., Aitchison, J.D., Suprapto, A., Hjertaas, K., Zhao, Y. and Chait, B.T. (2000) The yeast nuclear pore complex. composition, architecture, and transport mechanism. *J. Cell Biol.* **148**: 635–652.

Santoni, V., Molloy, M. and Rabilloud, T. (2000) Membrane proteins and proteomics: un amour impossible? *Electrophoresis* **21**: 1054–1070.

Schrotz-King, P., Wilm, M., Andersen, J.S., Ashman, K., Podtelejnikov, A.V., Bachi, A., King, A. and Mann, M. (1999) Use of mass spectrometric methods for protein identification in receptor research. *J. Recept. Signal Transduct. Res.* **19**: 659–672.

Schuler, E., Lin, F. and Worman, H.J. (1994) Characterization of the human gene encoding LBR, an integral protein of the nuclear envelope inner membrane. *J. Biol. Chem.* **269**: 11312–11317.

Seki, N., Ueki, N., Yano, K., Saito, T., Masuho, Y. and Muramatsu, M. (2000) cDNA cloning of a novel human gene NAKAP95, neighbor of A-kinase anchoring protein 95 (AKAP95) on chromosome 19p13.11-p13.12 region. *J. Human Gen.* **45**: 31–37.

Shumaker, D.K., Lee, K.K., Tanhehco, Y.C., Craigie, R. and Wilson, K.L. (2001) LAP2 binds to BAF.DNA complexes: requirement for the LEM domain and modulation by variable regions. *EMBO J.* **20**: 1754–1764.

Silve, S., Dupuy, P.H., Ferrara, P. and Loison, G. (1998) Human lamin B receptor exhibits sterol C14-reductase activity in Saccharomyces cerevisiae. *Biochim. Biophys. Acta* **1392**: 233–244.

Simos, G. and Georgatos, S.D. (1992) The inner nuclear membrane protein p58 associates in vivo with a p58 kinase and the nuclear lamins. *EMBO J.* **11**: 4027–4036.

Simos, G. and Georgatos, S.D. (1994) The lamin B receptor-associated protein p34 shares sequence homology and antigenic determinants with the splicing factor 2-associated protein p32. *FEBS Lett.* **346**: 225–228.

Siniossoglou, S., Santos-Rosa, H., Rappsilber, J., Mann, M. and Hurt, E. (1998) A novel complex of membrane proteins required for formation of a spherical nucleus. *EMBO J.* **17**: 6449–6464.

Soderqvist, H. and Hallberg, E. (1994) The large C-terminal region of the integral pore membrane protein, POM121, is facing the nuclear pore complex. *Eur. J. Cell Biol.* **64**: 186–191.

Sondermann, H., Ho, A.K., Listenberger, L.L., Siegers, K., Moarefi, I., Wente, S.R., Hartl, F.U. and Young, J.C. (2002) Prediction of novel Bag-1 homologs based on structure/function analysis identifies Snl1p as an Hsp70 co-chaperone in *Saccharomyces cerevisiae. J. Biol. Chem.* **277**: 33220–33227.

Soullam, B. and Worman, H.J. (1993) The amino-terminal domain of the lamin B receptor is a nuclear envelope targeting signal. *J. Cell Biol.* **120**: 1093–1100.

Spann, T.P., Goldman, A.E., Wang, C., Huang, S. and Goldman, R.D. (2002) Alteration of nuclear lamin organization inhibits RNA polymerase II-dependent transcription. *J. Cell Biol.* **156**: 603–608.

Starr, D.A. and Han, M. (2002) Role of ANC-1 in tethering nuclei to the actin cytoskeleton. *Science* **298**: 406–409.

Starr, D.A. and Han, M. (2003) ANChors away: an actin based mechanism of nuclear positioning. *J. Cell Sci.* **116**: 211–216.

Starr, D.A., Hermann, G.J., Malone, C.J., Fixsen, W., Priess, J.R., Horvitz, H.R. and Han, M. (2001) unc-83 encodes a novel component of the nuclear envelope and is essential for proper nuclear migration. *Development* **128**: 5039–5050.

Strambio-de-Castillia, C., Blobel, G. and Rout, M.P. (1995) Isolation and characterization of nuclear envelopes from the yeast *Saccharomyces. J. Cell Biol.* **131**: 19–31.

Strelkov, S.V., Herrmann, H. and Aebi, U. (2003) Molecular architecture of intermediate filaments. *BioEssays* **25**: 243–251.

Stuurman, N., Maus, N. and Fisher, P.A. (1995) Interphase phosphorylation of the Drosophila nuclear lamin: site-mapping using a monoclonal antibody. *J. Cell Sci.* **108**: 3137–3144.

Stuurman, N., Heins, S. and Aebi, U. (1998) Nuclear lamins: their structure, assembly, and interactions. *J. Struct. Biol.* **122**: 42–66.

Sullivan, T., Escalante-Alcalde, D., Bhatt, H., Anver, M., Bhat, N., Nagashima, K., Stewart, C.L. and Burke, B. (1999) Loss of A-type lamin expression compromises nuclear envelope integrity leading to muscular dystrophy. *J. Cell Biol.* **147**: 913–920.

Sutherland, H.G., Mumford, G.K., Newton, K., Ford, L.V., Farrall, R., Dellaire, G., Caceres, J.F. and Bickmore, W.A. (2001) Large-scale identification of mammalian proteins localized to nuclear sub-compartments. *Hum. Mol. Genet.* **10**: 1995–2011.

Tcheperegine, S.E., Marelli, M. and Wozniak, R.W. (1999) Topology and functional domains of the yeast pore membrane protein Pom152p. *J. Biol. Chem.* **274**: 5252–5258.

Tullio-Pelet, A., Salomon, R., Hadj-Rabia, S., *et al.* (2000) Mutant WD-repeat protein in triple-A syndrome. *Nat. Genet.* **26**: 332–335.

Ulitzur, N., Harel, A., Goldberg, M., Feinstein, N. and Gruenbaum, Y. (1997) Nuclear membrane vesicle targeting to chromatin in a Drosophila embryo cell-free system. *Mol. Biol. Cell* **8**: 1439–1448.

Washburn, M.P., Wolters, D. and Yates, J.R., 3rd. (2001) Large-scale analysis of the yeast proteome by multidimensional protein identification technology. *Nat. Biotechnol.* **19**: 242–247.

Wilkins, M.R., Pasquali, C., Appel, R.D., *et al.* (1996) From proteins to proteomes: large scale protein identification by two-dimensional electrophoresis and amino acid analysis. *Biotechnology (NY)* **14**: 61–65.

Worman, H.J. and Courvalin, J.C. (2002) The nuclear lamina and inherited disease. *Trends Cell. Biol.* **12**: 591–598.

Worman, H.J., Yuan, J., Blobel, G. and Georgatos, S.D. (1988) A lamin B receptor in the nuclear envelope. *Proc. Natl Acad. Sci. USA* **85**: 8531–8534.

Worman, H.J., Evans, C.D. and Blobel, G. (1990) The lamin B receptor of the nuclear envelope inner membrane: a polytopic protein with eight potential transmembrane domains. *J. Cell Biol.* **111**: 1535–1542.

Wozniak, R.W. and Blobel, G. (1992) The single transmembrane segment of gp210 is sufficient for sorting to the pore membrane domain of the nuclear envelope. *J. Cell Biol.* **119:** 1441–1449.

Wozniak, R.W., Bartnik, E. and Blobel, G. (1989) Primary structure analysis of an integral membrane glycoprotein of the nuclear pore. *J. Cell Biol.* **108:** 2083–2092.

Wozniak, R.W., Blobel, G. and Rout, M.P. (1994) POM152 is an integral protein of the pore membrane domain of the yeast nuclear envelope. *J. Cell Biol.* **125:** 31–42.

Yang, L., Guan, T. and Gerace, L. (1997) Lamin-binding fragment of LAP2 inhibits increase in nuclear volume during the cell cycle and progression into S phase. *J. Cell Biol.* **139:** 1077–1087.

Ye, Q. and Worman, H.J. (1994) Primary structure analysis and lamin B and DNA binding of human LBR, an integral protein of the nuclear envelope inner membrane. *J. Biol. Chem.* **269:** 11306–11311.

Ye, Q. and Worman, H.J. (1996) Interaction between an integral protein of the nuclear envelope inner membrane and human chromodomain proteins homologous to Drosophila HP1. *J. Biol. Chem.* **271:** 14653–14656.

Ye, Q., Callebaut, I., Pezhman, A., Courvalin, J.C. and Worman, H.J. (1997) Domain-specific interactions of human HP1-type chromodomain proteins and inner nuclear membrane protein LBR. *J. Biol. Chem.* **272:** 14983–14989.

Yu, J. and Wolfner, M.F. (2002) The Drosophila nuclear lamina protein YA binds to DNA and histone H2B with four domains. *Mol. Biol. Cell* **13:** 558–569.

Zhang, C., Jenkins, H., Goldberg, M.W., Allen, T.D. and Hutchison, C.J. (1996) Nuclear lamina and nuclear matrix organization in sperm pronuclei assembled in Xenopus egg extract. *J. Cell Sci.* **109:** 2275–2286.

Zhang, Q., Skepper, J.N., Yang, F., Davies, J.D., Hegyi, L., Roberts, R.G., Weissberg, P.L., Ellis, J.A. and Shanahan, C.M. (2001) Nesprins: a novel family of spectrin-repeat-containing proteins that localize to the nuclear membrane in multiple tissues. *J. Cell Sci.* **114:** 4485–4498.

Zhang, Q., Ragnauth, C., Greener, M.J., Shanahan, C.M. and Roberts, R.G. (2002) The nesprins are giant actin-binding proteins, orthologous to Drosophila melanogaster muscle protein MSP-300. *Genomics* **80:** 473–481.

Zhen, Y.Y., Libotte, T., Munck, M., Noegel, A.A. and Korenbaum, E. (2002) NUANCE, a giant protein connecting the nucleus and actin cytoskeleton. *J. Cell Sci.* **115:** 3207–3222.

Zheng, R., Ghirlando, R., Lee, M.S., Mizuuchi, K., Krause, M. and Craigie, R. (2000) Barrier-to-autointegration factor (BAF) bridges DNA in a discrete, higher-order nucleoprotein complex. *Proc. Natl Acad. Sci. USA* **97:** 8997–9002.

3

Nuclear envelope proteins and human disease

Howard J. Worman

1. Nuclear envelope: overall structure

The nuclear membranes, nuclear pore complexes and the nuclear lamina are the major structural components of the nuclear envelope (Worman and Courvalin, 2000). The nuclear membranes can be divided into three morphologically distinguishable continuous domains (*Figure 1*). The outer nuclear membrane is a direct continuation of the rough endoplasmic reticulum, sharing proteins with it and also containing ribosomes on its outer surface. The inner nuclear membrane is closest to the nucleoplasm and is associated with the nuclear lamina. The pore membranes connect the inner and outer nuclear membranes at numerous locations and are associated with the nuclear pore complexes.

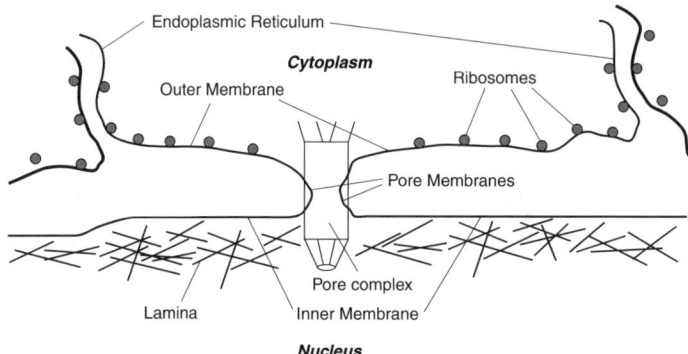

Figure 1. The nuclear envelope is composed of the nuclear membranes, nuclear pore complexes and nuclear lamina. This schematic diagram shows the different nuclear membrane domains and their associated structures. The outer nuclear membrane is continuous with the rough endoplasmic reticulum and has ribosomes on its outer surface. The inner nuclear membrane is associated with the nuclear lamina on its nuclear surface. The pore membranes connect the inner and outer nuclear membranes at numerous points and are associated with the nuclear pore complexes.

The Nuclear Envelope, edited by D.E. Evans, C. Hutchison & J.A. Bryant.
© 2004 Garland Science/BIOS Scientific Publishers

While the nuclear envelope is a relatively 'fixed' structure in interphase, it loses its morphological definition during cell division (Collas and Courvalin, 2000). Nuclear envelope breakdown in mitosis appears to be initiated by 'tearing' or the formation of 'holes' in the nuclear membranes that result from forces generated by microtubles (Beaudoin *et al.*, 2002; Salina *et al.*, 2002). The nuclear lamina and pore complexes disassemble into their protein building blocks. Integral proteins of the inner nuclear membrane associated with the lamina and chromatin in interphase become 'resorbed' into the endoplasmic reticulum (Ellenberg *et al.*, 1997; Yang *et al.*, 1997). As the chromosomes separate, nuclear envelopes reassemble in the two daughter cells in a step-wise process, with integral proteins of the inner nuclear membrane generally targeted first to the decondensing chromosomes, followed by pore complex proteins and, lastly, assembly of the nuclear lamina (Buendina and Courvalin, 1997; Chaudhary and Courvalin, 1993; Collas and Courvalin, 2000).

2. Integral proteins of the inner nuclear membrane of vertebrates

Although the membranes of the nuclear envelope and rough endoplasmic reticulum form a continuous network, the inner and pore membranes contain specific integral proteins, which are not present at significant steady-state concentrations in the outer membrane or endoplasmic reticulum in interphase. Many of the integral proteins of the inner nuclear membrane bind to the nuclear lamina and the chromatin (Worman and Courvalin, 2000). Several integral proteins have been definitively or presumptively localized to the inner nuclear membrane in interphase cells of vertebrates including lamin B receptor (LBR) (Worman *et al.*, 1990; Ye and Worman, 1994), three isoforms of lamina associated polypeptide (LAP 1) (Martin *et al.*, 1995; Senior and Gerace, 1988), several isoforms of LAP2 (Foisner and Gerace, 1993; Furukawa *et al.*, 1995; Harris *et al.*, 1994), emerin (Bione *et al.*, 1994; Manilal *et al.*, 1996; Nagano *et al.*, 1996), nurim (Rolls *et al.*, 1999), MAN1 (Lin *et al.*, 2000; Paulin-Levasseur *et al.*, 1996), nesprin-1/myne-1 (Mislow *et al.*, 2002; Zhang *et al.*, 2001), luma (Dreger *et al.*, 2001), a protein with sequence similarity to *Caenorhabditis elegans* Unc-84A (Dreger *et al.*, 2001) and a RING finger-binding protein (Mansharamani *et al.*, 2001).

3. Nuclear lamina and the human nuclear lamin gene family

The nuclear lamina is a meshwork of intermediate filaments concentrated at the inner surface of the inner nuclear membrane (Aebi *et al.*, 1986). The filaments also extend into the nuclear interior (Hozak *et al.*, 1995). The major protein building blocks of the nuclear lamina are called lamins (Gerace *et al.*, 1978). Complementary DNA cloning in 1986 showed that the lamins are intermediate filament proteins (Fisher *et al.*, 1986; McKeon *et al.*, 1986). Other intermediate filament proteins are cytoplasmic, such as keratins, vimentin, desmin and neurofilaments. Intermediate filament proteins have conserved alpha-helical rod domains and variable head and tail domains. They form homodimers and heterodimers that polymerize into higher-ordered filaments, usually approximately 10 nm in diameter, by still unclear mechanisms (Hermann and Aebi, 2000; Sturman *et al.*, 1998).

Lamins differ from cytoplasmic intermediate filament proteins in vertebrates in several ways (*Figure 2*). Lamins contain an additional 42 amino acids in their rod domains and nuclear localization signals in their tail domains. They are also phospho-ryated at specific sites in mitosis, which leads to disassembly of the filaments into dimers (Collas *et al.*, 1996; Heald and McKeon, 1990; Peter *et al.*, 1990; Ward and Kirschner, 1990). Mammalian lamins, except for lamin C and C2, contain 'CAAX boxes' at their carboxyl-termini that are modified by prenylation, specifically farnesy-lation (Beck *et al.*, 1988, 1990; Farnsworth *et al.*, 1989; Wolda and Glomset 1988). The last three amino acids are then removed by an endoprotease and the terminal cysteine is methylated on its carboxyl group by a methyltransferase. B-type lamins remain farnesylated and carboxymethylated when assembled in the lamina. Farnesylated prelamin A, the lamin A precursor, is recognized by an endoprotease that cleaves off the last 15 amino acids, including the prenylated cysteine (Beck *et al.*, 1990; Kilic *et al.*, 1997; Sinensky *et al.*, 1994). Metalloproteinase Zmpste24 is probably the endoprotease that removes the '-AAX' from prelamin A or possibly the endoprotease that removes the 15 amino acids (Bergo *et al.*, 2002; Leung *et al.*, 2001; Pendas *et al.*, 2002). Farnesylated prelamin A also binds to a nuclear protein called NARF, however, the significance of this interaction is not clear (Barton and Worman, 1999).

The near completion of the human genome sequence has confirmed that three genetic loci encode nuclear lamins (*Table 1*). *LMNA*, which encodes lamins A and C by alternative RNA splicing, was first characterized in 1993 (Lin and Worman, 1993). It is localized to chromosome 1q21.2-21.3 (Wydner *et al.*, 1996). Lamins A and C are expressed in most terminally differentiated somatic cells but lacking from early embryos, some undifferentiated cells and various cancers (Cance *et al.*, 1992; Guilly *et al.*, 1987; Rober *et al.*, 1989; Stewart and Burke, 1987; Worman *et al.*, 1988). *LMNA* also encodes a germ cell-specific isoform called lamin C2 (Fukuwara *et al.*, 1994) and a minor somatic cell isoform called lamin AΔ10 (Machiels *et al.*, 1996). *LMNB1* encodes lamin B1 (Lin and Worman, 1995), which appears to be expressed in all or most somatic cells (Cance *et al.*, 1992; Guilly *et al.*, 1987; Rober *et al.*, 1989; Stewart and Burke, 1987; Worman *et al.*, 1988). *LMNB1* is localized to chromosome 5q23.2-31.1 (Wydner *et al.*, 1996). *LMNB2* is localized to chromosome 19p13.3 and encodes lamin B2 (Biamonti *et al.*, 1992), which is also probably expressed in all or most somatic cells. A germ cell-specific isoform called lamin B3, first identified in mice, also arises from *LMNB2* by alternative RNA splicing (Furukawa and Hotta, 1993).

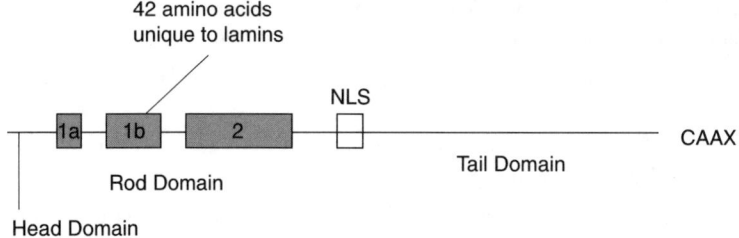

Figure 2. *Nuclear lamins are intermediate filament proteins with some unique features. Like all intermediate filament proteins, lamins have a conserved central rod domain divided into three alpha-helical regions 1b, 1b and 2 separated by short linkers. They also have variable amino-terminal head and carboxyl-terminal tail domains. Lamins differ from the cytoplasmic intermediate filament proteins in that they have an additional 42 amino acids in their rod domains and a nuclear localization signal (NLS) in their tail domains. B-type lamins and lamin A also have a 'CAAX-box' at their carboxyl-termini.*

Table 1. Human nuclear lamin gene family and their encoded proteins

Locus	Chromosome	Proteins	Cell types expressed
LMNA	1q21.2-21.3	Lamin A	Differentiated somatic cells
		Lamin C	Differentiated somatic cells
		Lamin C2	Germ cells
		Lamin A Δ10	Differentiated somatic cells
LMNB1	5q23.2-31.1	Lamin B1	Apparently all somatic cells
LMNB2	19p13.3	Lamin B2	All or most somatic cells
		Lamin B3	Germ cells

4. Invasion of the positional cloners

Few cell biologists studying the nuclear envelope thought much about its relevance to inherited diseases until the positional cloners invaded the field (*Table 2*). In 1994, Bione *et al.* (1994) used positional cloning methods to identify the gene responsible for X-linked Emery–Dreifuss muscular dystrophy, a disease characterized by regional skeletal muscle wasting, contractures, cardiac conduction abnormalities and cardiomyopathy. The encoded protein was named emerin after Alan Emery who, along with Fritz Dreifuss, first described the clinical condition (Emery and Dreifuss, 1966). Emerin contains one transmembrane segment and a LEM domain, which is also present in LAP2 and MAN1 (Lin *et al.*, 2000) at its amino-terminus. Northern analysis showed that emerin was expressed in virtually all tissues (Bione *et al.*, 1994). After the initial report that it was the X-linked Emery–Dreifuss muscular dystrophy protein,

Table 2. 'Invasion of the positional cloners': mutations in nuclear envelope proteins and the human diseases they cause

1994, Bione *et al.,* show that emerin mutations cause X-linked Emery-Dreifuss muscular dystrophy

1999, Bonne *et al.*, show that lamin A/C mutations cause autosomal dominant Emery-Dreifuss muscular dystrophy; others show mutations in related skeletal and cardiac muscle disorders

2000, Cao & Hegele, Shackleton *et al.*, and Speckman *et al.,* show lamin A/C mutations cause Dunnigan-type partial lipodystrophy

2002, De Sandre-Giovannoli *et al.*, show a lamin A/C mis-sense mutation cause recessive Charcot-Marie-Tooth disorder type 2

2002, Novelli *et al.*, show that a lamin A/C homozygous missense mutation causes mandibuloacral dysplasia

2002, Hoffmann *et al.*, show that mutations in LBR cause Pelger-Huët anomaly

2003, Waterham *et al.*, show that mutations in LBR cause autosomal recessive HEM/Greenberg skeletal dysplasia

2003, Eriksson *et al.*,and De Sandre-Giovannoli *et al.*, show lamin A splicing mutations in Hutchinson-Gilford progeria

two groups (Manilal *et al.*, 1996; Nagano *et al.*, 1996) made antibodies against emerin and reported that it was localized to the nuclear envelope and lacking from cells of subjects with the disease. This was the first demonstration that mutations in an inner nuclear membrane protein were responsible for a human disease.

A disease phenotypically indistinguishable from X-linked Emery–Dreifuss muscular dystrophy is inherited in an autosomal dominant manner. Bonne *et al.* (1999) used positional cloning to identify mutations in *LMNA*, the gene encoding lamins A and C, as the cause of autosomal dominant Emery–Dreifuss muscular dystrophy. Subsequently, Fatkin and collaborators (1999) also identified mutations in *LMNA* in familial cardiomyopathy with conduction deficit, a variant of autosomal dominant Emery–Dreifuss muscular dystrophy with identical heart abnormalities but minimal skeletal muscle involvement. Muchir *et al.* (2000) also showed that mutations in *LMNA* could cause a limb girdle muscular dystrophy that has different regional skeletal muscle involvement than Emery–Dreifuss muscular dystrophy but similar cardiomyopathy and cardiac conduction abnormalities.

The findings of Bione *et al.* (1994) and Bonne *et al.* (1999) showed that mutations in the genes encoding emerin and lamins A and C, different proteins of the inner nuclear membrane, cause the same clinical syndrome. Furthermore, they demonstrated that mutations in inner nuclear membrane proteins expressed in virtually all differentiated somatic cells cause a disease that affects primarily or exclusively striated muscle. Soon after, however, Cao and Hegele (2000), Shackleton *et al.* (2000) and Speckman *et al.* (2000) showed that mutations in *LMNA* also cause Dunnigan-type familial partial lipodystrophy, an autosomal dominantly inherited disease characterized by loss of peripheral fat after puberty, insulin resistance and usually overt diabetes mellitus. The large majority of the mutations causing Dunnigan-type partial lipodystrophy were located in *LMNA* exon 8, which encodes a portion of the tail domain common to lamins A and C. Hence, different mutations in lamins A and C can cause disease of either adipose tissue or striated muscle.

The story became even more complicated a couple of years later. In 2002, two autosomal recessive diseases, a rare Charcot–Marie–Tooth type 2 disorder, which is a peripheral neuropathy (De Sandre-Giovannoli *et al.*, 2002), and mandibuloacral dysplasia, a developmental disorder that has skeletal deformities and partial lipodystrophy as prominent features (Novelli *et al.*, 2002), were shown to be caused by mutations in lamins A and C. Intriguingly, mutations in the Zmpste24 protease that processes prelamin A also appear to cause mandibuloacral dysplasia (Agarwal *et al.*, 2003). Eriksson *et al.* (2003) and De Sandre-Giovanini *et al.* (2003) showed that a dominant mutation in *LMNA* causes Hutchinson–Gilford progeria syndrome, a syndrome characterized by features of premature aging in early childhood, including baldness, 'old-appearing' skin and cardiovascular disease. Cao and Hegele (2003) subsequently confirmed this finding. Atypical cases of Werner syndrome, which is also characterized by signs of premature aging, also appear to result from different mutations in *LMNA* (Chen *et al.*, 2003).

Positional cloners have also shown that mutations in LBR, an integral protein of the inner nuclear membrane, cause human diseases. Hoffmann *et al.* (2002) showed that mutations in LBR cause Pelger–Huët anomaly, an autosomal dominantly inherited condition characterized by ovoid-shaped neutrophil nuclei, which are normally hypersegmented. Subjects with Pelger–Huët anomaly generally have no other signs or symptoms. Rare individuals with homozygous mutations in LBR are born alive but suffer from HEM/Greenberg skeletal dysplasia (Waterham et al., 2003). This is a chondrodystrophy with a lethal course, characterized by fetal hydrops, short limbs, abnormal chondro-osseous calcification and a deficiency of 3-beta-hydroxysterol delta(14)-reductase activity (Waterham et al., 2003). The sterol reductase deficiency is consistent with the fact that the hydrophobic domain of LBR, with eight transmembrane segments, is very similar in sequence to other sterol reductases in humans, other animals, yeast and plants (Holmer et al., 1998; Schuler et al., 1994). At this time, it is not known how abnormal sterol metabolism, aberrant chromatin organization or both contribute to HEM/Greenberg skeletal dysplasia and the relatively benign Pelger–Huët anomaly.

5. Mouse models of human diseases caused by nuclear envelope protein mutations

Work by Colin Stewart and collaborators has led to two mouse models of human diseases with abnormalities in lamins A and C. In 1999, Stewart and collaborators reported on the characterization of *Lmna* (-/-) mice that develop regional skeletal muscle abnormalities and cardiomyopathy similar to human Emery–Dreifuss muscular dystrophy in humans (Sullivan et al., 1999). These mice also develop changes in peripheral nerves similar to those seen in human subjects with Charcot–Marie–Tooth type 2 (De Sandre-Giovannoli et al., 2002). *Lmna* (+/-) mice are apparently normal, whereas in humans with Emery–Dreifuss muscular dystrophy and *LMNA* mutations, the disease is autosomal dominant. Several mutant lamin A proteins that cause Emery–Dreifuss muscular dystrophy are as stable as wild-type lamin A (Östlund et al., 2001) and only a very few families have been described that are essentially null for an allele of *LMNA* (Bonne et al., 1999, 2000). Hence, partial loss of function of lamins A and C may not generally be the cause of Emery–Dreifuss muscular dystrophy and expression of the mutant protein may contribute to the pathogenesis. Experiments creating *Lmna* 'knock in' mice may help shed light on this issue.

While attempting to create a *Lmna* 'knock in' mouse with a mutation that causes autosomal dominant Emery–Dreifuss muscular dystrophy, Stewart and collaborators introduced a nucleotide change directing the substitution of proline for leucine at residue 530 (Mounkes et al., 2003). Mice heterozygous for this point mutation, however, did not show signs of muscular dystrophy but mice homozygous for the mutation showed a progeria phenotype. The 'knock in' mutation likely causes aberrant RNA splicing but how this relates to pathophysiology has not been determined. In humans, the G608G (GGC > GGT) mutation in *LMNA* that causes Hutchinson–Gilford progeria syndrome creates an abnormal RNA splice donor site that leads to expression of a truncated prelamin A lacking 50 amino acids nears its carboxyl-terminus (Eriksson et al., 2003). Based on the deleted amino acids, this prelamin A probably cannot be processed to lamin A. The *LMNA* mutation causing Hutchinson–Gilford progeria syndrome may also lead to lower levels of expression of wild-type lamins A and C in cells (De Sandre-Giovannoli et al., 2003).

Ichthyosis is a spontaneous mutant mouse, several strains of which are maintained at The Jackson Laboratory (Blake et al., 2002). Mice homozygous for deleterious alleles at the ichthyosis locus present with a blood phenotype similar to human

Pelger–Huët anomaly but also have other more significant abnormalities including alopecia, variable expression of syndactyly and hydrocephalus. Shultz *et al.* (2003) have shown that the ichthyosis mouse has homozygous mutations in *Lbr*. These mice could be useful in determining how mutations in inner nuclear membrane protein LBR may cause disease either by affecting chromatin organization or sterol metabolism.

6. How do mutations in emerin and lamins A and C cause the same disease?

Emery–Dreifuss muscular dystrophy is characterized by early contractures of the elbows, Achilles' tendons and posterior neck, slow progressive muscle wasting and cardiomyopathy with atrioventricular conduction block (Emery and Dreifuss, 1966; Rowland *et al.*, 1979). The X-linked form is caused by mutations in emerin (Bione *et al.*, 1994) and the autosomal dominant form by mutations in lamins A and C (Bonne *et al.*, 1999). How can mutations in emerin, an integral protein of the inner nuclear membrane, and lamins A and C, peripheral proteins of the inner nuclear membrane, both cause the same disease?

Most, but not all, mutations in X-linked Emery–Dreifuss muscular dystrophy lead to an absence of emerin from the nuclear envelope of striated muscle and other cells (Manilal *et al.*, 1996, 1998; Nagano *et al.*, 1996). Studies utilizing fluorescence loss in photobleaching demonstrate that emerin is normally restricted to the inner nuclear membrane without 'backwards' diffusion into the continuous endoplasmic reticulum (Östlund *et al.*, 1999). In cells from *Lmna* (-/-) mice, however, a considerable amount of emerin is localized to the endoplasmic reticulum, suggesting that binding to lamins A and C is required for efficient retention in the inner nuclear membrane (Sullivan *et al.*, 1999). Studies utilizing co-immunoprecipitation and direct binding assays have also shown that emerin binds to lamin A and C (Clements *et al.*, 2000; Fairley *et al.*, 1999). In transfected cells that express lamin A mutations that cause Emery–Dreifuss muscular dystrophy, some emerin is also mislocalized from the inner nuclear membrane and present in the endoplasmic reticulum (Holt *et al.*, 2003; Östlund *et al.*, 2001; Raharjo *et al.*, 2001). In contrast, in cells expressing mutant lamin A proteins that cause Dunnigan-type partial lipodystrophy, emerin remains in the inner nuclear membrane and is not significantly localized to the endoplasmic reticulum (Favreau *et al.*, 2003; Holt *et al.*, 2001, 2003; Östlund *et al.*, 2001; Raharjo *et al.*, 2001). Hence, a common feature of X-linked and autosomal dominant Emery–Dreifuss muscular dystrophy may be a loss of at least some emerin from the inner nuclear membrane or, possibly, an increase in the diffusional mobility of emerin in the membrane. How these changes in emerin could cause striated muscle abnormalities, however, remains to be elucidated.

7. How do mutations in widely expressed proteins cause tissue-specific diseases?

Mutations in emerin, a protein expressed in all or most somatic cells, causes abnormalities of striated muscle. Lamins A and C are also expressed in virtually all differentiated somatic cells but, in most instances, mutations in these proteins cause primarily tissue-specific diseases. Mutations in lamins A and C can cause Emery–Dreifuss muscular dystrophy and variant disorders, with problems mostly restricted to skeletal and heart muscle (Emery and Dreifuss, 1966; Rowland *et al.*, 1979), or Dunnigan-type partial lipodystrophy, which is characterized by loss of adipose tissue from the extremities, excess fat in the neck, face and trunk and insulin resistance and diabetes mellitus after the onset of puberty (Dunnigan *et al.*, 1974; Jackson *et al.*, 1997). While striated and cardiac muscle abnormalities have been reported in some subjects diagnosed with Dunnigan-type partial lipodystrophy (Garg *et al.*, 2002; Wildermuth *et al.*, 1996), it appears to be primarily a disease of peripheral adipose tissue. The Charcot–Marie–Tooth type 2 disorder caused by mutations in lamins A and C primarily affects peripheral nerves, with an onset usually in the second decade (Chaouch *et al.*, 2003). Hutchinson–Gilford progeria syndrome (DeBusk, 1972), atypical Werner syndrome (Chen *et al.*, 2003) and mandibuloacral dysplasia (Young *et al.*, 1971), which has lipodystrophy as a feature, demonstrate more generalized abnormalities but some tissues and organ systems are more preferentially affected than others.

Several investigators have put forth two different hypotheses for how mutations in inner nuclear membrane proteins expressed in virtually all somatic cells can cause tissue-specific abnormalities (Burke and Stewart, 2002; Goldman *et al.*, 2002; Hutchison, 2002; Östlund and Worman, 2003; Worman and Courvalin, 2000, 2002; Wilson, 2000). One hypothesis, which is generally referred to as the 'gene expression hypothesis', stipulates that mutations in nuclear envelope proteins cause cell-specific changes in gene expression that may be pathogenic. This is based in part on the 'gene gating' hypothesis of Blobel (1985), which initially proposed that the nuclear lamina and pore complexes function in gene regulation. The second hypothesis has been called the 'mechanical stress' hypothesis, which proposes that lamins and emerin are part of a cytoskeletal-karyoskeletal network that provides mechanical support in cells. Certain mutations in lamins or emerin can weaken this support network, making cells in tissues such as skeletal and cardiac muscle susceptible to damage by recurrent mechanical stress. This is somewhat analogous to the thinking regarding how mutations in proteins of the dystrophin–glycoprotein complex that cause muscular dystrophies may lead to loss of sarcolemmal integrity and render muscle fibres more susceptible to damage (Cohn and Campbell, 2000).

While both the 'gene expression' and 'mechanical stress' hypotheses provide appealing models to design future experiments on how mutations in nuclear envelope proteins cause tissue-specific diseases, little data are presently available to support either of them. So far, putative gene regulatory functions of inner nuclear membrane proteins have been inferred primarily from their interactions with DNA or with other proteins that function in gene regulation. Lamins A and C interact with DNA and histones (Goldberg *et al.*, 1999; Luderus *et al.*, 1992; Stierlé *et al.*, 2003; Taniura *et al.*, 1995) and also with retinoblastoma protein (Mancini *et al.*, 1994). LBR binds to mammalian orthologues of *Drosophila* heterochromatin protein 1 (Ye and Worman, 1996; Ye *et al.*, 1997) and also to DNA (Ye and Worman, 1994). Emerin and LAP2-beta bind to a transcriptional repressor called germ cell less (Holaska *et al.*, 2003; Nili *et al.*, 2001). Despite these protein–DNA and protein–protein interactions, however, there are few if any direct demonstrations that nuclear envelope proteins modulate gene regulation and experimental results demonstrating tissue-specific effects of mutations in these proteins on gene expression are presently lacking. Regarding the 'mechanical stress' hypothesis, there is at least one report showing that nuclei from fibroblasts of

subjects with Dunnigan-type partial lipodystrophy have abnormal mechanical properties, as judged from the extensive deformations in nuclei from heat-shocked cells
(Vigouroux *et al.*, 2001). A direct effect of lamin or emerin mutations on muscle fibre
fragility, however, has not been demonstrated.

8. How do different mutations in lamins A and C cause different diseases?

How do mutations in the same proteins, lamins A and C, cause different diseases?
Some insights can be gained from structural studies of a common region in the
carboxyl-terminal tail of the proteins. X-ray crystallographic (Dhe-Paganon *et al.*,
2002) and nuclear magnetic resonance (Krimm *et al.*, 2002) measurements have shown
that amino acids 430 to 545 in lamins A and C adopt a type S immunoglobulin-like
fold. Some mutations that cause striated muscle disease and almost 90% of the mutations that cause Dunnigan-type partial lipodystrophy create amino acid changes within
this fold. The amino acid changes in the immunoglobulin-like fold in that cause
striated muscle diseases affect hydrophobic residues or residues involved in intramolecular hydrogen bonding and are predicted to lead to overall disruption of fold
structure. In contrast, amino acid changes that cause Dunnigan-type partial lipodystrophy affect solvent-exposed residues and change the charge at a surface of the fold,
not affecting overall fold structure. Hence, mutations in lamins A and C that cause
striated muscle disease, including those within the immunoglobulin fold and the
remainder of the molecule, may cause significant overall disruption of protein
structure, possibly affecting lamina polymerization or interactions with many
protein partners. Mutation in lamins A and C that cause Dunnigan-type partial lipodystrophy appear to preserve overall protein structure but affect a relatively small
portion of the molecule, that may, for example, be involved in a specific
protein–protein or protein–nucleic acid interaction. Expression of lamin A and C with
amino acid differences in this portion of the immunoglobulin-like fold may dominantly interfere with such a specific putative interaction.

In Hutchinson–Gilford progeria, the lamin A mutation is very specific. It creates an
abnormal splice donor site and generation of a mRNA that encodes an abnormal prelamin missing 50 amino acids near its carboxyl-terminus (Eriksson *et al.*, 2003).
Expression of this abnormal protein may lead to the development of the
Hutchinson–Gilford progeria phenotype. Lamin C structure is not affected. Lamin A
and C mutations that cause a Charcot–Marie–Tooth type 2 disorder and mandibuloacral dysplasia, which are both autosomal recessive, are also very specific; however,
the consequences of these mutations on protein structure are not known.

9. Conclusions

Recent studies on the nuclear envelope have led to a coming together of cell biologists,
geneticists and clinicians. Mutations in nuclear envelope proteins have been shown to
be the cause of several different human diseases. However, how any of these mutations
are related to disease pathophysiology has not yet been clearly established. Two
hypotheses for how tissue-specific diseases arise from mutations in nuclear envelope
proteins implicate increased susceptibility to mechanical stress or alterations in

cell-specific gene expression that result from the mutations. These hypotheses provide a framework for the design of future experimental studies. Most importantly, more research is needed in this exciting area of medicine and cell biology.

References

Aebi, U., Cohn, J., Buhle, L. and Gerace, L. (1986) The nuclear lamina is a meshwork of inter-mediate-type filaments. *Nature* **323**: 560–564.

Agarwal, A.K., Fryns, J.P., Auchs, R.J. and Garg, A. (2003) Zinc metalloproteinase, ZMPSTE24, is mutated in mandibuloacral dysplasia. *Hum. Mol. Genet.* **12**: 1995–2001.

Barton, R.M. and Worman, H.J. (1999) Prenylated prelamin A interacts with Narf, a novel nuclear protein. *J. Biol. Chem.* **274**: 30008–30018.

Beaudouin, J., Gerlich, D., Daigle, N., Eils, R. and Ellenberg, J. (2002) Nuclear envelope breakdown proceeds by microtubule-induced tearing of the lamina. *Cell* **108**: 83–96.

Beck, L.A., Hosick, T.J. and Sinensky, M. (1988) Incorporation of a product of mevalonic acid metabolism into proteins of Chinese hamster ovary cell nuclei. *J. Cell Biol.* **107**: 1307–1316.

Beck, L.A., Hosick, T.J. and Sinensky, M. (1990) Isoprenylation is required for the processing of the lamin A precursor. *J. Cell Biol.* **110**: 1489–1499.

Bergo, M.O., Gavino, B., Ross, J., *et al.* (2002) Zmpste24 deficiency in mice causes spontaneous bone fractures, muscle weakness, and a prelamin A processing defect. *Proc. Natl. Acad. Sci. USA* **99**: 13049–13054.

Biamonti, G., Giacca, M., Perini, G., *et al.* (1992) The gene for a novel human lamin maps at a highly transcribed locus of chromosome 19 which replicates at the onset of S-phase. *Mol. Cell Biol.* **12**: 3499–3506.

Bione, S., Maestrini, E., Rivella, S., Mancini, M., Regis, S., Romeo, G. and Toniolo, D. (1994) Identification of a novel X-linked gene responsible for Emery-Dreifuss muscular dystrophy. *Nat. Genet.* **8**: 323–327.

Blake, J.A., Richardson, J.E., Bult, C.J., Kadin, J.A and Eppig, T.J. (2002) The Mouse Genome Database (MGD): the model organism database for the laboratory mouse. *Nucleic Acids Res.* **30**: 113–115.

Blobel, G. (1985) Gene gating: a hypothesis. *Proc. Natl. Acad. Sci. USA* **82**: 8527–8529.

Bonne, G., Di Barletta, M.R., Varnous, S., *et al.* (1999) Mutations in the gene encoding lamin A/C cause autosomal dominant Emery-Dreifuss muscular dystrophy. *Nat. Genet.* **21**: 285–288.

Bonne, G., Mercuri, E., Muchir, A., *et al.* (2000) Clinical and molecular genetic spectrum of autosomal dominant Emery-Dreifuss muscular dystrophy due to mutations of the lamin A/C gene. *Ann. Neurol.* **48**: 170–180.

Buendia, B. and Courvalin, J.C. (1997) Domain-specific disassembly and reassembly of nuclear membranes during mitosis. *Exp. Cell Res.* **230**: 133–144.

Burke, B. and Stewart, C.L. (2002) Life at the edge: the nuclear envelope and human disease. *Nat. Rev. Mol. Cell Biol.* **3**: 575–585.

Cance, W.G., Chaudhary, N., Worman, H.J., Blobel, G. and Cordon-Cardo, C. (1992) Expression of the nuclear lamins in normal and neoplastic human tissues. *J. Exp. Clin. Cancer Res.* **11**: 233–246.

Cao, H. and Hegele, R.A. (2003) *LMNA* is mutated in Hutchinson-Gilford progeria (MIM 176670) but not in Wiedemann-Rautenstrauch progeroid syndrome (MIM 264090). *J. Hum. Genet.* **48**: 271–274.

Cao, H. and Hegele, R.A. (2000) Nuclear lamin A/C R482Q mutation in canadian kindreds with Dunnigan-type familial partial lipodystrophy. *Hum. Mol. Genet.* **9**: 109–112.

Chaouch, M., Allal, Y., De Sandre-Giovannoli, A., *et al.* (2003) The phenotypic manifesta-tions of autosomal recessive axonal Charcot-Marie-Tooth due to a mutation in lamin A/C gene. *Neuromuscul. Disord.* **13**: 60–67.

Chaudhary, N. and Courvalin, J.C. (1993) Stepwise reassembly of the nuclear envelope at the end of mitosis. *J. Cell Biol.* **122**: 295–306.

Chen, L., Lee, L., Kudlow, B.A., *et al.* (2003) *LMNA* mutations in atypical Werner's syndrome. *Lancet* **362**: 440–445.

Clements, L., Manilal, S., Love, D.R. and Morris, G.E. (2000) Direct interaction between emerin and lamin A. *Biochem. Biophys. Res. Commun.* **267**: 709–714.

Cohn, R.D. and Campbell, K.P. (2000) Molecular basis of muscular dystrophies. *Muscle Nerve* **23**: 1456–1471.

Collas, I. and Courvalin, J.C. (2000) Sorting nuclear membrane proteins at mitosis. *Trends Cell Biol.* **10**: 5–8.

Collas, P., Courvalin, J.C. and Poccia, D. (1996) Targeting of membranes to sea urchin sperm chromatin is mediated by a lamin B receptor-like integral membrane protein. *J. Cell Biol.* **135**: 1715–1725.

De Sandre-Giovannoli, A., Bernard, R., Cau, P., *et al.* (2003) Lamin a truncation in Hutchinson-Gilford progeria. *Science* **300**: 2055.

De Sandre-Giovannoli, A., Chaouch, M., Kozlov, S., *et al.* (2002) Homozygous defects in *LMNA*, encoding lamin A/C nuclear-envelope proteins, cause autosomal recessive axonal neuropathy in human (Charcot-Marie-Tooth disorder type 2) and mouse. *Am. J. Hum. Genet.* **70**: 726–736.

DeBusk, F.L. (1972) The Hutchinson-Gilford progeria syndrome. Report of 4 cases and review of the literature. *J. Pediatr.* **80**: 697–724.

Dhe-Paganon, S., Werner, E.D., Chi, Y.I. and Shoelson, S.E. (2002) Structure of the globular tail of nuclear lamin. *J. Biol. Chem.* **277**: 17381–17384.

Dreger, M., Bengtsson, L., Schoneberg, T., Otto, H. and Hucho, F. (2001) Nuclear envelope proteomics: novel integral membrane proteins of the inner nuclear membrane. *Proc. Natl. Acad. Sci. USA* **98**: 11943–11948.

Duband-Goulet, I. and Courvalin, J.C. (2000) Inner nuclear membrane protein LBR preferentially interacts with DNA secondary structures and nucleosomal linker. *Biochemistry* **39**: 6483–6488.

Dunnigan, M.G., Cochrane, M.A., Kelly, A. and Scott, J.W. (1974) Familial lipoatrophic diabetes with dominant transmission. A new syndrome. *Q. J. Med.* **43**: 33–48.

Ellenberg, J., Siggia, E.D., Moreira, J.E., Smith, C.L., Presley, J.F., Worman, H.J. and Lippincott-Schwartz, J. (1997) Nuclear membrane dynamics and reassembly in living cells: targeting of an inner nuclear membrane protein in interphase and mitosis. *J. Cell Biol.* **138**: 1193–1206.

Emery, A.E. and Dreifuss, F.E. (1966) Unusual type of benign x-linked muscular dystrophy. *J. Neurol. Neurosurg. Psychiat.* **29**: 338–342.

Eriksson, M., Brown, W.T., Gordon, L.B., *et al.* (2003) Recurrent *de novo* point mutations in lamin A cause Hutchinson-Gilford progeria syndrome. *Nature* **423**: 293–298.

Fairley, E.A., Kendrick-Jones, J. and Ellis, J.A. (1999) The Emery-Dreifuss muscular dystrophy phenotype arises from aberrant targeting and binding of emerin at the inner nuclear membrane. *J. Cell Sci.* **112**: 2571–2582.

Farnsworth, C.C., Wolda, S.L., Gelb, M.H. and Glomset, J.A. (1989) Human lamin B contains a farnesylated cysteine residue. *J. Biol. Chem.* **264**: 20422–20429.

Fatkin, D., MacRae, C., Sasaki, T., *et al.* (1999) Missense mutations in the rod domain of the lamin A/C gene as causes of dilated cardiomyopathy and conduction-system disease. *N. Engl. J. Med.* **341**: 1715–1724.

Favreau, C., Dubosclard, E., Ostlund, C., Vigouroux, C., Capeau, J., Wehnert, M., Higuet, D., Worman, H.J., Courvalin, J.C. and Buendia, B. (2003) Expression of lamin A mutated in the carboxyl-terminal tail generates an aberrant nuclear phenotype similar to that observed in cells from patients with Dunnigan-type partial lipodystrophy and Emery-Dreifuss muscular dystrophy. *Exp. Cell Res.* **282**: 14–23.

Fisher, D.Z., Chaudhary, N. and Blobel, G. (1986) cDNA sequencing of nuclear lamins A and C reveals primary and secondary structural homology to intermediate filament proteins. *Proc. Natl. Acad. Sci. USA* **83**: 6450–6454.

Foisner, R. and Gerace, L. (1993) Integral membrane proteins of the nuclear envelope interact with lamins and chromosomes, and binding is modulated by mitotic phosphorylation. *Cell* **73:** 1267–1279.

Furukawa, K. and Hotta, Y. (1993) cDNA cloning of a germ cell specific lamin B3 from mouse spermatocytes and analysis of its function by ectopic expression in somatic cells. *EMBO J.* **12:** 97–106.

Furukawa, K., Inagaki, H. and Hotta, Y. (1994) Identification and cloning of an mRNA coding for a germ cell-specific A-type lamin in mice. *Exp. Cell Res.* **212:** 426–430.

Furukawa, K., Pante, N., Aebi, U. and Gerace, L. (1995) Cloning of a cDNA for lamina-associated polypeptide 2 (LAP2) and identification of regions that specify targeting to the nuclear envelope. *EMBO J.* **14:** 1626–1636.

Garg, A., Speckman, R.A. and Bowcock, A.M. (2002) Multisystem dystrophy syndrome due to novel missense mutations in the amino-terminal head and alpha-helical rod domains of the lamin A/C gene. *Am. J. Med.* **112:** 549–555.

Gerace, L., Blum, A. and Blobel, G. (1978) Immunocytochemical localization of the major polypeptides of the nuclear pore complex-lamina fraction. Interphase and mitotic distribution. *J. Cell Biol.* **79:** 546–566.

Goldberg, M., Harel, A., Brandeis, M., Rechsteiner, T., Richmond, T.J. and Weiss, A.M., and Gruenbaum, Y. (1999) The tail domain of lamin DmO binds histones H2A and H2B. *Proc. Natl. Acad. Sci. USA* **96:** 2852–2857.

Goldman, R.D., Gruenbaum, Y., Moir, R.D., Shumaker, D.K. and Spann, T.P. (2002) Nuclear lamins: building blocks of nuclear architecture. *Genes Dev.* **16:** 533–547.

Guilly, M.N., Bensussan, A., Bourge, J.F., Bornens, M. and Courvalin, J.C. (1987) A human T lymphoblastic cell line lacks lamins A and C. *EMBO J.* **6:** 3795–3799.

Harris, C.A., Andryuk, P.J., Cline, S., Chan, H.K., Natarajan, A., Siekierka, J.J. and Goldstein, G. (1994) Three distinct human thymopoietins are derived from alternatively spliced mRNAs. *Proc. Natl. Acad. Sci. USA* **91:** 6283–6287.

Heald, R. and McKeon, F. (1990) Mutations of phosphorylation sites in lamin A that prevent nuclear lamina disassembly in mitosis. *Cell* **61:** 579–589.

Herrmann, H and Aebi, U. (2000) Intermediate filaments and their associates: multi-talented structural elements specifying cytoarchitecture and cytodynamics. *Curr. Opin. Cell Biol.* **12:** 79–90.

Hoffmann, K., Dreger, C.K., Olins, A.L., *et al.* (2002) Mutations in the gene encoding the lamin B receptor produce an altered nuclear morphology in granulocytes (Pelger-Huët anomaly). *Nat. Genet.* **31:** 410–414.

Holaska, J.M., Lee, K.K., Kowalski, A.K. and Wilson, K.L. (2003) Transcriptional repressor germ cell-less (GCL) and barrier to autointegration factor (BAF) compete for binding to emerin in vitro. *J. Biol. Chem.* **278:** 6969–6975.

Holmer, L., Pezhman, A. and Worman, H.J. (1998) The human lamin B receptor/sterol reductase multigene family. *Genomics* **54:** 469–476.

Holt, I., Clements, L., Manilal, S., Brown, S.C. and Morris, G.E. (2001) The R482Q lamin A/C mutation that causes lipodystrophy does not prevent nuclear targeting of lamin A in adipocytes or its interaction with emerin. *Eur. J. Hum. Genet.* **9:** 204–208.

Holt, I., Östlund, C., Stewart, C.L., Man, N., Worman, H.J. and Morris, G.E. (2003) Effect of pathogenic mis-sense mutations in lamin A on its interaction with emerin *in vivo*. *J. Cell Sci.* **116:** 3027–3035.

Hozak, P., Sasseville, A.M., Raymond, Y. and Cook, P.R. (1995) Lamin proteins form an internal nucleoskeleton as well as a peripheral lamina in human cells. *J. Cell. Sci.* **108:** 635–644.

Hutchison, C.J. (2002) Lamins: building blocks or regulators of gene expression? *Nat. Rev. Mol. Cell Biol.* **3:** 848–858.

Jackson, S.N., Howlett, T.A., McNally, P.G., O'Rahilly, S. and Trembath, R.C. (1997) Dunnigan-Kobberling syndrome: an autosomal dominant form of partial lipodystrophy. *Q. J. Med.* **90:** 27–36.

Kilic, F., Dalton, M.B., Burrell, S.K., Mayer, J.P., Patterson, S.D. and Sinensky, M. (1997) *In vitro* assay and characterization of the farnesylation-dependent prelamin A endoprotease. *J. Biol. Chem.* **272**: 5298–5304.

Krimm, I., Östlund, C., Gilquin, B., *et al.* (2002) The Ig-like structure of the C-terminal domain of lamin A/C, mutated in muscular dystrophies, cardiomyopathy, and partial lipodystrophy. *Structure* **10**: 811–823.

Leung, G.K., Schmidt, W.K., Bergo, M.O., Gavino, B., Wong, D.H., Tam, A., Ashby, M.N., Michaelis, S. and Young, S.G. (2001) Biochemical studies of Zmpste24-deficient mice. *J. Biol. Chem.* **276**: 29051–29058.

Lin, F. and Worman, H.J. (1993) Structural organization of the human gene encoding nuclear lamin A and nuclear lamin C. *J. Biol. Chem.* **268**: 16321–16326.

Lin, F. and Worman, H.J. (1995) Structural organization of the human gene (*LMNB1*) encoding nuclear lamin B1. *Genomics* **27**: 230–236.

Lin, F., Blake, D.L., Callebaut, I., Skerjanc, I.S., Holmer, L., McBurney, M.W., Paulin-Levasseur, M. and Worman, H.J. (2000) MAN1, an inner nuclear membrane protein that shares the LEM domain with lamina-associated polypeptide 2 and emerin. *J. Biol. Chem.* **275**: 4840–4847.

Luderus, M.E., de Graaf, A., Mattia, E., den Blaauwen, J.L., Grande, M.A., de Jong, L. and van Driel, R. (1992) Binding of matrix attachment regions to lamin B1. *Cell* **70**: 949–959.

Machiels, B.M., Zorenc, A.H., Endert, J.M., Kuijpers, H.J., van Eys, G.J., Ramaekers, F.C. and Broers, J.L. (1996) An alternative splicing product of the lamin A/C gene lacks exon 10. *J. Biol. Chem.* **271**: 9249–9253.

Mancini, M.A., Shan, B., Nickerson, J.A., Penman, S. and Lee, W.H. (1994) The retinoblastoma gene product is a cell cycle-dependent, nuclear matrix-associated protein. *Proc. Natl. Acad. Sci. USA* **91**: 418–422.

Manilal, S., Nguyen, T.M., Sewry, C.A. and Morris, G.E. (1996) The Emery-Dreifuss muscular dystrophy protein, emerin, is a nuclear membrane protein. *Hum. Mol. Genet.* **5**: 801–808.

Manilal, S., Recan, D., Sewry, C.A., *et al.* (1998) Mutations in Emery-Dreifuss muscular dystrophy and their effects on emerin protein expression. *Hum. Mol. Genet.* **7**: 855–864.

Mansharamani, M., Hewetson, A. and Chilton, B.S. (2001) Cloning and characterization of an atypical type IV P-type ATPase that binds to the RING motif of RUSH transcription factors. *J. Biol. Chem.* **276**: 3641–3649.

Martin, L., Crimaudo, C., and Gerace, L. (1995) cDNA cloning and characterization of lamina-associated polypeptide 1C (LAP1C), an integral protein of the inner nuclear membrane. *J. Biol. Chem.* **270**: 8822–8828.

McKeon, F.D., Kirschner, M.W. and Caput, D. (1986) Homologies in both primary and secondary structure between nuclear envelope and intermediate filament proteins. *Nature* **319**: 463–468.

Mislow, J.M., Kim, M.S., Davis, D.B. and McNally, E.M. (2002) Myne-1, a spectrin repeat transmembrane protein of the myocyte inner nuclear membrane, interacts with lamin A/C. *J. Cell Sci.* **115**: 61–70.

Mounkes, L.C., Kozlov, S., Hernandez, L., Sullivan, T. and Stewart, C.L. (2003) A progeroid syndrome in mice is caused by defects in A-type lamins. *Nature* **423**: 298–301.

Muchir, A., Bonne, G., van der Kooi, A.J., van Meegen, M., Baas, F., Bolhuis, P.A., de Visser, M. and Schwartz, K. (2000) Identification of mutations in the gene encoding lamins A/C in autosomal dominant limb girdle muscular dystrophy with atrioventricular conduction disturbances (LGMD1B). *Hum. Mol. Genet.* **9**: 1453–1459.

Nagano, A., Koga, R., Ogawa, M., Kurano, Y., Kawada, J., Okada, R., Hayashi, Y.K., Tsukahara, T. and Arahata, K. (1996) Emerin deficiency at the nuclear membrane in patients with Emery-Dreifuss muscular dystrophy. *Nat. Genet.* **12**: 254–259.

Nili, E., Cojocaru, G.S., Kalma, Y., *et al.* (2001) Nuclear membrane protein LAP2beta mediates transcriptional repression alone and together with its binding partner GCL (germ-cell-less). *J. Cell Sci.* **114**: 3297–3307.

Novelli, G., Muchir, A., Sangiuolo, F., *et al.* (2002) Mandibuloacral dysplasia is caused by a mutation in *LMNA*-encoding lamin A/C. *Am. J. Hum. Genet.* **71**: 426–431.

Östlund, C. and Worman, H.J. (2003) Nuclear envelope proteins and neuromuscular diseases. *Muscle Nerve* **27**: 393–406.

Östlund, C., Ellenberg, J., Hallberg, E., Lippincott-Schwartz, J. and Worman, H.J. (1999) Intracellular trafficking of emerin, the Emery-Dreifuss muscular dystrophy protein. *J. Cell Sci.* **112**: 1709–1719.

Östlund, C., Bonne, G., Schwartz, K. and Worman, H.J. (2001) Properties of lamin A mutants found in Emery-Dreifuss muscular dystrophy, cardiomyopathy and Dunnigan-type partial lipodystrophy. *J. Cell Sci.* **114**: 4435–4445.

Paulin-Levasseur, M., Blake, D.L., Julien, M. and Rouleau, L. (1996) The MAN antigens are non-lamin constituents of the nuclear lamina in vertebrate cells. *Chromosoma* **104**: 367–379.

Pendas, A.M., Zhou, Z., Cadinanos, J., *et al.* (2002) Defective prelamin A processing and muscular and adipocyte alterations in Zmpste24 metalloproteinase-deficient mice. *Nat. Genet.* **31**: 94–99.

Peter, M., Nakagawa, J., Doree, M., Labbe, J.C. and Nigg, E.A. (1990) *In vitro* disassembly of the nuclear lamina and M phase-specific phosphorylation of lamins by *cdc2* kinase. *Cell* **61**: 591–602.

Raharjo W.H., Enarson, P., Sullivan, T., Stewart, C.L. and Burke, B. (2001) Nuclear envelope defects associated with LMNA mutations cause dilated cardiomyopathy and Emery-Dreifuss muscular dystrophy. *J. Cell Sci.* **114**: 4447–4457.

Rober, R.A., Weber, K. and Osborn, M. (1989) Differential timing of nuclear lamin A/C expression in the various organs of the mouse embryo and the young animal: a developmental study. *Development* **105**: 365–378.

Rolls, M.M., Stein, P.A., Taylor, S.S., Ha, E., McKeon, F. and Rapoport, T.A. (1999) A visual screen of a GFP-fusion library identifies a new type of nuclear envelope membrane protein. *J. Cell Biol.* **146**: 29–44.

Rowland, L.P., Fetell, M., Olarte, M., Hays, A., Singh, N. and Wanat, F.E. (1979) Emery-Dreifuss muscular dystrophy. *Ann. Neurol.* **5**: 111–117.

Salina, D., Bodoor, K., Eckley, D.M., Schroer, T.A., Rattner, J.B. and Burke, B. (2002) Cytoplasmic dynein as a facilitator of nuclear envelope breakdown. *Cell* **108**: 97–107.

Schuler, E., Lin, F. and Worman, H.J. (1994) Characterization of the human gene encoding LBR, an integral protein of the nuclear envelope inner membrane. *J. Biol. Chem.* **269**: 11312–11317.

Senior, A. and Gerace, L. (1988) Integral membrane proteins specific to the inner nuclear membrane and associated with the nuclear lamina. *J. Cell Biol.* **107**: 2029–2036.

Shackleton, S., Lloyd, D.J., Jackson, S.N., *et al.* (2000) *LMNA*, encoding lamin A/C, is mutated in partial lipodystrophy. *Nat. Genet.* **24**: 153–156.

Shultz, L.D., Lyons, B.L., Burzenski, L.M., *et al.* (2003) Mutations at the mouse ichthyosis locus are within the lamin B receptor gene: a single gene model for human Pelger-Huët anomaly. *Hum. Mol. Genet.* **12**: 61–69.

Sinensky, M., Fantle, K., Trujillo, M., McLain, T., Kupfer, A. and Dalton, M. (1994) The processing pathway of prelamin A. *J. Cell Sci.* **107**: 61–67.

Speckman, R.A., Garg, A., Du, F., Bennett, L., Veile, R., Arioglu, E., Taylor, S.I., Lovett, M. and Bowcock, A.M. (2000) Mutational and haplotype analyses of families with familial partial lipodystrophy (Dunnigan variety) reveal recurrent missense mutations in the globular C-terminal domain of lamin A/C. *Am. J. Hum. Genet.* **66**: 1192–1198.

Stewart, C. and Burke, B. (1987) Teratocarcinoma stem cells and early mouse embryos contain only a single major lamin polypeptide closely resembling lamin B. *Cell* **51**: 383–392.

Stierlé, V., Couprie, J., Ostlund, C., Krimm, I., Zinn-Justin, S., Hossenlopp, P., Worman, H.J., Courvalin, J.C. and Duband-Goulet, I. (2003) The carboxyl-terminal region common to lamins A and C contains a DNA binding domain. *Biochemistry* **42**: 4819–4828.

Stuurman, N., Heins, S. and Aebi, U. (1998) Nuclear lamins: their structure, assembly, and interactions. *J. Struct. Biol.* **122**: 42–66.

Sullivan, T., Escalante-Alcalde, D., Bhatt, H., Anver, M., Bhat, N., Nagashima, K., Stewart, C.L. and Burke, B. (1999) Loss of A-type lamin expression compromises nuclear envelope integrity leading to muscular dystrophy. *J. Cell Biol.* **147**: 913–920.

Taniura, H., Glass, C. and Gerace, L. (1995) A chromatin binding site in the tail domain of nuclear lamins that interacts with core histones. *J. Cell Biol.* **13**: 133–144.

Vigouroux, C., Auclair, M., Dubosclard, E., Pouchelet, M., Capeau, J., Courvalin, J.C. and Buendia, B. (2001) Nuclear envelope disorganization in fibroblasts from lipodystrophic patients with heterozygous R482Q/W mutations in the lamin A/C gene. *J. Cell Sci.* **114**: 4459–4468.

Ward, G.E. and Kirschner, M.W. (1990) Identification of cell cycle-regulated phosphorylation sites on nuclear lamin C. *Cell* **61**: 561–577.

Waterham, H.R., Koster, J., Mooyer, P., Noort, G. G., Kelley, R.I., Wilcox, W.R., Wanders, R.J., Hennekam, R.C. and Oosterwijk, J.C. (2003) Autosomal recessive HEM/Greenberg skeletal dysplasia is caused by 3 beta-hydroxysterol delta 14-reductase deficiency due to mutations in the lamin B receptor gene. *Am. J. Hum. Genet.* **72**: 1013–1017.

Wildermuth, S., Spranger, S., Spranger, M., Raue, F. and Meinck, H.M. (1996) Kobberling-Dunnigan syndrome: a rare cause of generalized muscular hypertrophy. *Muscle Nerve* **19**: 843–847.

Wilson, K.L. (2000) The nuclear envelope, muscular dystrophy and gene expression. *Trends Cell Biol.* **10**: 125–129.

Wolda, S.L. and Glomset, J.A. (1988) Evidence for modification of lamin B by a product of mevalonic acid. *J. Biol. Chem.* **263**: 5997–6000.

Worman, H.J. and Courvalin, J.C. (2000) The inner nuclear membrane. *J. Membr. Biol.* **177**: 1–11.

Worman, H.J. and Courvalin, J.C. (2002) The nuclear lamina and inherited disease. *Trends Cell Biol.* **12**: 591–598.

Worman, H.J., Lazaridis, I. and Georgatos, S.D. (1988) Nuclear lamina heterogeneity in mammalian cells. Differential expression of the major lamins and variations in lamin B phosphorylation. *J. Biol. Chem.* **263**: 12135–12141.

Worman, H.J., Evans, C.D. and Blobel, G. (1990) The lamin B receptor of the nuclear envelope inner membrane: a polytopic protein with eight potential transmembrane domains. *J. Cell Biol.* **111**: 1535–1542.

Wydner, K.L., McNeil, J.A., Lin, F., Worman, H.J. and Lawrence, J.B. (1996) Chromosomal assignment of human nuclear envelope protein genes *LMNA*, *LMNB1*, and *LBR* by fluorescence in situ hybridization. *Genomics* **32**: 474–478.

Yang, L., Guan, T. and Gerace, L. (1997) Integral membrane proteins of the nuclear envelope are dispersed throughout the endoplasmic reticulum during mitosis. *J. Cell Biol.* **137**: 1199–1210.

Ye, Q. and Worman, H.J. (1994) Primary structure analysis and lamin B and DNA binding of human LBR, an integral protein of the nuclear envelope inner membrane. *J. Biol. Chem.* **269**: 11306–11311.

Ye, Q. and Worman, H.J. (1996) Interaction between an integral protein of the nuclear envelope inner membrane and human chromodomain proteins homologous to *Drosophila* HP1. *J. Biol. Chem.* **271**: 14653–14656.

Ye, Q., Callebaut, I., Pezhman, A., Courvalin, J.C. and Worman, H.J. (1997) Domain-specific interactions of human HP1-type chromodomain proteins and inner nuclear membrane protein LBR. *J. Biol. Chem.* **272**: 14983–14989.

Young, L.W., Radebaugh, J.F., Rubin, P., Sensenbrenner, J.A., Fiorelli, G. and McKusick, V.A. (1971) New syndrome manifested by mandibular hypoplasia, acroosteolysis, stiff joints and cutaneous atrophy (mandibuloacral dysplasia) in two unrelated boys. *Birth Defects Orig. Artic. Ser.* **7**: 291–297.

Zhang, Q., Skepper, J.N., Yang, F., Davies, J.D., Hegyi, L., Roberts, R.G., Weissberg, P.L., Ellis, J.A. and Shanahan, C.M. (2001) Nesprins: a novel family of spectrin-repeat-containing proteins that localize to the nuclear membrane in multiple tissues. *J. Cell Sci.* **114**: 4485–4498.

4

Protein interactions, right or wrong, in Emery–Dreifuss muscular dystrophy

Glenn E. Morris

1. Introduction

The English idiom of 'not being able to see the wood for the trees' is somewhat ambiguous. It could mean that one cannot recognize the presence of a wood under bark and leaves, or, more often, that one spends so much time looking at individual trees that one cannot see the broader picture – the wood or forest – that the trees create. Either way, this is a common problem in trying to understand the molecular pathogenesis of Emery–Dreifuss muscular dystrophy (EDMD) and related laminopathies. So many surprising facts emerge that it is difficult to identify which are important clues and which are relatively trivial for understanding the disease process. Is it possible to account for all the disparate observations with a simple molecular hypothesis? Emery–Dreifuss muscular dystrophy is most commonly an X-linked form (X-EDMD) caused by complete absence of the nuclear membrane protein, emerin (Bione *et al.*, 1995; Manilal *et al.*, 1996; Nagano *et al.*, 1996). Rather less common is an autosomal dominant form (AD-EDMD) caused by missense mutations in lamin A/C (Bonne *et al.*, 1999). Why do these mutations affect skeletal and cardiac muscles and cause joint contractures while other tissues are functionally unaffected (Emery, 1993)? Is it significant that conducting cells in the heart are preferentially affected, compared to contractile cells? Some muscles in EDMD are only mildly affected while others may become severely wasted. Not unnaturally, EDMD muscle biopsies are usually taken from mildly affected muscles rather than severely wasted ones and they show a mild histopathology with evidence of regeneration but little or no necrosis (Sewry, 2000; Sewry *et al.*, 2001). There is also frequent clinical variability in EDMD, even within families with the same mutation. The variability may be in both the range of clinical features displayed and their severity and is usually attributed to individual genetic background (modifying genes), although environmental effects have not been ruled out (Morris, 2001). Variability has probably confused the clinical distinction between

The Nuclear Envelope, edited by D.E. Evans, C. Hutchison & J.A. Bryant.
© 2004 Garland Science/BIOS Scientific Publishers

the X-linked form of EDMD and the autosomal dominant form caused by missense mutations in lamin A/C. The two forms of the disease were once thought to be indistinguishable clinically, but it is now clear that the cardiac problem in AD-EDMD is more severe (Becane *et al.*, 2000). Sudden cardiac death is more common in AD-EDMD patients and they often develop a dilated cardiomyopathy that requires a heart transplant, whereas transplants are very rarely performed in X-EDMD. Systematic comparisons of skeletal muscle wasting and contracture severities in the two forms of the disease have not yet been reported. A large proportion of patients with the clinical features of EDMD have no mutations in either emerin or lamin A/C. There is at least one family with autosomal inheritance and this suggests that mutations in a third gene can cause EDMD. Most such cases, however, are sporadic with no family history of EDMD. This might suggest the possibility of a non-genetic origin in these cases, but a large number of sporadic cases do have lamin A/C mutations that are presumably responsible for the disease (Bonne *et al.*, 2000). How many of these observations will be critical clues in developing an explanation of the molecular pathogenesis and how many are irrelevant remains to be seen.

One man well able to see the wood for the trees is Alan Emery. In his recent book with Marcia Emery (Emery and Emery, 1995), he gives an authentic account of the discovery of EDMD as a separate disease from the Duchenne/Becker muscular dystrophies. When Emery was a Ph.D. student at Johns Hopkins in 1962, his supervisor, Victor McKusick, sent him to Virginia to study a large family with an X-linked muscular dystrophy. This family had been described the previous year by the eminent neurologist and epilepsy expert, Fritz Dreifuss, who had suggested a slight broadening of the definition of Duchenne muscular dystrophy to include this particular form (Dreifuss and Hogan, 1961). After a long weekend studying almost the whole family on the spot, Emery was able to define the essential features of a new disease. The data were published in 1966 (Emery and Dreifuss, 1996) but it was many years before EDMD was widely accepted as a distinct form of muscular dystrophy.

2. Nuclear foci of A-type lamins

When cells are transfected with lamin A or lamin C cDNAs carrying pathogenic mis-sense mutations, some of the mutants give rise to large aggregates or foci of recombinant protein in the nucleoplasm (Holt *et al.*, 2003; Östlund *et al.*, 2001; Raharjo *et al.*, 2001). These foci are produced in addition to the lamin A that goes to its normal location at the nuclear rim. When they occur, they are usually only found in 10–25% of all transfected cells. They are not observed in cells transfected with wild-type lamin A or lamin A carrying a lipodystophy mutation at R482, though a recent study using HeLa cells found nuclear aggregates formed by all A-type lamins (Bechert *et al.*, 2003). It is generally agreed that foci also contain B-type lamins but not LAP2 (Holt *et al.*, 2003; Östlund *et al.*, 2001; Raharjo *et al.*, 2001). Our recent observation that foci are usually produced by mutations in the helical rod region of lamin A, and not by mutations in the lamin A tail (Holt *et al.*, 2003), suggests that they may reflect a defect in normal assembly of lamin A into filaments due to the mutation.

Our recent studies of lamin A mutants (Holt *et al.*, 2003) and those of Raharjo *et al.* (2001) were performed by transfection of *lmna* -/- mouse embryonic fibroblasts (MEFs) from the lamin A/C knockout mouse. These differ from studies using lamin

A/C+ cell types, such as HeLa, COS or mouse C2C12 myoblasts (Bechert *et al.*, 2003; Östlund *et al.*, 2001), in three very important ways. Firstly, the transfected lamin A will be 100% mutant, like a homozygous mutation, instead of part mutant and part wild-type, more analogous to a heterozygote. Homozygous lamin A/C mutations produce a more severe phenotype in laminopathy patients. Secondly, lamin C remains completely absent from the transfected *lmna* -/- MEFs, so the A-type lamina produced from transfected lamin A alone is incomplete. Finally, reconstruction of the A-type lamina and its associated proteins, such as emerin, starts afresh after transfection of *lmna* -/- MEFs, whereas transfection into HeLa or similar cells introduces *extra* lamin A into an environment with a complete lamina complex already present. In *lmna* -/- MEFs, emerin is present in the endoplasmic reticulum (ER) of the cytoplasm because A-type lamins are required to trap it at the nuclear rim (Sullivan *et al.*, 1999). In normal cells, emerin is already at the nuclear rim and fluorescence photobleaching has shown that its mobility is greatly reduced by its interactions there (Östlund *et al.*, 1999). Our studies have shown that emerin is usually present in the large nuclear foci of *lmna* -/- MEFs tranfected with lamin A and that it is present *inside* a ring of recombinant lamin A (Holt *et al.*, 2003). *Figure 1* shows a schematic view of a hypothetical way in which nuclear foci might arise in *lmna* -/- MEFs. It takes account of the experimental observation that foci appear to grow from the top or the bottom of the nucleus into the nucleoplasm and may traverse the whole nucleus. In this hypothesis, much mutant lamin A assembly can occur apparently normally at the nuclear rim, but occasional misassembly results in the growth of aggregates which continue to trap emerin at the nuclear membrane. Some of these foci near the nuclear rim grow larger and invaginate, with emerin inside the outer layer of lamins. In general, large foci were only produced by mutations in the helical rod region involved in lamin A assembly, whereas mutations in the tail region of lamin A impaired its interaction with emerin (Holt *et al.*, 2003). It is possible to see from this model how transfection of C2C12 myoblasts might result in nuclear foci with much lower amounts of emerin (Östlund *et al.*, 2001), since emerin would already be trapped at the nuclear rim in these cells.

It is interesting to note that skin fibroblasts from patients with AD-EDMD also display nuclear foci with a similar frequency of up to 25%, depending on the lamin A mutation (author's unpublished observations). Although these AD-EDMD foci are usually smaller than those seen in transfected cells, their existence does show that focus formation is a normal biological process for the mutant lamins and not a mere artefact of recombinant transfection. Although skin fibroblasts can apparently function normally in spite of nuclear foci, this may not be true of affected tissues, like cardiac and skeletal muscle, so a role for foci in EDMD pathogenesis cannot be entirely ruled out. It is not yet clear, however, whether the origin and composition of nuclear foci in transfected cells and AD-EDMD fibroblasts is the same. It is interesting that nuclear foci of pre-lamin A have been reported in normal cell nuclei as a possible staging post before assembly at the nuclear rim (Sasseville and Raymond, 1995). Pre-lamin A is converted to mature lamin A by proteolytic removal of the last 15 amino-acids (Weber *et al.*, 1989). Although cDNA encoding pre-lamin A is commonly used for transient transfection experiments and the possibility of a relationship between mutant 'nuclear foci' and 'pre-lamin A foci' cannot be ruled out, a relationship seems unlikely because transfected mutant lamin C, which does not undergo proteolytic conversion, can also form nuclear foci (Raharjo *et al.*, 2001).

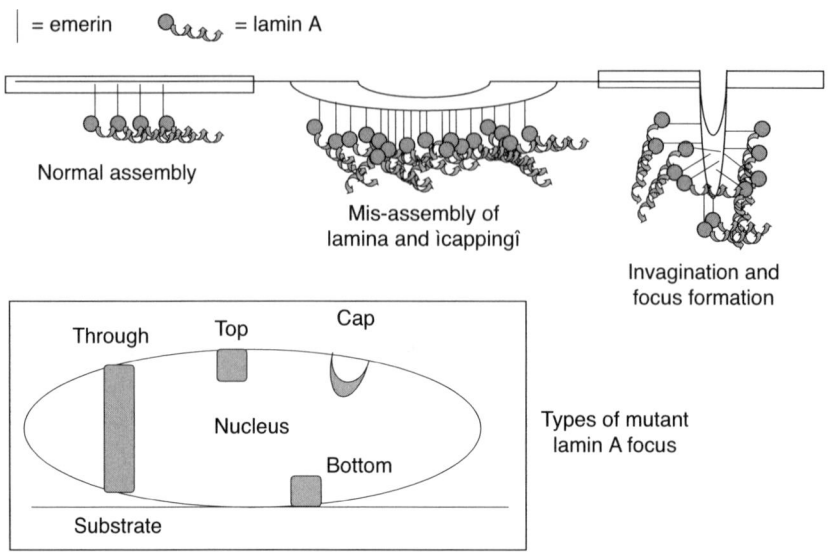

Figure 1. *Hypothetical mechanism for the formation of nuclear foci of mutant lamin A. The model is based on experimental evidence in Holt et al. (2003). Normal assembly of the emerin–lamin A complex involves assembly of lamin A into filaments via the rod domains and interaction with emerin by the globular tail domain. Mis-assembly caused by rod domain mutations is proposed to result in 'cap' formation (with emerin on the nuclear membrane side of the cap), followed by invagination as the focus enlarges (with emerin on the inside of the focus). Sometimes the focus runs from one side of the nucleus to the other, as illustrated.*

The need for caution before assuming that all nuclear foci are the same is emphasized by the quite different, though very large, nuclear aggregates produced by transfection of mutant lamin A cDNA (in plasmid vector pCS2+) into a special SS6 strain of HeLa cells (Bechert *et al.*, 2003). These very large foci were produced in up to 86% of all nuclei, even using the lipodystrophy mutant R482L, which does not produce foci at all in C2C12 myoblasts (Östlund *et al.*, 2001) or *lmna* -/- MEFs (Holt *et al.*, 2003; Raharjo *et al.*, 2001). Even wild-type lamin A produced foci in 40% of nuclei in this HeLa experimental system. Although these aggregates were also associated with the nuclear rim, they differ from nuclear foci in containing LAP2 and in having emerin, when present, on the outside, rather than the inside, of the aggregate (Bechert *et al.*, 2003). They are clearly formed by a different mechanism from that proposed in *Figure 1*.

Visible defects at the nuclear periphery were first reported in electron microscopy studies of skeletal muscle nuclei in both X-EDMD (Fidzianska *et al.*, 1998; Ognibene *et al.*, 1999) and AD-EDMD (Sewry *et al.*, 2001) and, once again, only 10–20% of nuclei were visibly affected (Ognibene *et al.*, 1999). Disruption of the nuclear lamina and detachment of peripheral heterochromatin from the nuclear rim were reported. This common feature of lamin A/C mis-sense mutations and emerin knockout mutations might suggest that it has some role in EDMD pathogenesis. However, these changes were seen both in an affected tissue (muscle biopsies) and in an unaffected tissue (skin fibroblast cultures), so there is no simple direct link between the nuclear abnormalities and pathogenesis. Furthermore, it has recently been suggested that the changes in nuclear morphology in X-EDMD and AD-EDMD are distinguishable from each other (Fidzianska and Hausmanowa-Petrusewicz, 2003), implying that they are not linked to a common defect in the emerin–lamin A/C complex. It is not known whether the morphological changes occur in nuclei of all tissues. Similar nuclear changes were seen in over 80% of nuclei in *lmna* -/- MEFs (Sullivan *et al.*, 1999), showing that they can be caused by absence of lamin A/C as well as by dominant-negative effects of heterozygous mis-sense mutations.

3. Cell division, degeneration and death

One might imagine that these nuclear defects interfere with cell division or cell survival in EDMD cell cultures, but this does not seem to be the case. Indeed, we have shown that absence of emerin actually causes the number of skin fibroblasts in culture to increase (G. Morris, Nguyen thi Man and C. Sewry, unpublished data). This experiment was done by counting emerin-positive and emerin-negative cells in a culture of skin fibroblasts from a female carrier of a null mutation in emerin. *Figure 2* shows emerin-negative and emerin-positive fibroblasts side-by-side in a culture from an EDMD carrier female. There is no nuclear rim staining in the EDMD cell nuclei. The advantage of using carrier fibroblasts, rather than comparing separate cultures of normal and EDMD fibroblasts, is three-fold. The genetic background (apart from the X-chromosome) is identical, as are the culture conditions and the culture history (e.g. passage number). We found that the proportion of emerin-positive cells decreased from 28% to 8% in five passages. We could not ascertain at that time whether this was due to faster cell division or better survival in the emerin-negative cells, though elsewhere in this volume Chris Hutchison has reported a higher rate of cell division in EDMD cells and a theory to explain it (Chapter 16). A third possibility, that the X-inactivation pattern is unstable in cell culture with a gradual switch of expression from the normal X-chromosome to the mutated X-chromosome, was eliminated by plating cells at clonal density. Each clone contained either emerin-positive cells or emerin-negative cells, but not a mixture of both. The faster growth rate of emerin-negative fibroblasts is clearly an artefact of cell culture because the ratio of emerin-negative to emerin-positive is about 50:50 in the epidermal layer of EDMD carrier skin (Manilal *et al.*, 1997), as predicted theoretically from random X-inactivation in females.

Although cultured EDMD fibroblasts may grow faster, it is clear that degeneration–regeneration cycles are occurring in EDMD skeletal muscle and that, in some specific muscles, the degeneration can lead to massive muscle wasting. A resemblance between the nuclear changes in EDMD and nuclear changes in apoptosis has been noted (Fidzianska *et al.*, 1998; Ognibene *et al.*, 1999; Sewry *et al.*, 2001) and loss of lamins, by proteolysis or otherwise, may accelerate apoptosis (Rao *et al.*, 1996). The absence of necrosis and near-normal serum CK levels in EDMD would also be consistent with apoptotic cell loss as a mechanism for muscle wasting. The late stages of apoptosis occur so rapidly that, in a slowly wasting EDMD muscle, the numbers of apoptotic nuclei might be too low to detect easily, even if apoptosis were entirely responsible for the on-going muscle loss (Morris, 2000).

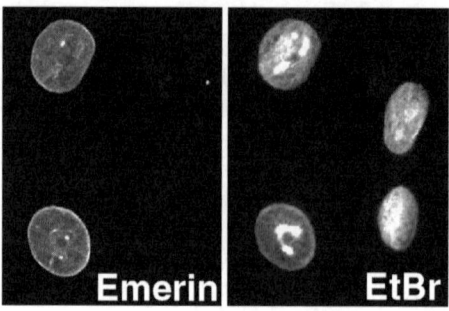

Figure 2. Skin fibroblast cultures from EDMD carrier females contain both emerin-positive and emerin-negative cells. Cells were stained with the anti-emerin mAb, MANEM5 [Manilal et al., 1996], and FITC rabbit anti-(mouse Ig). They were counterstained for nuclei using ethidium bromide (EtBr) and examined with a BioRadiance 2000 confocal microscope and a 60x objective.

4. Interactions of emerin with other proteins

The discoveries that lamin A/C interacts directly with emerin (Clements *et al.*, 2000) and that it is required for keeping emerin at the nuclear rim (Sullivan *et al.*, 1999) suggested that the distinctive EDMD clinical features result from a defect in some function of the emerin–lamin A/C complex. The identification of protein-binding partners for emerin and lamin A has provided some clues to what this function might be.

BAF (barrier to autointegration factor) is a small DNA-binding protein that also binds to the 'LEM' or 'thymopoietin' domain in emerin (Lee *et al.*, 2001). This LEM domain near the N-terminus of emerin (aa6-44) mediates chromatin binding and is also present in two other inner nuclear membrane proteins, LAP2β and MAN1 (Lin *et al.*, 2000). There is a clear relationship here between sequence homology in the three proteins and BAF-binding function, although the homology is not high (40% between emerin and LAP2β). Transfection studies showed that amino acids 107–175 of emerin (the Tsuchiya–Östlund sequence) are necessary (and sufficient when a transmembrane sequence is present) for targeting emerin to the nuclear rim (Östlund *et al.*, 1999; Tsuchiya *et al.*, 1999). Mutations within amino acids 78–178 of emerin affect its ability to bind lamin A *in vitro* (Lee *et al.*, 2001). This central region of emerin appears to interact with the globular tail domain common to lamins A and C (Vaughan *et al.*, 2001; Wilkinson *et al.*, 2003), the 3D structure of which has been established (Dhe Paganon *et al.*, 2002; Krimm *et al.*, 2002). These results are consistent with the lamin A/C–emerin interaction being one of the main mechanisms for targeting emerin to the nuclear rim. The fact that emerin locates to the ER, instead of the nuclear rim, in lamin A/C 'knockout' mouse embryonic fibroblasts indicates that the BAF–DNA inter-action is not retaining emerin in these nuclei. In some knockout mouse tissues, however, emerin is partially retained at the nuclear rim, as a result of an interaction with unidentified proteins other than lamin A/C (Sullivan *et al.*, 1999).

Nesprins (Zhang *et al.*, 2001) are type II inner nuclear membrane proteins, like emerin, and one isoform, nesprin-1-α, has been shown to interact with both emerin and A-type lamins (Mislow *et al.*, 2002). Although emerin and lamin A/C interact directly, it has been suggested that nesprin-1-α acts as an additional bridge linking these two proteins (Mislow *et al.*, 2002). Like emerin, nesprins may also require A-type lamins to capture them from the ER into the nuclear envelope. Nesprins are very large, dystrophin-like proteins that also have an actin-binding region (calponin homology domains) at their N-terminus, so they might also act as a bridge between nuclear actin and emerin-lamin A/C complexes. There is also evidence for a direct interaction between emerin and actin (Fairley *et al.*, 1999), in addition to that mediated by nesprin (see also Wilson, Chapter 20). Actin appears to have an important role in chromatin remodelling (Shumaker *et al.*, 2003), so a functional interaction with emerin might support a role for emerin in differentiation and gene expression.

Emerin also has a bi-partite binding site that binds both the GCL (germ cell less) transcription factor (Holaska *et al.*, 2003) and the YT521-B splicing-associated factor (Wilkinson *et al.*, 2003) and this provides a further link between emerin and regulation of gene expression. YT521-B has been shown to modulate alternative splice site selection *in vivo* (Hartman *et al.*, 1999); this activity requires the C-terminal Glu-Arg domain that our evidence suggests can also interact with emerin (Wilkinson *et al.*, 2003). If the normal functions of YT521-B are somehow regulated by interactions with emerin or the emerin–lamin A/C complex, some of the cardiac and skeletal muscle symptoms of EDMD may result from inappropriate splicing of tissue-specific mRNAs. This could explain how changes in widely expressed proteins like YT521-B and emerin might produce the tissue-specific effects characteristic of EDMD. Myotonic dystrophy is caused by RNA splicing defects that result from the accumulation of expanded CUG repeats in the nucleus (Savkur *et al.*, 2001) and, like EDMD, it also features cardiac conduction system defects and skeletal muscle wasting.

Figure 3 shows a speculative diagram of how emerin might function in organizing transcription and splicing factors in relation to the emerin–lamin A/C complex and chromatin associated with BAF. Because the nuclear periphery tends to be enriched for inactive chromatin, it is possible that the function of emerin is to sequester transcription and splicing factors, instead of promoting their normal functions on active chromatin. Inhibition of YT521-B splice-site selection activity by emerin (Wilkinson *et al.*, 2003) is also consistent with the latter view. On the other hand, a recent Xenopus study suggests that emerin promotes chromatin decondensation when it binds to BAF (Sugura-Totten, 2002). Although the mapped sites on emerin for BAF and GCL binding overlap only slightly, an inhibition of GCL binding by BAF has been demonstrated *in vitro* (Hartmann *et al.*, 1999) and this also seems inconsistent with a simple role for emerin in bringing GCL and chromatin together. It is impossible to say at present which of the currently known emerin-binding factors are involved in the pathogenesis of EDMD. However, the hypothesis that absence or disruption of the emerin–lamin A/C complex has downstream effects on gene expression that lead to specific degeneration of skeletal and cardiac muscle now deserves serious consideration.

There are other examples of transcription factor binding to lamina-associated proteins. The adipocyte transcription factor known as 'sterol response element binding protein 1' (SREBP1), binds to lamin A and this binding is affected by a lamin A/C mutation that causes lipodystrophy (Lloyd *et al.*, 2002). The retinoblastoma transcription factor, Rb, binds to the emerin-related LEM domain protein, LAP2α (Markiewicz *et al.*, 2002), although this particular isoform of LAP2 is nucleoplasmic.

Clearly, for such a small protein (254 aa), emerin appears to participate in a remarkable number of different interactions and it is difficult to say which, if any, are

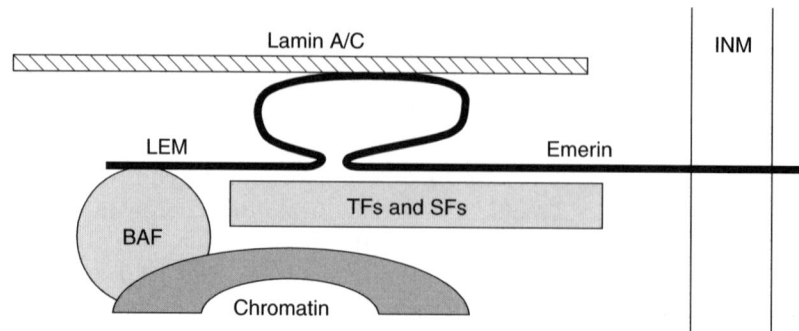

Figure 3. *Hypothetical model for interactions of emerin at the nuclear rim. Emerin is drawn diagrammatically as a partly folded structure to illustrate the bi-partite nature of the binding site for transcription factors (TFs) and splicing factors (SFs). The C-terminus of emerin is attached to the inner nuclear membrane (INM). BAF (barrier to autointegration factor) mediates binding of the N-terminal LEM (LAP2-Emerin-MAN1) domain to chromatin.*

likely to prove important in the pathogenesis of EDMD. It is also clear that emerin cannot bind to all of the proposed partners at the same time and this is one probable reason for the lack of complete co-localization of emerin with all its partners. BAF, for example, co-localizes with emerin only at one stage in the cell cycle (Haraguchi *et al.*, 2001). The interactions are probably dynamic and determined by the relative binding affinities of the partners for emerin. These interactions may be sensitive to post-translational modifications such as phosphorylation that alter the binding affinities, thus providing a possible mechanism for regulation. The interaction between emerin and lamin A/C is clearly important because mutations of either protein can cause EDMD. The other emerin interactors are candidate genes for those cases of EDMD that have neither emerin nor lamin A/C mutations. Identification of novel EDMD-causing genes will clearly be a major step in understanding both EDMD pathogenesis and the normal function of the emerin–lamin A/C complex.

Do lamin A/C mutations act by affecting emerin function or does the absence of emerin cause defects in lamina function mediated by lamin A/C? The former seems unlikely because emerin mutations produce three clinical features (cardiac conduction, muscle wasting and contractures) and lamin A/C mutations may produce only one or two of these (e.g. in dilated cardiomyopathy type 1A or limb-girdle muscular dystrophy type 1B) (Morris, 2001). Evidence for the latter is also rather limited, since changes in the lamina in X-EDMD may be secondary to the disease process. I prefer an alternative view that gains some support from the number of diseases unrelated to EDMD but caused by specific lamin A/C mutations (lipodystrophy (Shakleton *et al.*, 2000), Charcot–Marie–Tooth neuropathy (De Sandre-Giovannoli *et al.*, 2002), mandibuloacral dysplasia (Novelli *et al.*, 2002) and Hutchinson–Gifford progeria (Eriksson *et al.*, 2003; Mounkes *et al.*, 2003)). In this view, the EDMD-related disorders are caused by defective function of the emerin–lamin A/C complex, while lamin A/C mutations that cause non-EDMD diseases affect specific functions of lamin A/C unrelated to emerin. The precise function of the emerin–lamin A/C complex is unclear, but there is a growing realization that nuclear elements previously thought of as merely structural may have a role in gene expression, exerted through interactions with chromatin and with nuclear transcription/splicing factors. The possibility that the lamina or nuclear matrix is involved in fixing patterns of chromatin accessibility in post-mitotic nuclei (and therefore involved in cell differentiation) remains speculative at the present time. The transmembrane sequence near the C-terminus of emerin suggests that the influence of the emerin–lamin A/C complex would normally be restricted to the nuclear periphery, which is usually enriched in heterochromatin and inactive genes (Cohen et al., 2001), although invaginations of the nuclear membrane into the nuclear interior are observed (Fricker et al., 1997; Manilal et al., 1999).

Acknowledgements

I thank the British Heart Foundation, the EU 5th Framework and the Muscular Dystrophy Campaign for past and current grant support for muscular dystrophy research in my laboratory.

References

Becane, H.M., Bonne, G., Varnous, S., et al. (2000) High incidence of sudden death with conduction system and myocardial disease due to lamins A and C gene mutation. *Pacing Clin. Electrophysiol.* 23: 1661–1666.

Bechert, K., Lagos-Quintana, M., Harborth, J., Weber, L. and Osborn, M. (2003) Effects of expressing lamin A mutant protein causing Emery-Dreifuss muscular dystrophy and familial partial lipodystrophy in HeLa cells. *Exp. Cell Res.* 286: 75–86.

Bione, S., Small, K., Aksmanovic, V.M., et al. (1995) Identification of new mutations in the Emery-Dreifuss muscular dystrophy gene and evidence for heterogeneity of the disease. *Hum. Mol. Genet.* 4: 1859–1863.

Bonne, G., DiBarletta, M.R., Varnous, S., et al. (1999) Mutations in the gene encoding lamin A/C cause autosomal dominant Emery-Dreifuss muscular dystrophy. *Nature Genet.* 21: 285–288.

Bonne, G., Mercuri, E., Muchir, A., et al. (2000) Clinical and molecular genetic spectrum of autosomal dominant Emery-Dreifuss muscular dystrophy due to mutations of the lamin A/C gene. *Ann. Neurol.* 48: 170–180.

Clements, L., Manilal, S., Love, D.R. and Morris, G.E. (2000) Direct interaction between emerin and lamin A. *Biochem. Biophys. Res. Commun.* 267: 709–714.

Cohen, M., Lee, K.K., Wilson, K.L. and Gruenbaum, Y. (2001) Transcriptional repression, apoptosis, human disease and the functional evolution of the nuclear lamina. *Trends Biochem. Sci.* 26: 41–47.

De Sandre-Giovannoli, A., Chaouch, M., Kozlov, S., et al. (2002) Homozygous defects in LMNA, encoding lamin A/C nuclear-envelope proteins, cause autosomal recessive axonal neuropathy in human (Charcot-Marie-Tooth disorder type 2) and mouse. *Am. J. Hum. Genet.* 70: 726–736.

Dhe-Paganon, S., Werner, E.D., Chi, Y.I. and Shoelson, S.E. (2002) Structure of the globular tail of nuclear lamin. *J. Biol. Chem.* 277: 17381–17384.

Dreifuss, F.E. and Hogan, G.R. (1961) Survival in X-chromosomal muscular dystrophy. *Neurology* 11: 734–737.

Emery, A.E.H. (1993) *Duchenne muscular dystrophy*, 2nd edn, pp. 88–95. Oxford University Press, Oxford, UK.

Emery, A.E. and Dreifuss, F.E. (1966) Unusual type of benign X-linked muscular dystrophy. *J. Neurol. Neurosurg. Psychiatry* 29: 338–342.

Emery, A.E.H. and Emery, M.L.H. (1995) *The History of a Genetic Disease: Duchenne Muscular Dystrophy or Meryon's Disease.* Royal Society of Medicine Press, London.

Eriksson, M., Brown, W.T., Gordon, L.B., *et al.* (2003) Recurrent *de novo* point mutations in lamin A cause Hutchinson-Gilford progeria syndrome. *Nature* 423: 293–298.

Fairley, E.A., Kendrick-Jones, J. and Ellis, J.A. (1999) The Emery-Dreifuss muscular dystrophy phenotype arises from aberrant targeting and binding of emerin at the inner nuclear membrane. *J. Cell Sci.* 112: 2571–2582.

Fidzianska, A. and Hausmanowa-Petrusewicz, I. (2003) Architectural abnormalities in muscle nuclei. Ultrastructural differences between X-linked and autosomal dominant forms of EDMD. *J. Neurol. Sci.* 210: 47–51.

Fidzianska, A., Toniolo, D. and Hausmanowa-Petrusewicz, I. (1998) Ultrastructural abnormality of sarcolemmal nuclei in Emery-Dreifuss muscular dystrophy (EDMD). *J. Neurol. Sci.* 159: 88–93.

Fricker, M., Hollinshead, M., White, N. and Vaux, D. (1997) Interphase nuclei of many mammalian cell types contain deep, dynamic, tubular membrane-bound invaginations of the nuclear envelope. *J. Cell Biol.* 136: 531–544.

Haraguchi, T., Koujin, T., Segura-Totten, M., Lee, K.K., Matsuoka, Y., Yoneda, Y., Wilson, K.L. and Hiraoka, Y. (2001) BAF is required for emerin assembly into the reforming nuclear envelope. *J. Cell Sci.* 114: 4575–4585.

Hartmann, A.M., Nayler, O., Schwaiger, F.W., Obermeier, A. and Stamm, S. (1999) The interaction and colocalization of Sam68 with the splicing-associated factor YT521-B in nuclear dots is regulated by the src family kinase p59^{fyn}. *Mol. Biol. Cell* 10: 3909–3926.

Holaska, J.M., Lee, K.K., Kowalski, A.K. and Wilson, K.L. (2003) Transcriptional repressor germ cell-less (GCL) and barrier to autointegration factor (BAF) compete for binding to emerin in vitro. *J. Biol. Chem.* 278: 6969–6975.

Holt, I., Östlund, C., Stewart, C.L., Nguyenthi Man, Worman, H.J. and Morris, G.E. (2003) Effect of pathogenic mis-sense mutations in lamin A on its interaction with emerin in vivo. *J. Cell Sci.* 116: 3027–3035.

Krimm, I., Östlund, C., Gilquin, B., Couprie, J., Hossenlopp, P., Mornon, J.P., Bonne, G., Courvalin, J.C., Worman, H.J. and Zinn-Justin, S. (2002) The Ig-like structure of the C-terminal domain of lamin A/C, mutated in muscular dystrophies, cardiomyopathy, and lipodystrophy. *Structure (Camb)* 10: 811–823.

Lee, K.K., Haraguchi, T., Lee, R.S., Koujin, T., Hiraoka, Y. and Wilson, K. (2001) Distinct functional domains in emerin bind lamin A and DNA-bridging protein BAF. *J. Cell Sci.* 114: 4567–4573.

Lin, F., Blake, D.L., Callebaut, I., Skerjanc, I.S., Holmer, L., McBurney, M.W., Paulin-Levasseur, M. and Worman, H.J. (2000) MAN1, an inner nuclear membrane protein that shares the LEM domain with lamina-associated polypeptide 2 and emerin. *J. Biol. Chem.* 275: 4840–4847.

Lloyd, D.J., Trembath, R.C. and Shackleton, S. (2002) A novel interaction between lamin A and SREBP1: implications for partial lipodystrophy and other laminopathies. *Hum. Mol. Genet.* 11: 769–777.

Manilal, S., Nguyenthi Man, Sewry, C.A. and Morris, G.E. (1996) The Emery-Dreifuss muscular dystrophy protein, emerin, is a nuclear membrane protein. *Hum. Mol. Genet.* 5: 801–808.

Manilal, S., Sewry, C.A., Nguyenthi Man, Muntoni, F. and Morris, G.E. (1997) Diagnosis of X-linked Emery-Dreifuss muscular dystrophy by protein analysis of leucocytes and skin with monoclonal antibodies. *Neuromusc. Disord.* 7: 63–66.

Manilal, S., Sewry, C.A., Pereboev, A., Nguyenthi Man, Gobbi, P., Hawkes, S., Love, D.R. and Morris, G.E. (1999) Distribution of emerin and lamins in the heart and implications for Emery-Dreifuss muscular dystrophy. *Hum. Mol. Genet.* 8: 353–359.

Markiewicz, E., Dechat, T., Foisner, R., Quinlan, R.A. and Hutchison, C.J. (2002) Lamin A/C binding protein LAP2a is required for nuclear anchorage of retinoblastoma protein. *Mol. Biol. Cell.* **13**: 4401–4413.

Mislow, J.M.K., Holaska, J.M., Kim, M.S., Lee, K.K., SeguraTotten, M., Wilson, K.L. and McNally, E.M. (2002) Nesprin-1 self-associates and binds directly to emerin and lamin A in vitro. *FEBS Lett.* **263**: 1–6.

Morris, G.E. (2000) Nuclear proteins and cell death in inherited neuromuscular disease. *Neuromusc. Disord.* **10**: 217–227.

Morris, G.E. (2001) The role of the nuclear envelope in Emery-Dreifuss muscular dystrophy. *Trends Mol. Med.* **7**: 572–577.

Mounkes, L.C., Kozlov, S., Hernandez, L., Sullivan, T. and Stewart, C.L. (2003) A progeroid syndrome in mice is caused by defects in A-type lamins. *Nature* **423**: 298–301.

Nagano, A., Koga, R., Ogawa, M., Kurano, Y., Kawada, J., Okada, R., Hayashi, Y.K., Tsukahara, T. and Arahata, K. (1996) Emerin deficiency at the nuclear membrane in patients with Emery-Dreifuss muscular dystrophy. *Nat. Genet.* **12**: 254–259.

Novelli, G., Muchir, A., Sangiuolo, F., *et al.* (2002) Mandibuloacral dysplasia is caused by a mutation in LMNA-encoding lamin A/C. *Am. J. Hum. Genet.* **71**: 426–431.

Ognibene, A., Sabatelli, P., Petrini, S., Squarzoni, S., Riccio, M., Santi, S., Villanova, M., Palmeri, S., Merlini, L. and Maraldi, N.M. (1999) Nuclear alterations in a skeletal muscle biopsy and in skin cultured from one patient affected by X-linked Emery-Dreifuss muscular dystrophy. *Muscle Nerve* **22**: 864–869.

Östlund, C., Ellenberg, J., Hallberg, E., Lippincott-Schwartz, J. and Worman, H.J. (1999) Intracellular trafficking of emerin, the Emery-Dreifuss muscular dystrophy protein. *J. Cell Sci.* **112**: 1709–1719.

Östlund, C., Bonne, G., Schwartz, K. and Worman, H.J. (2001) Properties of lamin A mutants found in Emery-Dreifuss muscular dystrophy, cardiomyopathy and Dunnigan-type partial lipodystrophy. *J. Cell Sci.* **114**: 4435–4445.

Raharjo, W.H., Enarson, P., Sullivan, T., Stewart, C.L. and Burke, B. (2001) Nuclear envelope defects associated with LMNA mutations cause dilated cardiomyopathy and Emery-Dreifuss muscular dystrophy. *J. Cell Sci.* **114**: 4447–4457.

Rao, L., Perez, D. and White, E. (1996) Lamin proteolysis facilitates nuclear events during apoptosis. *J. Cell Biol.* **135**: 1441–1455.

Sasseville, A.M. and Raymond, Y. (1995) Lamin A precursor is localized to intranuclear foci. *J. Cell Sci.* **108**: 273–285.

Savkur, R.S., Philips, A.V. and Cooper, T.A. (2001) Aberrant regulation of insulin receptor alternative splicing is associated with insulin resistance in myotonic dystrophy. *Nature Genet.* **29**: 40–47.

Segura-Totten, M., Kowalski, A.K., Craigie, R. and Wilson, K.L. (2002) Barrier-to-autointegration factor: major roles in chromatin decondensation and nuclear assembly. *J. Cell Biol.* **158**: 475–485.

Sewry, C.A. (2000) Immunocytochemical analysis of human muscular dystrophy. *Microsc. Res. Tech.* **48**: 142–154.

Sewry, C.A., Brown, S.C., Mercuri, E., Bonne, G., Feng, L., Camici, G., Morris, G.E. and Muntoni, F. (2001) Skeletal muscle pathology in autosomal dominant Emery-Dreifuss muscular dystrophy with lamin A/C mutations. *Neuropathol. Appl. Neurobiol.* **27**: 281–290.

Shackleton, S., Lloyd, D.J., Jackson, S.N., *et al.* (2000) *LMNA,* encoding lamin A/C, is mutated in partial lipodystrophy. *Nat. Genet.* **24**: 153–156.

Shumaker, D.K., Kuczmarski, E.R. and Goldman, R.D. (2003) The nucleoskeleton: lamins and actin are major players in essential nuclear functions. *Curr. Opin. Cell Biol.* **15**: 358–366.

Sullivan, T., Escalante-Alcalde, D., Bhatt, H., Anver, M., Bhat, N., Nagashima, K., Stewart, C.L. and Burke, B. (1999) Loss of A-type lamin expression compromises nuclear envelope integrity leading to muscular dystrophy. *J. Cell Biol.* **147**: 913–919.

Tsuchiya, Y., Hase, A., Ogawa, M., Yorifuji, H. and Arahata, K. (1999) Distinct regions specify the nuclear membrane targeting of emerin, the responsible protein for Emery-Dreifuss muscular dystrophy. *Eur. J. Biochem.* **259**: 859–865.

Vaughan, A., Alvarez-Reyes, M., Bridger, J.M., Broers, J.L., Ramaekers, F.C., Wehnert, M., Morris, G.E., Whitfield, W.G.F. and Hutchison, C.J. (2001) Both emerin and lamin C depend on lamin A for localization at the nuclear envelope. *J. Cell Sci.* **114**: 2577–2590.

Weber, K., Plessmann, U. and Traub, P. (1989) Maturation of nuclear lamin A involves specific carboxy-terminal trimming, which removes the polyisoprenylation site from the precursor; implications for the structure of the nuclear lamina. *FEBS Lett.* **257**: 411–414.

Wilkinson, F.L., Holaska, J.M., Zhang, Z., Sharma, A., Manilal, S., Holt, I., Stamm, S., Wilson, K.L. and Morris, G.E. (2003) Emerin interacts in vitro with the splicing-associated factor, YT521-B. *Eur. J. Biochem.* **270**: 2459–2466.

Zhang, Q., Skepper, J.N., Yang, F., Davies, J.D., Hegyi, L., Roberts, R.G., Weissberg, P.L., Ellis, J.A. and Shanahan, C.M. (2001) Nesprins: a novel family of spectrin-repeat-containing proteins that localize to the nuclear membrane in multiple tissues. *J. Cell Sci.* **114**: 4485–4498.

5

Plant nuclear envelope proteins

Annkatrin Rose, Shalaka Patel and Iris Meier

1. Introduction

The nuclear envelope (NE) is the hallmark of all eukaryotic cells, separating the nucleoplasm from the cytoplasm during interphase. At the same time, the NE allows for the controlled exchange of macromolecules between the two compartments through nuclear pores and presents a surface for anchoring and organizing cytoskeletal components and chromatin. In order to provide these functions, the NE has evolved into a highly organized system of membranes and proteins. The outer nuclear membrane (ONM) is continuous with the endoplasmic reticulum, while the inner nuclear membrane (INM) contains a number of INM-specific integral and peripheral membrane proteins and is lined by the nuclear lamina in metazoan cells (reviewed by Holaska et al., 2002). While the protein composition of the NE has been studied in detail in animal cells, our knowledge of the components of the plant NE is still at an early stage. To date, only a handful of proteins have been cloned and characterized as components of the plant NE, while a few more have been detected using antibodies against NE components (see *Table 1*). Initial investigations have revealed similarities as well as striking deviations in the NE biology of the different kingdoms, providing intriguing insights into the evolution of this cell compartment.

2. The plant nuclear envelope as scaffold for macromolecular complexes

2.1 *Do plant cells have lamins?*

The lamins, a subgroup of the intermediate-filament proteins, form the protein meshwork of the nuclear lamina in animal cells. They are involved in attaching chromatin to the inner surface of the NE and in a number of additional functions such as NE assembly, DNA synthesis, transcription and apoptosis (Goldman et al., 2002; Holaska et al., 2002). Mutations in A-type lamins or associated proteins have been linked to at least six different inherited diseases in humans affecting skeletal and

The Nuclear Envelope, edited by D.E. Evans, C. Hutchison & J.A. Bryant.
© 2004 Garland Science/BIOS Scientific Publishers

Table 1. Antigens and proteins at the plant nuclear envelope

Protein	Immunodetection	Plant species	Size	pI	Domains/characteristics	Putative function	Reference
put. myosin	pAb against D.m. NMPCL fraction	Onion	n.d.	n.d.	Antiserum cross-reacts with animal non-muscle myosin	NPC-associated myosin-like ATPase	Berrios and Fisher, 1986
put. myosin	mAb S-1 against MHC	Tobacco	175 kDa[b]	n.d.	Only anti-MHC was found at NE, not anti-MLC	Myosin heavy chain	Tang et al., 1989
put. NM component	mAb JIM 63 against D.c. NM	Carrot	92 kDa[b]	5.5-5.9[b]	Wide cross-reactivity of antibody (tobacco, onion, Arabidopsis, pea, Aspergillus, rat)	Plant lamina component	Beven et al., 1991
put. nucleoporin	pAb against yeast Nsp1p, mAb against rat p62	Carrot, onion, brassica	100 kDa[b]	n.d.	Decorates nuclear pores	Nucleoporin	Scofield et al., 1992
put. MTOC component	mAb 6C6 against calf centrosomes	Onion, lily, ginko, tobacco	78 kDa[b] in onion	n.d.	Relocates from NE to telomeres/SC during meiotic prophase, localizes at kinetochores during mitosis, associated with NM	MT nucleation, NE-attachment of telomeres in meiosis	Chevrier et al., 1992; Schmit et al., 1994, 1996; Gindullis and Meier, 1999; Cowling et al., 2003
put. G-proteins	n.d.	Pea	27, 28, and 30 kDa[b]	n.d.	GTP-binding activity is regulated by phytochrome	Regulation of vesicle fusion, phytochrome signaling	Clark et al., 1993
put. lamin	pAb against chicken lamins	Onion	65 kDa[b]	6.8[b]	Also detectable in spots in the NM	Plant lamina component	Minguez and Morena Diaz de la Espina, 1993
gp40	pAb against N.t. gp40	Tobacco	357 aa, 39 kDa[a], 40-43 kDa[b]	9.45[a]	tGlcNAc glycoprotein, similar to aldose-1-epimerases	Nucleoporin	Heese-Peck and Raikhel, 1998b
LCA	pAb against L.e. LCA	Tomato	105/116 kDa[b]	n.d.	Transmembrane protein, relocalizes to spindle poles during mitosis	Ca^{2+}-pump	Downie et al., 1998

Table 1. continued

Protein	Immunodetection	Plant species	Size	pI	Domains/characteristics	Putative function	Reference
NMCP1	mAb CML-1 against D.c. NM	Carrot	1164 aa, 133.6 kDa[a], 96/103 kDa[b]	5.6-5.8[b]	Coiled-coil domain, NLS, kinase consensus sequences, associated with NM	IF protein of the plant lamina	Masuda et al., 1999
MAF1	pAb against L.e. MAF1	Tomato	152 aa, 16.2 kDa[a], 25 kDa[b]	4.2[a]	WPP domain for plant NE targeting, interacts with MFP1, associated with NM	Plant NPC- lamina component	Gindullis et al., 1999
put. spectrin	pAb S1390 against chicken spectrin	Pea	60 kDa, 220/240 k Da[b]	n.d.	Also detectable in spots and nucleolus-NE tracks inside the nucleus	Attachment of actin filaments	de Ruijter et al., 2000
RanGAP	pAb against A.t. RanGAP1	Arabidopsis	535-545 aa, 59 kDa[a], 60 kDa[b]	4.5[a]	WPP domain for plant NE targeting, conserved RanGAP domain (mammalian, yeast), relocates to spindle and phragmoplast during mitosis	Ran GTPase activation	Rose and Meier, 2001; Pay et al., 2002
Spc98p/GCP3	pAb against O.s. Spc98p	Arabidopsis, rice	838 aa, 95 kDa[a]	6.8[a]	Colocalizes with gamma-tubulin, also localizes to cytoplasmic and nuclear sites, conserved (animals, yeast)	MT nucleation	Erhardt et al., 2002

[a] predicted
[b] observed

A.t., *Arabidopsis thaliana*, D.c., *Daucus carota* (carrot), D.m., *Drosophila melanogaster* (fruitfly), L.e., *Lycopersicon esculentum* (tomato), N.t., *Nicotiana tabacum* (tobacco), O.s., *Oryza sativa* (rice); aa, amino acids, kDa, kilodalton, mAb, monoclonal antibody, MHC, myosin heavy chain, MLC, myosin light chain, MT, microtubule, n.d., not determined, NLS, nuclear localization signal, NM, nuclear matrix, NMPCL, nuclear matrix-pore complex-lamina, NPC, nuclear pore complex, pAb, polyclonal antibody, SC, synaptonemal complex, tGlcNAc, terminal N-acetylglucosamine.

cardiac muscle and/or the loss and redistribution of white fat (Mounkes *et al.*, 2003; see also Chapter 3), demonstrating the importance of NE function in certain types of animal cells.

Electron microscopy (EM) studies revealed that a structure similar to the vertebrate nuclear lamina is also found in the nuclei of higher plant cells (Galcheva-Gargova *et al.*, 1988; Moreno Díaz de la Espina *et al.*, 1991). However, there is no evidence for the existence of *bona fide* lamins in plants. Immunohistochemical studies suggested the existence of lamin-like proteins in plant nuclei (Li and Roux, 1992; McNulty and Saunders, 1992; Mínguez and Moreno Díaz de la Espina, 1993), but no sequence information from the antigens detected in these studies is available, and no lamin-coding genes seem to be present in the fully sequenced *Arabidopsis* genome. This might indicate that plants, like yeast, do not have this group of nuclear coiled-coil proteins. The questions remain as to which components make up the structures detected by EM and which plant proteins are detected by the anti-lamin antisera.

NMCP1 is a 134-kDa carrot nuclear matrix protein found exclusively at the periphery of the nucleus during interphase (Masuda *et al.*, 1997). NMCP1 has a central coiled-coil domain flanked by a non-helical short head and a larger tail domain and a pI of 5.6–5.8, similar to lamins. Although NMCP1 is roughly twice the size of lamins, it is at present the best candidate for a lamin-like protein in plants.

2.2 *A connection between the nuclear envelope and the actin cytoskeleton?*

The NE in animal cells contains large spectrin-related proteins of the α-actinin superfamily as integral proteins in both membranes. The nesprins localize to the INM and provide a connection to the lamina by interacting with the lamina components lamin A and emerin (Mislow *et al.*, 2002). The giant 796-kDa protein NUANCE is located at the ONM and might function in connecting the nucleus with the cytoskeleton through its interaction with actin (Zhen *et al.*, 2002). Attachment of NUANCE and nesprin-1 to the ONM and INM, respectively, is achieved by the presence of a C-terminal transmembrane domain (Zhang *et al.*, 2001; Zhen *et al.*, 2002). Members of this protein family differ from each other in the number of spectrin repeats, which fold into three-bundle helical coiled-coils forming rod-like structures. In animals, proteins containing the spectrin-repeat domain predominantly serve in organizing the cytoskeleton by crosslinking actin filaments, or linking actin filaments to membranes (Djinovic-Carugo *et al.*, 2002). Interestingly, an antiserum directed against chicken spectrin also detects an antigenic component at the plant NE, indicating the possible presence of this class of proteins in plants (de Ruijter *et al.*, 2000). Similar to the case of the lamins, no sequence homologues of spectrin-like proteins have been identified in plants to date.

2.3 *Microtubule nucleation at the plant nuclear envelope*

During nuclear division, the genetic material inside the nucleus is distributed into two daughter nuclei by the action of the microtubular spindle apparatus. Formation of the mitotic spindle originates at microtubule organizing centres (MTOCs), which are the cytoplasmic centrosomes in animal cells and the NE-embedded spindle pole bodies in yeast cells. One of the main differences in cell division between higher plant cells and other eukaryotic cells is the absence of centrosomes or spindle pole bodies in plant cells. Instead, the plant NE acts as the main site for plant microtubule nucleation. Early evidence pointed to the association of microtubules with the nuclear surface in plant cells and Stoppin *et al.* (1994) showed that isolated maize nuclei are capable of nucleating microtubules *in vitro*. While the nuclear surface might not be the only site for microtubule nucleation in plants cells, it is especially active in the G2 phase of the cell cycle before NE breakdown (reviewed by Canaday *et al.*, 2000), suggesting that its microtubule-nucleating activity might be under cell-cycle control. Microtubule-associated proteins (MAPs) modulate the nucleating activity of the plant nuclear surface (Stoppin *et al.*, 1996).

Significant progress has been made in recent years in identifying the molecular components of the microtubule nucleation machinery. In yeast cells, the spindle pole proteins Spc98p and Spc97p form a complex with γ-tubulin that binds to the large coiled-coil protein Spc110p of the spindle pole body at the NE (Knop and Schiebel, 1997). Homologues of these proteins have been identified in animals where they interact in a similar fashion (Murphy *et al.*, 1998; Oegema *et al.*, 1999). The complex of γ-tubulin and Spc98p/Spc97p constitutes the minimal core of microtubule nucleation units in yeast, also known as γ-tubulin ring complex or γ-TuRC (reviewed by Job *et al.*, 2003; Moritz and Agard, 2001; Schiebel, 2000). All three proteins can also be identified by sequence homology in plant genomes. Recently, plant homologues of Spc98p have been reported to co-localize with γ-tubulin at the nuclear surface in plant cells (Erhardt *et al.*, 2002). Microtubule nucleation in plant cells depends on the presence of both γ-tubulin and Spc98p, and GFP-labelled plant Spc98p localizes within the nucleus, at the NE and close to the plasma membrane, strengthening the hypothesis of multiple nucleation sites in plant cells (Seltzer *et al.*, 2003).

In addition, vertebrate γ-TuRC contains several proteins not present in yeast cells but conserved in plants. Human 76p, also known as GCP4 (γ-tubulin complex protein 4), was identified as a component of an isolated γ-tubulin complex and it localizes to the centrosomes and the spindle poles (Fava *et al.*, 1999). It has an orthologue in the higher plant *Medicago* and the moss *Physcomitrella*, but appears to be absent in yeast (Fava *et al.*, 1999). Murphy and colleagues reported the finding of the two remaining components of the γ-tubulin complex in mammalian cells, GCP5 and GCP6, and identified the *Arabidopsis* counterparts of h76p/GCP4 and the GCP5/6 family by sequence comparison (Murphy *et al.*, 2001). Although the functional and orthologue relationship of vertebrate GCP4, GCP5 and GCP6 to the GCP-like proteins in plants is not clear, their presence in plant genomes suggests that plant cells utilize the same basic molecular components as vertebrate cells for microtubule nucleation, even though the microscopically visible subcellular structures associated with microtubule nucleation differ significantly between higher plants (NE), animals (centrosomes) and yeast (spindle pole bodies).

The question remains how the association of microtubule nucleation with the NE is achieved in plant cells. In yeast, the coiled-coil proteins Spc72p and Spc110p of the spindle pole body bind the γ-tubulin/Spc98p/Spc97p complex and are thought to provide the attachment sites on the cytoplasmic and nucleoplasmic surfaces of the NE-embedded spindle pole body, respectively (Knop and Schiebel, 1997, 1998). Kendrin/pericentrin-B, the mammalian orthologue of Spc110p, forms a complex with other coiled-coil proteins that might serve as anchor for the γ-TuRC and regulatory proteins at the mammalian centrosome according to their protein interaction profiles

(Takahashi *et al.*, 2002). No homologues of Spc72p or Spc110p/Kendrin appear to be encoded by the fully sequenced *Arabidopsis* genome. However, it is tempting to speculate that the γ-TuRC might be anchored at the surface of the plant NE via interaction with other NE-associated coiled-coil proteins, such as NMCP1 (Masuda *et al.*, 1997), in analogy to the situation in yeast and animal cells.

2.4 *Meiotic tethering of telomeres to the nuclear envelope*

Homologous chromosome synapsis allowing for meiotic recombination is the hallmark of meiotic prophase and coincides with the clustering of telomeres at the NE in a bouquet arrangement that is unique to meiosis (reviewed by Scherthan, 2001). The telomere attachment at the NE occurs adjacent to the centrosome or, in yeast, the spindle pole body. As discussed above, plant cells do not have these organelles; however, plant telomeres also cluster into a bouquet structure during meiotic prophase (Bass *et al.*, 1997; Cowan *et al.*, 2001; Martínez-Pérez *et al.*, 1999). Microscopic evidence suggests a connection between the synaptonemal complex of meiotic chromosomes and their attachment site at the NE. Interestingly, an antibody raised against calf centrosomes decorates the NE as well as the synaptonemal complex in plant cells (Schmit *et al.*, 1996), suggesting a common molecular link between these structures. Asynaptic mutants have been characterized in rye and maize that are not only defective in formation of synaptonemal complexes but also in clustering telomeres into bouquets. The rye *sy1* mutant fails to form a bouquet in about half its nuclei (Mikhailova *et al.*, 2001). In the maize mutant *pam1* (plural abnormalities of meiosis 1), telomeres associate with the NE but their clustering is delayed or abolished (Golubovskaya *et al.*, 2002). In the maize *dsy1* (desynaptic 1) mutant, telomeres are misplaced during bouquet formation whereas the maize *dy* (desynaptic) mutant forms apparently normal bouquets but then fails to maintain the telomere–NE connection during middle prophase (Bass *et al.*, 2003). Interestingly, not all plant species show the typical bouquet arrangement at the NE. *Arabidopsis* telomeres were found to cluster inside the nucleus during meiosis, most likely attached to the periphery of the nucleolus (Armstrong *et al.*, 2001). This might indicate a higher diversity of patterns in nuclear organization during meiosis among plant species than in other eukaryotes.

How are telomeres attached to the NE during meiosis? The involvement of telomeric chromatin is one possibility. A sperm-specific histone H2B variant has been found as a component of the human sperm telomere-binding protein complex and has been suggested to contribute to NE attachment via binding to lamins (Gineitis *et al.*, 2000). However, analysis of a mouse knockout mutant deficient in telomerase has shown that most telomeric chromatin is dispensable for the attachment of the chromosome ends to the NE in mammals (Franco *et al.*, 2002). The nuclear pores have been discussed as candidate attachment sites (Tham and Zakian, 2000). However, meiotic telomere attachment to the NE occurs in regions that are distinct from nuclear pore-dense areas in mammalian cells (Scherthan *et al.*, 2000). An antibody raised against frog oocyte NE proteins interacting with telomere DNA recognized a 70-kDa outer NE component in frog cells as well as in mammalian cells (Podgornaya *et al.*, 2000) suggesting the involvement of its antigen in the anchoring process. The yeast proteins Taz1 (telomere-associated in *Schizosaccharomyces pombe*) and Ndj1p (nondisjunction 1) in *Saccharomyces cerevisiae* are required for the bouquet formation at the spindle pole body during meiosis, and it has been suggested that they might function in tethering meiotic chromosomes to the nuclear periphery (Cooper *et al.*, 1998; Trelles-Sticken *et al.*, 2000). No homologues of these proteins are currently known in either plants or animals. Future research will have to be undertaken to reveal the molecular nature of the meiotic telomere attachment sites on plant and vertebrate NEs.

3. Targeting across and into the plant nuclear envelope

3.1 *Gateways in the plant nuclear envelope*

The physical barrier of the NE needs to be permeable to a variety of macromolecules and signals in order for the cell to function. The only known gateways for the transport of macromolecules across the NE are the nuclear pore complexes (NPCs). The NPC is a 125-MDa multiprotein complex with a 9-nm aqueous pore through which molecules smaller than 30 kDa can readily diffuse (Gasiorowski and Dean, 2003; Stoffler *et al.*, 1999a).

The proteins forming the NPCs are called nucleoporins (Nups). Significant progress in the identification of animal and yeast Nups has been made recently by the use of proteomics (Allen *et al.*, 2001; Cronshaw *et al.*, 2002). It was assumed that the animal NPC would contain a larger number of distinct subunits than the yeast NPC because of its larger size. However, the proteomic analysis of mammalian NPC composition showed that it contains only 30 different nucleoporins, similar to the results from yeast and less complex than expected (Cronshaw *et al.*, 2002).

The composition of the plant NPC is largely unknown. Preliminary work done by Scofield *et al.* (1992) showed that an antibody against the yeast nucleoporin Nsp1p recognized a 100-kDa protein from the nuclear matrix of carrot suspension cells, indicating the possible existence of such a protein in plants. With a few exceptions, homologues to most animal and yeast Nups cannot be identified in plants using sequence homology searches. A putative 215-kDa orthologue of gp210, one of the integral membrane proteins of the NPC, has been identified in *Arabidopsis* and is predicted to contain a transmembrane domain at its C-terminus, consistent with its animal counterpart (Cohen *et al.*, 2001).

A number of yeast and vertebrate Nups have FG repeats (FXFG in vertebrates and GLFG in yeast), which are thought to provide transient, low-affinity binding sites for transport receptors (Mattaj and Englmeier, 1998). A nucleoporin-like FG-repeat-containing protein can be identified by sequence similarity searches with human Nup98 in the *Arabidopsis* database. Its FG repeats appear to be conserved with its animal Nup counterpart, indicating a possible role for this motif in plant NPCs (Patel S., Rose A., and Meier I., unpublished results).

A family of vertebrate Nups is modified with single N-acetylglucosamine (GlcNAc) residues (for review see Heese-Peck and Raikhel, 1998a). The function of this glycosy-lation is unclear, but it has been useful in purification of Nups by lectin affinity chromatography. The lectin wheat germ agglutinin (WGA) binds specifically to GlcNAc and inhibits nuclear protein import in vertebrates. It also appears to bind to plant nuclei; however, WGA only inhibits nuclear import of large complexes in plants indicating that direct interaction with the polysaccharides at the NPC is not sufficient to block nuclear import of smaller proteins (Hicks *et al.*, 1996). Heese-Peck and Raikhel (1998b) used the WGA affinity of GlcNAc to isolate the plant nucleoporin gp40. Plant gp40 shares similarity with bacterial aldose-1-epimerases and is a target of glycosylation, although these modifications are larger than the ones observed on mammalian Nups and consist of more than five GlcNAc residues (Heese-Peck and Raikhel, 1998b). This might provide an explanation for the lack of inhibition of nuclear transport by lectin due to the larger carbohydrates providing a longer 'linker' between the NPC and WGA, thus allowing more space for proteins to pass than at vertebrate NPCs. Gp40 has been hypothesized to play a role in the nuclear import of glyco-proteins (Heese-Peck and Raikhel, 1998b).

3.2 *Nuclear import and export*

The active transport of macromolecules across the nuclear pores involves a number of signals and transport factors. Although the NPC allows molecules up to about 60 kDa to passively diffuse across, molecules over 50 kDa diffuse at a very slow rate. Most proteins that need to be transported to the nucleus bear a nuclear localization signal (NLS) enabling active transport to the nucleus. The transport of small proteins that can diffuse easily through the nuclear pore can be improved significantly by the presence of an NLS (Gasiorowski and Dean, 2003). Proteins needing to be shuttled back from the nucleus into the cytoplasm often contain a nuclear export signal (NES). Animal and yeast NLS and NES sequences have been found to be functional in plants (Merkle, 2001; Smith *et al.*, 1997; Ward and Lasarowitz, 1999), and endogenous NLS and NES motifs have been identified on a variety of plant proteins. For example, the movement protein BR1 of the squash leaf curl virus has a functional NES with two basic NLSs (Ward and Lazarowitz, 1999). The NES on AtRanBP1a is functionally indistinguishable from the NES on the HIV-1 Rev protein (Haasen *et al.*, 1999).

NLS sequences are recognized by the import receptor importin α, a member of the karyopherin family. Hicks and Raikhel (1995) showed that NLS-binding proteins were also present in higher plants. Several importin α variants have been characterized in plants (Hübner *et al.*, 1999; Jiang *et al.*, 1998a, 2001; Shoji *et al.*, 1998; Smith *et al.*, 1997). Plant importin α appears to be strongly associated with the NE (Hicks *et al.*, 1996; Smith *et al.*, 1997). In contrast to animal systems where the importin α-mediated import is dependent on importin β, *Arabidopsis* importin At-IMPα binds NLS with high affinity and mediates nuclear import in the absence of importin β *in vitro* (Hübner *et al.*, 1999). Multiple isoforms of importin α could play an important role in plant signal transduction. For example, the nuclear import of the red light receptor phytochrome and several b-ZIP transcription factors is regulated by light (Hisada *et al.*, 2000; Yamamoto and Deng, 1999). Rice importin α1b selectively binds to a number of plant NLSs and is differentially expressed in different tissues and in response to light (Jiang *et al.*, 2001). It is involved in the nuclear import of constitutive photomor-phogenic 1 (COP1), a repressor of photomorphogenesis (Jiang *et al.*, 2001).

Another member of the karyopherin family that plays an important role in nuclear import is importin β. It facilitates the interaction of the importin/cargo complex with FG repeat components of the NPC (Bayliss *et al.*, 2002). Plant importin β homologues were isolated in rice by Jiang *et al.* (1998b) and are involved in NE docking of NLS-containing proteins and their subsequent nuclear localization. The *Arabidopsis* genome encodes at least 17 predicted importin β-like proteins (Bollman *et al.*, 2003).

NES sequences interact with another member of the karyopherin family, the nuclear export receptor CRM1/exportin 1. An *Arabidopsis* homologue of exportin 1,

AtXPO1, has been identified and functionally characterized (Haasen *et al.*, 1999). Similar to the animal exportin 1, it is inhibited by the antifungal antibiotic leptomycin B (Haasen *et al.*, 1999). The *Arabidopsis* homologue of exportin 5 (involved in double-stranded RNA export), HASTY, has also been identified (Bollman *et al.*, 2003). HASTY protein is located at the nuclear periphery and interacts with the Ran GTPase. Its loss causes a variety of developmental phenotypes (Bollman *et al.*, 2003). PAUSED, the *Arabidopsis* homologue of exportin T (involved in tRNA export), is able to rescue a tRNA export defective yeast mutant, indicating conservation of its function in plants (Hunter *et al.*, 2003). Like HASTY, mutants of PAUSED have pleiotropic effects in plant development (Hunter *et al.*, 2003; Li and Chen, 2003).

3.3 *The Ran cycle in plants*

The action of the karyopherins alone is not sufficient to drive transport across the NE. The small GTPase Ran (Ras-related nuclear protein) is required for nucleocytoplasmic transport. Ran exists in two forms, Ran GTP and Ran GDP, which are transformed into each other by the action of accessory proteins of the Ran cycle. The low intrinsic GTPase activity of Ran is enhanced by the cofactors Ran GTPase Activating Protein (RanGAP) and Ran Binding Proteins (RanBPs) on the cytoplasmic side of the NE, thus leading to the transformation of RanGTP to RanGDP outside the nucleus. The chromatin-bound Ran nucleotide exchange factor RCC1 (regulator of chromosome condensation 1) converts RanGDP to RanGTP inside the nucleus. The RanGDP vs. RanGTP gradient over the NE established by the spatial sequestering of the Ran accessory proteins is involved in maintaining the directionality of nucleocytoplasmic transport. The association of RanGTP with karyopherins inside the nucleus triggers the release of imported cargo proteins, whereas it stabilizes the interaction with export cargos. The hydrolysis of RanGTP to RanGDP outside the nucleus triggers the release of Ran from the karyopherins, thus dissociating export cargo complexes and recycling import receptors to the cytoplasm (for review see Gasiorowski and Dean, 2003; Merkle, 2001).

Apart from its function in nucleocytoplasmic transport, the Ran cycle plays a critical role in the regulation of spindle and NE assembly during mitosis (for review see Quimby and Dasso, 2003). Further evidence suggests that Ran might alter the conformation of NPCs (Goldberg *et al.*, 2000) and is involved in NPC assembly (Walther *et al.*, 2003).

Components of the Ran cycle have been identified in plants. Ran has been found in a variety of plant species (Ach and Gruissem, 1994; Merkle *et al.*, 1994; Saalbach and Christov, 1994) and is encoded as a small family of three Ran proteins in *Arabidopsis* (Haizel *et al.*, 1997). RanGAP also has been identified (Meier, 2000) and can complement a temperature-sensitive mutant of the yeast RanGAP homologue rna1p (Pay *et al.*, 2002). *Arabidopsis* RanGAP is concentrated at the plant NE during interphase (Rose and Meier, 2001; Pay *et al.*, 2002), consistent with the NE localization of its mammalian homologue, and is associated with spindle and phragmoplast microtubules during mitosis (Pay *et al.*, 2002). The phragmoplast is a unique feature of plant cells associated with the growing cell plate. These findings might be indicative of a role for the plant Ran cycle in microtubule organization during cell division also in plant cells.

Haizel *et al.* (1997) isolated genes encoding RanBPs (AtRanBP1a and AtRanBP1b) from *Arabidopsis*. Kim and Roux (2003) cloned RanBP1c and showed that suppression of its expression resulted in altered root development and hypersensitivity to auxin. They hypothesized that AtRanBP1c, by maintaining the RanGDP/RanGTP cycle, is involved in the control of proteins that regulate auxin sensitivity. AtRanBP1c stabilizes the RanGTP conformation and is a co-activator of RanGAP. Localization studies indicated that it is present in the cytoplasm and the nucleus, suggesting a possible role for plant RanBP1 inside the nucleus (Kim and Roux, 2003). The missing link in the plant Ran cycle is the Ran-specific guanine nucleotide exchange factor RCC1. So far, no RCC1 homologue has been identified in plants.

3.4 *Nuclear envelope targeting in plants*

MAF1, a small plant-specific NE-protein (Gindullis *et al.*, 1999), shows high sequence similarity to the N-terminal domain of plant RanGAPs (Meier, 2000). Based on immunocytochemistry and GFP-fusion protein localization, both MAF1 and RanGAP are localized at the plant NE, though not exclusively (see *Figure 1*). Plant RanGAP and MAF1 share a targeting domain termed the WPP domain, after a conserved tryptophan-proline-proline motif, that appears to be unique to plants (Meier, 2000; Rose and Meier, 2001). Mutation of the conserved WPP motif within this domain abolishes NE targeting (Rose and Meier, 2001). Vertebrate RanGAPs lack the WPP domain, but possess a C-terminal domain required for targeting mammalian RanGAP to the nuclear pore. The small ubiquitin-like modifier SUMO is attached to this domain, and the SUMOylated C-terminus of mammalian RanGAP binds to the nucleoporin RanBP2/Nup358 (Matunis *et al.*, 1998). This domain is not present on plant RanGAPs, and no homologues of RanBP2/Nup358 appear to be encoded by the *Arabidopsis* genome. Interestingly, human RanGAP1 does not localize to the NE in tobacco and *Arabidopsis* cells and *Arabidopsis* RanGAP1 does not localize to the NE in HeLa cells. (A. Rose, J. Joseph, M. Dasso and I. Meier, unpublished results). This suggests that plant and mammalian RanGAP not only have different targeting domains, but also utilize different binding partners for retention at the NE. The attachment site for plant RanGAP at the NE is currently not known.

   What could be the role of NE attachment of RanGAP in both mammalian and plant cells? *Saccharomyces cerevisiae* and *Schizosaccharomyces pombe* RanGAPs possess neither of the NE targeting domains of plant or animal RanGAPs and do not localize to the NE (Hopper *et al.*, 1990; Matynia *et al.*, 1996), indicating that the attachment of RanGAP to the NE might not be necessary for its function during interphase. However, yeast cells undergo closed mitosis without disassembly of the NE whereas higher plant and animal cells undergo open mitosis characterized by the breakdown of the NE during prophase. It is tempting to speculate that the sequestering of RanGAP to the NE and microtubular structures during nuclear division has become necessary with the development of open mitosis and has arisen independently during the evolution of plants and animals.

   To date, no integral membrane proteins of the NE have been cloned in plants. However, an N-terminal domain of the mammalian lamin B receptor (LBR) was targeted successfully to the NE in tobacco cells consistent with its animal cell localization pattern (Irons *et al.*, 2003). LBR is an integral membrane protein of the INM,

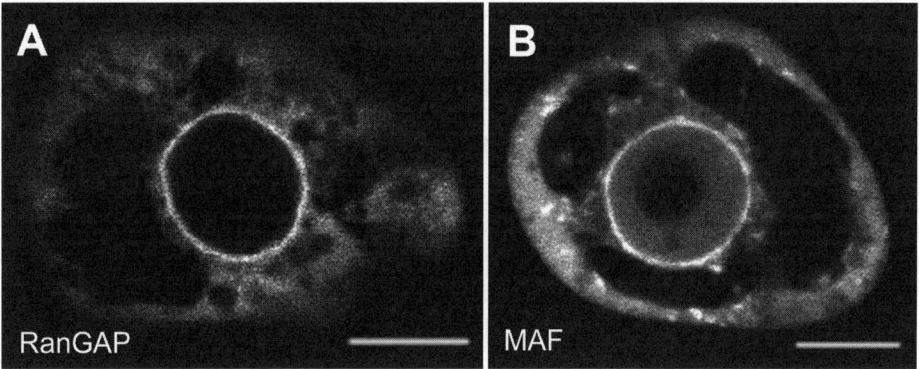

Figure 1. AtRanGAP1 and LeMAF1 are localized at the plant nuclear envelope. GFP fusion constructs of (A) AtRanGAP1 (Rose and Meier, 2001) and (B) LeMAF1 (Gindullis et al., 1999) were transiently expressed in tobacco BY-2 cells. Images of optical sections of a plane corresponding to the center of the nucleus were taken on a laser scanning confocal microscope. Bars equal 10 μm. Both AtRanGAP1 and LeMAF1 are distributed throughout the cytoplasm and concentrated at the nuclear envelope. MAF1-GFP also enters the nucleus, but not the nucleolus, and concentrates in speckles in the cytoplasm.

and protein and DNA database analysis showed no evidence for LBR homologues in plants. The N-terminal portion of LBR comprises the nucleoplasmic domain containing a bipartite NLS and a transmembrane domain. This domain is sufficient to anchor the protein in the INM (Smith and Blobel, 1993; Soullam and Worman, 1993) and can be used as a NE marker in mammalian cells (Ellenberg *et al.*, 1997). LBR not only targets to the NE in mammalian and plant cells, but also in yeast (Smith and Blobel, 1994), indicating the existence of a common mechanism for the targeting of INM proteins in eukaryotic cells.

One possibility is that the INM targeting signal might simply be comprised of an NLS in combination with a transmembrane segment to anchor the protein in the NE membrane. Soullam and Worman (1995) have shown that the nucleoplasmic portion of LBR can indeed function as an NLS for soluble proteins and as an INM localization signal when fused to membrane proteins. However, other nuclear localization signals tested failed to target membrane proteins to the NE (Soullam and Worman, 1995). The amino-terminal domain of LBR must therefore have different properties than nuclear localization signals in general. The INM targeting depended on the size of the nucleo-cytoplasmic domain, indicating that INM proteins reach the nucleus via migration through the nuclear pores (Soullam and Worman, 1995). The current model for targeting integral membrane proteins into the nucleus involves their synthesis in the ER, lateral diffusion into the outer NE membrane, diffusion through the nuclear pore membrane, and finally diffusion into the inner NE membrane where they are immobilized by interaction with the lamina or other intranuclear components (Ellenberg *et al.*, 1997). This INM targeting mechanism appears to be functional in animals, yeast and plants, and the usefulness of a tagged animal INM protein as NE marker in plant cells opens new possibilities for studies of plant NE dynamics.

4. Signal transduction at the plant nuclear envelope

4.1 *G proteins*

Heterotrimeric G proteins and other GTP-binding proteins are involved in a number of signalling pathways, including light, hormone and pathogen responses in plants (for review see Assmann, 2002; Yang, 2002). In animal systems, a number of small GTP-binding proteins and antigens of the heterotrimeric G-protein complex have been found at the NE (Cohen-Armon *et al.*, 1996; Rubins *et al.*, 1990; Saffitz *et al.*, 1994). Similar G proteins also are present at the plant NE. Clark *et al.* (1993) found small GTP-binding protein activities associated with the NE fraction in pea. Their ability to bind GTP appeared to be regulated by the red light receptor phytochrome. It was suggested that these small NE-associated G proteins could play a role in the regulation of vesicle fusions during NE reassembly in mitosis. In tobacco, a Gβ subunit has been detected in isolated nuclei (Peskan and Oelmüller, 2000). Carrot Gα subunit is detectable in membrane-free nuclear matrix preparations (Drøbak *et al.*, 1995). Similar findings have been obtained in animal systems. In rat neurons, GTP-binding proteins reacting with antibodies against Gα were found associated with the NE but resistant to extraction by detergent, indicating a possible connection with non-lipid NE components (Cohen-Armon *et al.*, 1996). While the role of the small GTPase Ran at the NE has been well established (see Section 3), the functions of heterotrimeric G-proteins at the nucleus are more speculative and might involve nuclear phospholipid and Ca^{2+} signalling (for review see Willard and Crouch, 2000).

4.2 *Ca^{2+} signalling*

A distinct role for the NE in controlling Ca^{2+} signalling in nuclear processes has been documented in animal cells. Recent evidence from mammalian systems suggests that a nuclear Ca^{2+} pool is stored in a nucleoplasmic reticulum formed by the INM of the NE (Echevarria *et al.*, 2003; Lui *et al.*, 1998). Ca^{2+} can be released from this pool in an IP_3-dependent fashion, and fluctuations in intranuclear Ca^{2+} levels trigger a variety of responses in animal nuclei, such as changes in gene expression (Hardingham *et al.*, 1997), the translocation of nuclear protein kinase C to the NE (Echevarria *et al.*, 2003), and regulation of nuclear transport via opening and closing of the nuclear basket of the NPC (Stoffler *et al.*, 1999b).

Electrophysiological studies performed on plant nuclei more than a decade ago indicated the presence of ion channel activity also in the plant NE (Matzke *et al.*, 1992). Similar experiments on isolated nuclei from red beet showed a voltage-dependent cation channel in the NE that could be activated by Ca^{2+} on the nucleoplasmic side (Grygorczyk and Grygorczyk, 1998). Downie *et al.* (1998) presented the first microscopic evidence for the NE localization of a Ca^{2+} pump in tomato. The import of Ca^{2+} across the NE into carrot nuclei occurs in an ATP-dependent fashion, and the use of an ionophore demonstrated that the nuclear Ca^{2+} is not irreversibly sequestered but accumulates in a releasable membrane-bound pool (Bunney *et al.*, 2000). In analogy to animal cells, the intramembrane space of the NE is a likely candidate to form this storage compartment. It remains to be seen whether plant nuclei contain a nucleoplasmic reticulum for Ca^{2+} storage as has been observed in animal cells. The role of intranuclear Ca^{2+} signalling in plant nuclei has not yet been established, but it is tempting to speculate that it might be similar to its function in animal cells where it is regulating processes involved in apoptosis, NE breakdown and mitosis.

4.3 *Proteasome-mediated proteolysis*

The 26S proteasome is the proteolysis machinery of the ubiquitin-mediated protein-degradation pathway. Selective degradation of regulatory proteins plays a crucial role in many signalling pathways and cell cycle control. Several studies in animal and yeast cells have shown a close relationship between the ubiquitin- and proteasome-mediated degradation pathway and the NE. Microscopic studies of GFP-labelled proteasomes showed their localization to the NE in both budding and fission yeast (Enenkel *et al.*, 1998; Wilkinson *et al.*, 1998). EM studies revealed a concentration at the nucleoplasmic side of the NE (Wilkinson *et al.*, 1998). It is still unclear how proteasomes pass the NE and how they are anchored (Enenkel *et al.*, 1999). The attachment of proteasomes at the NE could play a role in monitoring the protein traffic between the cytosol and the nucleus, in disassembly of the lamina during mitosis, and in signal transduction.

Recent evidence suggests that the NE in plant cells might also serve as a site for 26S proteasome-mediated proteolysis. Yanagawa *et al.* (2002) found the 26S proteasome associated with the NE and mitotic microtubular structures in tobacco BY-2 cell cultures. The NE localization of the protein-degradation machinery could facilitate its involvement in signal transduction pathways. The calcium-dependent protein kinase NtCDPK1 of tobacco has been found to co-localize with, and phosphorylate, NtRpn3, a regulatory subunit of the tobacco proteasome, in a Ca^{2+}-dependent manner in a variety of subcellular locations including the NE (Lee *et al.*, 2003). Both genes are preferentially expressed in proliferating cell types, and reduction of protein levels resulted in abnormal cell morphology and premature cell death. In another signal transduction cascade, COP1 has been hypothesized to suppress light-induced devel-opment by targeting transcription factors like HY5 for degradation in darkness (Osterlund *et al.*, 2000). COP1 has recently been shown to exhibit ubiquitin-ligase activity in plant nuclei, thus targeting the transcriptional activator LAF1 for degra-dation inside the nucleus (Seo *et al.*, 2003).

5. Summary

Compared to research in the animal field, the plant NE has been clearly under-investigated. The available data so far indicate similarities as well as striking differences that raise interesting questions about the function and evolution of the NE in different kingdoms. Despite a seemingly similar structure and organization of the NE, many of the proteins that are integral components of the animal NE appear to lack homologues in plant cells. The sequencing of the *Arabidopsis* genome has not led to the identifi-cation of homologues of animal NE components, but has indicated that the plant NE must have a distinct protein composition different from that found in metazoan cells.

Besides providing a selective barrier between the nucleoplasm and the cytoplasm, the plant NE functions as a scaffold for chromatin but the scaffolding components are not identical to those found in animal cells. The NE comprises an MTOC in higher plant cells, a striking difference to the organization of microtubule nucleation in other eukaryotic cells. Nuclear pores are present in the plant NE, but identifiable orthologues of most animal and yeast nucleoporins are presently lacking. The transport pathway through the nuclear pores via the action of karyopherins and the Ran cycle is conserved in plant cells. Interestingly, RanGAP is sequestered to the NE in plant cells and animal cells, yet the targeting domains and mechanisms of attachment are different between the two kingdoms.

At present, only a few proteins localized at the plant NE have been identified molecularly. Future research will have to expand the list of known protein components involved in building a functional plant NE.

Acknowledgements

We would like to thank Diane Furtney for expert manuscript editing. Financial support by the National Science Foundation (MCB-0079577 and MCB-0209399) and the U.S. Department of Agriculture (Plant Growth and Development no. 2001-01901) to I.M. is greatly acknowledged.

References

Ach, R. and Gruissem, W. (1994) A small nuclear GTP-binding protein from tobacco suppresses a *Schizosaccharomyces pombe* cell-cycle mutant. *Proc. Natl Acad. Sci. USA* 91: 5863–5867.

Allen, N.P., Huang, L., Burlingame, A. and Rexach, M. (2001) Proteomic analysis of nucleoporin interacting proteins. *J. Mol. Biol.* 276: 29268–29274.

Armstrong, S.J., Franklin, C.H. and Jones, G.H. (2001) Nucleolus-associated telomere clustering and pairing precede meiotic chromosome synapsis in *Arabidopsis thaliana*. *J. Cell Sci.* 114: 4207–4217.

Assmann, S.M. (2002) Heterotrimeric and unconventional GTP binding proteins in plant cell signaling. *Plant Cell Suppl.* 2002: S355–373.

Bass, H.W., Bordoli, S.J. and Foss, E.M. (2003) The desynaptic (dy) and desynaptic 1 (dsy1) mutations in maize (*Zea mays* L.) cause distinct telomere-misplacement phenotypes during meiotic prophase. *J. Exp. Bot.* 54: 39–46.

Bass, H.W., Marshall, W.F., Sedat, J.W., Agard, D.A. and Cande, W.Z. (1997) Telomeres cluster *de novo* before the initiation of synapsis: a three-dimensional spatial analysis of telomere positions before and during meiotic prophase. *J. Cell Biol.* 137: 5–18.

Bayliss, R., Littlewood, T., Strawn, L.A., Wente, S.R. and Stewart, M. (2002) GLFL and FxFG nucleoporins bind to overlapping sites on importin β. *J. Biol. Chem.* 52: 50597–50606.

Berrios, M. and Fisher, P.A. (1986) A myosin heavy-chain-like polypeptide is associated with the nuclear envelope in higher eukaryotic cells. *J. Cell Biol.* 103: 711–724.

Beven, A., Guan, Y., Peart, J., Cooper, C. and Shaw, P. (1991) Monoclonal antibodies to plant nuclear matrix reveal intermediate filament-related components within the nucleus. *J. Cell Sci.* 98: 293–302.

Bollman, K.M., Aukerman, M.J., Park, M.Y., Hunter, C., Berardini, T.Z. and Poethig, R.S. (2003) HASTY, the *Arabidopsis* ortholog of exportin 5/MSN5, regulates phase change and morphogenesis. *Development* 130: 1493–1504.

Bunney, T.D., Shaw, P.J., Watkins, P.A.C., Taylor, J.P., Beven, A.F., Wells, B., Calder, G.M. and Drøbak, B.K. (2000) ATP-dependent regulation of nuclear Ca^{2+} levels in plant cells. *FEBS Lett.* 476: 145–149.

Canaday, J., Stoppin-Mellet, V., Mutterer, J., Lambert, A.-M. and Schmit, A.-C. (2000) Higher plant cells: γ-tubulin and microtubule nucleation in the absence of centrosomes. *Microsc. Res. Tech.* 49: 487–495.

Chevrier, V., Komesli, S., Schmitt, A.C., Vantard, M., Lambert, A.M. and Job, D. (1992) A monoclonal antibody, raised against mammalian centrosomes and screened by recognition of plant microtubule organizing centers, identifies a pericentriolar component in different plant cells types. *J. Cell Sci.* **101**: 823–835.

Clark, G.B., Memon, A.R., Tong, C.-G., Thompson Jr., G.A. and Roux, S.J. (1993) Phytochrome regulates GTP-binding protein activity in the envelope of pea nuclei. *Plant J.* **4**: 399–402.

Cohen, M., Wilson, K.L. and Gruenbaum, Y. (2001) Membrane proteins of the nuclear pore complex: Gp210 is conserved in *Drosophila*, *C. elegans* and *A. thaliana*. *Gene Ther. Mol. Biol.* **6**: 47–55.

Cohen-Armon, M., Hammel, I., Anis, Y., Homburg, S. and Dekel, N. (1996) Evidence for endogenous ADP-ribosylation of GTP-binding proteins in neuronal cell nucleus. *J. Biol. Chem.* **271**: 26200–26208.

Cooper, J.P., Watanabe, Y. and Nurse, P. (1998) Fission yeast Taz1 protein is required for meiotic telomere clustering and recombination. *Nature* **392**: 828–831.

Cowan, C.R., Carlton, P.M. and Cande, W.Z. (2001) The polar arrangement of telomeres in interphase and meiosis. Rabl organization and the bouquet. *Plant Phys.* **125**: 532–538.

Cowling, R., Delichere, C., Morot-Sir, M., Wicker-Planquart, C., Gaillard, J., Chevrier, V., Job, D. and Vantard, M. (2003) Molecular and functional characterization of plant proteins involved in microtubule dynamic assembly. *Cell Biol. Int.* **27**: 185–186.

Cronshaw, J.M., Krutchinsky, A.N., Zhang, W., Chait, B.T. and Matunis, M.J. (2002) Proteomic analysis of the mammalian nuclear pore complex. *J. Cell Biol.* **158**: 915–927.

de Ruijter, N.C.A., Ketelaar, T., Blumenthal, S.S.D., Emons, A.M. and Schel, J.H.N. (2000) Spectrin-like proteins in plant nuclei. *Cell Biol. Int.* **24**: 427–438.

Djinovic-Carugo, K., Gautel, M., Ylänne, J. and Young, P. (2002) The spectrin repeat: a structural platform for cytoskeletal protein assembly. *FEBS Lett.* **513**: 119–123.

Downie, L., Priddle, J., Hawes, C. and Evans, D.E. (1998) A calcium pump at the higher plant nuclear envelope? *FEBS Lett.* **429**: 44–48.

Drøbak, B.K., Watkins, P.A.C., Bunney, T.D., Dove, S.K., Shaw, P.J., White, I.R. and Millner, P.A. (1995) Association of multiple GTP-binding proteins with the plant cytoskeleton and nuclear matrix. *Biochem. Biophys. Res. Comm.* **210**: 7–13.

Echevarria, W., Leite, M.F., Mateus, T.G., Zipfel, W.R. and Nathanson, M.H. (2003) Regulation of calcium signals in the nucleus by a nucleoplasmic reticulum. *Nature Cell Biol.* **5**: 440–446.

Ellenberg, J., Siggia, E.D., Moreira, J.E., Smith, C.L., Presley, J.F., Worman, H.J. and Lippincott-Schwartz, J. (1997) Nuclear membrane dynamics and reassembly in living cells: targeting of an inner nuclear membrane protein in interphase and mitosis. *J. Cell Biol.* **138**: 1193–1206.

Enenkel, C., Lehmann, A. and Kloetzel, P.-M. (1998) Subcellular distribution of proteasomes implicated a major location of protein degradation in the nuclear envelope-ER network in yeast. *EMBO J.* **17**: 6144–6154.

Enenkel, C., Lehmann, A. and Kloetzel, P.-M. (1999) GFP-labelling of 26S proteasomes in living yeast: insight into proteasomal functions at the nuclear envelope/rough ER. *Mol. Biol. Rep.* **26**: 131–135.

Erhardt, M., Stoppin-Mellet, V., Campagne, S., *et al.* (2002) The plant Spc98p homologue colocalizes with γ-tubulin at microtubule nucleation sites and is required for microtubule nucleation. *J. Cell Sci.* **115**: 2423–2431.

Fava, F., Raynaud-Messina, B., Leung-Tack, J., Mazzolini, L., Li, M., Guillemot, J.C., Cachot, D., Tollon, Y., Ferrara, P. and Wright, M. (1999) Human 76p: a new member of the γ-tubulin-associated protein family. *J. Cell Biol.* **147**: 857–868.

Franco, S., Alsheimer, M., Herrera, E., Benavente, R. and Blasco, M.A. (2002) Mammalian meiotic telomeres: composition and ultrastructure in telomerase-deficient mice. *Eur. J. Cell Biol.* **81**: 335–340.

Galcheva-Gargova, Z.L., Marinova, E.L. and Koleva, S.T. (1988) Isolation of nuclear shells from plant cells. *Plant Cell Environ.* 11: 819–825.

Gasiorowski, J.Z. and Dean, D.A. (2003) Mechanisms of nuclear transport and interventions. *Adv. Drug Del. Rev.* 55: 703–716.

Gindullis, F. and Meier, I. (1999) Matrix attachment region binding protein MFP1 is localized in discrete domains at the nuclear envelope. *Plant Cell* 11: 1117–1128.

Gindullis, F., Peffer, N.J. and Meier, I. (1999) MAF1, a novel plant protein interacting with matrix attachment region binding protein MFP1, is located at the nuclear envelope. *Plant Cell* 11: 1755–1767.

Gineitis, A.A., Zalenskaya, I.A., Yau, P.M. and Bradbury, E.M. (2000) Human sperm telomere-binding complex involves histone H2B and secures telomere membrane attachment. *J. Cell Biol.* 151: 1591–1597.

Goldberg, M.W., Rutherford, S.A., Hughes, M., Cotter, L.A., Bagley, S., Kiseleva, E., Allen, T.D. and Clarke, P.R. (2000) Ran alters nuclear pore complex conformation. *J. Mol. Biol.* 300: 519–529.

Goldman, R.D., Gruenbaum, Y., Moir, R.D., Shumaker, D.K. and Spann, T.P. (2002) Nuclear lamins: building blocks of nuclear architecture. *Genes Dev.* 16: 533–547.

Golubovskaya, I.N., Harper, L.C., Pawlowski, W.P., Schichnes, D. and Cande, W.Z. (2002) The pam1 gene is required for meiotic bouquet formation and efficient homologous synapsis in maize (*Zea mays* L.). *Genetics* 162: 1979–1993.

Grygorczyk, C. and Grygorczyk, R. (1998) A Ca^{2+}- and voltage-dependent cation channel in the nuclear envelope of red beet. *Biochim. Biophys. Acta* 1375: 117–130.

Haasen, D., Köhler, C., Neuhaus, G. and Merkle, T. (1999) Nuclear export of proteins in plants: AtXPO1 is the export receptor for leucine-rich nuclear export signals in *Arabidopsis thaliana*. *Plant J.* 20: 695–705.

Haizel, T., Merkle, T., Pay, A., Fejes, E. and Nagy, F. (1997) Characterization of proteins that interact with the GTP-bound form of the regulatory GTPase Ran in *Arabidopsis*. *Plant J.* 11: 93–103.

Hardingham, G.E., Chawla, S., Johnson, C.M. and Bading, H. (1997) Distinct functions of nuclear and cytoplasmic calcium in the control of gene expression. *Nature* 385: 260–265.

Heese-Peck, A. and Raikhel, N. (1998a) The nuclear pore complex. *Plant Mol. Biol.* 38: 145–162.

Heese-Peck, A. and Raikhel, N. (1998b) A glycoprotein modified with terminal N-acetylglu-cosamine and localized at the nuclear rim shows sequence similarity to aldose-1-epimerases. *Plant Cell* 10: 599–612.

Hicks, G.R. and Raikhel, N.V. (1995) Nuclear localization signal binding proteins in higher plant nuclei. *Proc. Natl Acad. Sci. USA* 92: 734–738.

Hicks, G.R., Smith, H.M., Lobreaux, S. and Raikhel, N.V. (1996) Nuclear import in permeabi-lized protoplasts from higher plants has unique features. *Plant Cell* 8: 1337–1352.

Hisada, A., Hanzawa, H., Weller, J.L., Nagatani, A., Reid, J.B. and Furuya, M. (2000) Light-induced nuclear translocation of endogenous pea phytochrome A visualized by immunocytochemcial procedures. *Plant Cell* 12: 1063–1078.

Holaska, J.M., Wilson, K.L. and Mansharamani, M. (2002) The nuclear envelope, lamins and nuclear assembly. *Curr. Op. Cell Biol.* 14: 357–364.

Hopper, A.K., Traglia, H.M. and Dunst, R.W. (1990) The yeast *RNA1* gene product necessary for RNA processing is located in the cytosol and apparently excluded from the nucleus. *J. Cell Biol.* 111: 309–321.

Hübner, S., Smith, H.M.S., Hu, W., Chan, C.K., Rihs, H.-P., Paschal, B.M., Raikhel, N.V. and Jans, D.A. (1999) Plant importin α binds nuclear localization sequences with high affinity and can mediate nuclear import independent of importin β. *J. Biol. Chem.* 274: 22610–22617.

Hunter, C.A., Aukerman, M.J., Sun, H., Fokina, M. and Poethig, R.S. (2003) PAUSED encodes the *Arabidopsis* exportin-t ortholog. *Plant Phys.* 132: 2135–2143.

Irons, S.L., Evans, D.E. and Brandizzi, F. (2003) The first 238 amino acids of the human lamin B receptor are targeted to the nuclear envelope in plants. *J. Exp. Bot.* **54**: 943–950.

Jiang, C.J., Imamoto, N., Matsuki, R., Yoneda, Y. and Yamamoto, N. (1998a) Functional characterization of a plant importin α homologue. Nuclear localization signal (NLS)-selective binding and mediation of nuclear import of NLS proteins *in vitro*. *J. Biol. Chem.* **273**: 24083–24087.

Jiang, C.J., Imamoto, N., Matsuki, R., Yoneda, Y. and Yamamoto, N. (1998b) *In vitro* characterization of rice importin β1: molecular interaction with nuclear transport factors and mediation of nuclear protein import. *FEBS Lett.* **437**: 127–130.

Jiang, C.J., Shoji, K., Matsuki, R., *et al.* (2001) Molecular cloning of a novel importin α homologue from rice, by which constitutive photomorphogenic 1 (COP1) nuclear localization signal (NLS)-protein is preferentially nuclear imported. *J. Biol. Chem.* **276**: 9322–9329.

Job, D., Valiron, O. and Oakley, B. (2003) Microtubule nucleation. *Curr. Op. Cell Biol.* **15**: 111–117.

Kim, S.H. and Roux, S.J. (2003) An *Arabidopsis* Ran-binding protein, AtRanBP1c, is a co-activator of Ran GTPase-activating protein and requires the C-terminus for its cytoplasmic localization. *Planta* **216**: 1047–1052.

Knop, M. and Schiebel, E. (1997) Spc98p and Spc97c of the yeast γ-tubulin complex mediate binding to the spindle pole body via their interaction with Spc110p. *EMBO J.* **23**: 6985–6995.

Knop, M. and Schiebel, E. (1998) Receptors determine the cellular localization of a γ-tubulin complex and thereby the site of microtubule formation. *EMBO J.* **17**: 3952–3997.

Lee, S.S., Cho, H.S., Yoon, G.M., Ahn, J.-W., Kim, H.-H. and Pai, H.-S. (2003) Interaction of NtCDPK1 calcium-dependent protein kinase with NtRpn3 regulatory subunit of the 26S proteasome in *Nicotiana tabacum*. *Plant J.* **33**: 825–840.

Li, J. and Chen, X. (2003) PAUSED, a putative exportin-t, acts pleiotropically in *Arabidopsis* development but is dispensable for viability. *Plant Phys.* **132**: 1913–1924.

Li, H. and Roux, S.J. (1992) Casein kinase II protein kinase is bound to lamina-matrix and phosphorylates lamin-like protein in isolated pea nuclei. *Proc. Natl Acad. Sci. USA* **89**: 8434–8438.

Lui, P.P.Y., Kong, S.K., Kwok, T.T. and Lee, C.Y. (1998) The nucleus of HeLa cell contains tubular structures for Ca^{2+} signaling. *Biochem. Biophys. Res. Comm.* **247**: 88–93.

Martínez-Pérez, E., Shaw, P., Reader, S., Aragón-Alcaide, L., Miller, T. and Moore, G. (1999) Homologous chromosome pairing in wheat. *J. Cell Sci.* **112**: 1761–1769.

Masuda, K., Haruyama, S. and Fujino, K. (1999) Assembly and disassembly of the peripheral architecture of the plant cell nucleus during mitosis. *Planta* **210**: 165–167.

Masuda, K., Xu, Z.-J., Takahashi, S., Ito, A., Ono, M., Nomura, K. and Inoue, M. (1997) Peripheral framework of carrot cell nucleus contains a novel protein predicted to exhibit a long α-helical domain. *Exp. Cell Res.* **232**: 173–181.

Mattaj, I.W. and Englmeier, L. (1998) Nucleocytoplasmic transport: the soluble phase. *Annu. Rev. Biochem.* **67**: 256–306.

Matunis, M.J., Wu, J. and Blobel, G. (1998) SUMO-1 modification and its role in targeting the Ran GTPase-activating protein, RanGAP1, to the nuclear pore complex. *J. Cell Biol.* **140**: 499–509.

Matynia, A., Dimitrov, K., Mueller, U., He, X. and Sazer, S. (1996) Perturbations in the spi1p GTPase cycle of *Schizosaccharomyces pombe* through its GTPase-activating protein and guanine nucleotide exchange factor components result in similar phenotypic consequences. *Mol. Cell. Biol.* **16**: 6352–6362.

Matzke, A.J.M., Behensky, C., Weiger, T. and Matzke, M.A. (1992) A large conductance ion channel in the nuclear envelope of a higher plant cell. *FEBS Lett.* **302**: 81–85.

McNulty, A.K. and Saunders, M.J. (1992) Purification and immunological detection of pea nuclear intermediate filaments: evidence for plant nuclear lamins. *J. Cell Sci.* **103**: 407–414.

Meier, I. (2000) A novel link between Ran signal transduction and nuclear envelope proteins in plants. *Plant Phys.* **124**: 1507–1510.

Merkle, T. (2001) Nuclear import and export of proteins in plants: a tool for the regulation of signaling. *Planta* **213**: 499–517.

Merkle, T., Haizel, T., Matsumoto, T., Harter, K., Dallmann, G. and Nagy, F. (1994) Phenotype of the fission yeast cell cycle regulatory mutant *pim1-46* is suppressed by a tobacco cDNA encoding a small, Ran-like GTP-binding protein. *Plant J.* **6**: 555–565.

Mikhailova, E.I., Sosnikhina, S.P., Kirillova, G.A., Tikholiz, O.A., Smirnov, V.G., Jones, R.N. and Jenkins, G. (2001) Nuclear dispositions of subtelomeric and pericentromeric chromosomal domains during meiosis in asynaptic mutants of rye (*Secale cereale* L.). *J. Cell Sci.* **114**: 1875–1882.

Mínguez, A. and Moreno Díaz de la Espina, S. (1993) Immunological characterization of lamins in the nuclear matrix of onion cells. *J. Cell Sci.* **106**: 431–439.

Mislow, J.M., Holaska, J.M., Kim, M.S., Lee, K.K., Segura-Totten, M., Wilson, K.L. and McNally, E.M. (2002) Nesprin-1α self-associates and binds directly to emerin and lamin A *in vitro*. *FEBS Lett.* **525**: 135–140.

Moreno Díaz de la Espina, S., Barthelemy, L. and Cerezuela, M.A. (1991) Isolation and ultra-structural characterization of the residual nuclear matrix in a plant cell system. *Chromosoma* **100**: 110–117.

Moritz, M. and Agard, D.A. (2001) γ-Tubulin complexes and microtubule nucleation. *Curr. Op. Struct. Biol.* **11**: 174–181.

Mounkes, L., Kozlov, S., Burke, B. and Steward, C.L. (2003) The laminopathies: nuclear structure meets disease. *Curr. Op. Genet. Dev.* **13**: 223–230.

Murphy, S.M., Preble, A.M., Patel, U.K., O'Connell, K.L., Dias, D.P., Moritz, M., Agard, D., Stults, J.T. and Stearns, T. (2001) GCP5 and GCP6: Two new members of the human γ-tubulin complex. *Mol. Biol. Cell* **12**: 3340–3352.

Murphy, S.M., Urbani, L. and Stearns, T. (1998) The mammalian γ-tubulin complex contains homologues of the yeast spindle pole body components Spc97p and Spc98p. *J. Cell Biol.* **141**: 663–674.

Oegema, K., Wiese, C., Martin, O.C., Milligan, R.A., Iwamatsu, A., Mitchison, T.J. and Zheng, Y. (1999) Characterization of two related *Drosophila* γ-tubulin complexes that differ in their ability to nucleate microtubules. *J. Cell Biol.* **144**: 721–733.

Osterlund, M.T., Hardtke, C.S., Wei, N. and Deng, X.W. (2000) Targeted destabilization of HY5 during light-regulated development of *Arabidopsis*. *Nature* **405**: 462–466.

Pay, A., Resch, K., Frohnmeyer, H., Fejes, E., Nagy, F. and Nick, P. (2002) Plant RanGAPs are localized at the nuclear envelope in interphase and associated with microtubules in mitotic cells. *Plant J.* **30**: 699–709.

Peskan, T. and Oelmüller, R. (2000) Heterotrimeric G-protein β-subunit is localized in the plasma membrane and nuclei of tobacco leaves. *Plant Mol. Biol.* **42**: 915–922.

Podgornaya, O.I., Bugaeva, E.A., Voronin, A.P., Gilson, E. and Mitchell, A.R. (2000) Nuclear envelope associated protein that binds telomeric DNAs. *Mol. Reprod. Dev.* **57**: 16–25.

Quimby, B.B. and Dasso, M. (2003) The small GTPase Ran: interpreting the signs. *Curr. Op. Cell Biol.* **15**: 338–344.

Rose, A. and Meier, I. (2001) A domain unique to plant RanGAP is responsible for its targeting to the plant nuclear rim. *Proc. Natl Acad. Sci. USA* **98**: 15377–15382.

Rubins, J.B., Benditt, J.O., Dickey, B.F. and Riedel, N. (1990) GTP-binding proteins in rat liver nuclear envelopes. *Proc. Natl Acad. Sci. USA* **87**: 7080–7084.

Saalbach, G. and Christov, V. (1994) Sequence of a plant cDNA from *Vicia faba* encoding a novel Ran-related GTP-binding protein. *Plant Mol. Biol.* **24**: 969–972.

Saffitz, J.E., Nash, J.A., Green, K.G., Luke, R.A., Ransnas, L.A. and Insel, P.A. (1994) Immunoelectron microscopic identification of cytoplasmic and nuclear Gsα in S49 lymphoma cells. *FASEB J.* **8**: 252–258.

Scherthan, H. (2001) A bouquet makes ends meet. *Nature Rev.* **2**: 621–627.

Scherthan, H., Jerratsch, M., Li, B., Smith, S., Hultén, M., Lock, T. and de Lange, T. (2000) Mammalian meiotic telomeres: protein composition and redistribution in relation to nuclear pores. *Mol. Biol. Cell* **11**: 4189–4208.

Schiebel, E. (2000) γ-tubulin complexes: binding to the centrosome, regulation and micro-tubule nucleation. *Curr. Op. Cell Biol.* **12**: 113–118.

Schmit, A.C., Endle, M.C. and Lambert, A.M. (1996) The perinuclear microtubule-organizing center and the synaptonemal complex of higher plants share a common antigen: its putative transfer and role in meiotic chromosomal ordering. *Chromosoma* **104**: 405–413.

Schmit, A.C., Stoppin, V., Chevrier, V., Job, D. and Lambert, A.M. (1994) Cell cycle dependent distribution of a centrosomal antigen at the perinuclear MTOC or at the kineto-chores of higher plant cells. *Chromosoma* **103**: 343–351.

Scofield, G.N., Beven, A.F., Shaw, P.J. and Doonan, J.H. (1992) Identification and localization of a nucleoporin-like protein component of the plant nuclear matrix. *Planta* **187**: 414–420.

Seltzer, V., Pawlowski, T., Campagne, S., Canaday, J., Erhardt, M., Evrard, J.-L., Herzog, E. and Schmit, A.-C. (2003) Multiple microtubule nucleation sites in higher plants. *Cell Biol. Int.* **27**: 267–269.

Seo, H.S., Yang, J.-Y., Ishikawa, M., Bolle, C., Ballesteros, M.L. and Chua, N.-H. (2003) LAF1 ubiquitination by COP1 controls photomorphogenesis and is stimulated by SPA1. *Nature* **423**: 995–999.

Shoji, K., Iwasaki, T., Matsuki, R., Miyao, M. and Yamamoto, N. (1998) Cloning of a cDNA encoding an importin-α and down-regulation of the gene by light in rice leaves. *Gene* **212**: 279–286.

Smith, H.M., Hicks, G.R. and Raikhel, N.V. (1997) Importin α from *Arabidopsis thaliana* is a nuclear import receptor that recognizes three classes of import signals. *Plant Physiol.* **114(2)**: 411–417.

Smith, S. and Blobel, G. (1993) The first membrane spanning region of the lamin B receptor is sufficient for sorting to the inner nuclear membrane. *J. Cell Biol.* **120**: 631–637.

Smith, S. and Blobel, G. (1994) Colocalization of vertebrate lamin B and lamin B receptor (LBR) in nuclear envelopes and in LRB-induced membrane stacks of the yeast *Saccharomyces cerevisiae*. *Proc. Natl Acad. Sci. USA* **91**: 10124–10128.

Soullam, B. and Worman, H.J. (1993) The amino-terminal domain of the lamin B receptor is a nuclear envelope targeting signal. *J. Cell Biol.* **120**: 1093–1100.

Soullam, B. and Worman, H.J. (1995) Signals and structural features involved in integral membrane protein targeting to the inner nuclear membrane. *J. Cell Biol.* **130**: 15–27.

Stoffler, D., Fahrenkrog, B. and Aebi, U. (1999a) The nuclear pore complex: from molecular architecture to functional dynamics. *Curr. Op. Cell Biol.* **11**: 391–401.

Stoffler, D., Goldie, K.N., Feja, B. and Aebi, U. (1999b) Calcium-mediated structural changes of native nuclear pore complexes monitored by time-lapse atomic force microscopy. *J. Mol. Biol.* **287**: 741–752.

Stoppin, V., Lambert, A.-M. and Vantard, M. (1996) Plant microtubule-associated proteins (MAPs) affect microtubule nucleation and growth at plant nuclei and mammalian centro-somes. *Eur. J. Cell Biol.* **69**: 11–23.

Stoppin, V., Vantard, M., Schmit, A.-C. and Lambert, A.-M. (1994) Isolated plant nuclei nucleate microtubule assembly: the nuclear surface in higher plants has centrosome-like activity. *Plant Cell* **6**: 1099–1106.

Takahashi, M., Yamagiwa, A., Tamako, N., Mukai, H. and Ono, Y. (2002) Centrosomal proteins CG-NAP and Kendrin provide microtubule nucleation sites by anchoring γ-tubulin ring complex. *Mol. Biol. Cell* **13**: 3235–3245.

Tang, X.J., Hepler, P.K. and Scordilis, S.P. (1989) Immunochemical and immunocytochemical identification of a myosin heavy chain polypeptide in *Nicotiana* pollen tubes. *J. Cell Sci.* **92**: 569–574.

Tham, W.-H. and Zakian, V.A. (2000) Telomeric tethers. *Nature* **403**: 34–35.

Trelles-Sticken, E., Dresser, M.E. and Scherthan, H. (2000) Meiotic telomere protein Ndj1p is required for meiosis-specific telomere distribution, bouquet formation and efficient homo-logue pairing. *J. Cell Biol.* **151**: 95–106.

Walther, T.C., Askjaer, P., Gentzel, M., Habermann, A., Griffiths, G., Wilm, M., Mattaj, I.W. and Hetzer, M. (2003) RanGTP mediates nuclear pore complex assembly. *Nature* **424:** 689–694.

Ward, B.M. and Lazarowitz, S.G. (1999) Nuclear export in plants: use of geminivirus movement proteins for a cell-based export assay. *Plant Cell* **11:** 1267–1276.

Wilkinson, C.R.M., Wallace, M., Morphew, M., Perry, P., Allshire, R., Javerzat, J.-P., McIntosh, J.R. and Gordon, C. (1998) Localization of the 26S proteasome during mitosis and meiosis in fission yeast. *EMBO J.* **17:** 6465–6476.

Willard, F.S. and Crouch, M.F. (2000) Nuclear and cytoskeletal translocation and localization of heterotrimeric G-proteins. *Immun. Cell Biol.* **78:** 387–394.

Yamamoto, N. and Deng, X.-W. (1999) Protein nucleocytoplasmic transport and its light regulation in plants. *Genes Cells* **4:** 489–500.

Yanagawa, Y., Hasezawa, S., Kumagai, F., *et al.* (2002) Cell-cycle dependent dynamic changes of 26S proteasome distribution in tobacco BY-2 cells. *Plant Cell Physiol.* **43:** 604–613.

Yang, Z. (2002) Small GTPases: versatile signaling switches in plants. *Plant Cell Suppl.* **2002:** S375–S388.

Zhang, Q., Skepper, J.N., Yang, F., Davies, J.D., Hegyi, L., Roberts, R.G., Weissberg, P.L., Ellis, J.A. and Shanahan, C.M. (2001) Nesprins: a novel family of spectrin-repeat-containing proteins that localize to the nuclear membrane in multiple tissues. *J. Cell Sci.* **114:** 4485–4498.

Zhen, Y.-Y., Libotte, T., Munck, M., Noegel, A.A. and Korenbaum, E. (2002) NUANCE, a giant protein connecting the nucleus and actin cytoskeleton. *J. Cell Sci.* **115:** 3207–3222.

6

Structure, function and assembly of the nuclear pore complex

Sheona Drummond and Terry Allen

1. Introduction

The compartmentalization of cellular functions presents a physical barrier between the cytoplasm and the contents of the membrane-enveloped organelles. The enclosure of the genome within a nuclear envelope has necessitated the evolution of nuclear membrane-associated structures termed nuclear pore complexes (NPCs) through which molecules can be transported. Communication between the cytoplasm and nucleoplasm also requires biochemical processes that facilitate and regulate such exchange. NPCs are large proteinaceous assemblies present within pores formed by fusion of the inner and outer nuclear membrane bilayers (INM and ONM, respectively). NPCs are the sites of both active transport (energy and 'transporter-molecule' dependent) and passive diffusion of molecules between the cytoplasm and nucleoplasm. Several NPC-specific proteins (nucleoporins) and other NPC-associated proteins also function in a diverse range of cellular processes in addition to their previously characterized roles in the nucleocytoplasmic trafficking. For example, in yeast cells, nucleoplasmic filamentous nucleoporins play a role in the structural and functional organization of chromatin (Galy *et al.*, 2000). Yeast nucleoporins have also been found to be components of the spindle pole body (Chial *et al.*, 1998), while others play a role in cell cycle regulation (Iouk *et al.*, 2002).

Many techniques and model systems have been employed to investigate the composition, dynamics and the protein–protein interactions involved in establishing and maintaining NPC structure and function. Here we discuss the structural components of the NPC and the protein–protein interactions that are required for the assembly and structural maintenance of the NPC. In addition, theories concerning the process of translocation through the pore shall be discussed.

The Nuclear Envelope, edited by D.E. Evans, C. Hutchison & J.A. Bryant.
© 2004 Garland Science/BIOS Scientific Publishers

2. NPC structure

NPCs are highly ordered modular structures. They have eight-fold rotational symmetry, although they are asymmetrical about the plane of the NE (see Chapter 7). Mature NPCs comprise filaments that radiate into the cytoplasm from outer nuclear membrane-associated annular structures, a central compartment through which molecules are translocated, and a nucleoplasmic compartment composed of further annular components that have associated filaments organized into a basket structure (see colour *Plate 1*).

The structure of the NPC has been studied extensively in yeasts, primarily the budding yeast *Saccharomyces cerevisiae* and the fission yeast *Schizosaccharomyces pombe* and in vertebrates, including the clawed toad *Xenopus laevis*, rat and human. Some of the most detailed description of the global structure of the NPC, and its component sub-structures, has been possible through analyses of *Xenopus* oocyte nuclei (Akey, 1989; Akey and Radermacher, 1993; Goldberg and Allen, 1993; Goldberg *et al.*, 1997; Hinshaw and Milligan, 2003; Jarnik and Aebi, 1991; Stoffler *et al.*, 2003; Unwin and Milligan, 1982). These nuclei are easily isolated and visualized as they are large (approximately 100 μm in diameter) and contain approximately 50 NPCs per μm² of nuclear envelope.

2.1 *Cytoplasmic filaments and cytoplasmic annular components*

Upon the cytoplasmic face of the nuclear envelope, at the borders of each NPC fusion pore, there is a 'thin ring' (colour *Plate 1*). This 'thin ring' forms the structural base for eight equidistantly positioned cytoplasmic particles each 'anchoring' a single filament that projects in to the cytoplasm (colour *Plate 1*). The cytoplasmic particles appear to have the ability to pivot inwards towards the centre of the NPC (Akey and Radermacher, 1993; Goldberg and Allen, 1996; Ris, 1997), thereby changing the orientation of the associated cytoplasmic filaments. It has been demonstrated that the cytoplasmic filaments have the ability to bend away from the central channel of the pore possibly to avoid physically impeding the passage of large transport cargoes (Goldberg *et al.*, 2000). Below the thin ring is a 'star ring' composed of eight triangular sub-units that are juxtaposed with the outer nuclear membrane. The cytoplasmic filaments of the NPC have been proposed to be the initial docking sites for specific transport molecules and their cargo prior to their import in to the nucleus (reviewed in Chook and Blobel, 2001; Steggerda and Paschal, 2002; Strom and Weis, 2001). However, it has been shown that nuclei devoid of cytoplasmic filaments, or where binding to these filaments has been blocked through incubation with gold-conjugated antibodies, remain competent for importin α/β and transportin-dependent nuclear import (Walther *et al.*, 2002).

2.2 *Nuclear pore membrane proteins*

The nuclear envelope can be divided into three functionally and biochemically distinct domains: (i) the outer nuclear envelope (ONM) which is continuous with the endoplasmic reticulum; (ii) the inner nuclear membrane (INM) which binds the nuclear lamina and both directly and indirectly associates with chromatin; and (iii) the nuclear pore membrane domain. The pore membrane can be defined as the regions of continuity between the INM and ONM produced by fusion of these membrane bilayers. Subsequent dilation of the fusion pore coupled with continued sequential recruitment of nucleoporins leads to the assembly of a mature NPC bordered by the pore membrane. Integral membrane proteins unique to this region of the nuclear envelope are thought to be necessary for the assembly and anchoring of the NPC within the nuclear envelope. Additional proteins localized to this region of the NPC assemble to form a lumenally disposed spoke ring which is contiguous with radial arms of the pore complex that traverse the membrane and most likely serve to maintain the structural integrity of the NPC.

At present there are only two nuclear pore domain-specific integral membrane proteins identified in vertebrates; gp210 (Greber *et al.*, 1990; Wozniak and Blobel, 1989) and POM121 (Hallberg *et al.*, 1993) and one in yeast; POM152 (Wozniak *et al.*, 1994).

Topological studies have shown that gp210 has a large lumenal domain (approximately 95% of its mass), a single transmembrane domain and a short cytoplasmic 'tail' that faces the NPC (Greber *et al.*, 1990). Gp210 is a glycoprotein containing N-linked oligosaccharide moieties (Greber *et al.*, 1990; Wozniak *et al.*, 1992). A recent study has implicated this protein in the earliest events in NPC assembly (Drummond and Wilson, 2002).

POM121 has a short domain within the nuclear envelope lumen and a single transmembrane domain with the majority of the protein abutting the NPC (Soderqvist and Hallberg, 1994). It has been shown that a short sequence, amino acids 129–618, within the carboxy-terminus of POM121 is sufficient for targeting to NPCs suggesting that this part of the protein interacts with other proteins of the pore complex.

The sole integral membrane protein unique to the pore membrane domain in yeast is POM152. Interestingly, deletion mutants of this gene were still viable and their growth was indistinguishable from wild-type yeast cells (Wozniak *et al.*, 1994). This demonstrated that POM152, albeit an abundant nucleoporin, is non-essential for life and suggests that there is a degree of redundancy of protein function within the yeast NPC.

Two studies in yeast have indicated that other membrane-associated proteins are associated with the nuclear transport function of the NPC. In a genetic screen for suppressors of the lethal nup116 deletion, Snl1p (suppressor of nup116 lethality 1) was found to be necessary and sufficient to rescue the lethal phenotype. Snl1p is an ~18-kDa protein with limited homology to regions of POM153. Snl1p has a single transmembrane domain and is found at both the NE and the ER (Ho *et al.*, 1998). Snl1p has been shown to interact with Heat shock protein 70 (Hsp70) (Sondermann *et al.*, 2002) that shuttles bi-directionally through the NPC and is required for nuclear transport activity (Shi and Thomas, 1992; Shulga *et al.*, 1996).

Although not a *bone fide* nucleoporin, Brr6p was identified in a genetic screen for cs mRNA transport mutants in *S. cerevisiae*. This protein is ~23 kDa and has been localized to the nuclear envelope of yeast cells at, or near, NPCs. Deletion of the Brr6 gene results in altered nuclear morphology and NPC distribution coupled with mRNA export defects (de Bruyn Kops and Guthrie, 2001). This suggests the Brr6 protein is required for spatial organization and transport function(s) of NPC.

2.3 *Central compartment of the NPC*

The central compartment of the NPC comprises a central 'transporter' channel, through which import and export complexes are actively translocated, and proteinaceous structures that extend into the bordering lumenal domain of the nuclear envelope. In addition, there are aqueous channels between the pore complex and the adjoining membrane that allow the passive diffusion of small molecules of diameter of 9 nm or less (Feldherr and Akin, 1990; Hinshaw *et al.*, 1992). It is generally believed that this central channel acts as a molecular sieve that, due to its composition of natively unfolded, hydrophobic, phenylalanine-rich nucleoporins, allows only receptor-mediated active transport and limited passive diffusion. The method(s) by which nucleocytoplasmic trafficking may be regulated and mediated through this channel are discussed below.

2.4 *Nucleoplasmic annular components*

At, and below, the level of the inner nuclear membrane the NPC is composed of further annular structures. The 'nucleoplasmic ring' is comprised of eight triangular sub-units. In addition SEM studies have identified a peripheral structure, or 'globule' which appears to close the nucleoplasmic entrance to the central transporter channel (Goldberg and Allen, 1993). However it should be noted that similar globular structures have been observed at the cytoplasmic face of the NPC and it is unclear whether this constitutes a *bona fide* NPC structure or transport cargo in transit through the pore.

2.5 *Nuclear pore basket and associated nucleoplasmic filaments*

Eight filaments, with diameters of approximately 7–10 nm, extend from between each of the sub-units of the nucleoplasmic ring into the nucleoplasm. These filaments come together at a putative distal ring ~50 nm from the inner nuclear membrane to form a basket structure (reviewed in Goldberg *et al.*, 1999). It has been shown that filaments extend from these baskets into the nucleoplasm and they can also link neighbouring NPCs (Ris, 1997). These filaments are composed, at least in part, of previously described nucleoporins that are components of the nuclear basket such as Tpr (Cordes *et al.*, 1993, 1997; Paddy, 1998; Zimowska *et al.*, 1997), nup153 (Pant, *et al.*, 1994) and nup98 (Powers *et al.*, 1997). Mlp1p and Mlp2p, the conserved yeast homologues of vertebrate Tpr (Cordes *et al*, 1997), comprise filaments extending from NPCs into the nucleus and have been shown to mediate the structural and functional organization of chromatin in close proximity to the nuclear membrane (Galy *et al.*, 2000).

3. Annulate lamellae

Annulate lamellae (AL) are arrays of stacked membrane cisternae found in the cytoplasm. These membrane arrays contain a high density of both soluble and membrane-associated nucleoporins. Investigation of the morphology of these membrane stacks has shown that they contain closely packed complexes very similar to mature NPCs. However there is still considerable debate as to whether AL NPCs have biochemically or structurally similar nuclear basket structures.

AL are especially found in rapidly proliferating embryonic or tumour cells and AL production can be induced by treatment with sub-lethal doses of the anti-mitotic drugs colchicines and vinblastine sulfate (Chemnitz et al., 1977; reviewed in Kessel, 1992).

AL are thought to be storage organelles for NPC components, therefore the assembly of nucleoporins into NPC precursors may also act as a qualitative checkpoint for these proteins prior to their recruitment to nascent NPCs at the nuclear envelope. AL NPCs are significantly more mobile than NPCs present in the NE (Daigle et al., 2001) and this may be due to the absence of a stabilizing underlying lamina.

4. Identification of the NPC proteome

Proteomic analyses of the mammalian and yeast NPC (Cronshaw et al., 2002; Rout et al., 2000, respectively) have revealed that both structures contain ~30 distinct proteins. Many of these proteins display degrees of structural and functional conservation between species although there are several proteins unique to either yeast or mammals (see *Table 1*). Studies of yeast and mammalian NPCs have shown that they share a great many similar morphological features. The yeast NPC is markedly smaller with a calculated mass of 44 MDa compared to vertebrate NPC mass of 60 MDa (Cronshaw et al., 2002; Rout et al., 2000). The mass difference between the yeast and vertebrate NPC is reflected in measured differences in both height (~350 A versus ~800 A) and diameter (~960 A versus 1450 A) (Yang et al., 1998). These disparities may be explained in a number of ways. Structural studies of the yeast NPC suggest that they lack specific sub-structures, such as the thin cytoplasmic and nucleoplasmic rings, cytoplasmic particles and the lumenal spoke ring (Yang et al., 1998). In addition, vertebrate nucleoporins are often larger than their yeast homologues, for example nup107 compared with its yeast homologue nup84. It has been suggested that the larger size and mass of the vertebrate pore may be due to post-translational modifications, additional structural domains and/or phosphorylation of constituent proteins rather than differences in the number of copies of each protein incorporated into the NPCs (Cronshaw et al., 2002).

4.1 *FG-repeat, FXFG-repeat and GLFG-repeat nucleoporins*

Nucleoporins can be further classified into families of proteins that share specific sequence motifs. Identification of shared motifs in novel proteins has often given suggestion of protein structure and function.

Many nucleoporins are enriched in phenylalanine-rich repeat motifs (see colour *Plate 2*). These domains are characterized by short stretches of hydrophobic residues, such as FG, FxFG (where x = any hydrophobic amino acid) or GLFG separated by very hydrophilic linker sequences. Specific nucleoporins have been shown to contain up to 30 phenylalanine-rich domains and it can be estimated that there are approximately 10 000 of these hydrophobic motifs per NPC. The bulk of the identified transport factors bind these Phe-rich domains directly (Allen et al., 2001; Bachi et al., 2000; Clarkson et al., 1996; Pashal and Gerace, 1995; Radu et al., 1995; Strasser et al., 2000) and these interactions are essential for translocation through the pore (Bayliss et al., 1999, 2000; Fribourg et al., 2001; Strasser et al., 2000).

Table 1. NPC proteins in vertebrates and their yeast homologues

Vertebrate	localization	component of/ interacts with	sequence motif	yeast homologue(s)
Tpr	nuclear[a,b]	nup98, nup153		Mlp1 Mlp2
nup214/CAN	cytoplasmic[c]	nup84[v]/nup88 complex	FG	nup159
nup358/RanBP2	cytoplasmic[d]	nup205 complex	FXFG	
gp210	pore membrane[e]	?	TMD	
nup205	symmetrical[f]	nup205 complex		nup192
nup188[v]	symmetrical[f]	nup205 complex		nup188[y]
nup160	nuclear γ	nup107 complex/nup98/nup153		
nup155	symmetrical[h]	?		nup157
nup153	nuclear[i]	Tpr, nup98	FXFG	nup1
POM121	pore membrane[j]	?	TMD	
Seh1	symmetrical[f]	?	WD	Seh1
nup133	nuclear[g]	nup107 complex		nup133
nup96	nuclear[k]	nup107 complex		nup145C
nup107	symmetrical[l]	nup107 complex		nup84[y]
nup98	nuclear[m,n(mobile nup)]	nup153/nup107 complex	GLFG[v], FG[y]	nup145N
	cytoplasmic-bias[f]	nup205 complex	GLFG	nup100
	cytoplasmic-bias[f]		GLFG	nup116
nup93	nuclear[o]	nup205 complex		Nic96
nup88	cytoplasmic[p]	nup88 complex		nup82
nup84[v]	cytoplasmic[q]	nup214/CAN		
nup75	?	?		
nup62	central[r]	p62 complex	FXFG[v],FG[y]	Nsp1
nup58	central*	p62 complex	FG[v], GLFG[y]	nup49
ALADIN	cytoplasmic[s]	?	WD	
nup54	central*	p62 complex	FG[v], GLFG[y]	nup57
nup50/NPAP60	nuclear[t]	?		
nup45	central*	p62 complex	FG[v], GLFG[y]	nup49
p43	?	?	WD	
hGle1	cytoplasmic-bias[f]	?		Gle1
RAE1/Gle2	symmetrical[f]	?	WD	Gle2
Sec13-like		?		
Sec13-related		?		
p37	?	?	WD	
p35	?	?		
p30 ?	?	?	WD	

[a]Byrd et al., 1994; [b]Cordes et al., 1997; [c]Kraemer et al., 1994; [d]Wu et al., 1995; [e]Greber et al., 1990; [f]localization of vertebrate nups based on localization of yeast homologue, Rout et al., 2000; [g]Vasu et al., 2001; [h]Radu et al., 1993; [i]Sukegawa and Blobel, 1993; [j]Soderqvist and Hallberg, 1994; [k]Fontoura et al., 1999; [l]Radu et al., 1995; [m]Fontoura et al., 2001; [n]Griffis et al., 2002; [o]Grandi et al., 1997; [p]Fornerod et al., 1997; [q]Bastos et al., 1997; [r]Grote et al., 1995; *localization based on nup62 localization; [s]Cronshaw and Matunis, 2002; [t]Guan et al., 2000; [v]vertebrate [y]yeast
Nup62 complex = nup62, nup58, nup54, nup45
Nup88 complex = nup88, nup62, nup214, nup98
Nup170 complex = nup107, nup160, nup133, nup96, nup98 (= vertebrate homologue of yeast nup84 complex)
Nup205 complex = nup205, nup188, nup93

4.2 WD-repeat nucleoporins

Recent proteomic analysis of the vertebrate NPC (Cronshaw et al., 2002) has identified a family of nucleoporins that share repeated WD (Trp-Asp) residues. Three of these proteins were novel components of the NPC; namely p37, p43 and p30. Three others were putative vertebrate homologues of previously characterized nucleoporins namely yeast nup85, yeast nup53 and human Seh1p. Another was ALADIN a 60-kDa protein encoded for by a gene whose mutation is implicated in the human disease triple-A syndrome (AAAS) (Huebner et al., 2002; Tullio-Pelet et al., 2000). Although there is little information on the structure or function of these proteins within the NPC, proteins containing WD repeats have been extensively investigated in other cellular contexts (reviewed in Li and Roberts, 2001; Smith et al., 1999). From these studies a WD repeat has been defined as a 44–60-residue sequence that typically contains a GH di-peptide 11–24 residues from its amino-terminus and a WD motif at its carboxy-terminus. The crystal structure of only two WD repeat proteins has been resolved, the Gβ subunit of heterotrimeric G proteins and Aip1p, a protein involved in actin depolymerisation (Sondek et al., 1996; Voegtli et al., 2003, respectively). These proteins adopt a highly symmetrical propeller-like configuration where each WD-sequence repeat corresponds to three-quarters of one blade and a quarter of the following blade of the propeller. It has been presumed that most, if not all, other proteins containing this sequence motif are likely to adopt similar surface structures. However, WD-repeat proteins have been implicated in a wide range of diverse cellular processes and it remains unclear if this motif confers a specific function (reviewed in Smith et al., 1999). Interestingly, several WD-repeat proteins have been characterized through their association with human diseases (reviewed in Li and Roberts, 2001).

4.3 Post-translation modifications of nucleoporins

Finite genomic sequence is not necessarily reflected in a limited number of resultant proteins, either structurally or functionally. Post-translational modification of proteins is a method by which a cell can increase the complexity, and thereby the uniqueness, of its proteome without the requirement for increased complexity of the DNA template. In addition, as the association of modifying phosphate and carbo-hydrate moieties is dynamic this association can serve to regulate biochemical pathways within the cell.

At least ten vertebrate nucleoporins are glycosylated by addition of monomeric O-linked N-acetylglucosamine (O-GlcNAc) to serine and/or threonine residues (Holt et al., 1987; Miller et al., 1999; reviewed in Hanover, 2001). These proteins are important for NPC structure and transport functions, although covalent masking of the O-GlcNAc with galactose residues does not interfere with NPC morphology or transport competency (Miller and Hanover, 1994). It has been suggested that glycosy-lation and phosphorylation may compete for sites of modification. Cells that have been treated with O-GlcNAcase inhibitors (thereby 'locking' O-GlcNAc modifications onto proteins) are hyperglycosylated and hypophosphorylated (Haltiwanger et al., 1998).

Protein phosphorylation is a common and essential post-translational modification as, for example, progression through the cell cycle is regulated and dependent upon dynamic interplay between phosphatases and kinases (reviewed in John et al., 2001). Upon entry into mitosis, the nuclear envelope of higher eukaryotes disassembles (reviewed in Foisner, 2003) and concomitantly there is disassembly of NPCs (Collas, 1998; Cotter et al., 1998; Kiseleva et al., 2001). Phosphorylation of NPC components governs their disassembly. Glycosylated nucleoporins such as nup153, nup214 and RanBP2/nup358 are phosphorylated throughout the cell cycle and become hyper-phosphorylated during mitosis thereby disrupting the protein–protein interactions that anchor them at the nuclear pore. Similarly, the pore membrane domain glyco-protein gp210 is phosphorylated during mitosis but not during interphase (Favreau et al., 1996; Macaulay et al., 1995).

Post-translational modification, in the form of proteolytic cleavage of a protein precursor, is required for the biogenesis of specific nucleoporins in both yeast and vertebrates. For example, nup98 and nup96 are generated by the cleavage of a 186-kDa precursor and this cleavage event is essential for the correct localization of both proteins (Fontoura et al., 1999). Nup98 is also generated through autoproteolysis of another distinct precursor to yield nup98 and a 6-kDa peptide (Griffis et al., 2003; Hodel et al., 2002). It seems likely that this biogenesis pathway is conserved as the putative yeast homologues of nups 98 and 96, N-nup145p and C-nup145p are also generated by the site-specific auto-proteolysis of a single precursor polypeptide (Teixeira et al., 1997).

Ubiquitinization is another post-translational modification and it involves the covalent attachment of ubiquitin (Ub), a 76-amino-acid protein, to lysine residues within the target substrate. This modification can direct the fate of the protein in a number of different ways; mono-ubiqinization (the addition of a single Ub molecule) is a relatively stable modification whereas addition of numerous Ub molecules affects a number of cellular processes and is relatively unstable. The best-characterized function of the multi-Ub signal is ATP-dependent proteolysis by the 26S proteome (reviewed in Hochstrasser, 1996). Similarly, a family of proteins, small ubiquitin-like modifiers (SUMOs), are processed and covalently conjugated to substrate proteins by mecha-nisms analogous to ubiquinization (Boddy et al., 1996; Mahajan et al., 1997; Matunis et al., 1996; Okura et al., 1996; Shen et al., 1996). It has been shown that SUMO-1, the first identified member of the SUMO family, acts as a covalent modification of RanGAP1, the sole Ran-GTPase activating protein which is located on the cytoplasmic domain of the NPC (Mahajan et al., 1997; Matunis et al., 1996). This modification, termed SUMOylation, has been proposed to be necessary for the targeting of RanGAP1 to the cytoplasmic filaments of the NPC by revealing a domain within the carboxy-terminus of RanGAP1 that binds to Nup358 (Matunis et al., 1998). In addition it has been shown that nup358 has SUMO1 E3 ligase activity meaning that nup358 has the ability to stim-ulate the conjugate of SUMO1 to proteins and suggesting that the SUMOylation and nuclear transport of, at least some substrates, are associated processes (Pichler et al., 2002). In addition, investigation of the behaviour of RanGTP during mitosis has suggested that its conjugation to SUMO1 is necessary for localization to spindle poles and kinetochores during mitosis (Joseph et al., 2002).

5. NPC protein complexes and their functions

Analyses of component proteins of the NPC have given insight into the ordered affinities of many nucleoporins thereby identifying specific protein sub-complexes that comprise the NPC (see *Table 1*). These studies have provided information on the protein assemblages that play roles in nuclear pore complex assembly, transport through the pore, and establishment and maintenance of structure within the plane of the nuclear envelope. In addition, elegant immunological and structural studies of the yeast p84 complex (Lutzmann et al., 2002) serve as a paradigm for the modular compo-sition of the NPC.

5.1 *p62 complex*

The p62 complex has been localized at the central channel of the NPC (Hu *et al.*, 1996). This complex comprises 4 nucleoporins, all of which are O-linked glycoproteins with phenylalanine-glycine (FG) repeat motifs; p62, p58, p54 and p45. Comparison of the peptide profiles of p58 and p45 generated by tryptic digestion suggests that these proteins are alternate splicing variants of the same gene. P58 has FG repeats at both its N and C termini flanking a predicted α-helical coil coiled region, whereas p45 is truncated and lacks the C-terminal FG repeats. P54 also contains FG repeats at its N-terminus with a predicted C-terminal coil coiled region. It has been suggested that the role of this protein complex is to achieve accumulation of transport complexes near the central channel prior to their translocation across the pore (Hu *et al.*, 1996).

5.2 *Nup107–160 complex and yeast nup84 complex*

The nup107–160 complex is the vertebrate equivalent of the p84 complex in the yeast *S. cerevisiae* and consists of nup107, nup133, nup96, nup160/120 and Sec13. The nup107–160 complex has been implicated as a key player in NPC assembly (Walther *et al.*, 2003) and shall be discussed in this context below.

The nup84 complex was first isolated from yeast, under native conditions, as a 375-kDa complex as judged by quantitative and qualitative TEM and ultracentrifugation (Siniossoglou *et al.*, 2000). This mass estimate is consistent with a complex composed of monomers of each component protein. Co-expression of these proteins has shown that when all components are present this complex adopts a 'Y-shaped' structure with a diameter of ~25 nm (Lutzmann *et al.*, 2002). Although it is not clear exactly which sub-structure of the yeast NPC the nup84 complex corresponds to, or is a component of, this study gives the first direct, biochemically defined, insight into the modular organization of nucleoporins and their immediate binding partners.

5.3 *Nup93–nup205–nup188 complex*

Identification of the nucleoporin nup93 revealed it to be the vertebrate homologue of the yeast nucleoporin Nic96p localized at the distal end of the nucleoplasmic basket of the vertebrate NPC. Grandi *et al.* (1997) demonstrated that the majority of nup93 interacts with a previously uncharacterized 205-kDa protein. Studies of the yeast homologue of this 205-kDa protein (Nup192p; Kosova *et al.*, 1999) showed it to be localized to the nucleoplasmic aspect of nuclear pores and to interact with the yeast nup93 homologue Nic96p. Further co-purification experiments identified another nucleoporin that complexes with nup93 and nup205, termed *Xenopus* nup188 (Miller *et al.*, 2000). Nuclei assembled in the absence of the nup93 complex were competent for nuclear transport although they were significantly smaller and contained fewer complete NPCs (Grandi *et al.*, 1997).

As the vertebrate nup93–nup205–nup188 complex and the yeast Nic96p–nup192p–nup188p complexes show a high degree of similarity both in protein sequence and protein–protein interactions it has been suggested that these sub-complexes may have the same structural and functional role in vertebrate and yeast NPC, respectively.

5.4 *Nup98–nup88 and nup98–nup96 complexes*

Nup98 was first identified as an FG, GLFG and FxFG containing NPC component involved in the translocation of transport complexes through the nuclear pore (Radu *et al.*, 1995). Initial studies suggested that nup98 was present only on the nucleoplasmic side of the NPC upon the nucleoplasmic filaments that comprise the nuclear pore basket (Radu *et al.*, 1995). However, more recent investigations indicate that nup98 interacts *in vivo* with nucleoporins on both the nucleoplasmic and cytoplasmic faces of the NPC (Frosst *et al.*, 2002; Griffis *et al.*, 2003; Vasu *et al.*, 2001). Nup98 interacts with nup96 on the nucleoplasmic side of the NPC and can also interact with nup88 on the cytoplasmic side of the NPC. The same domain within nup98 mediates the interaction with both differentially localized binding partners. In addition, nup98 is also present as a highly mobile nucleoporin within the nucleoplasm and it has been suggested that nup98 may serve to direct RNAs for export through the NPC into the cytoplasm for translation (Griffis *et al.*, 2002).

5.5 *Nup88 complex*

Together with Nup214 and p62, Nup88 forms a sub-complex that is found only on the cytoplasmic face of the pore (Bastos *et al.*, 1997; Fornerod *et al.*, 1997). Nup88 binds directly to nup214/CAN and is dependent upon this protein for its localization at the cytoplasmic face of the NPC (Fornerod *et al.*, 1997). The yeast homologue of nup88, termed nup82, is also cytoplasmically disposed (Rout *et al.*, 2000). Recent work by Griffis *et al.* (2003) has revealed that the mobile nucleoporin Nup98 has two potential binding partners on the cytoplasmic face of the NPC, namely Nup96 and Nup88, and it can also bind nup96 located in the nucleoplasmic domain of the NPC. From this it has been suggested that Nup98 may acts as a bridge between the Nup107–160 complex and the Nup88–Nup214–Nup62 complex (Griffis *et al.*, 2003).

6. Traversing the nuclear pore

The central channel of the NPC acts as a conduit for the active transport of molecules, and individual NPCs are thought to be capable of both import and export during nucleocytoplasmic trafficking (Feldherr *et al.*, 1984). A number of studies have sought to define the operational parameters of the NPC and combinatorial studies involving computational simulation and experimental data acquisition have proposed differing estimated rates of translocation. For example, it has been suggested that each NPC performs in the region of ~120 (Smith *et al.*, 2002) to 10^3 (Ribbeck and Gorlich, 2001) translocation events per second.

6.1 *Transport-cargo chaperones and the Ran gradient*

Nucleocytoplasmic trafficking requires chaperone transport proteins, which bind to the designated cargo molecule. This binding is mediated through sequence motifs within the cargo molecule that designate the destination of the cargo, i.e. nuclear import sequences (NISs) are present in molecules destined for the nucleoplasm and conversely nuclear export sequences (NESs) are found within molecules that require transportation out of the nucleus to the cytoplasm. The transport molecules (termed importins, exportins and transportins depending on the direction of transport that they mediate) have the ability to shuttle back to the compartment in which they meet their cargo in order to facilitate further rounds of translocation through the NPC. The small GTPase Ran determines the direction of transport afforded by these chaperone molecules. Ran is found in two states in the cell, either as a GTP-bound form in the nucleus or GDP-bound form in the cytoplasm and a steep RanGTP/RanGDP gradient has been measured across the nuclear envelope (Görlich et al., 2003; Kalab et al., 2002; Smith et al., 2002). In the case of nuclear import, RanGDP stimulates the binding of cargo to import factors that, upon entry to the nucleus, are triggered to release their cargo by the nuclear concentration of RanGTP. Conversely, RanGTP activates the binding of exportins to their cargo within the nucleus and, upon entry into the cytoplasm, RanGDP disrupts this association.

6.2 *Translocation through the centre of the pore*

Several recent studies have focused on the mechanism(s) by which molecules are translocated through the central channel of the NPC. Although these investigations have generated several differing models to explain this phenomenon, they are all based on the necessity of transport factors to bind the FG-repeat domains of nups present in the central channel (*Figure 1*).

6.3 *Filamentous size exclusion model*

Following proteomic analyses of the yeast NPC Rout et al. (2000) suggested a model for translocation through the NPC whereby Brownian diffusion drives such transport. In their model the narrow bore of the pore acts as a size-exclusion barrier to the inappropriate trafficking of molecules, and the binding of transport molecules to FG repeat nucleoporins increases the residency time at the pore and facilitates passage through the NPC. This model was further refined, and somewhat modified, by characterization of the structural properties of the nucleoporins present at the central channel of the NPC (Denning et al., 2003).

They specifically focused on the properties of nucleoporins containing FG repeat motifs, as they are the NPC components that facilitate the NPC-binding and subsequent nucleocytoplasmic trafficking of transport chaperone–cargo complexes. Biophysical and amino acid composition analyses revealed that FG-repeat-containing nucleoporins were highly disordered within the context of the nuclear pore suggesting that FG-repeat nups exist and function in a native unfolded state. However, the small non-FG-repeat regions do adopt folded structures and these are, in the main, regions involved in binding other NPC components. Denning et al. (2003) have proposed a model for translocation through the NPC whereby the filamentous nups composing the central channel of the NPC create a highly dynamic meshwork of differing sizes of filaments that serve to define the size limit exclusion barrier of the NPC. In such a way small molecules could easily diffuse between these filaments whereas larger cargo molecules would require the specific FG-binding properties conferred by importin/exportin/transportin proteins (see *Figure 1a*).

6.4 *Hydrophobic partitioning model*

Studies by Ribbeck and Görlich (2002) have shown that nuclear transport factors can be distinguished from other soluble proteins due to their high degree of surface

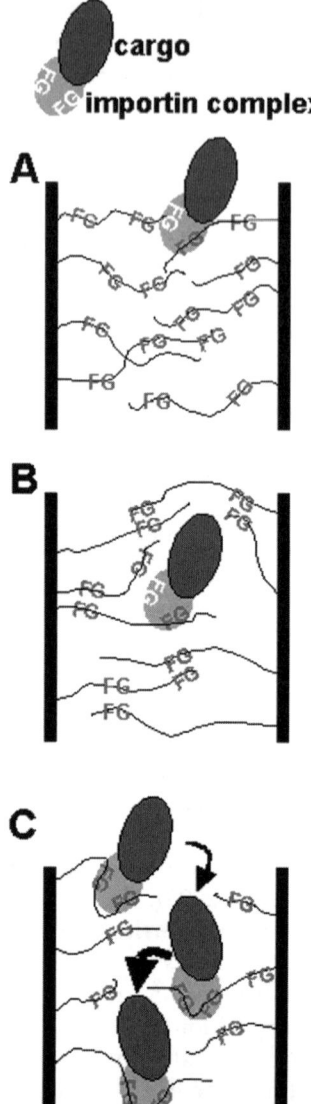

Figure 1. Models for translocation through the nuclear pore complex. Recent studies of the process by which molecules are actively transported through the NPC have led to the proposal of three models. (A) Filamentous size exclusion model. Nucleoporins containing FG repeats, within the central channel of the NPC, are the key components of a filamentous network that determines the size-exclusion barrier of the NPC. Direct interaction of these nups with FG-binding transport factors allows passage through the pore (Denning et al., 2003). (B) Hydrophobic partitioning model. The central channel of the pore is defined by the highly hydrophobic nature of FG-repeat nups. Molecules are excluded and passage through the pore is facilitated through the binding of hydrophobic transporter molecules that navigate through the pore as they bind FG-repeat moieties. In addition the exclusion barrier maybe tightened by association of FG repeats between nups (Ribbeck and Gorlich, 2002; Shulga and Goldfarb, 2003). (C) Affinity gradient model. The direction of transport may be defined by increasing affinity of specific transport molecules for nups within distinct regions of the NPC (Ben-Efraim and Gerace, 2001).

hydrophobicity. They propose that this characteristic of transport chaperone molecules allows their passage through the hydrophobic environment of the central channel of the NPC. Their studies showed that the FG-repeats, characteristic of nucleoporins within the central channel of the NPC, render this channel a highly hydrophobic environment and these proteins were also capable of 'tightening' this exclusion barrier through weak hydrophobic interactions between adjacent FG-repeat proteins (see *Figure 1b*). Furthermore, they analysed the effects upon translocation of these transport proteins when bound to cargo molecules that lack such hydrophobicity and they found that bound, relatively hydrophilic cargo could drastically impede NPC passage. Thus the larger the cargo the greater the number of transport molecules required for efficient transport through the NPC. The Ribbeck and Görlich (2002) model for translocation through the NPC suggests that the hydrophobic central channel excludes passage of hydrophilic material, and the binding of transport proteins overcomes this exclusion by virtue of their hydrophobicity and their ability to bind FG-repeat nucleoporins. This model was supported by independent studies showing that treatment of cells with aliphatic alcohols disrupt the size-exclusion/permeability barrier of the NPC by reversibly dissociating neighbouring FG-repeat nucleoporins (Shulga and Goldfarb, 2003).

6.5 *Affinity gradient model*

Biochemical investigation of the importin β-mediated nuclear import pathway has suggested a model for translocation across the NPC in which transport directionality is facilitated by the increasing affinity of importin β for nucleoporins through the NPC (Ben-Efraim and Gerace, 2001). This study showed that importin β first binds nup358/RanBP2 on the cytoplasmic filaments of the NPC then, via the p62 complex and possibly other unidentified nucleoporins, it binds nup153 and this translocation follows a gradient of increasing binding affinity of importin β for these nucleoporins (see *Figure 1c*). In addition, Ben-Efraim and Gerace (2001) showed that importin β had very similar binding affinities for nucleoporins localized to specific compartments of the NPC; therefore they suggest that the affinity of import factors for specific regions of the NPC may be important for conferring directionality to their translocation pathway. The 'affinity gradient model' for translocation through the NPC was tested in yeast and it was shown that one specific import cargo–transporter complex (Kap95p–Kap60p–cargo) has progressively higher affinities for nucleoporins from the cytoplasm through to the nuclear pore basket (Pyhtila and Rexach, 2003). However, this model of increasing affinity of transport factors for nucleoporins appears to conflict with the ability of these molecules, after dissociation of cargo proteins, to traverse the pore in the opposite direction prior to another round of transport.

Research into the mechanisms that mediate and regulate nucleocytoplasmic trafficking remains a field of much active investigation.

7. NPC assembly

All higher eukaryotes undergo an open mitosis whereby the nuclear envelope loses its tethering at the chromatin–lamina interface and retreats back into the endoplasmic reticulum (ER) (Ellenberg and Lippincott-Schwartz, 1999; Yang *et al.*, 1997). Other organisms undergo partial membrane dissociation during mitosis, e.g. the fruit fly

Drosophila melanogaster (Stafstrom and Staehelin, 1984), or a closed mitosis where the NE remains intact. Irrespective of the nature of mitosis there is the necessity for assembly of new NPCs. In yeast, which undergoes a closed mitosis, new NPCs are assembled throughout the cell cycle and NPC density peaks during S-phase (Winey *et al.*, 1997). In higher eukaryotes the number of NPCs doubles during S-phase (Maul, 1977) and then, at telophase, a nuclear envelope and associated NPCs must assemble around both daughter chromosome populations. A structural pathway for the assembly of NPCs has been proposed on the basis of studies involving chemical agents that arrest nuclear assembly at defined timepoints (Macaulay and Forbes, 1996). Complementary study of NPC assembly during nuclear formation was performed by scanning electron microscopy (SEM) examination of nascent NPCs in the *Xenopus* cell free system (Gant *et al.*, 1998; Goldberg *et al.*, 1997; see *Figure 1a–f*). The binding of membranes to the surface of decondensing chromatin and subsequent fusion and flattening of these membranes is a prerequisite for the formation of NPCs (Holaska *et al.*, 2002; Macaulay and Forbes, 1996; reviewed in Vasu and Forbes, 2001). At the initiation of pore assembly 'dimples' or concavities, with no associated structures, were visible on the ONM (*Figure 2a*). As SEM visualizes only the surface topology of the sample the corresponding morphology of the INM remains unresolved. These dimples then form holes, or empty pores, and these nascent pores appear to adopt a more regular 'stabilized' surface morphology, often with material within the pore itself (*Figure 2b,c*). Subsequently there is assembly of intermediate annular structures, specifically the cytoplasmic star-ring and thin-ring (*Figure 2e–g*) and then there is the assembly of the eight filaments that project into the cytoplasm (*Figure 2g,h*). This structural pathway for NPC assembly has been supported by *in vivo* observation of NPC formation in *Drosophila* embryos (Kiseleva *et al.*, 2001).

During mitosis, once the NPCs have disassembled, many nucleoporins maintain their associations with their resident complex members in the cytoplasm and resultantly upon exit from mitosis recruitment of nucleoporins is thought to involve the step-wise recruitment of specific protein complexes rather than relocation of each NPC component individually.

Immuno-localization studies have afforded insight into the ordered return of nucleoporins to nascent NPCs and have given further evidence for the step-wise assembly of NPC intermediates (Bodoor *et al.*, 1999; Ellenberg *et al.*, 1997). In addition, analysis of NPC assembly in mutant yeast strains has suggested proteins that are involved in NPC formation (Bucci and Wente, 1998; Gomez-Ospina *et al.*, 2000; Ryan and Wente, 2002) and the maintenance of NPC structure (Wente and Blobel, 1993).

8.   The molecular requirements for NPC assembly

Although putative structural NPC assembly pathways have been proposed through chemical manipulations of NPC assembly in model systems, only recently has there been insight into the molecular requirements of this essential cellular process.

8.1 *Hierarchical recruitment of nups to the reforming nuclear envelope*

Recent studies of the molecular requirements of NPC assembly have utilized biochemical depletion, RNAi and siRNA technologies to determine the roles and

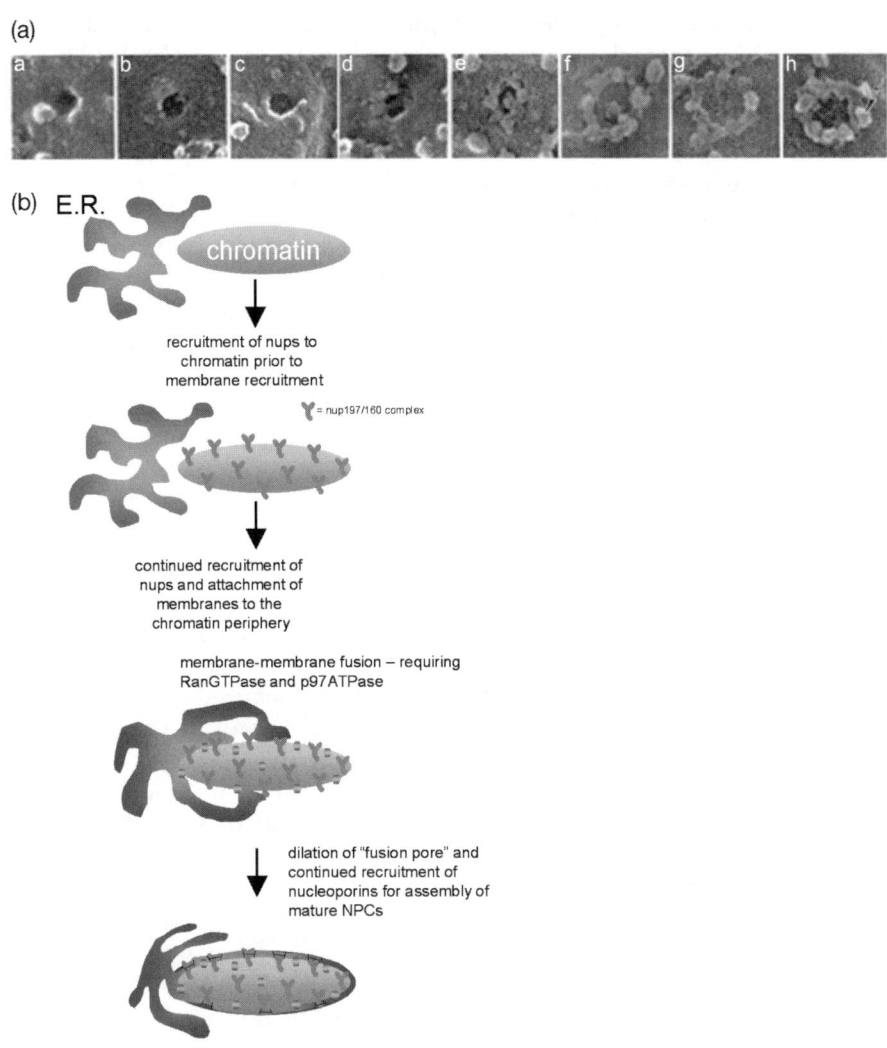

Figure 2. Assembly of the vertebrate nuclear pore complex. (a) NPC assembly in the Xenopus cell-free system. (a–h) A structural pathway for NPC assembly has been proposed on the basis of SEM observations of nuclear assembly in cell free extracts of Xenopus eggs (Goldberg et al., 1997). (b) A model for nuclear pore complex assembly. During mitosis in vertebrates the nuclear envelope retracts into the endoplasmic reticulum (ER) therefore upon exit from mitosis there is first recruitment of soluble nucleoporins (i.e. the nup107/160 complex; Walther et al., 2003) to the chromatin surface then membranes are targeted to the decondensing chromatin and fuse to enclose the chromatin. Further recruitment of membrane associated and soluble nucleoporins supports assembly of mature NPCs.

necessity of specific nucleoporins in NPC function and assembly. This has revealed that the nup107–160 complex is essential for assembly of NPCs and its absence results in the formation of nuclear membranes devoid of NPCs and nuclear pore intermediates (Harel *et al.*, 2003; Walther *et al.*, 2003). This is further reflected in an absence of recruitment of other nucleoporins to the nascent nuclear envelope (Boehmer *et al.*, 2003). These studies provide strong evidence that the nup107–160 complex is a key determinant of NPC assembly. In addition, this complex is recruited to chromatin prior to the binding of membranes to chromosome surfaces and thereby, at sites of chromatin binding, may act as the foundation for subsequent NPC formation. Upon formation of the nuclear pore there is the necessity for dilation of the pore with further recruitment of nucleoporins, both soluble and membrane-associated. Gp210, one of only two integral membrane proteins present exclusively at the interphase vertebrate pore membrane domain, has been implicated to play a role in dilation of the nuclear pore (Drummond and Wilson, 2002).

8.2 *The role of Ran in NPC assembly*

The small GTPase Ran has been well characterized as the enzyme that plays a central role in mediating the directional transport of molecules between the nucleoplasm and cytoplasm (see previous section) and a wealth of recent work has shown it to have an essential role in nuclear assembly in vertebrates (Hughes *et al.*, 1998; Walther *et al.*, 2003; Zhang and Clarke, 2000; Zhang *et al.*, 1999), *Caenorhabditis elegans* (Askjaer *et al.*, 2002) and yeast (Ryan *et al.*, 2003). The potential role for Ran as a mediator of NPC assembly was investigated in nuclear assembly extracts of *Xenopus* eggs (Walther *et al.*, 2003). This study revealed that inhibition of RanGTP production, by incubation with mutant Ran, depletion of RCC1 that converts RanGDP to RanGTP or depletion of Ran itself, inhibits the recruitment of nucleoporins to chromatin and thereby leads to the complete inhibition of NPC assembly.

9. Nucleoporins and disease

Nucleocytoplasmic transport is clearly an essential process and therefore defects in the composition and resultant changes in the function of the NPC can result in severe disease phenotypes. Cell-cycle progression is regulated by proteins that by definition require translation in the cytoplasm and are then transported into the nucleus via NPCs, therefore defects in this transport process can have catastrophic effects upon cell viability and proliferative status (reviewed in Hood and Silver, 2000).

Several nucleoporins are involved in chromosomal rearrangements that induce oncogenic transformation of the cells in which they are expressed. These fusion proteins have been implicated in human leukaemias (mainly acute myelogenous leukaemia; AML) and myelodysplastic syndrome (MDS) (reviewed in Scandura *et al.*, 2002) and a diverse range of human cancers.

9.1 *Nup98 and Nup214/CAN*

Nup98 is the target of at least 12 distinct and recurrent chromosomal translocation events that produce oncogenic fusion proteins (see Scandura *et al.*, 2002 and references therein). In the majority of these situations the Nup98 gene becomes fused to a homeobox gene such as HOXA9, PMX1 or HOXD13 (Borrow *et al.*, 1996; Nakamura *et al.*, 1996, 1999; Nishiyama *et al.*, 1999; Raza-Egilmez *et al.*, 1998). Homeobox genes are classified as a family of transcription factors that share a conserved DNA-binding motif termed the homeodomain. This domain determines the DNA-binding specificity of the gene products and thereby is pivotal in determining the transcriptional regulation of genes downstream of the protein-binding site. As mentioned previously, Nup98 is generated by the proteolytic cleavage of a precursor polypeptide that, when cleaved, generates both Nup98 and Nup96 or Nup98 and a 6- or 8-kDa peptide (Hodel *et al.*, 2002; see Post-translational modification of nucleoporins). In all of the Nup98 fusion proteins identified to date the N-terminal portion of the Nup98 gene product is fused to the carboxy-terminal portion of its fusion partner and these fusions lack the portions encoding Nup96 or the smaller polypeptide (see Hussey and Dobrovic, 2002, and references therein). The effect that this exerts on Nup96 expression and localization, or NPC function has not been determined. In addition, the direct effect the production of these protein fusions has on Nup98 dynamics, interactions or functions is not well understood. It has been suggested that the FG repeat region of Nup98, present in the HOX–Nup98 fusions, specifically acts as a potent trans-activator of gene transcription and together with inappropriate association of Nup98 with transcription-regulating proteins leads to oncogenic cellular proliferation (Kasper *et al.*, 1999; Kroon *et al.*, 2001).

9.2 *Nup214/CAN*

Nup214 was first characterized as the human protein CAN: a putative oncogene associated with leukaemogenesis. Oncogenic proliferation is induced by chromosomal rearrangements that fuse Nup214/CAN to DEK: phospho-protein with site-specific DNA-binding ability involved in regulation of DNA architecture and replication (Alexiadis *et al.*, 2000; Fu *et al.*, 1997), or SET: a phospho-protein that plays a role in regulating protein phosphatase 2A activity (Li *et al.*, 1996). The Nup214/CAN–DEK or Nup214/CAN–SET chromosomal rearrangements comprise the C-terminus of the Nup214/CAN gene fused to the N-terminal portion of the DEK gene (a translocation event) or the complete SET gene (a gene fusion event) (Von Lindern *et al.*, 1992a, 1992b). In both cases the expressed chimeric proteins share the C-terminus of Nup214/CAN and therefore it was suggested that this component of the fusion protein is the oncogenic determinant. As Nup214 has been localized to the cytoplasmic filaments of the NPC and has a role in nucleocytoplasmic trafficking it is possible that aberrant regulation of traffic between the nucleus and cytoplasm may confer a proliferative advantage to the cell. In addition, it has been shown that the SET protein is a potent specific inhibitor of protein phosphatase 2A (PP2A) and therefore alterations in this function of SET, due to fusion with Nup214/CAN, may contribute to the cellular transformation to a leukaemic state (Li *et al.*, 1996). Similarly, the DEK protein has been shown to exert an inhibitory effect on DNA replication (Alexiadis *et al.*, 2000) and this activity may be altered when expressed as part of a fusion protein.

9.3 *Nup88*

Nup88 is specifically over-expressed in cells derived from a range of human cancers (Gould *et al.*, 2002), especially colorectal and ovarian malignancies (Emterling *et al.*,

2003; Martinez *et al.*, 1999, respectively). As a result it has been suggested that detection and quantitation of Nup88 protein expression levels may be a useful, broad-based histological indicator of tumorogenesis and tumour progression. Interestingly Nup88 interacts directly with Nup214 and therefore the oncogenic fusion of these proteins may ablate their cognate binding domains thereby resulting in altered NPC structure and/or function, and may contribute to oncogenic transformation.

References

Akey, C.W. (1989) Interactions and structure of the nuclear pore complex revealed by cryo-electron microscopy. *J. Cell Biol.* **109**(3): 955–970.

Akey, C.W. and Radermacher, M. (1993) Architecture of the Xenopus nuclear pore complex revealed by three-dimensional cryo-electron microscopy. *J. Cell Biol.* **1122**(1): 1–19.

Alexiadis, V., Waldmann, T., Andersen, J., Mann, M., Knippers, R. and Gruss, C. (2000) The protein encoded by the proto-oncogene DEK changes the topology of chromatin and reduces the efficiency of DNA replication in a chromatin-specific manner. *Genes Dev.* **14**: 1308–1312.

Allen, N.P., Huang, L., Burlingame, A. and Rexach, M. (2001) Proteomic analysis of nucleoporin interacting proteins. *J. Biol. Chem.* **276**(31): 29268–29274.

Askjaer, P., Galy, V., Hannak, E. and Mattaj, I.W. (2002) Ran GTPase cycle and importins alpha and beta are essential for spindle formation and nuclear envelope assembly in living *Caenorhabditis elegans* embryos. *Mol. Biol. Cell.* **13**(12): 4355–4370.

Bachi, A., Braun, I.C., Rodrigues, J.P., Pante, N., Ribbeck, K., von Kobbe, C., Kutay, U., Wilm, M., Gorlich, D., Carmo-Fonseca, M. and Izaurralde, E. (2000) The C-terminal domain of TAP interacts with the nuclear pore complex and promotes export of specific CTE-bearing RNA substrates. *RNA* **6**(1): 136–158.

Bastos, R., Ribas de Pouplana, L., Enarson, M., Bodoor, K. and Burke, B. (1997) Nup84, a novel nucleoporin that is associated with CAN/Nup214 on the cytoplasmic face of the nuclear pore complex. *J. Cell Biol.* **137**(5): 989–1000.

Bayliss, R., Ribbeck, K., Akin, D., Kent, H.M., Feldherr, C.M., Gorlich, D. and Stewart, M. (1999) Interaction between NTF2 and xFxFG-containing nucleoporins is required to mediate nuclear import of RanGDP. *J. Mol. Biol.* **293**(3): 579–593.

Bayliss, R., Kent, H.M., Corbett, A.H. and Stewart, M. (2000) Crystallization and initial X-ray diffraction characterization of complexes of FxFG nucleoporin repeats with nuclear transport factors. *J. Struct. Biol.* **131**(3): 240–247.

Ben-Efraim, I. and Gerace, L. (2001) Gradient of increasing affinity of importin beta for nucleoporins along the pathway of nuclear import. *J. Cell Biol.* **152**(2): 411–417.

Boddy, M.N., Howe, K., Etkin, L.D., Solomon, E. and Freemont, P.S. (1996) PIC 1, a novel ubiquitin-like protein which interacts with the PML component of a multiprotein complex that is disrupted in acute promyelocytic leukemia. *Oncogene* **13**(5): 971–982.

Bodoor, K., Shaikh, S., Salina, D., Raharjo, W.H., Bastos, R., Lohka, M. and Burke, B. (1999) Sequential recruitment of NPC proteins to the nuclear periphery at the end of mitosis. *J. Cell. Sci.* **112** (Pt 13): 2253–2264.

Boehmer, T., Enninga, J., Dales, S., Blobel, G. and Zhong, H. (2003) Depletion of a single nucleoporin, Nup107, prevents the assembly of a subset of nucleoporins into the nuclear pore complex. *Proc. Natl Acad. Sci. USA* **3**: 981–985.

Borrow, J., Shearman, A.M., Stanton, Jr. V.P., *et al.* (1996) The t(7:11)(p15:p15) translocation in acute myeloid leukemia fused the genes for nucleoporin NUP98 and class I homeoprotein HOXA9. *Nature Genet.* **12**: 159–167.

Bucci, M. and Wente, S.R. (1998) A novel fluorescence-based genetic strategy identifies mutants of *Saccharomyces cerevisiae* defective for nuclear pore complex assembly. *Mol. Biol. Cell.* **9**(9): 2439–2461.

Byrd, D.A., Sweet, D.J., Pante, N., Konstantinov, K.N., Guan, T., Saphire, A.C., Mitchell, P.J., Cooper, C.S., Aebi, U. and Gerace, L. (1994) Tpr, a large coiled coil protein whose amino terminus is involved in activation of oncogenic kinases, is localized to the cytoplasmic surface of the nuclear pore complex. *J. Cell. Biol.* **127**(6 Pt 1): 1515–1526.

Chemnitz, J., Salmberg, K. and Bierring, F. (1977) Observations on the association of annulate lamellae with vinblastine-induced paracrystals in tumour cells in vitro. *Arch. B Cell. Pathol.* **24**(2): 147–156.

Chial, H.J., Rout, M.P., Giddings, T.H. and Winey, M. (1998) *Saccharomyces cerevisiae* Ndc1p is a shared component of nuclear pore complexes and spindle pole bodies. *J. Cell. Biol.* **143**(7): 1789–1800.

Chook, Y.M. and Blobel, G. (2001) Karyopherins and nuclear import. *Curr. Opin. Struct. Biol.* **11**(6): 703–715.

Clarkson, W.D., Kent, H.M. and Stewart, M. (1996) Separate binding sites on nuclear transport factor 2 (NTF2) for GDP-Ran and the phenylalanine-rich repeat regions of nucleoporins p62 and Nsp1p. *J. Mol. Biol.* **263**(4): 517–524.

Collas, P. (1998) Nuclear envelope disassembly in mitotic extract requires functional nuclear pores and a nuclear lamina. *J. Cell. Sci.* **111**(Pt 9): 1293–1303.

Cordes, V.C., Reidenbach, S., Kohler, A., Stuurman, N., van Driel, R. and Franke, W.W. (1993) Intranuclear filaments containing a nuclear pore complex protein. *J. Cell. Biol.* **123**(6 Pt 1): 1333–1344.

Cordes, V.C., Reidenbach, S., Rackwitz, H.R. and Franke, W.W. (1997) Identification of protein p270/Tpr as a constitutive component of the nuclear pore complex-attached intranuclear filaments. *J. Cell. Biol.* **136**(3): 515–529.

Cotter, L.A., Goldberg, M.W. and Allen, T.D. (1998) Nuclear pore complex disassembly and nuclear envelope breakdown during mitosis may occur by both nuclear envelope vesicularisation and dispersion throughout the endoplasmic reticulum. *Scanning* **20**(3): 250–251.

Cronshaw, J.M. and Matunis, M.J. (2003) The nuclear pore complex protein ALADIN is mislocalized in triple A syndrome. *Proc. Natl Acad. Sci. USA* **100**(10): 5823–5827.

Cronshaw, J.M., Krutchinsky, A.N., Zhang, W., Chait, B.T. and Matunis, M.J. (2002) Proteomic analysis of the mammalian nuclear pore complex. *J. Cell. Biol.* **158**(5): 915–927.

Daigle, N., Beaudouin, J., Hartnell, L., Imreh, G., Hallberg, E., Lippincott-Schwartz, J. and Ellenberg, J. (2001) Nuclear pore complexes form immobile networks and have a very low turnover in live mammalian cells. *J. Cell. Biol.* **154**(1): 71–84.

de Bruyn Kops, A. and Guthrie, C. (2001) An essential nuclear envelope integral membrane protein, Brr6p, required for nuclear transport. *EMBO J.* **20**(15): 4183–4193.

Denning, D.P., Patel, S.S., Uversky, V., Fink, A.L. and Rexach, M. (2003) Disorder in the nuclear pore complex: the FG repeat regions of nucleoporins are natively unfolded. *Proc. Natl Acad. Sci. USA* **100**(5): 2450–2455.

Drummond, S.P. and Wilson, K.L. (2002) Interference with the cytoplasmic tail of gp210 disrupts "close apposition" of nuclear membranes and blocks nuclear pore dilation. *J. Cell. Biol.* **158**(1): 53–62.

Ellenberg, J. and Lippincott-Schwartz, J. (1999) Dynamics and mobility of nuclear envelope proteins in interphase and mitotic cells revealed by green fluorescent protein chimeras. *Methods* **19**(3): 362–372.

Ellenberg, J., Siggia, E.D., Moreira, J.E., Smith, C.L., Presley, J.F., Worman, H.J. and Lippincott-Schwartz, J. (1997) Nuclear membrane dynamics and reassembly in living cells: targeting of an inner nuclear membrane protein in interphase and mitosis. *J. Cell. Biol.* **138**(6): 1193–1206.

Emterling, A., Skoglund, J., Arbman, G., Schneider, J., Eversson, S., Carstensen, J., Zhang, H. and Sun, X.F. Clinopathological significance of Nup88 expression in patients with colorectal cancer. *Oncology* **64**: 361–369.

Favreau, C., Worman, H.J., Wozniak, R.W., Frappier, T. and Courvalin, J.C. (1996) Cell cycle-dependent phosphorylation of nucleoporins and nuclear pore membrane protein Gp210. *Biochemistry* **35**(24): 8035–8044.

Feldherr, C.M. and Akin, D. (1990) EM visualization of nucleocytoplasmic transport processes. *Electron Microsc. Rev.* **3**(1): 73–86.

Feldherr, C.M., Kallenbach, E. and Schultz, N. (1984) Movement of a karyophilic protein through the nuclear pores of oocytes. *J. Cell. Biol.* **99**(6): 2216–2222.

Foisner, R. (2003) Cell cycle dynamics of the nuclear envelope. *Scientific World Journal* **17**;3(3): 1–20.

Fontoura, B.M., Blobel, G. and Matunis, M.J. (1999) A conserved biogenesis pathway for nucleoporins: proteolytic processing of a 186-kilodalton precursor generates Nup98 and the novel nucleoporin, Nup96. *J. Cell. Biol.* **144**(6): 1097–112.

Fontoura, B.M., Dales, S., Blobel, G. and Zhong, H. (2001) The nucleoporin Nup98 associates with the intranuclear filamentous protein network of TPR. *Proc. Natl Acad. Sci. USA* **98**(6): 3208–3213.

Fornerod, M, van Deursen, J., van Baal, S., Reynolds, A., Davis, D., Murti, K.G., Fransen, J. and Grosveld, G. (1997) The human homologue of yeast CRM1 is in a dynamic subcomplex with CAN/Nup214 and a novel nuclear pore component Nup88. *EMBO J.* **16**(4): 807–816.

Fribourg, S., Braun, I.C., Izaurralde, E. and Conti, E. (2001) Structural basis for the recognition of a nucleoporin FG repeat by the NTF2-like domain of the TAP/p15 mRNA nuclear export factor. *Mol. Cell.* **8**(3): 645–656.

Frosst, P., Guan, T., Subauste, C., Hahn, K. and Gerace, L. (2002) Tpr is localized within the nuclear basket of the pore complex and has a role in nuclear protein export. *J. Cell. Biol.* **156**(4): 617–630.

Fu, G.K., Grosveld, G. and Markovitz, D.M. (1997) DEK, an autoantigen involved in a chromosomal translocation in acute myelogenous leukemia, binds to the HIV-2 enhancer. *Proc. Natl Acad. Sci. USA* **94**: 1811–1815.

Galy, V., Olivo-Marin, J.C., Scherthen, H., Doye, V., Rascalou, N. and Nehrbass, U. (2000) Nuclear pore complexes in the organisation of silent telomeric chromatin. *Nature* **403**: 108–112.

Gant, T.M., Goldberg, M.W. and Allen, T.D. (1998) Nuclear envelope and nuclear pore assembly: analysis of assembly intermediates by electron microscopy. *Curr. Opin. Cell. Biol.* **10**(3): 409–415.

Gerace, L., Ottaviano, Y. and Kondor-Koch, C. (1982) Identification of a major polypeptide of the nuclear pore complex. *J. Cell. Biol.* **95**(3): 826–837.

Goldberg, M.W. and Allen, T.D. (1993) The nuclear pore complex: three-dimensional surface structure revealed by field emission, in-lens scanning electron microscopy, with underlying structure uncovered by proteolysis. *J. Cell. Sci.* **106**: 261–274.

Goldberg, M.W. and Allen, T.D. (1996) The nuclear pore complex and lamina: three-dimensional structures and interactions determined by field emission in-lens scanning electron microscopy. *J. Mol. Biol.* **257**(4): 848–865.

Goldberg, M.W., Wiese, C., Allen, T.D. and Wilson, K.L. (1997) Dimples, pores, star-rings, and thin rings on growing nuclear envelopes: evidence for structural intermediates in nuclear pore complex assembly. *J. Cell. Sci.* **110**: 409–420.

Goldberg, M.W., Rutherford, S.A., Hughes, M., Cotter, L.A., Bagley, S., Kiseleva, E., Allen, T.D. and Clarke, P.R. (2000) Ran alters nuclear pore complex conformation. *J. Mol. Biol.* **300**(3): 519–529.

Gomez-Ospina, N., Morgan, G., Giddings, T.H. Jr., Kosova, B., Hurt, E. and Winey, M. (2000) Yeast nuclear pore complex assembly defects determined by nuclear envelope reconstruction. *J. Struct. Biol.* **132**(1): 1–5.

Görlich, D., Seewald, M.J. and Ribbeck, K. (2003) Characterization of Ran-driven cargo transport and the RanGTPase system by kinetic measurements and computer simulation. *EMBO J.* **22**(5): 1088–1100.

Gould, V.E., Oruceivic, A., Zentgraf, H., Gattuso, P., Martinez, N. and Alonso, A. (2002) Nup88 (karyoporin) in human malignant neoplasms and dysplasias: correlations of immunostaining of tissue sections, cytologic smears and immunoblot analysis. *Hum. Pathol.* **33**: 536–544.

Grandi, P., Dang, T., Pane, N., Shevchenko, A., Mann, M., Forbes, D. and Hurt, E. (1997) Nup93, a vertebrate homologue of yeast Nic96p, forms a complex with a novel 205-kDa protein and is required for correct nuclear pore assembly. *Mol. Biol. Cell.* **8(10)**: 2017–2038.

Greber, U.F., Senior, A. and Gerace, L. (1990) A major glycoprotein of the nuclear pore complex is a membrane-spanning polypeptide with a large lumenal domain and a small cytoplasmic tail. *EMBO J.* **9(5)**: 1495–1502.

Griffis, E.R., Altan, N., Lippincott-Schwartz, J. and Powers, M.A. (2002) Nup98 is a mobile nucleoporin with transcription-dependent dynamics. *Mol. Biol. Cell.* **13(4)**: 1282–1297.

Griffis, E.R., Xu, S. and Powers, M.A. (2003) Nup98 localizes to both nuclear and cytoplasmic sides of the nuclear pore and binds to two distinct nucleoporin subcomplexes. *Mol. Biol. Cell.* **14(2)**: 600–610.

Grote, M., Kubitscheck, U., Reichelt, R. and Peters, R. (1995) Mapping of nucleoporins to the center of the nuclear pore complex by post-embedding immunogold electron microscopy. *J. Cell. Sci.* **108 (Pt 9)**: 2963–2972.

Guan, T., Kehlenbach, R.H., Schirmer, E.C., Kehlenbach, A., Fan, F., Clurman, B.E., Arnheim, N. and Gerace, L. (2000) Nup50, a nucleoplasmically oriented nucleoporin with a role in nuclear protein export. *Mol. Cell. Biol.* **20(15)**: 5619–5630.

Hallberg, E., Wozniak, R.W. and Blobel, G. (1993) An integral membrane protein of the pore membrane domain of the nuclear envelope contains a nucleoporin-like region. *J. Cell. Biol.* **122(3)**: 513–521.

Haltiwanger, R.S., Grove, K. and Philipsberg, G.A. (1998) Modulation of O-linked N-acetylglucosamine levels on nuclear and cytoplasmic proteins in vivo using the peptide O-GlcNAc-beta-N-acetylglucosaminidase inhibitor O-(2-acetamido-2-deoxy-D-glucopyranosylidene)amino-N-phenylcarbamate. *J. Biol. Chem.* **273(6)**: 3611–3617.

Hanover, J.A. (2001) Glycan-dependent signaling: O-linked N-acetylglucosamine. *FASEB J.* **15(11)**: 1865–1876.

Harel, A., Orjalo, A.V., Vincent, T., Lachish-Zalait, A., Vasu, S., Shah, S., Zimmerman, E., Elbaum, M. and Forbes, D.J. (2003) Removal of a single pore subcomplex results in vertebrate nuclei devoid of nuclear pores. *Mol. Cell.* **11(4)**: 853–864.

Hinshaw, J.E. and Milligan, R.A. (2003) Nuclear pore complexes exceeding eightfold rotational symmetry. *J. Struct. Biol.* **141(3)**: 259–268.

Hinshaw, J.E., Carragher, B.O. and Milligan, R.A. (1992) Architecture and design of the nuclear pore complex. *Cell* **69(7)**: 1133–1141.

Ho, A.K., Raczniak, G.A., Ives, E.B. and Wente, S.R. (1998) The integral membrane protein snl1p is genetically linked to yeast nuclear pore complex function. *Mol. Biol. Cell.* **9(2)**: 355–373.

Hochstrasser, M. (1996) Ubiquitin-dependent protein degradation. *Ann. Rev. Genet.* **30**: 405–439.

Hodel, A.E., Hodel, M.R., Griffis, E.R., Hennig, K.A., Ratner, G.A., Xu, S. and Powers, M.A. (2002) The three-dimensional structure of the autoproteolytic, nuclear pore-targeting domain of the human nucleoporin Nup98. *Mol. Cell.* **10(2)**: 347–358.

Holaska, J.M., Wilson, K.L. and Mansharamani, M. (2002) The nuclear envelope, lamins and nuclear assembly. *Curr. Opin. Cell. Biol.* **14(3)**: 357–364.

Holt, G.D., Snow, C.M., Senior, A., Haltiwanger, R.S., Gerace, L. and Hart, G.W. (1987) Nuclear pore complex glycoproteins contain cytoplasmically disposed O-linked N-acetylglucosamine. *J. Cell. Biol.* **104(5)**: 1157–1164.

Hood, J.K. and Silver, P.A. (2000) Diverse nuclear transport pathways regulate cell proliferation and oncogenesis. *Biochim. Biophys. Acta* **1471**: M31–41.

Hu, T., Guan, T. and Gerace, L. (1996) Molecular and functional characterization of the p62 complex, an assembly of nuclear pore complex glycoproteins. *J. Cell. Biol.* **134(3)**: 589–601.

Huebner, A., Kaindl, A.M., Braun, R. and Handschug, K. (2002) New insights into the molecular basis of the triple A syndrome. *Endocr. Res.* **28(4):** 733–739.

Hughes, M., Zhang, C., Avis, J.M., Hutchison, C.J. and Clarke, P.R. (1998) The role of the ran GTPase in nuclear assembly and DNA replication: characterisation of the effects of Ran mutants. *J. Cell. Sci.* **111(Pt 20):** 3017–3026.

Iouk, T., Kerscher, O., Scott, R.J., Basrai, M.A. and Wozniak, R.W. (2002) The yeast nuclear pore complex functionally interacts with components of the spindle assembly checkpoint. *J. Cell. Biol.* **159(5):** 807–819.

Jarnik, M. and Aebi, U. (1991) Toward a more complete 3D structure of the nuclear pore complex. *J. Struct. Biol.* **107(3):** 291–308.

John, P.C., Mews, M. and Moore, R. (2001) Cyclin/Cdk complexes: their involvement in cell cycle progression and mitotic division. *Protoplasma* **216(3–4):** 119–142.

Joseph, J., Tan, S.H., Karpova, T.S., McNally, J.G. and Dasso, M. (2002) SUMO-1 targets RanGAP1 to kinetochores and mitotic spindles. *J. Cell. Biol.* **156(4):** 595–602.

Hussey, D.J. and Dubrovic, A. (2002) Recurrent coiled-coil motifs in NUP98 fusion partners provide a clue to leukemogenesis. *Blood* **99:** 1097–1098.

Kalab, P., Weis, K. and Heald, R. (2002) Visualization of a Ran-GTP gradient in interphase and mitotic *Xenopus* egg extracts. *Science* **295(5564):** 2452–2456.

Kasper, L.H., Brindle, P.K., Schnabel, C.A., Pritchard, C.E.J., Cleary, M.L. and Van Deursen, J.M.A. (1999) CREB binding protein interacts with nucleoporin-specific FG repeats that activate transcription and mediate NUP98-HOXA9 oncogenicity. *Mol. Cell. Biol.* **19(1):** 764–776.

Kessel, R.G. (1992) Annulate lamellae: a last frontier in cellular organelles. *Int. Rev. Cytol.* **133:** 43–120.

Kiseleva, E., Rutherford, S., Cotter, L.M., Allen, T.D. and Goldberg, M.W. (2001) Steps of nuclear pore complex disassembly and reassembly during mitosis in early Drosophila embryos. *J. Cell. Sci.* **114(Pt 20):** 3607–3618.

Kosova, B., Pante, N., Rollenhagen, C. and Hurt, E. (1999) Nup192p is a conserved nucleoporin with a preferential location at the inner site of the nuclear membrane. *J. Biol. Chem.* **274(32):** 22646–22651.

Kraemer, D., Wozniak, R.W., Blobel, G. and Radu, A. (1994) The human CAN protein, a putative oncogene product associated with myeloid leukemogenesis, is a nuclear pore complex protein that faces the cytoplasm. *Proc. Natl Acad. Sci. USA* **91(4):** 1519–1523.

Kroon, E., Thorsteinsdottir, U., Mayotte, N., Nakamura, T. and Sauvageau, G. (2001) NUP98-HOXA9 expression in hemopoietic stem cells induceschronic and acute myeloid leukemias in mice. *EMBO J.* **20:** 350–361.

Li, D. and Roberts, R. (2001) WD-repeat proteins: structure characteristics, biological function, and their involvement in human diseases. *Cell. Mol. Life Sci.* **58(14):** 2085–2097.

Li, M., Makkinje, A. and Damuni, Z. (1996) The myeloid leukemia-associated protein SET is a potent inhibitor of protein phosphatase 2A. *J. Biol. Chem.* **271:** 11059–11062.

Lutzmann, M., Kunze, R., Buerer, A., Aebi, U. and Hurt, E. (2002) Modular self-assembly of a Y-shaped multiprotein complex from seven nucleoporins.
EMBO J. **21(3):** 387–397.

Macaulay, C. and Forbes, D.J. (1996) Assembly of the nuclear pore: biochemically distinct steps revealed with NEM, GTP gamma S, and BAPTA. *J. Cell. Biol.* **132(1–2):** 5–20.

Macaulay, C., Meier, E. and Forbes, D.J. (1995) Differential mitotic phosphorylation of proteins of the nuclear pore complex. *J. Biol. Chem.* **270(1):** 254–262.

Mahajan, R., Delphin, C., Guan, T., Gerace, L. and Melchior, F. (1997) A small ubiquitin-related polypeptide involved in targeting RanGAP1 to nuclear pore complex protein RanBP2. *Cell* **88(1):** 97–107.

Martinez, N., Alonso, A., Moragues, M.D., Poton, J. and Schneider, J. (1999) The nuclear pore complex protein Nup88 is overexpressed in tumour cells. *Cancer Res.* **59:** 5408–5411.

Matunis, M.J., Coutavas, E. and Blobel, G. (1996) A novel ubiquitin-like modification modulates the partitioning of the Ran-GTPase-activating protein RanGAP1 between the cytosol and the nuclear pore complex. *J. Cell. Biol.* **135(6 Pt 1):** 1457–1470.

Matunis, M.J., Wu, J. and Blobel, G. (1998) SUMO-1 modification and its role in targeting the Ran GTPase-activating protein, RanGAP1, to the nuclear pore complex. *J. Cell. Biol.* **140(3):** 499–509.

Maul, G.G. (1977) Nuclear pore complexes. Elimination and reconstruction during mitosis. *J. Cell. Biol.* **74(2):** 492–500.

Miller, B.R., Powers, M., Park, M., Fischer, W. and Forbes, D.J. (2000) Identification of a new vertebrate nucleoporin, Nup188, with the use of a novel organelle trap assay. *Mol. Biol. Cell.* **11(10):** 3381–3396.

Miller, M.W. and Hanover, J.A. (1994) Functional nuclear pores reconstituted with beta 1-4 galactose-modified O-linked N-acetylglucosamine glycoproteins. *J. Biol. Chem.* **269(12):** 9289–9297.

Miller, M.W., Caracciolo, M.R., Berlin, W.K. and Hanover, J.A. (1999) Phosphorylation and glycosylation of nucleoporins. *Arch. Biochem. Biophys.* **367(1):** 51–60.

Nakaruma, T., Largaespada, D.A., Lee, M.P., *et al.* (1996) Fusion of the nucleoporin gene Nup98 to HOXA9 by the chromosome translocation t(7:11)(p15:p15) in human myeloid leukemia. *Nature Genet.* **12:** 154–158.

Nakamura, T., Yamazaki, Y., Hatano, Y. and Miura, I. (1999) Nup98 is fused to PMX1 homeobox gene in human acyte myeloid leukemia with chromosomal translocation t(1:11)(q23:p15). *Blood* **94:** 741–747.

Nishiyama, M., Arai, T., Tsunematsu, Y., *et al.* (1999) 11p15 translocations involving the NUP98 gene in childhood therapy-related acute myeloid leukemia/myelodysplastic syndrome. *Genes Chrom. Cancer* **26:** 215–220.

Okura, T., Gong, L., Kamitani, T., Wada, T., Okura, I., Wei, C.F., Chang, H.M. and Yeh, E.T. (1996) Protection against Fas/APO-1- and tumor necrosis factor-mediated cell death by a novel protein, sentrin. *J. Immunol.* **157(10):** 4277–4281.

Paddy, M.R. (1998) The Tpr protein: linking structure and function in the nuclear interior? *Am. J. Hum. Genet.* **63(2):** 305–310.

Pante, N., Bastos, R., McMorrow, I., Burke, B. and Aebi, U. (1994) Interactions and three-dimensional localization of a group of nuclear pore complex proteins. *J. Cell. Biol.* **126(3):** 603–617.

Paschal, B.M. and Gerace, L. (1995) Identification of NTF2, a cytosolic factor for nuclear import that interacts with nuclear pore complex protein p62. *J. Cell. Biol.* **129(4):** 925–937.

Pichler, A., Gast, A., Seeler, J.S., Dejean, A. and Melchior, F. (2002) The nucleoporin RanBP2 has SUMO1 E3 ligase activity. *Cell* **108(1):** 109–120.

Powers, M.A., Forbes, D.J., Dahlberg, J.E. and Lund, E. (1997) The vertebrate GLFG nucleoporin, Nup98, is an essential component of multiple RNA export pathways. *J. Cell. Biol.* **136(2):** 241–250.

Pyhtila, B. and Rexach, M. (2003) A gradient of affinity for the karyopherin Kap95p along the yeast nuclear pore complex. *J. Biol. Chem.* (Q2)

Radu, A., Blobel, G. and Wozniak, R.W. (1993) Nup155 is a novel nuclear pore complex protein that contains neither repetitive sequence motifs nor reacts with WGA. *J. Cell. Biol.* **121(1):** 1–9.

Radu, A., Moore, M.S. and Blobel, G. (1995) The peptide repeat domain of nucleoporin Nup98 functions as a docking site in transport across the nuclear pore complex. *Cell* **81(2):** 215–222.

Raza-Egilmez, S.Z., Jani-Sait, S.N., Gross, M., Higgins, M.J., Shows, T.B. and Aplan, P.D. (1998) Nup98-HOXD13 gene fusion in therapy-related acute myelogenous leukemia. *Cancer Res.* **58:** 4269–4273.

Ribbeck, K. and Gorlich, D. (2001) Kinetic analysis of translocation through nuclear pore complexes. *EMBO J.* **20(6):** 1320–1330.

Ribbeck, K. and Gorlich, D. (2002) The permeability barrier of nuclear pore complexes appears to operate via hydrophobic exclusion. *EMBO J.* **21(11)**: 2664–2671.

Ris, H. (1997) High-resolution field-emission scanning electron microscopy of nuclear pore complex. *Scanning* **19(5)**: 368–375.

Rout, M.P., Aitchison, J.D., Suprapto, A., Hjertaas, K., Zhao, Y. and Chait, B.T. (2000) The yeast nuclear pore complex: composition, architecture, and transport mechanism. *J. Cell. Biol.* **148(4)**: 635–651.

Ryan, K.J. and Wente, S.R. (2002) Isolation and characterization of new *Saccharomyces cerevisiae* mutants perturbed in nuclear pore complex assembly. *BMC Genet.* **3(1)**: 17.

Ryan, K.J., McCaffery, J.M. and Wente, S.R. (2003) The Ran GTPase cycle is required for yeast nuclear pore complex assembly. *J. Cell. Biol.* **160(7)**: 1041–1053.

Scandura, J.M., Boccuni, P., Cammenga, J. and Nimer, S.D. (2002) Transcription factor fusions in acute leukemia: variations on a theme. *Oncogene* **21**: 3422–3444.

Shen, Z., Pardington-Purtymun, P.E., Comeaux, J.C., Moyzis, R.K. and Chen, D.J. (1996) UBL1, a human ubiquitin-like protein associating with human RAD51/RAD52 proteins. *Genomics* **36(2)**: 271–279.

Shi, Y. and Thomas, J.O. (1992) The transport of proteins into the nucleus requires the 70-kilodalton heat shock protein or its cytosolic cognate. *Mol. Cell. Biol.* **12(5)**: 2186–2192.

Shulga, N. and Goldfarb, D.S. (2003) Binding dynamics of structural nucleoporins govern nuclear pore complex permeability and may mediate channel gating. *Mol. Cell. Biol.* **23(2)**: 534–542.

Shulga, N., Roberts, P., Gu, Z., Spitz, L., Tabb, M.M., Nomura, M. and Goldfarb, D.S. (1996) In vivo nuclear transport kinetics in *Saccharomyces cerevisiae*: a role for heat shock protein 70 during targeting and translocation. *J. Cell. Biol.* **135(2)**: 329–339.

Siniossoglou, S., Lutzmann, M., Santos-Rosa, H., Leonard, K., Mueller, S., Aebi, U. and Hurt, E. (2000) Structure and assembly of the Nup84p complex. *J. Cell. Biol.* **149(1)**: 41–54.

Smith, A.E., Slepchenko, B.M., Schaff, J.C., Loew, L.M. and Macara, I.G. (2002) Systems analysis of Ran transport. *Science* **295(5554)**: 488–491.

Smith, T.F., Gaitatzes, C., Saxena, K. and Neer, E.J. (1999) The WD repeat: a common architecture for diverse functions. *Trends Biochem. Sci.* **24(5)**: 181–185.

Soderqvist, H. and Hallberg, E. (1994) The large C-terminal region of the integral pore membrane protein, POM121, is facing the nuclear pore complex. *Eur. J. Cell. Biol.* **64(1)**: 186–191.

Sondek, J., Bohm, A., Lambright, D.G., Hamm, H.E. and Sigler, P.B. (1996) Crystal structure of a G-protein beta gamma dimer at 2.1A resolution. *Nature* **379(6563)**: 369–374.

Sondermann, H., Ho, A.K., Listenberger, L.L., Siegers, K., Moarefi, I., Wente, S.R., Hartl, F.U. and Young, J.C. (2002) Prediction of novel Bag-1 homologs based on structure/function analysis identifies Snl1p as an Hsp70 co-chaperone in *Saccharomyces cerevisiae*. *J. Biol. Chem.* **277(36)**: 33220–33227.

Stafstrom, J.P. and Staehelin, L.A. (1984) Dynamics of the nuclear envelope and of nuclear pore complexes during mitosis in the Drosophila embryo. *Eur. J. Cell. Biol.* **34(1)**: 179–189.

Steggerda, S.M. and Paschal, B.M. (2002) Regulation of nuclear import and export by the GTPase Ran. *Int. Rev. Cytol.* **217**: 41–91.

Stoffler, D., Feja, B., Fahrenkrog, B., Walz, J., Typke, D. and Aebi, U. (2003) Cryo-electron tomography provides novel insights into nuclear pore architecture: implications for nucleocytoplasmic transport. *J. Mol. Biol.* **328(1)**: 119–130.

Strasser, K., Bassler, J. and Hurt, E. (2000) Binding of the Mex67p/Mtr2p heterodimer to FXFG, GLFG, and FG repeat nucleoporins is essential for nuclear mRNA export. *J. Cell. Biol.* **150(4)**: 695–706.

Strom, A.C. and Weis, K. (2001) Importin-beta-like nuclear transport receptors. *Genome Biol.* **2(6)**: REVIEWS3008.Epub.

Sukegawa, J. and Blobel, G. (1993) A nuclear pore complex protein that contains zinc finger motifs, binds DNA, and faces the nucleoplasm. *Cell* **72**(1): 29–38.

Teixeira, M.T., Siniossoglou, S., Podtelejnikov, S., *et al.* (1997) Two functionally distinct domains generated by in vivo cleavage of Nup145p: a novel biogenesis pathway for nucleoporins. *EMBO J.* **16**(16): 5086–5097.

Tullio-Pelet, A., Salomon, R., Hadj-Rabia, S., *et al.* (2000) Mutant WD-repeat protein in triple-A syndrome. *Nat. Genet.* **26**(3): 332–335.

Unwin, P.N. and Milligan, R.A. (1982) A large particle associated with the perimeter of the nuclear pore complex. *J. Cell. Biol.* **93**(1): 63–75.

Vasu, S.K. and Forbes, D.J. (2001) Nuclear pores and nuclear assembly. *Curr. Opin. Cell. Biol.* **13**(3): 363–375.

Vasu, S., Shah, S., Orjalo, A., Park, M., Fischer, W.H. and Forbes, D.J. (2001) Novel vertebrate nucleoporins Nup133 and Nup160 play a role in mRNA export. *J. Cell. Biol.* **155**(3): 339–354.

Voegtli, W.C., Madrona, A.Y. and Wilson, D.K. (2003) The structure of Aip1p, a WD repeat protein that regulates cofilin-mediated actin depolymerization. *J. Biol. Chem.* **278**(36): 34373–34379.

Von Lindern, M., Fornerod, M., Soekarman, N., Van Baal, S., Jaegle, A., Hagemeijer, A., Bootsma, D. and Grosveld, G. (1992a) Translocation t(6:9) in acute nonlymphocytic leukemia results in the formation of a DEK-CAN fusion gene. *Baillieres Clin. Haematol.* **5**: 857–879.

Von Lindern, M., Van Baal, S., Wiegent, J., Raap, A., Hagemeijer, A. and Grosveld, G. (1992b) Can, a putative oncogene associated with myeloid leukemogenesis, may be activated by fusion of its 3' half to different genes: characterisation of the set gene. *Mol. Cell. Biol.* **12**: 3346–3355.

Walther, T.C., Pickersgill, H.S., Cordes, V.C., Goldberg, M.W., Allen, T.D., Mattaj, I.W. and Fornerod, M. (2002) The cytoplasmic filaments of the nuclear pore complex are dispensable for selective nuclear protein import. *J. Cell. Biol.* **158**(1): 63–77.

Walther, T.C., Alves, A., Pickersgill, H., *et al.* (2003a) The conserved Nup107-160 complex is critical for nuclear pore complex assembly. *Cell* **113**(2): 195–206.

Walther, T.C., Askjaer, P., Gentzel, M., Habermann, A., Griffiths, G., Wilm, M., Mattaj, I.W. and Hetzer, M. (2003b) RanGTP mediates nuclear pore complex assembly. *Nature* **424**(6949): 689–694.

Wente, S.R. and Blobel, G. (1993) A temperature-sensitive NUP116 null mutant forms a nuclear envelope seal over the yeast nuclear pore complex thereby blocking nucleocytoplasmic traffic. *J. Cell. Biol.* **123**(2): 275–284.

Winey, M., Yarar, D., Giddings, T.H., Jr. and Mastronarde, D.N. (1997) Nuclear pore complex number and distribution throughout the *Saccharomyces cerevisiae* cell cycle by three-dimensional reconstruction from electron micrographs of nuclear envelopes. *Mol. Biol. Cell.* **8**(11): 2119–2132.

Wozniak, R.W. and Blobel, G. (1992) The single transmembrane segment of gp210 is sufficient for sorting to the pore membrane domain of the nuclear envelope. *J. Cell. Biol.* **119**(6): 1441–1449.

Wozniak, R.W., Blobel, G. and Rout, M.P. (1994) POM152 is an integral protein of the pore membrane domain of the yeast nuclear envelope. *J. Cell. Biol.* **125**(1): 31–42.

Wu, J., Matunis, M.J., Kraemer, D., Blobel, G. and Coutavas, E. (1995) Nup358, a cytoplasmically exposed nucleoporin with peptide repeats, Ran-GTP binding sites, zinc fingers, a cyclophilin A homologous domain, and a leucine-rich region. *J. Biol. Chem.* **270**(23): 14209–14203.

Yang, L., Guan, T. and Gerace, L. (1997) Integral membrane proteins of the nuclear envelope are dispersed throughout the endoplasmic reticulum during mitosis. *J. Cell. Biol.* **137**(6): 1199–1210.

Yang, Q., Rout, M.P. and Akey, C.W. (1998) Three-dimensional architecture of the isolated yeast nuclear pore complex: functional and evolutionary implications. *Mol. Cell.* **1**(2): 223–234.

Zhang, C. and Clarke, P.R. (2000) Chromatin-independent nuclear envelope assembly induced by Ran GTPase in *Xenopus* egg extracts. *Science* **288**(5470): 1429–1432.

Zhang, C., Hughes, M. and Clarke, P.R. (1999) Ran-GTP stabilises microtubule asters and inhibits nuclear assembly in Xenopus egg extracts. *J. Cell. Sci.* 112(Pt 14): 2453–2461.

Zimowska, G., Aris, J.P. and Paddy, M.R. (1997) A Drosophila Tpr protein homolog is localized both in the extrachromosomal channel network and to nuclear pore complexes. *J. Cell. Sci.* 110(Pt 8): 927–944.

7

Import and export at the nuclear envelope

Martin Goldberg

1. Introduction

For a eukaryotic cell to function, its genome has to communicate with its cytoplasm. The genome however is surrounded by a double membrane which is impermeable to the necessary molecules, which either regulate nuclear functions or are a result of nuclear activities. Assembly of a nuclear pore complex (NPC) causes the fusion of the inner and outer membranes of the nuclear envelope (NE) resulting in a route of passage across the membranes. This 'hole' is however rapidly plugged with a beautiful, highly symmetrical protein complex (*Figure 1*), consisting of about 30 different proteins in multiple copies (Cronshaw *et al.*, 2002; Rout *et al.*, 2000). The number of proteins is in fact rather smaller than expected considering the gargantuan size of the NPC which is 90–120 nm in diameter (Akey and Radermacher, 1993; Goldberg and Allen, 1996; Hinshaw *et al.*, 1992, Kiseleva *et al.*, 1998; Yang *et al.*, 1998) and over 100 000 kDa in vertebrates (Reichelt *et al.*, 1990). This relative compositional simplicity can be explained by the NPC's symmetrical character as it appears to consist of eight identical radially arranged subunits. There is also symmetry within each subunit, something that is emphasized by immuno-gold electron microscopy, where the majority of nucleoporins (nuclear pore proteins) appear to be located on both nucleoplasmic and cytoplasmic sides of the NPC (Rout *et al.*, 2000).

Nuclear transport, on the other hand, is highly asymmetric: it is necessarily a one-way process, either into or out of the nucleus. Although many molecules, such as transport factors, shuttle back and forth, the processes of import and export are distinct. Some asymmetry is afforded by the peripheral components of the NPC, such as the cytoplasmic filaments and nucleoplasmic baskets, but it is generally agreed that the nucleotide state of the small GTPase, Ran, marks one compartment as the nucleus (RanGTP) and the other as the cytoplasm (RanGDP), and it is this biochemical asymmetry that is used to determine the direction of transport (Nachury and Weis, 1999) (*Figure 2*). In other words, import is driven towards a high RanGTP concentration, whereas export is driven away from it. This is achieved because, in general, import receptors bind to their cargo in the absence of RanGTP (i.e. in the cytoplasm) and are

The Nuclear Envelope, edited by D.E. Evans, C. Hutchison & J.A. Bryant.
© 2004 Garland Science/BIOS Scientific Publishers

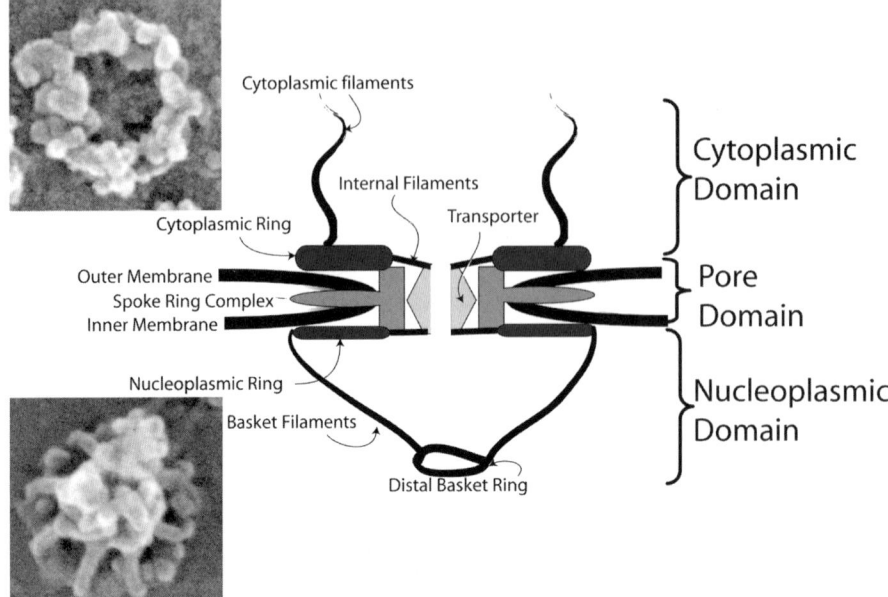

Figure 1. *Structural components of the NPC which consists of a region within the 'pore' containing the spoke ring complex and the transporter, a cytoplasmic domain with the cytoplasmic ring and filaments and the nucleoplasmic domain comtaining the nucleoplasmic ring and basket. Images are FESEM micrographs of the cytoplasmic and nucleoplasmic faces of the NPC.*

dissociated in the presence of RanGTP (i.e. in the nucleus). Export receptors, on the other hand, bind their cargo in the presence of RanGTP (in the nucleus) and dissociate when Ran hydrolyses its GTP to become RanGDP (in the cytoplasm). To establish the so-called 'Ran gradient', the Ran GTPase activating protein (RanGAP) is located in the cytoplasm, whereas the nucleotide exchange factor (RanGEF) is in the nucleus (for some recent reviews see: Azuma and Dasso, 2000; Chook and Blobel, 2001; Görlich and Kutay, 1999; Quimby and Dasso, 2003; Steggerda and Paschal, 2002).

Although we have a comprehensive understanding of the soluble factors involved in numerous transport pathways (see Weis, 2003 for review), our understanding of the unidirectional movement of a transport complex through the NPC itself is much less certain. It is known that transport factors interact with particular regions of certain nucleoporins, but how this drives movement in the correct direction is open to some speculation and has spawned several related models (Ben Efraim and Gerace, 2001; Ribbeck and Görlich, 2002; Rout et al., 2000; Srambio-de-Caatillia and Rout, 2002). Electron microscopy data, however, suggest a role for structural alterations of the NPC in the transport process (Akey, 1990, 1995; Goldberg et al., 2000; Panté and Aebi, 1996; Rutherford et al., 1997). However, calculations on the speed and capacity of a NPC during transport, which are extremely high (Ribbeck and Görlich, 2001; Smith et al., 2002), and the findings that energy utilization is not required for a single round of import (Schwoebel et al., 1998), casts doubt on the possibility that the NPC is a biomechanical pump. It is possible however, that aspects of certain transport pathways could be controlled by regulating NPC conformation.

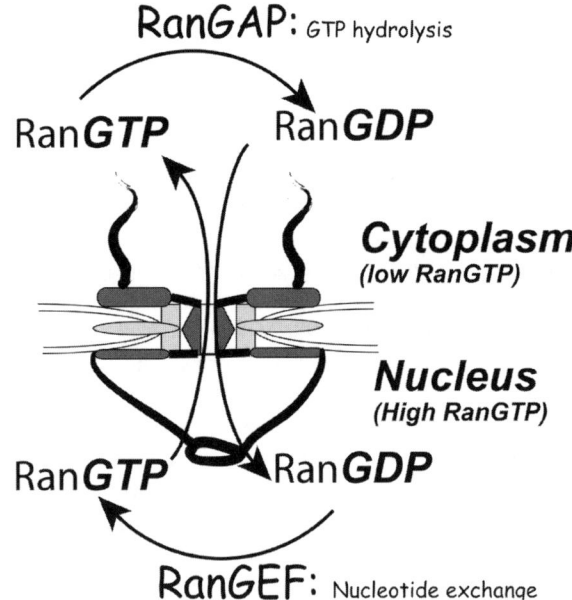

Figure 2. *The Ran cycle. GTP hydrolysis and nucleotide exchange appear to be separated from each other by the nuclear envelope during interphase, creating a high concentration of RanGTP in the nucleus.*

2. The soluble phase of transport

Colour *Plate 3* illustrates a widely accepted model for the import of a protein containing a classical nuclear localization sequence (cNLS). This is just one particular biochemical pathway across the NE, amongst several. However part of this pathway is the recycling of transport factors back to the cytoplasm, which illustrates the process of export as well. This model essentially represents the 'soluble phase' of transport (Mattaj and Englmeier, 1998) and it will used here to establish the principles of this phase. Thereafter, however, we shall focus on the more controversial role that the NPC plays in transport. As the NPC appears to be a spatially fixed component, this is known as the 'stationary phase' (Ohno *et al.*, 1998).

2.1 *Import of cargo*

For direct import of a protein, it must contain an NLS (colour *Plate 3*). The primary structures of NLSs are rather diverse and ill defined, but in the case of the cNLS (Dingwall *et al.*, 1982; Kalderon *et al.*, 1984; Lanford and Butel, 1984; Robbins *et al.*, 1991) these are single short basic sequences, such as that of the SV40 T-antigen (PKKKRKV), or similar bipartite sequences separated by a spacer. The cNLS binds to an import factor called importin α (also known as karyopherin α). Importin α in turn binds to importin β, which can bind to nucleoporins. Importin α therefore is the cNLS receptor, whereas importin β is the nucleoporin receptor and they work co-operatively to carry the cNLS-containing cargo to the NPC. Other import pathways, using non-classical NLSs, have the role of NLS receptor and NPC receptor rolled into one of a large family of importin β-like proteins (see Chook and Blobel, 2001; Görlich and Kutay, 1999, for reviews).

The importin α protein has two cNLS-binding sites, although in the free protein one of these is occupied by a segment of the flexible N-terminal domain of itself (Kobe, 1999). The flexible N-terminal domain is also the region that interacts with importin β and is known as the importin β binding (IBB) domain (Görlich *et al.*, 1996a; Weis *et al.*, 1996). When importin α binds importin β this region moves from the cNLS binding site in order to bind importin β. Therefore when importin β binds importin α, the cNLS binding site is revealed and importin α becomes an active cNLS receptor (Cingolani *et al.*, 1999; Conti *et al.*, 1998; Moroianu *et al.*, 1996). The IBB domain of importin α is highly basic and interacts with the acidic inner surface of importin β (Cingolani *et al.*, 1999).

The trimeric import complex then finds its way (probably by diffusion) to the NPC, where the nucleoporin-binding regions of importin β bind to certain nucleoporins. It has been shown that importin β binds specifically to regions of nucleoporins containing FG repeats (Bayliss *et al.*, 2002; Rexach and Blobel, 1995). Hydrophobic patches on the outer surface of importin β at the N-terminal portion are thought to bind to these hydrophobic repeat regions (Bayliss *et al.*, 2000), although C-terminal regions have also been shown to be important (Bednenko *et al.*, 2003). Approximately a third of all nucleoporins contain FG repeats (Cronshaw *et al.*, 2002; Rout *et al.*, 2000), so there are predicted to be large numbers of potential binding sites for transport complexes within the NPC. Most current models for the translocation step of nuclear transport assign a central role to the interaction of β importins with the FG nucleoporins.

The final step of this phase of the process is the dissociation of all these proteins from each other. Dissociation is driven by the multi-talented small GTPase, Ran. Ran was predicted and recently shown (Kalab *et al.*, 2002) to be predominantly in the active GTP-bound form in the nucleus, due, presumably, to the exclusively nuclear location of the RanGEF (Ohtsubo *et al.*, 1989). A high concentration of RanGTP in the nucleus, means that all in-coming import complexes rapidly come into contact with RanGTP. RanGTP binds to importin β (RanGDP does not) at its acidic inner surface (Chook and Blobel, 1999; Vetter *et al.*, 1999). This is the same region that binds to the IBB domain of importin α. Therefore there is competition between importin α and RanGTP for binding to this region, which partially explains how RanGTP causes substrate release. However, other members of the importin β family do not bind their substrates in this region (Chook and Blobel, 1999) despite having a similar structure, suggesting a role for conformational alteration of importin β in the substrate release mechanism (Chook and Blobel, 1999). Structural studies (Bayliss *et al.*, 2000) also suggest that conformational changes in importin β upon binding to RanGTP distorts the FG-repeat binding regions on its outer surface causing release from nucleoporins and hence the NPC.

2.2 *Export of importin β*

Importin β is now in a non-functional state for another round of import: not only is it in the wrong compartment, but it also has RanGTP bound to it, which would prevent the binding of cargo (importin α; colour *Plate 3b*). Exactly how importin β gets to the cytoplasm and looses RanGTP is uncertain. As RanGTP is required in other export pathways involving other members of the importin β superfamily, it would make sense for Ran GTP to be involved in importin β export (as shown in colour *Plate 3b*). Indeed there is evidence that the importin β–RanGTP complex does move intact from the nucleus to the cytoplasm (Hieda *et al.*, 1999). However it is also clear that when importin β is bound to RanGTP it does not bind to the FG repeats of nucleoporins (Bayliss *et al.*, 2000), so there would have to be some other mechanism for the complex to pass through the NPC, as it is too big to diffuse across. The other possibility is that importin β is dissociated from Ran in the nucleus and is exported independently from Ran. Evidence that an importin β mutant that cannot bind RanGTP is efficiently exported supports such a possibility (Kose *et al.*, 1999). However, depletion of GTP-bound Ran in the nucleus does significantly reduce importin β export (Izaurralde *et al.*, 1997), suggesting that there may be alternative importin β export pathways depending on whether Ran is bound or not.

In order to release Ran from importin β, Ran must hydrolyse its GTP (Görlich *et al.*, 1996b; Rexach and Blobel, 1995). This is stimulated by the Ran-GTPase-activating protein (RanGAP) (Bischoff *et al.*, 1994). However, the bound importin β inhibits this GTPase activation (Floer and Blobel, 1996). In order to overcome this inhibition RanGTP has to bind Ran-binding protein 1 (RanBP1) (Lounsbury and Macara, 1997) or similar Ran-binding domains of Nup358 (also called RanBP2). RanBP1 is a soluble protein that shuttles across the NE (Plafker and Macara, 2000) but is found largely in the cytoplasm. The cytoplasmic location is maintained by a leucine-rich nuclear export signal (NES) and export can be inhibited by leptomycin B, which inhibits CRM1-dependent export (CRM1 is an importin β-like export factor). RanBP1 therefore could bind to RanGTP (whether bound to importin β or not) anywhere in the cytoplasm, but also, potentially, in and around the NPC. RanGAP also appears to harbour an NLS and NESs, suggesting that it may also shuttle (Matunis and Blobel, 1998), making it possible that Ran could hydrolyse its GTP and release importin β on either side of the NPC, although there is no published evidence for this. On the other hand Nup358 (which also binds Ran and co-activates Ran GTPase activity with RanGAP) is located exclusively on the cytoplasmic filaments of the NPC. RanGAP is also targeted specifically to this location when it is covalently attached to the ubiquitin-like protein SUMO1 (Matunis *et al.*, 1998). Such an arrangement could specifically capture export complexes (including importin β–RanGTP) on their way out of the NPC and induce GTP hydrolysis and dissociation.

2.3 *Recycling of importin α*

In principle, the export of importin α appears to be very similar to the export of importin β, except that importin α cannot bind to RanGTP and apparently has no intrinsic export activity. Instead it has to first bind to a member of the importin β superfamily, an export factor called CAS (Hood and Silver, 1998; Kunzler and Hurt, 1998; Kutay *et al.*, 1997; Solsbacher *et al.*, 1998). Importin α binds strongly to CAS in the presence of RanGTP (i.e. in the nucleus), forming a trimeric complex which is exported (colour *Plate 3c*). As discussed above, this complex is then dissociated when Ran hydrolyses its GTP in association with RanGAP and RanBP1 or Nup358. This would be expected to occur on the cytoplasmic side of the NPC.

2.4 *Import and recharging of Ran (colour Plate 3d)*

The above process would result in an accumulation of RanGDP in the cytoplasm (after GTP hydrolysis and dissociation of export complexes). Ran is in fact small enough to diffuse back into the nucleus, but because of the extremely high demand for active Ran in the nucleus (at least ~500 transport events per NPC per second (Ribbeck and Görlich, 2001; Smith *et al.*, 2002)), this is unlikely to be efficient enough. Instead, RanGDP is actively imported by its own import factor NTF2 (Quimby *et al.*, 2000; Ribbeck *et al.*, 1998). NTF2 only binds to Ran in its GDP-bound state (Ribbeck *et al.*, 1998). One of the major structural alterations of Ran when it goes from the GTP state to the GDP state is a shift in the position of two loop regions called Switch I and Switch II. These loops are on the outer surface of the protein and they are structurally adjacent. Both are involved in the binding to NTF2, explaining how it can discriminate between RanGTP and RanGDP (Stewart *et al.*, 1998). Like importin β, NTF2 has a hydrophobic patch which interacts with FG-repeat regions of nucleoporins (Bayliss *et al.*, 2000; Quimby *et al.*, 2001) and presumably facilitates the passage of RanGDP-NTF2 through the NPC.

Once in the nucleus Ran has to be released from NTF2 and reactivated by exchanging the GDP for GTP. This is achieved by the Ran guanine nucleotide exchange factor (RanGEF), which was originally named RCC1 (Bischoff and Ponstingl., 1991; Ohtsubo *et al.*, 1989). RanGEF has an exclusively nuclear local-ization, which appears to be achieved by binding to the core regions of the histones H2A and H2B (Nemergut *et al.*, 2001). Binding of Ran to RanGEF appears to affect the nucleotide-binding site on Ran and cause the release of both GDP and GTP (Renault *et al.*, 2001), increasing the exchange rate by five orders of magnitude (Klebe *et al.*, 1995). Ran is then released from RanGEF and is then free to rebind to nucleotide, which because it is at a higher concentration is usually GTP (rather then GDP). It has been shown that when Ran binds to RanGEF, the binding of RanGEF to chromatin is greatly enhanced (Li *et al.*, 2003). The nucleotide exchange activity of RanGEF is also enhanced on binding to chromatin (Nemergut *et al.*, 2001). RanGEF can therefore probably pick up RanGDP almost anywhere, such as in the NPC (Iborra *et al.*, 2000), before binding to chromatin and generating RanGTP. This guarantees that RanGTP is only generated at a high concentration in the region of the chromatin, even when the NE is absent such as at mitosis.

3.   Role of NPC in transport

Studies using electron microscopy have found that various aspects of NPC structure can be found in differing conformations and that these can possibly be related to transport activities.

3.1 *Translocation and the transporter*

The central transporter is a controversial cylindrical component which traverses the NPC within the central channel (Akey and Radermacher, 1993). It is controversial because, sometimes, using some techniques it is not always observed and it seems to have a variable appearance. This can be explained by suggesting that the 'transporter' is in fact simply material caught in transit (Stoffler *et al.*, 2003) and not really part of the structure. On the other hand other studies have found that central structures are found in all NPCs (Goldberg and Allen, 1996) and that these can be classified into a limited number of conformations (Akey, 1990). Using cryoEM, image processing and classification analysis, Akey (1990) found that the central structure could be found in basically two conformations: a 'closed' conformation with a 9-nm central aperture (see transporter in *Figure 3a*) or an 'open' conformation with a 21-nm aperture (*Figure 3b*). Both had a defined structure with the eight-fold symmetry, typical of NPC components and unlikely to represent highly variable in-transit material. Consistent with this, earlier EM transport assays had shown that the upper limit for the size of a transport substrate was 19–26 nm depending on the cell (Feldherr and Akin, 1990; Feldherr *et al.*, 1984).

Field effect scanning electron microscopy (FESEM) also showed central structures that were either 'compact' or 'expanded' (Goldberg and Allen, 1996). In the compact conformation (illustrated in *Figure 3a*) putative channels were observed between the transporter and the inner spoke ring, which were about 10 nm in diameter and consistent with the upper limit for the diffusion of non-transported molecules across the NE. In the expanded conformation however, these channels were 'filled' (*Figure 3b*), suggesting the possibility that when the central transport channel is open this closes the diffusion channels. However, statistical analysis of the route of diffusion through the NPC (Feldherr and Akin, 1997) in TEM thin sections showed that

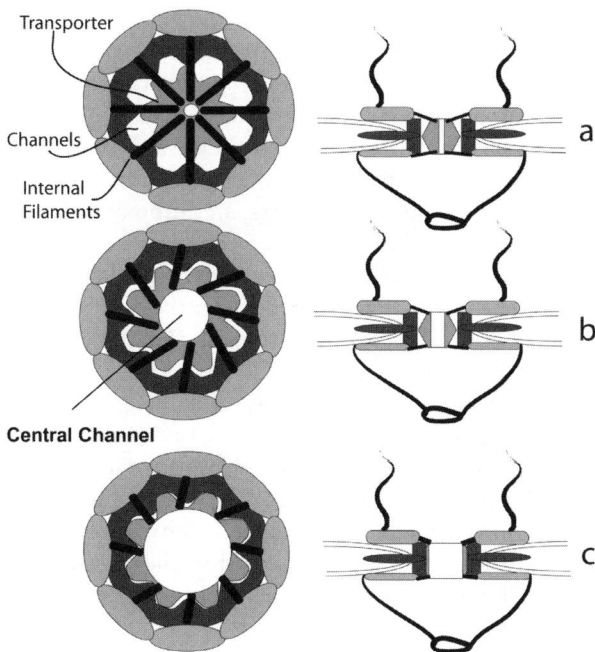

Figure 3. Model for transporter dynamics. The transporter can be either 'closed' (a) or 'dilated' (b) (Akey, 1990). Putative diffusion channels are between the transporter and the inner spoke ring (Akey and Radermacher, 1993) (a), but may be filled when the transporter 'opens' (Goldberg and Allen, 1996). Internal filaments could drive opening and closing. (c) hypothetical transporter configuration to accommodate large substrates.

diffusion of small particles occurred only partially at the periphery of the transporter. However, much of the diffusion was down the central axis, presumably through the transporter, which was exclusively the route for larger particles. Active transport (import and export), on the other hand, appears to be only centrally located (Akey, 1989; 1990; Feldherr and Akin, 1997; Panté et al., 1997; Rutherford et al., 1997; Stewart et al., 1990), suggesting that it is more restricted in its path.

In contrast, it was also shown (Panté and Kann, 2002) that particles as large as 39 nm could be transported through the NPC. As the central channel in which the transporter resides is not much bigger than this, the transporter, as we know it, could not be present in the NPC at the same time. On the other hand even 5 nm gold particles coated with nucleoplasmin pass through the NPC in single file at the central axis (Rutherford et al., 1997), suggesting that they are moving through a narrow channel.

It is possible that the transporter might not be a *bona fide* structural component of the NPC. In this case there must be some other means to restrict the size of molecules able to diffuse across the NPC and selectively facilitate the passage of larger molecules. In the Selective Phase Model (Ribbeck and Görlich, 2001) it is suggested that the central channel, instead of containing a defined structure (the transporter), is 'filled' with a meshwork of the FG-repeat-containing regions of several nucleoporins located near the channel. Because these are hydrophobic and relatively unstructured (Bayliss et al., 2000), they could interact with each other to form a loose mesh, presumably with gaps large enough to allow diffusion of smaller macromolecules. However, larger molecules could only enter the mesh if they contained hydrophobic regions able to interact with FG-repeat regions of nucleoporins, allowing them to effectively break into the mesh by competing for interactions with the FG-repeat regions. Such molecules would include, for instance, importin-β-containing complexes. It could be that transport occurs preferentially down the central axis because the meshwork would be loosest there. However, this model does not explain the apparently defined structures seen in some EM studies. It also assumes that the mesh simply fills the ~40 nm wide central channel so that the mechanism for the translocation of large and smaller transported molecules is the same. Although molecules as wide as this channel can be translocated (Panté and Kann, 2002), there is evidence that movement of smaller substrates is energy independent, as accounted for in the Selective Phase Model, but translocation of large substrates, such as this, requires the hydrolysis of GTP (Lyman et al., 2002), which is not accounted for.

The Brownian Affinity Gate Model (Rout et al., 2000) is in principle similar in that it proposes that FG-repeat regions, protruding as fine disordered filaments from either side of the NPC, are used to capture transport substrates and repel non-interacting molecules. Diffusion through the NPC is then facilitated by FG regions in the NPC core. Directionality is achieved by high affinity binding sites for transport complexes on the destination side of the NPC, whereat the transport complex is dissociated by the binding of RanGTP (import) or hydrolysis of GTP by Ran (export). In this model the transporter is still a component of the structure but it is not the selective mechanical gate that was originally proposed (Akey, 1990), but rather a channel of restricted size that acts as a diffusional barrier to molecules that are not concentrated at the NPC by binding to FG nucleoporins.

If the transporter is part of the structure it must be explained how particles almost as large as the central channel in which it resides can pass through the NPC. There are two possible explanations: firstly that the transporter is highly deformable and can accommodate large objects by massive conformational changes (*Figure 3c*), or secondly that in some circumstances the transporter can temporarily move out of the NPC, effectively unplugging it. There is some evidence from *Chironomus thummi* salivary glands that the transporter distorts considerably during the passage of the giant mRNP complex, the Balbiani ring granule (Kiseleva *et al.*, 1998). However, this particle unwinds into a ~25-nm wide rod before feeding into the NPC (Kiseleva *et al.*, 1996; Mehlin *et al.*, 1995), so whether the transporter can distort further to accommodate a 39-nm particle is uncertain. Whether the transporter does dissociate from the NPC during transport of large particles awaits any experimental evidence.

As the transporter appears to be dynamic, there must be some mechanism to drive these conformational changes. It is possible that it is simply very flexible and passively accommodates transport complexes. Such flexibility could be induced by the binding of transport factors to the FG regions of its constituent nucleoporins, possibly by breaking inter-nucleoporin interactions and loosening the structure. In many ways this model is compatible with the Selective Phase Model (Ribbeck and Görlich, 2001) but allows for a defined but flexible central structure. Here the purpose of the transporter would be to prevent the entry of 'unlicensed' molecules above a certain size and facilitate the diffusion of transport complexes through the transporter central channel.

In principle, the size of the diffusion channel, as well as the maximum diameter of the transport channel, could be varied by altering which nucleoporins are present, which could affect not only the transporter's dimensions but also its deformability properties. Although this has not been shown, it is known that the size of the diffusion channel varies over the cell cycle (Feldherr and Akin, 1990), that the functional pore size for signal-mediated transport decreases when cells become quiescent (Feldherr and Akin, 1991) and that this decrease is due to changes in the NPC itself (Feldherr and Akin, 1993). More specifically, it was shown in yeast (Shulga and Goldfarb, 2003) that agents which increase the permeability of the NPCs also caused the release of several nucleoporins, such as Nup170p, which is very centrally located and likely to be part of the transporter (Rout *et al.*, 2000). Therefore the role of some nucleoporins could be to restrict the functional size of the central aperture and releasing them allows a dilation of the transporter.

It has also been found that statistically most FG nucleoporins are found equally distributed between the cytoplasmic and nucleoplasmic faces, with only a very small number being biased or exclusively located to one side or the other (Rout *et al.*, 2000). This has been attributed to the symmetrical nature of the NPC. However, a growing number of nucleoporins are being found to be 'mobile' (Griffis *et al.*, 2002; Nakielny *et al.*, 1999). This makes it difficult to distinguish, by immuno-gold EM, proteins that are genuinely symmetrically distributed, with ones that are either shuttling continuously across the NPC, or ones that are located in different positions in different NPCs. The former can be shown by other means such as photobleaching experiments (e.g. Griffis *et al.*, 2002), and the latter by classifying NPCs in EM labelling experiments. Therefore it is possible that the association and dissociation of particular mobile nucleoporins from parts of the NPC (such as the transporter) could be used to control the functional size of the NPC or to open and close it.

Another function of nucleoporin mobility could be to define and/or change the transport pathway that an individual NPC is involved in. In other words the nucleoporin composition could define an 'export NPC' or an 'import NPC', for instance. It has been argued (Ribbeck and Görlich, 2001) that because the direction of nuclear transport across the NPC is fully reversible (by reversing the RanGTP gradient – Nachury and Weis, 1999), then the directionality of transport across the NPC cannot be determined by differential affinities of transport factors to specific nucleoporins in specific positions within the NPC as described in the Affinity Gate Model (Ben-Efraim and Gerace, 2001). This model is based on data showing that there is an increasing affinity of importin β (involved in import) for nucleoporins that are located nearer to the nucleoplasmic interior. However, it is possible that by reversing the RanGTP gradient there is also a concomitant rearrangement of nucleoporins within the NPC. This would be especially feasible for mobile nucleoporins, such as Nup153 (Nakielny et al., 1999) which is thought to be the terminal binding site for import complexes prior to dissociation and has a high affinity for importin β. Little is currently known about whether different NPCs may be involved in different transport pathways. However, it has been shown (Iborra et al., 2000) that NPCs can be categorized into those associated with NTF2 (the RanGDP import factor) and those associated with molecules involved in export, such as poly A containing mRNA and SR proteins.

The dynamics of the transporter may, additionally or alternatively, be controlled in a more mechanical way. Filaments have been observed by field emission scanning electron microscopy (FESEM) that connect the cytoplasmic and nucleoplasmic faces of the transporter to the more rigid coaxial rings (Goldberg and Allen, 1996). These 'internal filaments' have been found to vary in length in different NPCs (Goldberg et al., 1997), raising the possibility that they could alter their conformation (e.g. contract) to pull open the transporter aperture. These various dynamics are illustrated in the hypothetical model shown in *Figure 3*.

3.2 The cytoplasmic filaments

The initial encounter of import complexes with the NPC, and the terminal binding site for export complexes, is with the cytoplasmic filaments. In their simplest form these appear to be rod-shaped particles, attached to the cytoplasmic ring and extending into the cytoplasm (Goldberg and Allen, 1996; Ris, 1991). A major constituent of the filament, in vertebrates, is the large nucleoporin Nup358 which may in fact be the filament itself (Delphin et al., 1997; Walther and Pickersgill et al., 2002; Wu et al., 1995). This is a complex multifunctional protein, which is dispensable in certain transport assays (Walther and Pickersgill et al., 2002). Nup358 contains FG repeats (Wu et al., 1995; Yokoyama et al., 1995) and so binds transport factors. It also contains Ran-binding domains which bind RanGTP and functionally co-activate Ran's GTPase activity with RanGAP. RanGAP is also specifically targeted to Nup358 when it is covalently modified by a ubiquitin-like protein called SUMO1 (Matunis et al., 1998). Nup358 itself may be involved in SUMO1 ligation (Pichler et al., 2002). The Ran binding domains also bind RanGDP as well as RanGTP, but there is also a zinc finger domain that just binds RanGDP in a zinc-dependent manner (Yaseen and Blobel, 1999a). Several other factors involved in transport also bind Nup358. For instance, exportin-t (involved in tRNA export) binds in a RanGTP-dependent manner (Kuersten et al., 2002), whereas exportin 1 (also called CRM1), which is involved in exporting NES containing proteins, binds to the zinc finger domain and is insensitive to the nucleotide state of Ran (Singh et al., 1999).

Because Nup358 is dispensable, at least in some import pathways (Walther *et al.*, 2002), it is likely to have a regulatory role or a role in increasing efficiency. It was shown that Nup358 may be important for coupling the export/import cycles for importin β (Yaseen and Blobel, 1999b) because when RanGTP/importin β bind to Nup358, hydrolysis of GTP by Ran, stimulated by RanGAP, only occurs when importin α binds. Therefore, subsequent to export, importin β is only released from the cytoplasmic filaments for the next round of import when importin α interacts with it. Nup358 also has a cyclophilin domain which is involved in processing red/green opsin in the retina (Ferreira *et al.*, 1996, 1997) and a cyclophilin-like domain that interacts with proteosomes (Ferreira *et al.*, 1998), suggesting a role for the filaments in coupling transport and processing.

EM studies have indicated that the cytoplasmic filaments are dynamic and these dynamics may be related to transport. It was shown that during import of nucleo-plasmin-gold particles the cytoplasmic filaments extended into the centre of the NPC, almost as if the filaments were delivering the substrate to the central channel (Panté and Aebi, 1996; Rutherford *et al.*, 1997). It was also found that when RanGTP was injected into the cytoplasm, this caused the filaments to extend in the same way (Goldberg *et al.*, 2000), whereas injection of excess RanGDP caused them to fold up onto the cytoplasmic ring. Consistently, it was found that purified Nup358 is a rod-shaped protein that could also be found in a curled up conformation (Delphin *et al.*, 1997). The significance of these dynamics is uncertain. It could be that they have a mechanical role to modulate the physical access to the NPC central channel. Alternatively, there may be a biochemical role in masking or uncovering binding sites.

3.3 *The basket*

The nucleoplasmic face of the NPC consists of the eight-subunit nucleoplasmic coaxial ring. At a position between each subunit on the outer periphery of this ring a filament is attached. Each of these eight filaments extends into the nucleoplasm about 40 nm and then branches (Goldberg and Allen, 1996). The branches are then woven together to form a basket- or fishtrap-like structure (Ris, 1991). The branches are woven in a specific way so that the right-hand branch from one filament passes over the left-hand branch from the adjacent filament and connects to the left-hand branch of the next filament (Goldberg and Allen, 1996), which results in two interlocking squares as illustrated in *Figure 4*.

Of all the NPC components, the basket has most strikingly been shown to be a dynamic structure. The salivary glands of the midge larva *Chironomus thummi* produces a large morphologically distinctive mRNP particle called the Balbiani ring granule (as it is produced from the Balbiani ring puff of the polytene chromosome) (Mehlin and Daneholt, 1993). When NEs are isolated from the salivary glands and examined by FESEM (Kiseleva *et al.*, 1996) the Balbiani ring granule is observed in different positions on the basket and the position appears to affect the conformation of the basket. Firstly, there are apparently closed baskets with no Balbiani ring granule associated with them (*Figure 5a*). Then there are baskets where the mRNP particle has docked and the basket has opened (*Figure 5b*). Next the granule moves into the open basket (*Figure 5c*) and, as shown by EM tomography (Mehlin *et al.*, 1995), it unfolds into a rod-shaped particle that can then feed into the central channel of the NPC.

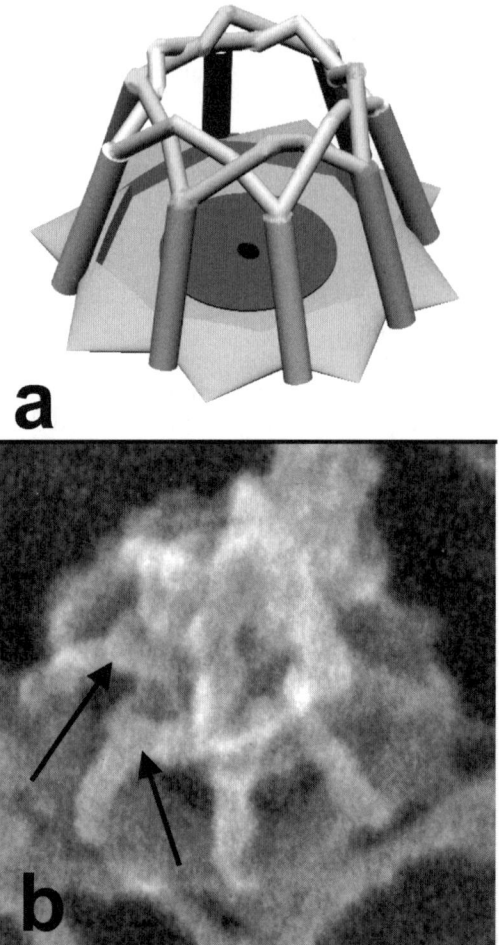

Figure 4. Model of the basket. (a) Basket filaments attach to the outer periphery of the nucleoplasmic ring, extend into the nucleoplasm and branch. The branches appear to be woven together. Branching of the filaments is indicated by the arrows in the FESEM micrograph (b).

The actual purpose or composition of the basket remains uncertain. Superficially, it has the appearance of a sieve and indeed large transport complexes such as the Balbiani ring granule and other mRNP particles could not pass through a closed basket. This then offers the chance, for instance, for an mRNA-processing checkpoint, where the signal to open the basket is only activated when the mRNA is fully processed and is ready to leave the nucleus. Therefore the elaborate structural conformation and dynamics of the basket suggests that it controls the passage of certain complexes, such as mRNP particles, on the basis of size. This is unlikely to be so for all transport substrates as things such as small nucleophilic proteins, even complexed with import factors, would be too small to be physically retained by the basket. Interestingly, it was found that the transport of large substrates requires the use of energy in the form of the hydrolysis of GTP, whereas the transport of small substrates does not (Lyman *et al.*, 2002). It is possible, though untested, that the energy is being used to open the basket.

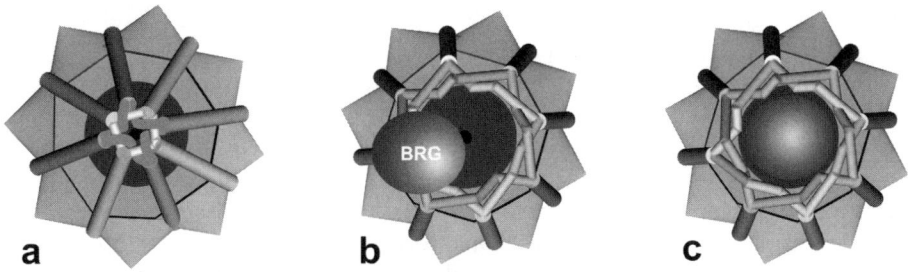

Figure 5. The basket is dynamic. (a) Baskets not involved in Balbiani ring granule export have a compact conformation. (b) The basket opens when the mRNP particle binds, and (c) moves into the NPC.

One problem with understanding the basket is that we do not know what proteins make up the structure. Immuno-EM has provided some candidates and clues, but all of these remain highly controversial. In vertebrates two of the most promising candidates were Nup153 and Tpr.

Nup153 was originally located to the distal basket ring (Panté *et al.*, 1994), but later using domain-specific antibodies (Fahrenkrog *et al.*, 2002) labelling was found in several positions on the basket and at a position which was consistent with the nucleoplasmic ring, as well as the cytoplasmic face of the NPC. Hence it was suggested that the Nup153 protein spanned the entire basket structure. Consistently, immuno-gold labelling in FESEM and TEM, using an antibody to the N-terminal domain, found Nup153 at the nucleoplasmic ring (Walther *et al.*, 2001). However when the basket was completely removed the labelling remained in the same position, showing that Nup153 cannot be a major structural constituent of the basket. In addition Nup153 is a 'mobile' nucleoporin which appears to shuttle across the NPC (Nakielny *et al.*, 1999). There is also a rapid exchange of Nup153 between the NPC and an intranuclear pool of the protein (Daigle *et al.*, 2001). As it is most likely that a structural component of the basket would be stably located at the basket, Nup153 does not seem like a good candidate for this role. It seems more likely that Nup153 is a dynamic protein that binds to several regions of the NPC including the distal basket ring (Fahrenkrog *et al.*, 2002) as well as to the nucleoplasmic coaxial ring (Walther *et al.*, 2001). Nup153 clearly has a role in nuclear transport and is particularly implicated in the cNLS pathway (Shah and Forbes, 1998; Shah *et al.*, 1998; Walther *et al.*, 2001), as well as several export pathways (Bastos *et al.*, 1996; Ullman *et al.*, 1999). It also interacts with lamins, which form a network of proteins on the inner membrane surface (Smythe *et al.*, 2000) and when Nup153 is depleted from nuclear assembly extracts, the NPCs become highly mobile (Walther *et al.*, 2001), suggesting that it may be involved in anchoring the NPC to the nuclear lamina. Such a role would be consistent with it being bound to the nucleoplasmic ring. Nup153 is also located close to (Frosst *et al.*, 2002) and interacts directly with Tpr (Hase and Cordes, 2003). Tpr is located at the distal basket ring and in intranuclear filaments (Cordes *et al.*, 1993) and nuclear bodies (Zimowska and Paddy, 2002). Like Nup153, Tpr may also be dynamic, especially the intranuclear pool and it has been implicated in the processing or transport of mRNA (Zimowska and Paddy, 2002) and protein export (Frosst *et al.*, 2002). Therefore, Nup153 may be a protein that binds to several parts of the NPC and facilitates the binding of other proteins to it. These may be structural proteins such as lamins or possibly Tpr or transport factors. Therefore where Nup153 binds on the NPC may depend on what is bound to it.

Although baskets are unlikely to physically retain smaller transport complexes, they could play a role as binding sites and therefore the basket could act as a biochemical sieve rather than a physical one. As transport complexes have to pass through the basket to either gain entry to or escape from the NPC (during export or import, respectively), any binding sites located on the basket or nucleoplasmic ring would be highly accessible to these complexes. Therefore when proteins such as Nup153 are located somewhere on the basket they would have a very high chance of coming into contact and binding to importin α/importin β/NLS protein complex (Shah *et al.*, 1998). This would capture the complex before it entered the nuclear interior, allowing RanGTP to bind and dissociate the complex (Shah *et al.*, 1998). Likewise the binding of export complexes to Nup153 at the basket could either be an important export checkpoint or a place where other factors are recruited prior to export.

4. Summary

Eukaryotic cells transport a myriad of molecules between the nucleus and cytoplasm and have evolved a number of related biochemical pathways to achieve this, many of which have been elucidated in recent years. One central and common component to all the pathways is the NPC. NPC components appear to play vital roles in transport and the NPC is structurally dynamic, but whether its role is as a facilitator, a controller or both is yet to be decided and awaits further analysis on the role of individual components in specific pathways.

References

Akey, C.W. (1989) Interactions and structure of the nuclear pore complex revealed by cryo-electron microscopy. *J. Cell Biol.* **109**: 955–970.

Akey, C.W. (1990) Visualization of transport-related configurations of the nuclear pore transporter. *Biophys. J.* **58**: 341–355.

Akey, C.W. (1995) Structural plasticity of the nuclear pore complex. *J. Mol. Biol.* **248**: 273–293.

Akey, C.W. and Radermacher, M. (1993) Architecture of the Xenopus nuclear pore complex revealed by three-dimensional cryo-electron microscopy. *J. Cell. Biol.* **122**: 1–19.

Azuma, Y. and Dasso, M. (2000) The role of Ran in nuclear function. *Curr. Opin. Cell. Biol.* **12**: 302–307.

Bastos, R., Lin, A., Enarson, M. and Burke, B. (1996) Targeting and function in mRNA export of nuclear pore complex protein Nup153. *J. Cell. Biol.* **134**: 1141–1156.

Bayliss, R., Ribbeck, K., Akin, D., Kent, H.M., Feldherr, C.M., Gorlich, D. and Stewart, M. (1999) Interaction between NTF2 and xFxFG-containing nucleoporins is required to mediate nuclear import of RanGDP. *J. Mol. Biol.* **293**: 579–593.

Bayliss, R., Littlewood, T. and Stewart, M. (2000) Structural basis for the interaction between FxFG nucleoporin repeats and importin-beta in nuclear trafficking. *Cell* **102**: 99–108.

Bayliss, R., Leung, S.W., Baker, R.P., Quimby, B.B., Corbett, A.H. and Stewart, M. (2002) Structural basis for the interaction between NTF2 and nucleoporin FxFG repeats. *EMBO J.* **21**: 2843–2853.

Bednenko, J., Cingolani, G. and Gerace, L. (2003) Importin beta contains a COOH-terminal nucleoporin binding region important for nuclear transport. *J. Cell. Biol.* **162**: 391–401.

Ben-Efraim, I. and Gerace, L. (2001) Gradient of increasing affinity of importin beta for nucleoporins along the pathway of nuclear import. *J. Cell. Biol.* **152**: 411–417.

Bischoff, F.R. and Ponstingl, H. (1991) Catalysis of guanine nucleotide exchange on Ran by the mitotic regulator RCC1. *Nature* **354**: 80–82.

Bischoff, F.R., Klebe, C., Kretschmer, J., Wittinghofer, A. and Ponstingl, H. (1994) RanGAP1 induces GTPase activity of nuclear Ras-related Ran. *Proc. Natl Acad. Sci. USA* **91**: 2587–2591.

Chook, Y.M. and Blobel, G. (1999) Structure of the nuclear transport complex karyopherin-beta2-Ran.GppNHp. *Nature* **399**: 230–237.

Chook, Y.M. and Blobel, G. (2001) Karyopherins and nuclear import. *Curr. Opin. Struct. Biol.* **11**: 703–715.

Cingolani, G., Petosa, C., Weis, K. and Muller, C.W. (1999) Structure of importin-beta bound to the IBB domain of importin-alpha. *Nature* **399**: 221–229.

Conti, E., Uy, M., Leighton, L., Blobel, G. and Kuriyan, J. (1998) Crystallographic analysis of the recognition of a nuclear localization signal by the nuclear import factor karyopherin alpha. *Cell* **94**: 193–204.

Cordes, V.C., Reidenbach, S., Kohler, A., Stuurman, N., van Driel, R. and Franke, W.W. (1993) Intranuclear filaments containing a nuclear pore complex protein. *J. Cell. Biol.* **123**: 1333–1344.

Cronshaw, J.M., Krutchinsky, A.N., Zhang, W., Chait, B.T. and Matunis, M.J. (2002) Proteomic analysis of the mammalian nuclear pore complex. *J. Cell. Biol.* **158**: 915–927.

Daigle, N., Beaudouin, J., Hartnell, L., Imreh, G., Hallberg, E., Lippincott-Schwartz, J. and Ellenberg, J. (2001) Nuclear pore complexes form immobile networks and have a very low turnover in live mammalian cells. *J. Cell. Biol.* **154**: 71–84.

Delphin, C., Guan, T., Melchior, F. and Gerace, L. (1997) RanGTP targets p97 to RanBP2, a filamentous protein localized at the cytoplasmic periphery of the nuclear pore complex. *Mol. Biol. Cell* **8**: 2379–2390.

Dingwall, C., Sharnik, S.V. and Laskey, R.A. (1982) A polypeptide domain that specifies migration of nucleoplasmin into the nucleus. *Cell* **30**: 449–458.

Fahrenkrog, B., Maco, B., Fager, A.M., Koser, J., Sauder, U., Ullman, K.S. and Aebi, U. (2002) Domain-specific antibodies reveal multiple-site topology of Nup153 within the nuclear pore complex. *J. Struct. Biol.* **140**: 254–267.

Feldherr, C.M. and Akin, D. (1990) The permeability of the nuclear envelope in dividing and nondividing cell cultures. *J. Cell. Biol.* **111**: 1–8.

Feldherr, C.M. and Akin, D. (1991) Signal-mediated nuclear transport in proliferating and growth-arrested BALB/c 3T3 cells. *J. Cell. Biol.* **115**: 933–939.

Feldherr, C.M. and Akin, D. (1993) Regulation of nuclear transport in proliferating and quiescent cells. *Exp. Cell. Res.* **205**: 179–186.

Feldherr, C.M. and Akin, D. (1997) The location of the transport gate in the nuclear pore complex. *J. Cell. Sci.* **110**: 3065–3070.

Feldherr, C.M., Kallenbach, E. and Schultz, N. (1984) Movement of a karyophilic protein through the nuclear pores of oocytes. *J. Cell. Biol.* **99**: 2216–2222.

Ferreira, P.A., Nakayama, T.A., Pak, W.L. and Travis, G.H. (1996) Cyclophilin-related protein RanBP2 acts as chaperone for red/green opsin. *Nature* **383**: 637–640.

Ferreira, P.A., Nakayama, T.A. and Travis, G.H. (1997) Interconversion of red opsin isoforms by the cyclophilin-related chaperone protein Ran-binding protein 2. *Proc. Natl Acad. Sci. USA* **94**: 1556–1561.

Ferreira, P.A., Yunfei, C., Schick, D. and Roepman, R. (1998) The cyclophilin-like domain mediates the association of Ran-binding protein 2 with subunits of the 19 S regulatory complex of the proteasome. *J. Biol. Chem.* **273**: 24676–24682.

Floer, M. and Blobel, G. (1996) The nuclear transport factor karyopherin beta binds stoichiometrically to Ran-GTP and inhibits the Ran GTPase activating protein. *J. Biol. Chem.* **271**: 5313–5316.

Frosst, P., Guan, T., Subauste, C., Hahn, K. and Gerace, L. (2002) Tpr is localized within the nuclear basket of the pore complex and has a role in nuclear protein export. *J. Cell. Biol.* **156**: 617–630.

Goldberg, M.W. and Allen, T.D. (1996) The nuclear pore complex and lamina: three-dimensional structures and interactions determined by field emission in-lens scanning electron microscopy. *J. Mol. Biol.* **257**: 848–865.

Goldberg, M.W., Solovei, I.I. and Allen, T.D. (1997) Nuclear pore complex structure in birds. *J. Struct. Biol.* **119**: 284–294.

Goldberg, M.W., Rutherford, S.A., Hughes, M., Cotter, L.A., Bagley, S., Kiseleva, E., Allen, T.D. and Clarke, P.R. (2000) Ran alters nuclear pore complex conformation. *J. Mol. Biol.* **300**: 519–529.

Görlich, D. and Kutay, U. (1999) Transport between the cell nucleus and the cytoplasm. *Ann. Rev. Cell. Dev. Biol.* **15**: 607–660.

Görlich, D., Henklein, P., Laskey, R.A. and Hartmann, E. (1996a) A 41 amino acid motif in importin-alpha confers binding to importin-beta and hence transit into the nucleus. *EMBO J.* **15**: 1810–1817.

Görlich, D., Panté, N., Kutay, U., Aebi, U. and Bischoff, F.R. (1996b) Identification of different roles for RanGDP and RanGTP in nuclear protein import. *EMBO J.* **15**: 5584–5594.

Griffis, E.R., Altan, N., Lippincott–Schwartz, J. and Powers, M.A. (2002) Nup98 is a mobile nucleoporin with transcription-dependent dynamics. *Mol. Biol. Cell.* **13**: 1282–1297.

Hase, M.E. and Cordes, V.C. (2003) Direct interaction with nup153 mediates binding of tpr to the periphery of the nuclear pore complex. *Mol. Biol. Cell.* **14**: 1923–1940.

Hieda, M., Tachibana, T., Yokoya, F., Kose, S., Imamoto, N. and Yoneda, Y. (1999) A monoclonal antibody to the COOH-terminal acidic portion of Ran inhibits both the recycling of Ran and nuclear protein import in living cells. *J. Cell. Biol.* **144**: 645–655.

Hinshaw, J.E., Carragher, B.O. and Milligan, R.A. (1992) Architecture and design of the nuclear pore complex. *Cell* **69**: 1133–1141.

Hood, J.K. and Silver, P.A. (1998) Cse1p is required for export of Srp1p/importin-alpha from the nucleus in Saccharomyces cerevisiae. *J. Biol. Chem.* **273**: 35142–35146.

Iborra, F.J., Jackson, D.A. and Cook, P.R. (2000) The path of RNA through nuclear pores: apparent entry from the sides into specialized pores. *J. Cell. Sci.* **113**: 291–302.

Izaurralde, E., Kutay, U., von Kobbe, C., Mattaj, I.W. and Görlich, D. (1997) The asymmetric distribution of the constituents of the Ran system is essential for transport into and out of the nucleus. *EMBO J.* **16**: 6535–6547.

Kalab, P., Weis, K. and Heald, R. (2002) Visualization of a Ran-GTP gradient in interphase and mitotic Xenopus egg extracts. *Science* **295**: 2452–2456.

Kalderon, D., Richardson, W.D., Markham, A.F. and Smith, A.E. (1984). Sequence requirements for nuclear location of Simian Virus 40 large T-antigen. *Nature* **311**: 33–38.

Kiseleva, E., Goldberg, M.W., Daneholt, B. and Allen, T.D. (1996) RNP export is mediated by structural reorganization of the nuclear pore basket. *J. Mol. Biol.* **260**: 304–311.

Kiseleva, E., Goldberg, M.W., Allen, T.D. and Akey, C.W. (1998). Active nuclear pore complexes in Chironomus: visualization of transporter configurations related to mRNP export. *J. Cell. Sci.* **111**: 223–236.

Klebe, C., Bischoff, F.R., Ponstingl, H. and Wittinghofer, A. (1995) Interaction of the nuclear GTP-binding protein Ran with its regulatory proteins RCC1 and RanGAP1. *Biochemistry* **34**: 639–647.

Kobe, B. (1999) Autoinhibition by an internal nuclear localization signal revealed by the crystal structure of mammalian importin alpha. *Nat. Struct. Biol.* **6**: 388–397.

Kose, S., Imamoto, N., Tachibana, T., Yoshida, M. and Yoneda, Y. (1999) beta-subunit of nuclear pore-targeting complex (importin-beta) can be exported from the nucleus in a Ran-independent manner. *J. Biol. Chem.* **274**: 3946–3952.

Kuersten, S., Arts, G.J., Walther, T.C., Englmeier, L. and Mattaj, I.W. (2002) Steady-state nuclear localization of exportin-t involves RanGTP binding and two distinct nuclear pore complex interaction domains. *Mol. Cell. Biol.* **22**: 5708–5720.

Kunzler, M. and Hurt, E.C. (1998) Cse1p functions as the nuclear export receptor for importin alpha in yeast. *FEBS Lett.* **433**: 185–190.

Kutay, U., Bischoff, F.R., Kostka, S., Kraft, R. and Görlich, D. (1997) Export of importin alpha from the nucleus is mediated by a specific nuclear transport factor. *Cell* **90**: 1061–1071.

Lanford, R.E. and Butel, R.S. (1984) Construction and characterization of an SV40 mutant defective in nuclear transport of T antigen. *J. Cell* **37**: 801–813.

Li, H.Y., Wirtz, D. and Zheng, Y. (2003) A mechanism of coupling RCC1 mobility to RanGTP production on the chromatin in vivo. *J. Cell. Biol.* **160**: 635–644.

Lounsbury, K.M. and Macara, I.G. (1997) Ran-binding protein 1 (RanBP1) forms a ternary complex with Ran and karyopherin beta and reduces Ran GTPase-activating protein (RanGAP) inhibition by karyopherin beta. *J. Biol. Chem.* **272**: 551–555.

Lyman, S.K., Guan, T., Bednenko, J., Wodrich, H. and Gerace, L. (2002) Influence of cargo size on Ran and energy requirements for nuclear protein import. *J. Cell. Biol.* **159**: 55–67.

Mattaj, I.W. and Englmeier, L. (1998) Nucleocytoplasmic transport: the soluble phase. *Ann. Rev. Biochem.* **67**: 265–306.

Matunis, M.J., Wu, J. and Blobel, G. (1998) SUMO-1 modification and its role in targeting the Ran GTPase-activating protein, RanGAP1, to the nuclear pore complex. *J. Cell. Biol.* **140**: 499–509.

Mehlin, H. and Daneholt, B. (1993) The Balbiani ring particle: a model for the assembly and export of RNPs from the nucleus? *Trends Cell. Biol.* **3**: 443–447.

Mehlin, H., Daneholt, B. and Skoglund, U. (1995) Structural interaction between the nuclear pore complex and a specific translocating RNP particle. *J. Cell. Biol.* **129**: 1205–1216.

Moroianu, J., Blobel, G. and Radu, A. (1996) The binding site of karyopherin alpha for karyopherin beta overlaps with a nuclear localization sequence. *Proc. Natl Acad. Sci. USA* **93**: 6572–6576.

Nachury, M.V. and Weis, K. (1999) The direction of transport through the nuclear pore can be inverted. *Proc. Natl Acad. Sci. USA* **96**: 9622–9627.

Nakielny, S., Shaikh, S., Burke, B. and Dreyfuss, G. (1999) Nup153 is an M9-containing mobile nucleoporin with a novel Ran-binding domain. *EMBO J.* **18**: 1982–1995.

Nemergut, M.E., Mizzen, C.A., Stukenberg, T., Allis, C.D. and Macara, I.G. (2001) Chromatin docking and exchange activity enhancement of RCC1 by histones H2A and H2B. *Science* **292**: 1540–1543.

Ohno, M., Fornerod, M. and Mattaj, I.W. (1998) Nucleocytoplasmic transport: the last 200 nanometers. *Cell* **92**: 327–336.

Ohtsubo, M., Okazaki, H. and Nishimoto, T. (1989) The RCC1 protein, a regulator for the onset of chromosome condensation locates in the nucleus and binds to DNA. *J. Cell. Biol.* **109**: 1389–1397.

Panté, N. and Aebi, U. (1996) Sequential binding of import ligands to distinct nucleopore regions during their nuclear import. *Science* **273**: 1729–1732.

Panté, N. and Kann, M. (2002) Nuclear pore complex is able to transport macromolecules with diameters of about 39 nm. *Mol. Biol. Cell.* **13**(2): 425–434.

Panté, N., Bastos, R., McMorrow, I., Burke, B. and Aebi, U. (1994) Interactions and three-dimensional localization of a group of nuclear pore complex proteins. *J. Cell. Biol.* **126**: 603–617.

Panté, N., Jarmolowski, A., Izaurralde, E., Sauder, U., Baschong, W. and Mattaj, I.W. (1997) Visualizing nuclear export of different classes of RNA by electron microscopy. *RNA* **3**: 498–513.

Pichler, A., Gast, A., Seeler, J.S., Dejean, A. and Melchior, F. (2002) The nucleoporin RanBP2 has SUMO1 E3 ligase activity. *Cell* **108**: 109–120.

Plafker, K. and Macara, I.G. (2000) Facilitated nucleocytoplasmic shuttling of the Ran binding protein RanBP1. *Mol. Cell. Biol.* **20**: 3510–3521.

Quimby, B.B. and Dasso, M. (2003) The small GTPase Ran: interpreting the signs. *Curr. Opin. Cell. Biol.* **15**: 338–344.

Quimby, B.B., Lamitina, T., L'Hernault, S.W. and Corbett, A.H. (2000) The mechanism of ran import into the nucleus by nuclear transport factor 2. *J. Biol. Chem.* **275**: 28575–28582.

Quimby, B.B., Leung, S.W., Bayliss, R., Harreman, M.T., Thirumala, G., Stewart, M. and Corbett, A.H. (2001) Functional analysis of the hydrophobic patch on nuclear transport factor 2 involved in interactions with the nuclear pore in vivo. *J. Biol. Chem.* **276**: 38820–38829.

Reichelt, R., Holzenburg, A., Buhle, E.L., Jr, Jarnik, M., Engel, A. and Aebi, U. (1990) Correlation between structure and mass distribution of the nuclear pore complex and of distinct pore complex components. *J. Cell. Biol.* **110**: 883–894.

Renault, L., Kuhlmann, J., Henkel, A. and Wittinghofer, A. (2001) Structural basis for guanine nucleotide exchange on Ran by the regulator of chromosome condensation (RCC1). *Cell* **105**: 245–255.

Rexach, M. and Blobel, G. (1995) Protein import into nuclei: association and dissociation reactions involving transport substrate, transport factors, and nucleoporins. *Cell* **83**: 683–692.

Ribbeck, K. and Görlich, D. (2001) Kinetic analysis of translocation through nuclear pore complexes. *EMBO J.* **20**: 1320–1330.

Ribbeck, K. and Görlich, D. (2002) The permeability barrier of nuclear pore complexes appears to operate via hydrophobic exclusion. *EMBO J.* **21**: 2664–2671.

Ribbeck, K., Lipowsky, G., Kent, H.M., Stewart, M. and Görlich, D. (1998) NTF2 mediates nuclear import of Ran. *EMBO J.* **17**: 6587–6598.

Ris, H. (1991) The 3D structure of the nuclear pore complex as seen by high voltage electron microscopy and high resolution low voltage scanning electron microscopy. *EMSA Bull.* **21**: 54–56.

Robbins, J., Dilworth, S.M., Lasket, R.A. and Dingwall, C. (1991) Two interdependent basic domains in nucleoplasmin nuclear targeting sequence: identification of a class of bipartite nuclear targeting sequences. *Cell* **64**: 615–623.

Rout, M.P., Aitchison, J.D., Suprapto. A., Hjertaas, K., Zhao, Y. and Chait, B.T. (2000) The yeast nuclear pore complex: composition, architecture, and transport mechanism. *J. Cell. Biol.* **148**: 635–651.

Rutherford, S.A., Goldberg, M.W. and Allen, T.D. (1997) Three-dimensional visualization of the route of protein import: the role of nuclear pore complex substructures. *Exp. Cell. Res.* **232**: 146–160.

Shah, S. and Forbes, D.J. (1998) Separate nuclear import pathways converge on the nucleoporin Nup153 and can be dissected with dominant-negative inhibitors. *Curr. Biol.* **8**: 1376–1386.

Shah, S., Tugendreich, S. and Forbes, D. (1998) Major binding sites for the nuclear import receptor are the internal nucleoporin Nup153 and the adjacent nuclear filament protein Tpr. *J. Cell. Biol.* **141**: 31–49.

Shulga, N. and Goldfarb, D.S. (2003) Binding dynamics of structural nucleoporins govern nuclear pore complex permeability and may mediate channel gating. *Mol. Cell. Biol.* **23**: 534–542.

Schwoebel, E.D., Talcott, B., Cushman, I. and Moore, M.S. (1998) Ran-dependent signal-mediated nuclear import does not require GTP hydrolysis by Ran. *Biol. Chem.* **273**: 35170–35175.

Singh, B.B., Patel, H.H., Roepman, R., Schick, D. and Ferreira, P.A. (1999) The zinc finger cluster domain of RanBP2 is a specific docking site for the nuclear export factor, exportin-1. *J. Biol. Chem.* **274**: 37370–37378.

Solsbacher, J., Maurer, P., Bischoff, F.R. and Schlenstedt, G. (1998) Cse1p is involved in export of yeast importin alpha from the nucleus. *Mol. Cell. Biol.* **18**: 6805–6815.

Smith, A.E., Slepchenko, B.M., Schaff, J.C., Loew, L.M. and Macara, I.G. (2002) Systems analysis of Ran transport. *Science* **295**: 488–491.

Smythe, C., Jenkins, H.E. and Hutchison, C.J. (2000) Incorporation of the nuclear pore basket protein nup153 into nuclear pore structures is dependent upon lamina assembly: evidence from cell-free extracts of Xenopus eggs. *EMBO J.* **19**: 3918–3931.

Steggerda, S.M. and Paschal, B.M. (2002) Regulation of nuclear import and export by the GTPase Ran. *Int. Rev. Cytol.* **217**: 41–91.

Stewart, M., Whytock, S. and Mills, A.D. (1990) Association of gold-labelled nucleoplasmin with the centres of ring components of Xenopus oocyte nuclear pore complexes. *J. Mol. Biol.* **213**: 575–582.

Stewart, M., Kent, H.M. and McCoy, A.J. (1998) The structure of the Q69L mutant of GDP-Ran shows a major conformational change in the switch II loop that accounts for its failure to bind Nuclear Transport Factor 2 (NTF2). *J. Mol. Biol.* **284**: 1517–1527.

Stoffler, D., Feja, B., Fahrenkrog, B., Walz, J., Typke, D. and Aebi, U. (2003) Cryo-electron tomography provides novel insights into nuclear pore architecture: implications for nucleo-cytoplasmic transport. *J. Mol. Biol.* **328**: 119–130.

Strambio-de-Castillia, C. and Rout, M.P. (2002) The structure and composition of the yeast NPC. *Results Probl. Cell. Differ.* **35**: 1–23.

Ullman, K.S., Shah, S., Powers, M.A. and Forbes, D.J. (1999) The nucleoporin nup153 plays a critical role in multiple types of nuclear export. *Mol. Biol. Cell.* **10**: 649–664.

Vetter, I.R., Arndt, A., Kutay, U., G"rlich, D. and Wittinghofer, A. (1999) Structural view of the Ran-Importin beta interaction at 2.3 A resolution. *Cell* **97**: 635–646.

Walther, T.C., Fornerod, M., Pickersgill, H., Goldberg, M., Allen, T.D. and Mattaj, I.W. (2001) The nucleoporin Nup153 is required for nuclear pore basket formation, nuclear pore complex anchoring and import of a subset of nuclear proteins *EMBO J.* **20**: 5703–5714.

Walther, T.C., Pickersgill, H.S., Cordes, V.C., Goldberg, M.W., Allen, T.D., Mattaj, I.W. and Fornerod, M. (2002) The cytoplasmic filaments of the nuclear pore complex are dispensable for selective nuclear protein import. *J. Cell. Biol.* **158**: 63–77.

Weis, K. (2003) Regulating access to the genome: nucleocytoplasmic transport throughout the cell cycle. *Cell* **112**: 441–451.

Weis, K., Ryder, U. and Lamond, A.I. (1996) The conserved amino-terminal domain of hSRP1 alpha is essential for nuclear protein import. *EMBO J.* **15**: 1818–1825.

Wu, J., Matunis, M.J., Kraemer, D., Blobel, G. and Coutavas, E. (1995) Nup358, a cytoplasmi-cally exposed nucleoporin with peptide repeats, Ran-GTP binding sites, zinc fingers, a cyclophilin A homologous domain, and a leucine-rich region. *J. Biol. Chem.* **270**: 14209–14213.

Yang, Q., Rout, M.P. and Akey, C.W. (1998) Three-dimensional architecture of the isolated yeast nuclear pore complex: functional and evolutionary implications. *Mol. Cell.* **1**: 223–234.

Yaseen, N.R. and Blobel, G. (1999a) Two distinct classes of Ran-binding sites on the nucleo-porin Nup-358. *Proc. Natl Acad. Sci. USA* **96**: 5516–5521.

Yaseen, N.R. and Blobel, G. (1999b) GTP hydrolysis links initiation and termination of nuclear import on the nucleoporin nup358. *J. Biol. Chem.* **274**: 26493–26502.

Yokoyama, N., Hayashi, N., Seki, T., *et al.* (1995) A giant nucleopore protein that binds Ran/TC4. *Nature* **376**: 184–188.

Zimowska, G. and Paddy, M.R. (2002) Structures and dynamics of Drosophila Tpr inconsistent with a static, filamentous structure. *Exp. Cell. Res.* **276**: 223–232.

8

Regulating gene expression in mammalian cells: how nuclear architecture influences mRNA synthesis and export to the cytoplasm

Dean A. Jackson

1. Introduction

The molecular mechanisms that regulate patterns of gene expression have been studied intensively over the past 20 years and many of the critical interactions have been characterized in detail (Bulger *et al.*, 2002; Butler and Kadonaga, 2002; Emerson, 2002; Jenuwein and Allis, 2001; Maniatis and Reed, 2002; Orphanides and Reinberg, 2002). The process is inevitably complex and recent reviews focus on the complex interplay of networks that somehow integrate different types of information that influence levels and patterns of gene expression. Indeed, the control of this process is so sophisticated that for the sake of understanding it is helpful to consider specific features that impact on the behaviour of these regulatory networks. For simplicity, five broad areas will be considered here:

1. Transcriptions factors – function as adaptors to decode the genetic information within DNA and then recruit the transcription machinery to gene promoters to engage gene expression. Transcription factors bind within the promoter and at more distal upstream activator elements and sequences such as enhancers that stimulate levels of gene expression. The behaviour of transcription factors, their concentrations in the nucleus and the structure and stability of the promoter-associated complexes are likely to play a critical role in determining levels of mRNA synthesis. Note that the concentration of transcription factors within nuclei and their affinity for different promoters can vary enormously.

The Nuclear Envelope, edited by D.E. Evans, C. Hutchison & J.A. Bryant.
© 2004 Garland Science/BIOS Scientific Publishers

2. Chromatin structure – serves to define regions of a genome that are accessible to transcription factors and subsequently the transcription machinery. Specific histone modifications correlate with and stabilize the active and inactive chromatin forms. Broadly speaking, active genes are found in open or accessible chromatin (euchromatin) and inactive genes, as well as many repetitive DNA elements are located in more condensed heterochromatin.

3. Chromosome structure and global nuclear architecture – provide a variety of mechanisms that influence the behaviour of components involved in gene expression.

4. Post-transcriptionally – the primary RNA transcript is processed at the 3' and 5' ends and spliced to remove non-coding introns. The mature mRNA–protein complex (mRNP) can then leave the transcription site and diffuse to the nuclear periphery, through the nuclear pores and subsequently engage sites of protein synthesis in the cytoplasm.

5. In the cytoplasm – various activities will also influence the stability of each mRNP and the amount of mature proteins that might be generated by each mRNA that leaves the nucleus.

This chapter sets out to evaluate how chromatin structure and global nuclear architecture collaborate to regulate the early stages of gene expression. Particular emphasis will be placed on the architecture of the transcription centres and mechanisms that control how the components involved in gene expression are able to interact. The chapter presents a global overview that emphasizes the relative significance of the different regulatory mechanisms. This inevitably raises many fascinating questions, for a deeper treatment of which the reader is referred to the more specialized reviews that are cited herein.

2. Initiating gene expression

The critical feature of gene expression in eukaryotes is that the transcription machinery is recruited to genes through its interaction with transcription factors and adaptor proteins that interact with these factors (Butler and Kadonaga, 2002; Emerson, 2002; Maniatis and Reed, 2002; Orphanides and Reinberg, 2002). Genetic elements within gene promoters define where transcription will begin. General transcription factors play a fundamental role in this process. For most eukaryotic promoters, the process is activated by association of the factor TFIID with DNA. TFIID is a large multi-protein complex, one sub-unit of which, the TATA sequence-binding protein (TBP) can recognize and associate with the canonical promoter element TATA. TFIID bound to the promoter can then activate the sequential recruitment of other factors as follows: TFIIB; TFIIE with TFIIF and RNA polymerase; TFIIH. Once assembly of this pre-initation complex is complete, the TFIIH complex directs phosphorylation of a domain of the RNA polymerase complex called the C-terminal domain (CTD). The CTD is conserved in eukaryotic RNA polymerase II proteins. The mammalian enzyme has 52 repeats of the amino acid (consensus) sequence YSPTSPS, in which the serine (S) and threonine (T) residues can be phosphorylated. Phosphorylation of the CTD correlates with release of the synthetic complex from the promoter and the initiation of RNA synthesis. During synthesis the DNA template is used to generate a complementary RNA molecule.

The association of transcription factors within a promoter precedes the onset of RNA synthesis (Butler and Kadonaga, 2002; Emerson, 2002). Transcription factors are generally small proteins that might be perceived to diffuse through chromatin as they seek binding sites in DNA. But while this is partly true – the factors are generally dynamic and most remain bound to promoters for only seconds (McNally *et al.*, 2000) – a 'snapshot' view will often show them to be concentrated at discrete nuclear sites (Grande *et al.*, 1997). An excellent example of this is provided by the spatial distribution of the aryl hydrocarbon receptor (AhR) and its binding partner, the AhR nuclear translocator, in human cells (Elbi *et al.*, 2002). The heterodimer formed between these two proteins is a transcription factor complex that binds to specific sites in the regulatory domains of numerous target genes. Using confocal microscopy, it is clear that AhR is recruited to transcription sites from the nuclear receptor complex and that its binding partner is required for this to occur; the complex is recruited to discrete sub-nuclear compartments, suggesting that these sites reflect the location of AhR target genes.

The location of transcription factors within promoters is known to define where the transcription complex is positioned and where transcription then begins. However, numerous other genetic elements have been described that contribute to the initiation process and play a role in defining the genetic units or domains (Bell *et al.*, 2001; Blackwood and Kadonaga 1998; Bode *et al.*, 2000; Bulger *et al.*, 2002). These include: upstream activating elements within promoters, enhancers, locus control regions (LCRs), scaffold/matrix attachment regions (S/MARs) and insulators. The combined action of promoters, enhancer, S/MAR and insulator elements will together dictate levels of gene expression. In addition, through their interaction with proteins involved in gene expression, these elements will play an important part in determining how DNA is organized and packaged within the nuclear space (Bulger *et al.*, 2002; Jackson 2003a, b).

3. Chromatin structure and gene expression

It is self-evident that chromatin organization might have profound effects on gene expression by modulating the accessibility of genetic sequences to activating transcription factors (Dillon and Festenstein, 2002; Grewal and Moazed, 2003). Indeed, as the chromatin fibre is the template for functions such as DNA replication and RNA transcription the importance of this DNA–protein complex cannot be over-stated. Broadly speaking, chromatin within a mammalian nucleus can be characterized as euchromatin or heterochromatin (Cremer and Cremer, 2001; Dillon and Festenstein, 2002). The former is ostensibly composed of chromosomal regions that have a high density of transcriptionally active genes. Euchromatin corresponds to chromosomal regions that can be classified as R-bands using cytological criteria. These are slightly GC-rich (because of GC islands associated with housekeeping genes), have a high density of Alu-repeats and are duplicated in the early part of S phase. Heterochromatin, in contrast, has many fewer transcribed genes, occupies chromosomal G and C-bands and is replicated in the second half of S phase. Euchromatin and heterochromatin have distinct features that correlate with these characteristics. For example, euchromatin is readily digested with enzymes such as DNase – it is classified as being DNase sensitive – whereas heterochromatin is relatively insensitive; digestion reflects the accessibility of DNA within the more open euchromatic structures.

4. Chromatin modification

These basic chromatin states correlate with functional status and reflect how transcribed genes are packed into chromatin that is modified to allow transcription to occur. As the basic structural unit of chromatin, the organization of nucleosomes is well known and will not be discussed here. However, in discussing the regulation of gene expression it is essential to understand how histone modifications such as acetylation, methylation, phosphorylation, ADP-ribosylation and ubiquitination impact on chromatin architecture (Jenuwein and Allis, 2001; Narlikar *et al.*, 2002). These modifications are used to generate and stabilize the different classes of chromatin, modulate chromatin structure throughout the cell cycle and control histone turnover. In terms of gene expression, acetylation of the N-terminal domains of histones – particularly H3 and H4 – is of greatest importance. These lysine-rich domains each have numerous lysine residues that are targets for acetylation. Acetylation influences the stability of the nucleosome complex so that DNA in chromatin with acetylated histones is more readily accessible to the transcription machinery.

The mechanisms by which histone acetylation is controlled are extremely complex. Levels of acetylation at any locus are dictated by the combined activities of histone acetyl transferases (HATs) and histone deacetylase complexes (HDACs). The activity of these large protein complexes is determined, in turn, by mechanisms that control their recruitment to different nuclear sites. The protein p300/CBP is a global transcriptional regulator that binds to enhancers within many gene loci and contains a HAT activity that is capable of acetylating specific sites in all the core histones and other transcription factors to stimulate transcription. In addition, complexes such as PCAF interact with p300/CBP, SCR1/ACTR and other DNA-binding activators to stimulate transcription by histone acetylation, predominantly in histone H3. Components of the multi-subunit general transcription factors TFIID and TFIIIC also have HAT activities that modify histones H3 and H4 to stimulate transcription and other HAT activities are associated with the elongating polymerase holoenzyme complexes. The patterns of histone acetylation that might be developed in response to these different activities can be complex; different HATs have numerous potential acetylation targets and different preferences for the various sites. Major modifications that correlate with an active chromatin status include acetylation of histone H3 at lysines 9 and 14 and H4 at lysine 5. Specificity for particular sites in chromatin is a product of the mechanisms of association of each HAT with chromatin. The bromodomain of these proteins is thought to play a role in this process.

Local changes in the levels of histone modification reflect patterns of gene activity within different chromosomal loci. For example, clear transitions from active to inactive chromatin suggest that chromatin domains are demarcated by elements that determine the boundaries of functional genetic units. This is clearly seen in the chicken β-globin locus, where very strong constitutive foci of hyperacetylation correspond with insulator elements that define the globin domain. This chromatin structure is in marked contrast to the very low levels of histone acetylation that are maintained in inactive regions of the locus and intermediate levels seen throughout transcribed gene domains (Litt *et al.*, 2001).

Histone acetylation in euchromatin correlates with gene activity. In contrast, inactive heterochromatin has lower levels of acetylation and very much higher level of histone methylation and phosphorylation. The human suppressor of variegation protein SUV39H1 encodes a histone methyltransferase (HMT) that selectively methylates histone H3 at lysine 9. This activity is dependent on a SET domain within the protein. This particular histone modification induces high-affinity binding of the heterochromatin protein HP1, through chromodomains. The other major group of heterochromatin proteins the polycomb group (Pc-G) proteins are known to recruit protein complexes with histone deacetylase activity. Pc-G proteins and antagonizing proteins of the trithorax group (trx-G) together play a role in modulating the dynamic transition between inactive and active chromatin states.

Chromatin status and the activity of complexes that catalyse modification of the histones are also influenced by the activity of a variety of protein complexes that perform ATP-dependent chromatin remodelling (Becker, 2002). These complexes were first described in studies to understand the control of mating type switching (SWI) and sucrose fermentation (SNF for sucrose non-fermenting) in yeast. Chromatin remodelling was recognized as a major factor in these two processes and has since been shown to be a fundamental regulator of gene expression. Examples of the remodelling machines in human cells include the human SWI/SNF complex, NURD and RSF. These multi-protein complexes operate through different chromatin binding domains. hSWI/SNF has a bromodomain, NURD chromodomains and RSF a SANT domain. These chromatin-remodelling machines serve to increase the local dynamic properties of chromatin. DNA and histones in chromatin make so many contacts that the nucleosomes they form are inherently stable structures. Nucleosomes can form on almost all stretches of DNA of sufficient length, though the need to fold the DNA duplex over the nucleosome surface does impose constraints on the way chromatin forms. In particular, the centre of the dyad axis in a nucleosome has a region of DNA that is distorted or kinked in order to make the necessary contacts with the histones of the nucleosome core. AT bases in DNA are preferred at this location. Other mechanisms exist to position nucleosomes in a specific way – the binding of a factor with DNA prior to establishing local chromatin structure would be an obvious mechanism. The ATP-dependent chromatin-remodelling activities serve to enhance the fluid properties of nucleosomes; in essence they allow nucleosomes to slide on DNA. The mechanism of this process is not known in detail, but is assumed to reduce the activation energy needed to reposition a nucleosome core.

Hence, the chromatin-remodelling machines provide a means by which nucleosomes can be repositioned to allow access to previously inaccessible sites in DNA. This activity will function co-operatively with the histone modification systems described above to modulate and stabilize different chromatin states. The combination of these activities adds a huge complexity to the epigenetic control of chromatin function. Indeed, present estimates suggest that something in the region of 50 chromatin-modifying complexes will collaborate to ensure that chromatin is an extremely complex and structurally dynamic substrate for gene expression.

5. Spatial nuclear architecture

Though the mechanisms might be complex, the principle that chromatin fluidity might determine the accessibility of transcription factors to DNA is rather obvious. But it is also clear that many other layers of control impact on the regulation of gene expression. For example, it has been known for many years that the local chromosome structure influences gene expression so that genes introduced into unnatural (ectopic) chromosomal sites might display unpredictable patterns and levels of gene expression (Wallace *et al.*, 2000; Wilson *et al.*, 1990). This is clearly a complex phenomenon, which incorporates various aspects of chromosome and nuclear architecture.

6. Transcription centres

When they are transcribed, protein-coding genes are generally found in euchromatin and expressed at nuclear sites that are located at the borders of condensed chromatin in association with perichromatin fibrils (Jackson, 2003a). Proliferative mammalian cells have 500–1000 active sites of RNA polymerase II activity for each haploid chromosome set (Elbi *et al.*, 2002; Jackson *et al.*, 1998; Pombo *et al.*, 1999). However, as cells have at least five times this number of active RNA polymerase II holoenzyme complexes (Jackson *et al.*, 1998; Kimura *et al.*, 1999) each active centre must represent a nuclear compartment where groups of transcripts are generated and processed together. This spatial co-ordination of the different steps required to produce mature mRNAs at specific nuclear sites forms the basis of the concept of transcription factories (Cook, 1999). Within these sites, transcripts are polymerised, processed and assembled into the required mRNA–protein complex before being released to engage the downstream export pathway. For a typical transcript, events occurring at the transcription site take roughly 15 minutes to complete (Femino *et al.*, 1998; Iborra *et al.*, 1998).

Transcription centres represent a major class of functional nuclear compartment. But the organization of these sites provides other insights into the mechanisms that regulate gene expression. Two basic features warrant further analysis. First, early studies using high-resolution analysis of transcription sites labelled with Br-UTP showed that the nascent RNA occupied a much smaller volume of the nucleus than the euchromatin of active genes (Iborra *et al.*, 1996). This observation is consistent with the view that the nascent transcript and chromatin clouds are somehow spatially compartmentalized (Colour *Plate 4*). Second, related studies demonstrated that most chromatin could be removed from human nuclei without disrupting the spatial organization of the transcription centres (Jackson *et al.*, 1993). This was interpreted to show that some architectural feature, but not the chromatin, must play a dominant role in regulating the distribution of the active sites. A nucleoskeleton or nuclear matrix is proposed to fulfil the necessary role.

How then does the idea of a spatially static active centre impact on our perception of gene expression? If the transcription sites are fixed, the chromatin, inevitably, must move during transcription (Colour *Plate 4*); the chromatin loops must be locally dynamic as RNA synthesis forces the gene to progressively associate (i.e. from promoter to 3' end) with the synthetic centre. The idea that genes might be associated with the active centre during synthesis and displaced from it thereafter defines a type of spatially determined 'transcription cycle' (Kimura *et al.*, 1999, 2002) within which the interplay between chromatin dynamics and promoter-bound factors will dictate levels of gene expression. The dynamic behaviour of transcription factors will also be central to this control. Interestingly, some classes of transcription factor can be shown to associate with chromatin only transiently (McNally *et al.*, 2000), whereas others are more stable, and might even remain bound to chromatin throughout mitosis (Chen *et al.*, 2002). Factors such as TBP (TATA binding protein), which fall in the second

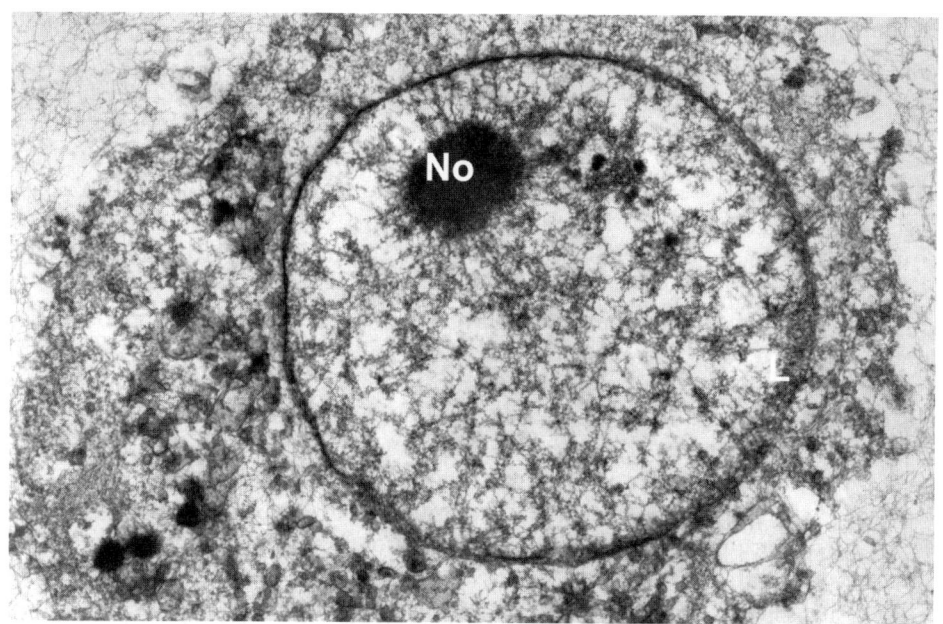

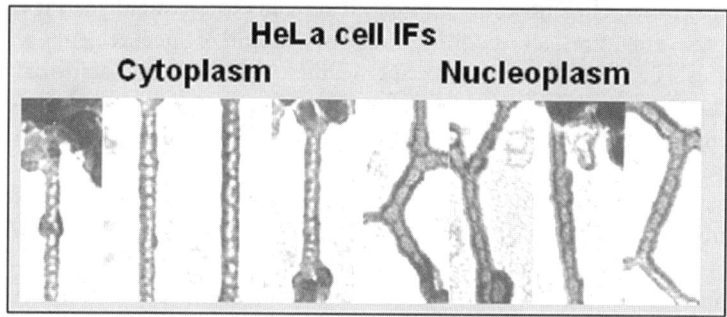

Figure 1. The nucleoskeleton in mammalian cells. To reveal the nucleoskeleton, HeLa cells encapsulated in agarose microbeads were permeabilized and chromatin cut with nucleases so that ~90% DNA could be removed before preparing a resinless electron micrograph. In this ~250 nm thick section, note that a diffuse agarose mesh surrounds the densely stained cell. The spherical, central nucleus is demarcated by a cage-like structure that forms from the highly branched nuclear lamin (L). Note that nuclear architecture is preserved even though most chromatin – half of the nuclear mass – is removed. The residual nucleoskeleton can be visualized as a diffuse network of coated filaments that pervade the nuclear volume and provide a 'solid phase' that seems to support nuclear compartments such as nucleoli (No), replication and transcription factories and inter chromatin granule clusters. Reproduced from Hozak et al. (1994). Replication factories and nuclear bodies: the ultrastructural characterization of replication sites during the cell cycle. J. Cell Sci. 107: 2191–2202 with permission of the Company of Biologists. The image below shows a sample of intermediate filaments from this type of preparation. In HeLa cells, the cytoplasmic vimentin filaments are unbranched and generally smooth, though some associated protein complexes are seen. The nuclear filaments, in contrast, are commonly branched. Staining and shadowing used in this technique reveals the 23–25 nm axial repeat that is characteristic of the intermediate filaments in mammalian cells. See Jackson and Cook (1988) for details.

category, provide ideal candidates to mark active genes and regulate the association of these genes with transcription sites.

7. Nuclear architecture and chromosome structure

The analysis of nuclear architecture begins to provide some clues as to the mechanisms that might influence the spatial organization of transcription sites. Chromosome structure and nuclear architecture clearly impact on both local and long-range chromatin organization (Cremer and Cremer, 2001) and the dynamic properties of chromatin (Gasser, 2002). Mammalian chromosomes occupy discrete regions within interphase nuclei. These chromosome 'territories' appear to have distinct borders with little mixing of chromatin from adjacent territories. Territories do not occupy specific positions, though gene activity can influence interphase chromosome location; chromosomes with a high density of active genes tend to be located towards the nuclear centre whereas those with a low density tend to lie closer to the nuclear periphery (Mahy et al., 2002a; Tanabe et al., 2002). In some instances, chromatin can be seen to escape from the confines of an individual territory to form loops that might spread 1 or 2 μm into the surrounding nucleoplasm (Mahy et al., 2002b; Williams et al., 2002). Chromatin loops of this sort most probably reflect the transcriptional status of highly expressed gene-rich chromosomal regions. It remains to be established if this type of chromatin movement reflects any functional association of specific genes with remote active centres – i.e. specialized nuclear sites – which lie outside the territorial boundary. It is worth noting that the activation of gene expression might correspond with the spatial changes in gene location that correspond with the transition for inert to active chromatin compartments (Francastel et al., 2001).

In discussing gene regulation it is important to remember that chromosome territories are not homogeneous structures but composed of a collection of DNA-rich foci that appear to be discrete structures that are separated by the RNA-rich interchromatin compartment (Colour *Plate 4*). Various studies confirm that transcription sites occupy this interchromatin compartment (Mahy et al., 2002a; Verschure et al., 2002). Hence it can be demonstrated that active chromatin (using specific antibodies for modified histones found in transcribed chromatin) is in spatially distinct compartments from transcription components such as the transcription factor TFIIH, RNA polymerase II and hnRNP-U (Verschure et al., 2002). The analysis of an active gene locus using the FISH (fluorescence *in situ* hybridization) approach also shows that DNA within a territory maintains a distinct spatial organization and that active genes can be transcribed from loci located throughout the territory and not only from the territory borders. Intriguingly, this technique also indicates that chromatin from transcribed regions of a locus might be located in apparent 'holes' in the chromosome paint, implying that the transcribed chromatin is highly decondensed, consistent with the architecture seen using high-resolution electron microscopy (Colour *Plate 4*).

In certain cases, chromatin location and, by association, global chromatin movement, also regulates gene expression. In the lymphocyte lineage, gene expression has been shown to correlate with gene location so that transcriptionally silent genes re-locate adjacent to centromeric heterochromatin in association with the DNA-binding protein Ikaros (Baxter et al., 2002). During B lymphocyte development the immunoglobulin H (IgH) and IgK loci are located at the nuclear periphery in haematopoietic progenitors and pro-T cells and in the nuclear interior in pro-B nuclei

(Kosak *et al.*, 2002). The inactive loci are associated with the nuclear lamina and must move to active sites within the nuclear interior before recombination and transcription of the IgH and IgK loci can occur. Observations of this type emphasize that certain nuclear compartments are permissive for gene expression while others are not.

8. The nuclear matrix

While the structure of chromosome territories provides one level of organization that impacts on gene expression it is clear that local features of nuclear structure can also influence gene expression (Jackson, 2003b). This level of organization provides a role for the nuclear matrix and related scaffold. Nuclear scaffold/matrix attachment regions (S/MARs) are DNA elements that clearly augment transcription and may achieve this using AT-rich sequences to modulate superhelical stress that arises during transcription (Bode *et al.*, 2000). The importance of these elements in gene regulation is confirmed by the fact that a classical MAR element is known to play a critical role in orchestrating the temporal and spatial expression of many genes during T-cell development (Alvarez *et al.*, 2000). In addition to this, efficient MAR elements have been shown to support long-term gene expression from ectopic chromosomal sites (Zan-Zabal *et al.*, 2001), suggesting that these elements might play a role in targeting gene domains to specific nuclear sites, such as matrix-associated transcription centres, prior to gene expression. It is difficult to define the relative importance of the different layers of control that regulate gene expression. Even so, at this point in time it will suffice to state that the layers exist and that the interplay between genetic elements in DNA, chromatin structure and spatial architecture will be critical to the process.

However one looks at gene expression, it is inevitable that some components of the transcription complex must move. The analysis of co-localization between transcription factors and transcription sites sheds some light on this by demonstrating that transcription factors are recruited to nuclear sites that are competent to perform gene expression. Only when this platform is established is a relevant gene recruited to the site (Choi *et al.*, 2001; Harrington *et al.*, 2002; Zaidi *et al.*, 2002). This type of model is supported by studies on the runt-related transcription factors (RUNX/CBFA/AML) that play essential roles in cellular differentiation and fetal development. A characterized domain of RUNX transcriptional activators targets the protein to discrete sub-nuclear foci. Removal of the targeting domain results in lethal haematopoietic and skeletal phenotypes, and implies that for this class of protein the correct nuclear location is critical for function (Choi *et al.*, 2001).

When RUNX1 and RUNX2 were studied in the human osteosarcoma cell line SaOS-2 (Harrington *et al.*, 2002) EGFP–RUNX fusion proteins were seen to occupy discrete foci that coincided with transcription sites, visualized by Bromo-UTP incorporation in permeabilized cells. An analysis of RUNX foci in living cells demonstrated that these structures moved only slightly over a period of 30 minutes, consistent with their association with the low-salt nuclear matrix. However, while the RUNX-rich foci themselves are spatially constrained their components are dynamic. Using fluorescence recovery after photobleaching (FRAP) the estimated half-life for recovery of EGFP–RUNX proteins in foci was ~10 seconds, ~20-fold slower that EGFP alone or the RUNX deletion without the nuclear matrix targeting domain. The phenotype of the RUNX2 deletion in transgenic animals and behaviour of proteins in cultured cells demonstrates that the functional activity of RUNX proteins is dependent on their location at specific nuclear sites. RUNX proteins also interact with Smads, a family of signalling proteins that regulate various developmental and biological processes in response to the transforming growth factor β/bone morphogenic protein family of growth factors. Critically, RUNX proteins are required to target Smads to nuclear sites where transcription is performed. Again this suggests that gene expression involves the *in situ* integration of critical signals through the assembly of regulatory complexes at transcriptionally active sub-nuclear sites (Zaidi *et al.*, 2002). This work and other studies (Francastel *et al.*, 2001; Schaufele *et al.*, 2002) suggest that activating factors engage appropriate nuclear sites before gene expression can occur.

9. Nuclear compartments

Even a cursory appreciation of the complexities of nuclear architecture touched on above leads us to consider how nuclear space might be controlled and if the organization of this space might play an active or passive role in gene regulation. We have seen that nuclei are highly structured and that different layers of structure – chromatin structure, chromosome architecture and global nuclear organization – might influence gene expression. It is now appropriate to develop the organizational theme by looking in a little more detail at mechanisms that might control spatial architecture in mammalian cells.

Many studies using GFP-tagged proteins analysed in living cells demonstrate that nuclear components are often extremely dynamic (Hendzel *et al.*, 2001). But this does not mean that they will be distributed evenly throughout the accessible nuclear space. Proteins of the so-called splicing speckles – also called the interchromatin granule clusters – emphasize this point. The speckles were named because of their appearance following immuno-labelling; in mammalian cells, many proteins involved in RNA metabolism are concentrated in ~20 large irregular foci that give an overall speckled appearance. But despite the fact that these sites are spatially structured and associated with the nuclear matrix the components within them are very dynamic. Hence, fluorescence loss in photobleaching shows that a particular splicing protein that is localized to speckles might be sufficiently dynamic that all molecules of the protein pass through the soluble nuclear compartment within a timeframe of about 10 minutes (Phair and Misteli, 2000). The speckled appearance represents a steady state and presumably reflects nuclear sites where interactions between the speckle components lead to their accumulation. While the components are dynamic the sites themselves might change little in structure over long periods of time. They are capable of changing however, and particularly dramatic changes in the shape and number (they lose surface texture to become spherical and the number decreases) of the sites is seen if transcription is inhibited.

But while speckles can be shown to represent classical nuclear matrix components the nature of their association remains controversial. In particular, it is notable that when the speckle components are dispersed by manipulating their phosphorylation status *in vivo*, the spaces that are revealed show no residual structures that might be related to a matrix network (Sacco-Bubulya and Spector, 2002). Partly because of this type of observation, the nature of the nuclear matrix remains a matter for debate. But while these compartments might be plastic it is also interesting to note that the spatial organization of sites such as transcription and replication factories and nuclear speckles is unaffected if nuclei are depleted of chromatin. This suggests that the compartments are spatially constrained and raises the possibility that a 'solid phase' exists within nuclei to provide a platform upon which these sites are assembled. A 'nucleoskeleton' has been proposed to fulfil his role (Jackson, 2003b).

By analogy with the cytoskeleton, which is known to play essential roles in many aspects of cytoplasmic function, it appears reasonable to assume that a nucleoskeleton might rely on the same architectural principles (Nickerson, 2001). The basic features would incorporate a skeleton of linked protein filaments and interacting adaptor proteins that dictate function. Core filaments that could provide a form of continuity throughout nuclei have been described. These filaments have characteristic morphological features of the intermediate filaments seen throughout the cytoplasm (Jackson and Cook, 1988; *Figure 1*). The best-characterized intermediate filaments within nuclei are undoubtedly the nuclear lamins (Goldman *et al.*, 2002). However, in addition to roles played by the nuclear lamins at the nuclear periphery, recent studies using GFP-lamin proteins and classical high-resolution immuno-staining have suggested that a veil of lamin filaments spreads throughout the nucleus (Moir *et al.*, 2000). These observations support the idea that a lamin-based network might pervade the nucleoplasm to provide a framework to co-ordinate the organization of various nuclear compartments. EM studies have also provided evidence for a network of lamin filaments in mammalian cells (Barboro *et al.*, 2003). This idea, however, remains controversial and it is still a matter of debate as to whether the lamins might form a contiguous internal nucleoskeleton. As an alternative, the lamins could feasibly play a structural role in the local organization of nuclear compartments such as transcription centres. Properties of the lamins are considered later in this chapter.

10. Proteins of the nuclear matrix – links to function

The nuclear matrix is classically described as an amorphous fibro-granular structure that can be isolated from nuclei by hypertonic treatment following nuclease digestion (Nickerson, 2001; Pederson, 1998). By this definition the matrix is perceived as being a product of the nucleoskeleton/nuclear lamina networks and associated proteins. The matrix typically contains many hundreds of different proteins, many of which have been studied in detail. Two of the best characterized – SAF-A (scaffold attachment factor A) and ARBP (attachment region binding protein) – were discovered because of their association with matrix-attached DNA elements (Romig *et al.*, 1992; Stratling and Yu, 1999). SAF-A turns out to be a major RNA-binding protein (hnRNP-U) and ARBP a protein that binds methylated DNA (MeCP2). SAF-A has been shown to bind p300, a major transcriptional co-activator, and so recruit active genes to the nuclear matrix (Martens *et al.*, 2002). ARBP, in contrast, interacts with methylated DNA in MAR elements and through an intermediary that contains Sin3A protein recruits a histone deacetylase to generate a silenced chromatin state. Another protein, SAF-B, (scaffold attachment factor B) specifically binds to S/MAR regions, interacts with RNA polymerase II (RNA pol II) and a subset of serine-/arginine-rich RNA-processing factors (SR proteins). It was proposed that these interactions allow SAF-B to provide a surface for the assembly of the transcription apparatus (Nayler *et al.*, 1998). As with the vast majority of characterized matrix proteins these are involved in different aspects of chromatin function. Extending this idea implies that the nucleoskeleton and nuclear matrix (that forms during salt extraction) are an expression of different processes that are performed within the interchromatin space.

Recent studies on the nuclear matrix protein SATB1 (special AT-rich sequence binding 1) emphasize this view. SATB1 is a nuclear matrix component that is found predominantly in thymocytes (Alvarez *et al.*, 2000). In these cells, a nuclear network of SATB1 binds chromatin to form chromatin domains that are believed to play important roles in gene regulation. SATB1 in association with MARs acts as a transcriptional repressor to regulate gene expression during T-cell development. This protein has the classical properties of a nuclear matrix component and deletion of SATB1 results in defects in the temporal and spatial regulation of many lymphocyte genes, leading to an arrest of T-cell development. SATB1 is known to target chromatin-remodelling factors to specific chromatin domains and so provide a mechanism for the long-range regulation of gene expression (Yasui *et al.*, 2002). SATB1 recruits histone deacetylase complexes containing the NURD chromatin-remodelling complex to SATB1-bound sites in the interleukin-2 receptor gene to elicit specific deacetylation of histones within the locus. In addition, SATB1 was shown to target CHRAC and ACF chromatin-remodelling complexes to regulate nucleosome positioning over several kb DNA. SATB1 might also play a direct role in recruiting RNA polymerase II to specific nuclear sites.

11. The nucleoskeleton and nuclear lamina – common themes

A small numbers of reports have used rather advanced techniques to reveal a network of core filaments that might form the basis of the nucleoskeleton. A seminal study in this regard demonstrated that the nucleoskeleton was composed of a branched network of intermediate filament proteins (Jackson and Cook, 1988). In view of this observation, the nuclear lamins provide the best candidates to form any internal nucleoskeleton. With this in mind, it is reasonable to argue that a detailed understanding of the structure and function of the lamin filaments is likely to reveal whether related structures might play any critical roles in the nuclear interior. As the behaviour of the nuclear lamina has been review extensively (Goldman *et al.*, 2002; Wilson *et al.*, 2001; and references therein) only features pertinent to the present theme will be considered here.

During the 1960s it was recognized that nuclei from mammalian cells had a protein network between the nuclear envelope and peripheral chromatin. This network or lamina is composed of a branched network of polymerized lamin proteins. The lamins are classified as type V intermediate filaments (IFs), and like all intermediate filament proteins have a specific domain with an α-helical coiled coil region flanked by non-helical domains. The central coiled coil domain is essential for assembly of the filaments. This process occurs initially through dimers and then higher multimers, such that a mature IF of 10 nm in cross-section has ~32 monomers in its cross-section. During the assembly process, specific alignment of the IF protein monomers gives a repeat structure with 23–25 nm periodicity. Major IF proteins in the cytoplasm of mammalian cells, such as vimentin, form very regular 10-nm (diameter) filaments that do not branch.

The nuclear lamins, in contrast, are highly branched and branches of smaller cross-section (i.e. suggesting filaments with fewer monomers in cross-section) are often seen. The lamin proteins contain nuclear localization signals and are subjected to post-translational modifications such as phosphorylation and most importantly are modified by isoprenylation, a modification which targets the lamin filaments to the nuclear envelope. This modification takes place on a special CaaX motif (not found in lamin C). Following cleavage after the cysteine residue the new terminal amino acid is modified by isoprenylation and methyl esterification. In addition to the peripheral nuclear lamina, nuclear lamin proteins are also found within the nuclear interior. *In vivo* studies support the view that these, like the peripheral proteins, are assembled into structures that have rather slow rates of exchange. However, it is still unclear if this represents a lamin filament network that pervades the nucleus or local aggregates of lamin proteins that might serve a function that does not demand a contiguous 'nucleoskeleton'.

12. The lamin genes

The lamin proteins represent a small protein family that has increased in complexity during metazoan evolution. Simple eukaryotes, such as the yeast *S. cerevisiae*, do not have lamin proteins; perhaps the lamina evolved during the transition from a closed to open mitosis; plants, however, which show an open mitosis, also lack lamin genes (discussed in Chapters 1 and 5). The simple multicellular organism *Caenorhabditis elegans* has a single lamin gene, which is expressed in all somatic cells. Mammals have three lamin genes (*LMNA*, *LMNB1* and *LMNB2*) that are processed to yield seven iso-forms. RNA processing – alternative splicing – of the *LMNA* primary transcript generates four A-type lamins – lamins A, AΔ10 (a splice variant without exon 10), C and C2. Three B-type lamins, called lamins B1–B3, are encoded by the *LMNB1/2* genes. All vertebrate cells express at least one variant of the B-type lamin proteins. The A-type lamins, in contrast, are expressed primarily in differentiated cells so that patterns of lamins A, AΔ10 and C expression are developmentally regulated. Lamins C2 and B3 are found only in germ cells.

13. Lamin function

Various studies identify structural features as a major role for the lamin proteins. It is notable, for example, that nuclei maintain their shape even after most of the nuclear contents have been removed. This and many other experiments suggest a role for the lamins in maintaining the mechanical stability of nuclei. The importance of maintaining a particular nuclear shape is not clear. Perhaps more important functionally is the possibility that the nuclear lamina can potentially define nuclear sites that have particular functional roles. Many proteins are known that act as adaptors that link the lamina to the nuclear envelope and components within chromatin to the lamina. Heterochromatin – which is transcriptionally inert – is commonly associated with the nuclear periphery and this location might have important implications for the regulation of gene expression. A-type lamins bind the retinoblastoma protein and other transcriptional repressors and may play important roles in regulating gene expression through spatial organization.

Other experiments support the fascinating possibility that nuclear lamins might be involved in the processes of DNA and RNA synthesis. The experimental evidence involves the introduction into cells – by micro-injection – of bacterially expressed lamin proteins that have amino-terminal deletions. The micro-injected proteins are engineered so that they interact with the endogenous proteins to disrupt stable filament formation. It has been reported that when the lamin filaments are disrupted in this way, DNA synthesis and RNA polymerase II-dependent RNA synthesis are also disrupted. One explanation for this is that lamin filaments or perhaps smaller local aggregates of the lamin proteins play some structural role within the active centres of DNA and RNA synthesis.

Many proteins are known to interact with the lamins to influence lamin function. These might be classified into two broad groups. Integral proteins of the inner nuclear membrane are the first group and include various isoforms of lamin-associated protein 2 (LAP2), LAP1, lamin B receptor (LBR), emerin, otefin, MAN1, Nurim, nesprin, RING finger-binding protein, A-kinase anchoring protein 149, and p19 (an isoquinoline-binding protein that is also found in the endoplasmic reticulum). The second group contains proteins that associate with the lamins but are not membrane components. This group includes germ cell less (GCL), young arrest (YA), PP1 phosphatase and the transcription factor Oct1.

The nuclear lamina is known to play a role in chromatin organization and the lamins can be shown to bind DNA elements called nuclear matrix and scaffold attachment regions (S/MARs) *in vitro*. The LAP2 isoforms LAP2α and LAP2β bind to chromatin. Chromatin binding involves two motifs on the LAP proteins, a chromatin-binding domain and a LEM domain. The LEM domain is a region of about 40 amino acids that is also found in emerin, otefin and MAN1. The LEM domain interacts with other chromatin-associated proteins such as BAF – and might then influence chromatin architecture and function.

14. Internal nuclear lamins

The lamina appears to form a single sheet of inter-connected filaments that lies opposed to the nuclear face of the nuclear membrane. Electron microscopy shows the vertebrate lamina to be an interlocked network of filaments that appears to have a 'woven' appearance. However, the most compelling images describing the structure of the lamina come from the extracted membranes of *Xenopus* oocytes and it is not clear to what extent these structures reflect those found in somatic mammalian cells. Unpublished observations by the author show that the lamina from mammalian cells is a highly branched network containing regions with filament diameter of 10, ~8 and ~5 nm. The structure of the filaments reflect the branch density, with the smaller filaments found at four-way junctions (author's unpublished data). Nucleo-filaments found deep in the nucleus are also branched (Jackson and Cook, 1988; *Figure 1*) and have filament diameters of 10 and ~8 nm. However, in appearance these internal filaments are more similar to intermediate filaments of the cytoskeleton than the nuclear lamina. The nucleo-filaments have a branch density of <3 branches/μm whereas the lamina filaments have ~25 branches/μm.

Taken together, this morphological evidence and electron microscope immunocytochemistry studies support the view that a lamin-based filament network spreads throughout the nucleus. Differences in appearance of the lamina and internal network warrant attention. However, it is not clear how different lamin proteins contribute to the structure of the lamina filaments or how the organization of the filaments might vary when assembled in different nuclear environments. The proximity of the nuclear membrane in particular might be expected to have a profound effect on the structure and appearance of the filaments as lamin B (and perhaps lamin C prior to its carboxy-terminal processing) are linked through their post-translational modifications to the nuclear membrane. Importantly, it is known that the LAP proteins are found throughout the nucleus and these and other lamin-associated proteins could provide a means of linking chromatin and classical nuclear matrix proteins to a lamin-based nucleoskeleton.

15. The pathway of gene expression – from gene to cytoplasm

The preceding discussion describes the early stages of the gene expression pathway, focusing on the conditions that are required to generate the primary transcript and how chromatin structure, chromosome architecture and global nuclear organization might regulate this process. Concerning gene expression, it is self-evident that activating transcription is a crucial first step in the process – without the transcript no post-transcription regulation would be possible. For protein-coding genes, once transcription is activated the train of events that will eventually lead to protein synthesis is set in motion. Critical aspects of regulation take place at the transcription site. For example, during transcription the transcript is modified at the 5' and 3' termini by special structures – the 5' Cap and 3' polyA tail – that regulate mRNA stability. Transcription is also accompanied by the process of RNA splicing, which removes intronic DNA sequences to generate a mature mRNA with protein coding potential. This process has two additional consequences for gene expression. First, most genes have many introns and exons and the process of differential splicing contributes to the complexity of the proteome – while the human genome appears to have roughly 35 000 genes at least half of these are thought to be differentially spliced to produce many more proteins than the number of primary transcription units. Second, the process of splicing is important for maturation of the protein complexes that are associated with the nascent RNA. In particular, splicing deposits a protein complex at the exon–exon junction and this complex is used to ensure that the final product is capable of generating an intact polypeptide. Only when splicing is complete are the mature mRNA–protein complexes able to move from the transcription sites and into the inter-chromatin channels en route to the cytoplasm.

16. RNA export pathways

Basic details of the primary mechanisms that control mRNP export have been elucidated (Reed and Hurt, 2002). The molecular interactions that underlie this process are dependent on events that take place at the site of mRNA synthesis. The critical components are small proteins that are part of a much larger complex that is deposited on the mRNA during splicing. The complex – called the –20/24 complex because of its location at about 20 nucleotides upstream of the exon–intron junction – contains a number of proteins including the protein Aly. Aly interacts specifically with TAP/mex67, which serves a major function for mRNAs. It is important to recognize that mRNA export is coupled to many of the steps of gene expression. In broad terms, the pathway that delivers mature mRNAs to the cytoplasm is influenced by promoter strength, the RNA polymerase complex that performs synthesis, transcriptional elongation, as well as downstream processing events such as RNA splicing and polyadenylation. A significant aspect of this pathway reflects the fact that mRNAs are retained at the transcription site until the required processing events have been performed. This is likely to be part of the 'quality control' process that ensures that only authentic mature mRNAs are exported to the cytoplasm.

17. Nuclear transport – facilitated diffusion

Once they have been released from the transcription site, mature mRNP complexes are believed to move to the nuclear periphery by diffusion through the inter-chromatin channels. Indeed, while the nucleus is known to contain significant amounts of actin and tubulin, these proteins do not appear to be assembled into extensive filamentous networks like those seen in the cytoplasm. At present there is no compelling evidence to support the view that nuclear complexes rely on motor proteins that operate on a filament network for their movement to the cytoplasm. At the nuclear periphery, export cargoes dock on the pore filaments that emanate from the nucleoplasmic face of the pore complex (Suntharalingam and Wente 2003; Vasu and Forbes, 2001). Transport then appears to occur by a process of directed diffusion – or facilitated translocation. Models of facilitated translocation suppose that selectivity – gating – is achieved through interactions between the transport complex and particular protein motifs in the pore proteins – called FG repeats. The FG repeats are proposed to provide a sort of hydrophobic plug that severely reduces the efficiency with which molecules that are not destined for transport are able to pass through nuclear pores. The receptor/cargo complex, in contrast, by virtue of its structure and interaction with other pore components is able to diffuse through this region of the pore. Though the details by which selectivity is achieved remain to be refined, changes in pore permeability following treatment with organic solvents are consistent with a role for a hydrophobic meshwork (provided by the FG repeats) inside the pore channel.

Some insight into the translocation mechanism is provided by a detailed analysis of the location of mRNP complexes during export. Interestingly, while a classical view holds that translocation occurs through the centre of the pore, high-resolution studies show that the export complexes are associated with the periphery of the pore – as seen in cross-section (Iborra, 2002). This implies that movement through the pore relies on a series of interactions between the export complex and pore components in the axial ring of the pore complex. Following translocation, the mRNP complex becomes associated with ribosomes to provide a template for protein synthesis.

18. Gene regulation and transcript stability

mRNPs released from transcription sites reach the cytoplasm in the timescale of minutes (Femino et al., 1998). Within the cytoplasm, many factors regulate mRNA stability and turnover; however these complex processes fall beyond the scope of this review. Intriguingly, one aspect of regulation that appears to have a nuclear component is the degradation of defective mRNAs by nonsense mediated decay (NMD). NMD of mRNA is a process by which RNA molecules with translational stop codons upstream of introns are recognized as being aberrant and destined for destruction (Hentze and Kulozik, 1999; Wilkinson and Shyu, 2002). This process is linked to the generation of the post-splicing complex. The molecular mechanism of NMD involves the UPF proteins (Culbertson and Leeds, 2003). UPF1 is associated with the exon junction complex and UPF2 and 3 associate to activate decay. As stop codon recognition is a critical feature of this process it is assumed that scanning by a ribosome is required. In many cells it appears that some level of NMD can take place within nuclei (Buhler et al., 2002; Wilkinson and Shyu, 2002), which may account for recent reports that some ribosome-based polypeptide synthesis occurs within the nucleus of mammalian cells (Iborra et al., 2001).

19. Conclusions

The nuclear features that contribute to the control of gene expression are extremely complex. This chapter sets our to provide a brief overview of the different features that contribute to this process. The pathway begins at the gene promoter where transcription factors bind to recruit the transcription machinery so that RNA synthesis can begin. In addition to promoters, many other DNA elements – such as enhancers – contribute to the activation of gene expression. Enhancers, like promoters, bind transcription factors and the complexes formed appear to augment the activation of gene expression. Recent evidence suggests that the enhancer and promoter complexes form a tertiary complex that recruits the transcription machinery (Carter *et al.*, 2002; Tolhuis *et al.*, 2002).

Events that precede RNA synthesis rely on simple DNA–protein interactions, and though this process is complicated by chromatin structure its molecular basis is easy to understand. Other features that regulate gene expression are less easy to define in molecular detail. For example, it is known that both chromosome structure and nuclear location can influence gene expression. However, the behaviour of these epigenetic features is often subtle and at present we do not know enough about the molecular details to predict how different locations might alter gene expression. One aspect of this level of organization relates to the structure of the active transcription sites. Here, the traditional view holds that individual transcription complexes engage the promoter and then scan along the gene as transcription proceeds. We now know that the components needed to generate mature mRNAs are so bulky and the volume occupied by transcription centres so small (Jackson *et al.*, 1998) that it seems inconceivable that the synthetic machinery and nascent transcript might move along the euchromatin template. Instead, centres of gene expression appear to be organized so that the chromatin template, transcription machinery and nascent product each occupy distinct zones within dedicated sites of gene expression. In this case, nuclear structure dictates that transcription sites are spatially constrained so that the template must become the dynamic participant during the transcription process. This is consistent with transcription occurring at the interface between the chromatin and inter-chromatin compartment and ensures that mature mRNAs pass directly into the inter-chromatin channels for efficient export to the cytoplasm. This arrangement is epitomised by the architecture of the transcription centres within nucleoli (Dundr *et al.*, 2002; Koberna *et al.*, 2002) and applies equally to sites of gene expression throughout the nucleoplasm (Jackson *et al.*, 1998; Mahy *et al.*, 2002b). This view of the active centres also means that their anatomy is not dictated by the functional requirements of a single transcription complex. Consequently, each active site can be very complex, simultaneously performing both the synthesis and processing of many transcripts from groups of genes. The architecture of gene clusters and their mechanisms of interaction with the transcription centres must be crucial to the control of gene expression in eukaryotic cells.

As with nucleoli, many components of the nucleoplasmic transcription sites must be incorporated *de novo*, though some factors remain bound to chromatin throughout the cell cycle to mark genes that are committed for expression (Chen *et al.*, 2002; Martinezbalbas *et al.*, 1995). Other transcription factors required to activate gene expression accumulate at discrete nuclear sites as foci, presumably through their association with other nuclear sites, perhaps a nucleoskeleton. These sites might subsequently provide a platform upon which the transcription machinery and appropriate genes interact to engage gene expression. Precisely how the spatial organization of these sites is maintained remains a matter for speculation. However, it is interesting to note that under some circumstances disruption of the nuclear network formed from nuclear lamins is accompanied by a dramatic reduction in transcription by RNA polymerase II (Spann *et al.*, 2002). This raises the possibility that lamin filaments throughout the nucleus provide a framework upon which the active sites of RNA polymerase II transcription are assembled.

Events occurring at the transcription centres activate gene expression. However, many other regulatory mechanisms contribute to this extremely complex process. At the transcription site, the primary transcripts are processed and assembled into an RNA–protein complex that is competent to move from the transcription site, through the inter-chromatin channels of the nucleus and through nuclear pores to the cytoplasm. Each of these steps is regulated to ensure that only authentic mature mRNAs reach their destination in the cytoplasm where protein synthesis completes the process of gene expression.

References

Alvarez, J.D., Yasui, D.H., Niida, H., Joh, T., Loh, D.Y. and Kohwi-Shigematsu, T. (2000) The MAR-binding protein SATB1 orchestrates temporal and spatial expression of multiple genes during T-cell development. *Gene Dev.* **14**: 521–535.

Barboro, P., D'Arrigo, C., Diaspro, A., Mormino, M., Alberti, I., Parodi, S., Patrone, E. and Balbi, C. (2002) Unraveling the organization of the internal nuclear matrix: RNA-dependent anchoring of NuMA to a lamin scaffold. *Exp. Cell. Res.* **279**: 202–218.

Baxter, J., Merkenschlager, M. and Fisher, A.G. (2002) Nuclear organisation and gene expression. *Curr. Opin. Cell. Biol.* **14**: 372–376.

Becker, P.B. (2002) Nucleosome sliding: facts and fiction. *EMBO J.* **21**: 4749–4753.

Bell, A.C., West, A.G. and Felsenfeld, G. (2001) Gene regulation – Insulators and boundaries: Versatile regulatory elements in the eukaryotic genome. *Science* **291**: 447–450.

Blackwood, E.M. and Kadonaga, J.T. (1998) Going the distance: a current view of enhancer action. *Science* **281**: 60–63.

Bode, J., Benham, C., Knopp, A. and Mielke, C. (2000) Transcriptional augmentation: modulation of gene expression by scaffold/matrix attached regions (S/MAR elements). *Crit. Rev. Eukaryotic Gene Expr.* **10**: 73–90.

Buhler, M., Wilkinson, M.F. and Muhlemann, O. (2002) Intranuclear degradation of nonsense codon-containing mRNA. *EMBO Rep.* **3**: 646–651.

Bulger, M., Sawado, T., Schubeler, D. and Groudine, M. (2002) ChIPs of the beta-globin locus: unraveling gene regulation within an active domain. *Curr. Opin. Genet. Dev.* **12**: 170–177.

Butler, J.E.F. and Kadonaga, J.T. (2002) The RNA polymerase II core promoter: a key component in the regulation of gene expression. *Gene Dev.* **16**: 2583–2592.

Carter, D., Chakalova, L., Osborne, C.S., Dai, Y.F. and Fraser, P. (2002) Long-range chromatin regulatory interactions in vivo. *Nat. Genet.* **32**: 623–626.

Chen, D.Y., Hinkley, C.S., Henry, R.W. and Huang, S. (2002) TBP dynamics in living human cells: Constitutive association of TBP with mitotic chromosomes. *Mol. Biol. Cell.* **13**: 276–284.

Choi, J.Y., Pratap, J., Javed, A., *et al.* (2001) Subnuclear targeting of Runx/Cbfa/AML factors is essential for tissue-specific differentiation during embryonic development. *Proc. Natl Acad. Sci. USA* **98**: 8650–8655.

Cook, P.R. (1999) The organization of replication and transcription. *Science* **284**: 1790–1795.

Cremer, T. and Cremer, C. (2001) Chromosome territories, nuclear architecture and gene regulation in mammalian cells. *Nat. Rev. Genet.* **2**: 292–301.

Culbertson, M.R. and Leeds, P.F. (2003) Looking at mRNA decay pathways through the window of molecular evolution. *Curr. Opin. Genet. Dev.* **13**: 207–214.

Dillon, N. and Festenstein, R. (2002) Unravelling heterochromatin: competition between positive and negative factors regulates accessibility. *Trends Genet.* **18**: 252–258.

Dundr, M., Hoffmann-Rohrer, U., Hu, Q.Y., Grummt, I., Rothblum, L.I., Phair, R.D. and Misteli, T. (2002) A kinetic framework for a mammalian RNA polymerase in vivo. *Science* **298**: 1623–1626.

Elbi, C., Misteli, T. and Hager, G.L. (2002) Recruitment of dioxin receptor to active transcription sites. *Mol. Biol. Cell.* **13**: 2001–2015.

Emerson, B.M. (2002) Specificity of gene regulation. *Cell* **109**: 267–270.

Femino, A.M., Fay, F.S., Fogarty, K. and Singer, R.H. (1998) Visualization of single RNA transcripts in situ. *Science* **280**: 585–590.

Francastel, C., Magis, W. and Groudine, M. (2001) Nuclear relocation of a transactivator subunit precedes target gene activation. *Proc. Natl Acad. Sci. USA* **98**: 12120–12125.

Gasser, S.M. (2002) Visualizing chromatin dynamics in interphase nuclei. *Science* **296**: 1412–1416.

Goldman, R.D., Gruenbaum, Y., Moir, R.D., Shumaker, D.K. and Spann, T.P. (2002) Nuclear lamins: building blocks of nuclear architecture. *Gene Dev.* **16**: 533–547.

Grande, M.A., van der Kraan, I., de Jong, L. and van Driel, R. (1997) Nuclear distribution of transcription factors in relation to sites of transcription and RNA polymerase II. *J. Cell Sci.* **110**: 1781–1791.

Grewal, S.I.S. and Moazed, D. (2003) Heterochromatin and epigenetic control of gene expression. *Science* **301**: 798–802.

Harrington, K.S., Javed, A., Drissi, H., McNeil, S., Lian, J.B., Stein, J.L., van Wijnen, A.J., Wang, Y.-L. and Stein, G.S. (2002) Transcription factors RUNX1/AML1 and RUNX2/Cbfa1 dynamically associate with stationary subnuclear domains. *J. Cell. Sci.* **115**: 4167–4176.

Hendzel, M.J., Kruhlak, M.J., MacLean, N.A.B., Boisvert, F.M., Lever, M.A. and Bazett-Jones, D.P. (2001) Compartmentalization of regulatory proteins in the cell nucleus. *J. Steroid Biochem.* **76**: 9–21.

Hentze, M.W. and Kulozik, A.E. (1999) A perfect message: RNA surveillance and nonsense-mediated decay. *Cell* **96**: 307–310.

Hozak, P., Jackson, D.A. and Cook, P.R. (1994) Replication factories and nuclear bodies: the ultrastructural characterization of replication sites during the cell cycle. *J. Cell. Sci.* **107**: 2191–2202.

Iborra, F.J. (2002) The path that RNA takes from the nucleus to the cytoplasm: a trip with some surprises. *Histochem. Cell. Biol.* **118**: 95–103.

Iborra, F.J., Pombo, A., Jackson, D.A. and Cook, P.R. (1996) Active RNA polymerases are localized within discrete transcription factories in human nuclei. *J. Cell. Sci.* **109**: 1427–1436.

Iborra, F.J., Jackson, D.A. and Cook, P.R. (1998) The path of transcripts from extra-nucleolar synthetic sites to nuclear pores: transcripts in transit are concentrated in discrete structures containing SR proteins. *J. Cell. Sci.* **111**: 2269–2282.

Iborra, F.J., Jackson, D.A. and Cook, P.R. (2001) Sites of translation in mammalian cell nuclei. *Science* **293**: 1139–1142.

Jackson, D.A. (2003a) The anatomy of transcription sites. *Curr. Opin. Cell. Biol.* **15**: 311–317.

Jackson, D.A. (2003b) Principles of nuclear structure. *Chromosome Res.* **11**: 387–401.

Jackson, D.A. and Cook, P.R. (1988) Visualization of a nucleoskeleton. *EMBO J.* **7**: 3667–3677.

Jackson, D.A., Hassan, A.B., Errington, R.J. and Cook, P.R. (1993) Visualization of focal sites of transcription within human nuclei. *EMBO J.* **12**: 1059–1065.

Jackson, D.A., Iborra, F.J., Manders, E.M. and Cook, P.R. (1998) Numbers and organization of RNA polymerases, nascent transcripts, and transcription units in HeLa nuclei. *Mol. Biol. Cell.* **9**: 1523–1536.

Jenuwein, T. and Allis, C.D. (2001) Translating the histone code. *Science* **293**: 1074–1080.

Kimura, H., Tao, Y., Roeder, R.G. and Cook, P.R. (1999) Quantitation of RNA polymerase II and its transcription factors in an HeLa cell: Little soluble holoenzyme but significant amounts of polymerases attached to the nuclear substructure. *Mol. Cell. Biol.* **19**: 5383–5392.

Kimura, H., Sugaya, K. and Cook, P.R. (2002) The transcription cycle of RNA polymerase II in living cells. *J. Cell. Biol.* **159**: 777–782.

Koberna, K., Malinsky, J., Pliss, A., Masata, M., Vecerova, J., Fialova, M., Bednar, J. and Raska, I. (2002) Ribosomal genes in focus: new transcripts label the dense fibrillar components and form clusters indicative of "Christmas trees" in situ. *J. Cell. Biol.* **157**: 743–748.

Kosak, S.T., Skok, J.A., Medina, K.L., Riblet, R., Le Beau, M.M., Fisher, A.G. and Singh, H. (2002) Subnuclear compartmentalization of immunoglobulin loci during lymphocyte development. *Science* **296**: 158–162.

Litt, M.D., Simpson, M., Recillas-Targa, F., Prioleau, M.N. and Felsenfeld, G. (2001) Transitions in histone acetylation reveal boundaries of three separately regulated neighboring loci. *EMBO J.* **20**: 2224–2235.

Mahy, N.L., Perry, P.E., Gilchrist, S., Baldock, R.A. and Bickmore, W.A. (2002a) Spatial organization of active and inactive genes and noncoding DNA within chromosome territories. *J. Cell. Biol.* **157**: 579–589.

Mahy, N.L., Perry, P.E. and Bickmore, W.A. (2002b) Gene density and transcription influence the localization of chromatin outside of chromosome territories detectable by FISH. *J. Cell. Biol.* **159**: 753–763.

Maniatis, T. and Reed, R. (2002) An extensive network of coupling among gene expression machines. *Nature* **416**: 499–506.

Martens, J.H.A., Verlaan, M., Kalkhoven, E., Dorsman, J.C. and Zantema, A. (2002) Scaffold/matrix attachment region elements interact with a p300-scaffold attachment factor A complex and are bound by acetylated nucleosomes. *Mol. Cell. Biol.* **22**: 2598–2606.

Martinezbalbas, M.A., Dey, A., Rabindran, S.K., Ozato, K. and Wu, C. (1995) Displacement of sequence-specific transcription factors from mitotic chromatin. *Cell* **83**: 29–38.

McNally, J.G., Muller, W.G., Walker, D., Wolford, R. and Hager, G.L. (2000) The glucocorticoid receptor: rapid exchange with regulatory sites in living cells. *Science* **287**: 1262–1265.

Moir, R.D., Yoon, M., Khuon, S. and Goldman, R.D. (2000) Nuclear lamins A and B1: Different pathways of assembly during nuclear envelope formation in living cells. *J. Cell. Biol.* **151**: 1155–1168.

Narlikar, G.J., Fan, H.Y. and Kingston, R.E. (2002) Cooperation between complexes that regulate chromatin structure and transcription. *Cell* **108**: 475–487.

Nayler, O., Stratling, W., Bourquin, J.P., Stagljar, I., Lindemann, L., Jasper, H., Hartmann, A.M., Fackelmayer, F.O., Ullrich, A. and Stamm, S. (1998) SAF-B protein couples transcription and pre-mRNA splicing to SAR/MAR elements. *Nucl. Acids Res.* **26**: 3542–3549.

Nickerson, J.A. (2001) Experimental observations of a nuclear matrix. *J. Cell. Sci.* **114**: 463–474.

Orphanides, G. and Reinberg, D. (2002) A unified theory of gene expression. *Cell* **108**: 439–451.

Pederson, T. (1998) Thinking about a nuclear matrix. *J. Mol. Biol.* **227**: 147–159.

Phair, R.D. and Misteli, T. (2000) High mobility of proteins in the mammalian cell nucleus. *Nature* **404**: 604–609.

Pombo, A., Jackson, D.A., Hollinshead, M., Wang, Z., Roeder, R.G. and Cook, P.R. (1999) Regional specialization in human nuclei: visualization of discrete sites of transcription by RNA polymerase III. *EMBO J.* **18**: 2241–2253.

Reed, R. and Hurt, E. (2002) A conserved mRNA export machinery coupled to pre-mRNA splicing. *Cell* **108**: 523–531.

Romig, H., Fackelmayer, F.O., Renz, A., Ramsperger, U. and Richter, A. (1992) Characterization of SAF-A, a novel nuclear DNA binding protein from HeLa cells with high affinity for nuclear matrix scaffold attachment DNA elements. *EMBO J.* **11**: 3431–3440.

Sacco-Bubulya, P. and Spector, D.L. (2002) Disassembly of interchromatin granule clusters alters the coordination of transcription and pre-mRNA splicing. *J. Cell. Biol.* **156**: 425–436.

Schaufele, F., Enwright, J.F., Wang, X., Teoh, C., Srihari, R., Erickson, R., MacDougald, O.A. and Day, R.N. (2002) CCAAT/enhancer binding protein alpha assembles essential cooperating factors in common subnuclear domains. *Mol. Endocrinol.* **15**: 1665–1676.

Spann, T.E., Goldman, A.E., Wang, C., Huang, S. and Goldman, R.D. (2002) Alteration of nuclear lamin organization inhibis RNA polymerase II-dependent transcription. *J. Cell. Biol.* **156**: 603–608.

Stratling, W.H. and Yu, F. (1999) Origin and roles of nuclear matrix proteins. Specific functions of the MAR-binding protein MeCP2/ARBP. *Crit. Rev. Eukaryotic Gene Expr.* **9**: 31–318.

Suntharalingam, M. and Wente, S.R. (2003) Peering through the pore: Nuclear pore complex structure, assembly, and function. *Dev. Cell* **4**: 775–789.

Tanabe, H., Muller, S., Neusser, M., von Hase, J., Calcagno, E., Cremer, M., Solovei, I., Cremer, C. and Cremer, T. (2002) Evolutionary conservation of chromosome territory arrangements in cell nuclei from higher primates. *Proc. Natl Acad. Sci. USA* **99**: 4424–4429.

Tolhuis, B., Palstra, R.J., Splinter, E., Grosveld, F. and de Laat, W. (2002) Looping and interaction between hypersensitive sites in the active beta-globin locus. *Mol. Cell.* **10**: 1453–1465.

Vasu, S.K. and Forbes, D.J. (2001) Nuclear pores and nuclear assembly. *Curr. Opin. Cell. Biol.* **13**: 363–375.

Verschure, P.J., van der Kraan, I., Enserink, J.M., Mone, M.J., Manders, E.M.M. and van Driel, R. (2002) Large-scale chromatin organization and the localization of proteins involved in gene expression in human cells. *J. Histochem. Cytochem.* **50**: 1303–1312.

Wallace, H., Ansell, R., Clark, J. and McWhir, J. (2000) Pre-selection of integration sites imparts repeatable transgene expression. *Nucl. Acids Res.* **28**: 1455–1464.

Wilkinson, M.F. and Shyu, B.-A. (2002) RNA surveillance by nuclear scanning? *Nat. Cell. Biol.* **4**: E144–E147.

Williams, R.R.E., Broad, S., Sheer, D. and Ragoussis, J. (2002) Subchromosomal positioning of the epidermal differentiation complex (EDC) in keratinocyte and lymphoblast interphase nuclei. *Exp. Cell. Res.* **272**: 163–175.

Wilson, C., Bellen, H.J. and Gehring, W.J. (1990) Position effects on eukaryotic gene expression. *Annu. Rev. Cell. Biol.* **6**: 679–714.

Wilson, K.L., Zastrow, M.S. and Lee, K.K. (2001) Lamins and disease: Insights into nuclear infrastructure. *Cell* **104**: 647–650.

Yasui, D., Miyano, M., Varga-Weisz, P. and Kohwi-Shigematsu, T. (2002) SATB1 regulates gene expression over long distances via chromatin remodelling. *Nature* **419**: 641–645.

Zahn-Zabal, M., Kobr, M., Girod, P.A., Imhof, M., Chatellard, P., de Jesus, M., Wurm, F. and Mermod, N. (2001) Development of stable cell lines for production or regulated expression using matrix attachment regions. *J. Biotechnol.* **87**: 29–42.

Zaidi, S.K., Sullivan, A.J., van Wijnen, A.J., Stein, J.L., Stein, G.S. and Lian, J.B. (2002) Integration of Runx and Smad regulatory signals at transcriptionally active subnuclear sites. *Proc. Natl Acad. Sci. USA* **99**: 8048–8053.

9

Nuclear shuttling in plant cells

Sondra G. Lazarowitz, Roisin C. McGarry, Yoshimi D. Barron and Miguel F. Carvalho

1. Introduction

The past decade has witnessed major advances in our understanding of the mechanism and regulation of nucleo-cytoplasmic transport in vertebrates and yeast (reviewed in Macara, 2001; Ossareh-Nazari *et al.*, 2001; Weis, 2002). This transport occurs through nuclear pore complexes (NPC), large supramolecular structures that span the double membrane of the nuclear envelope (NE). Ions and small molecules less than ~40 kDa can passively diffuse through aqueous channels that cross the NPC, however larger macromolecules (most proteins and ribonucleoprotein complexes) are translocated through the central channel of the NPC in a saturable and sequence-mediated process that is rapid and highly selective (Kuersten *et al.*, 2001; Macara, 2001; Mattaj and Englmeier, 1998). This nucleo-cytoplasmic traffic has evolved to be functionally and mechanistically diversified: beyond being needed for basal replication, transcription and RNA processing, it also serves to regulate the cell cycle, transcriptional activation and repression in response to a variety of physiological and environmental stimuli, circadian rhythms and a number of other processes (Macara, 2001; Ossareh-Nazari *et al.*, 2001; Weis, 2002; Yamamoto and Deng, 1999). The small GTPase Ran, a member of the ras superfamily, is key in this process, both in establishing compartmental identity and in providing directionality to transport.

Studies have begun to investigate mechanisms of nucleo-cytoplasmic transport in plants, but our knowledge lags behind that for animal and yeast cells due to technical difficulties in doing cell-based *in vivo* and *in vitro* studies in plants. Much of what we currently know is based on analogies to what is known in animal cells and yeast. Sequences that resemble nuclear targeting and export sequences in plant-encoded and plant pathogen-encoded proteins have been identified, and homologues of nuclear import and export receptors, adaptors and Ran have been cloned (reviewed in Heese-Peck and Raikhel, 1998; Hicks and Raikhel, 1995; Meier, 2001). While studies have begun to characterize the functions of these sequences and proteins in plant cells,

The Nuclear Envelope, edited by D.E. Evans, C. Hutchison & J.A. Bryant.
© 2004 Garland Science/BIOS Scientific Publishers

what have been lacking are unbiased experimental approaches to investigate the mechanism and regulation of nucleo-cytoplasmic transport in plant cells. The demonstration that one of the movement proteins encoded by the DNA-containing geminiviruses is a nuclear shuttle protein now affords an opportunity to do this (Ward and Lazarowitz, 1999).

Most basic differences between plant and animal viruses can be traced to the plant cell wall being an effective barrier to virus entry and exit. One critical difference is the mechanism for local virus spread. Most animal viruses go through an extracellular phase (the virion) to initiate infection in neighbouring cells, as well as at distant sites in the host. In contrast, plant virus local cell-to-cell movement within a leaf is direct without an extracellular phase. An extracellular form of the virus will enter the vascular system (phloem for most viruses) to move long distances and systemically invade the host. For local spread, plant viruses overcome the cell wall barrier by encoding movement proteins. Viral mutants that are null for movement protein function replicate and encapsidate progeny genomes, but do not exit from the inoculated cell (Atabekov and Taliansky, 1990; Dawson and Hilf, 1992). This can be directly visualized in plants inoculated with virus mutants that express GUS or GFP (Carrington *et al.*, 1996; Oparka *et al.*, 1996). Thus, movement proteins act to eliminate the cell wall barrier, which suggests that they can profoundly alter cell structure.

Movement proteins target a common pathway to transport the viral genome across the cell wall: they alter plasmodesmata, trans-wall channels that connect adjacent cells within a plant (reviewed in Carrington *et al.*, 1996; Lazarowitz, 1999; Tzfira *et al.*, 2000). However, prior to the viral genome reaching the cell wall, movement proteins also play a key role in co-ordinating replication of the viral genome with its directed movement towards the plant cell wall. In doing this, movement proteins utilize the cell machinery for trafficking macromolecules. The final act of crossing the cell wall is preceded by a series of regulated events in which movement proteins interact not only with replicating viral genomes, but with the endomembrane system, the cytoskeleton and, in geminiviruses, the nuclear import and export machinery. It is this property of movement proteins that make them robust models for investigating the regulation of macromolecular transport within and between plants cells.

2.   Nuclear import and export

2.1 *Lessons from animals and yeast*

Our current view of nucleo-cytoplasmic transport is based mainly on genetic and biochemical studies using yeast (*S. cerevisiae* and *S. pombe*), *Xenopus laevis* oocytes and cultured mammalian cells, and recent structural studies of transport receptors and their cargoes or regulators. Transported cargoes have nuclear import (NLS) or export (NES) targeting sequences that are recognized by specific soluble transport receptors. Three classes of transport receptors have been identified: (i) the importin β-like family of proteins (importins and exportins), representing the largest group; (ii) NTF2/p10[1] which imports Ran into the nucleus; and (iii) the Tap/Mex67 family which transports mRNA. These receptors, all but two of which appear to function

---

[1] Factors are named according to their animal/yeast designations. The export receptor Crm1 (chromosome region maintenance based on its phenotype as first described in *S. pombe*) is now also termed Xpo1, and has been proposed to be named Exportin 1 in higher eukaryotes. For simplicity, it will be referred to as Crm1.

exclusively for import or for export, shuttle between the cytoplasm and nucleus, and carry their cargo through the NPC by specifically interacting with a subset of nucleoporins that contain phenylalanine-glycine (FG)-rich repeat motifs. The best characterized of these, in terms of the regulation of cargo pick-up and delivery, is the importin β receptor family, four members of which are exportins (Crm1, exporting proteins with a leucine-rich NES; CAS/CSE1, which exports importin α; exportin-t, for tRNA export; and exportin 4/Msn5, which exports multiple phosphoproteins but imports yeast replication protein (RP)A) (Macara, 2001; Strom and Weis, 2001; Weis, 2002). Ran acts as a positional marker for cargo delivery in interphase cells and during mitosis. A steep RanGDP (cytoplasm)/Ran GTP (nucleus) gradient, created by the asymmetric distribution of Ran effectors, is responsible for directionality in transport. In interphase cells, the Ran guanine nucleotide exchange factor (RanGEF) RCC1 is exclusively nuclear, being bound to chromatin *via* histones H2A and H2B. However, the Ran GTPase activating protein RanGAP, is exclusively cytoplasmic. For nuclear import, RanGTP induces cargo release by binding to importins to destablize the import complex on the nucleoplasmic side of the NE. In contrast, binding of RanGTP to exportins is required to assemble export complexes in the nucleoplasm, which are disassembled in the cytoplasm by RanGTP hydrolysis induced by RanGAP. Translocation across the NPC occurs by a process of facilitated diffusion: it does not require energy and is reversible. Energy, provided by hydrolysis of Ran GTP to Ran GDP, is required for cargo accumulation against a concentration gradient.

This basic model has become more complex as additional adaptor proteins, which bridge transport receptors and specific cargoes, and Ran binding proteins (RanBPs), which themselves are asymmetrically or symmetrically located in relation to the NE, have been identified (Kunzler and Hurt, 2001). In particular, the RanGTP-binding partners RanBP1 and RanBP2 (a nucleoporin) act as cytoplasmic co-factors of RanGAP in the dissociation of export complexes and the recycling of import receptors. This may, in part, be facilitated by a pool of RanGAP1 (the only RanGAP identified in animals) being associated with RanBP2 through a post-translational modification by the ubitquitin-related protein SUMO-1. RanBP3, homologous to RanBP1, acts as a co-factor for the exportin Crm1 to stabilize cargo binding to Crm1–RanGTP complexes. Nxt1 and RanBP1 have been reported as additional co-factors that facilitate dissociation of the export complex from the cytoplasmic side of the NPC (Ossareh-Nazari *et al.*, 2001). Functional assays have recently expanded this view of Ran targeting and regulation, showing that Ran also functions in microtubule aster formation and NE formation during the process of open mitosis that occurs in higher eukaryotes, with importin β being one important target in at least mitotic spindle formation (Hetzer *et al.*, 2002; Sazer and Dasso, 2000; Wiese *et al.*, 2001; Zhang and Clarke, 2000). As in nucleo-cytoplasmic transport during interphase, the mitotic Ran gradient is proposed to regulate spindle assembly: RCC1, which remains associated with chromatin during mitosis, causes local generation of RanGTP in the vicinity of chromosomes and thus local release of importin cargoes. In support of this model, mitotic cargoes of importin β, including microtubule-associated proteins NuMA and TPX2, have been shown to regulate mitotic spindle assembly when released around chromatin (see Kuersten *et al.*, 2001; Weis, 2002). Both RCC1 and RanGAP1 acting on Ran seem to be required for NE assembly, as demonstrated in extracts of *Xenopus* eggs or human somatic cells (Zhang and Clarke, 2000, 2001). The relevant Ran targets in this process remain to be identified.

The best-characterized nuclear import targeting sequences are the two forms of the short basic NLS first described in SV40 large T antigen (monopartite) and in nucleoplasmin (bipartite) (Dingwall *et al.*, 1982; Kalderon *et al.*, 1984). Proteins containing this so-called classical cNLS specifically bind to the import adaptor importin α, which heterodimerizes with importin β, the import receptor for translocation through the NPC. In the past 5 years it has become clear that there are multiple additional signal-dependent nuclear import pathways. Besides six isoforms of importin α in animals, other importin-β-dependent adaptors in vertebrates include snurportin 1 (import of m3G-capped snRNPs), and XRIPα (import of replication protein A, RPA) and importin β can also complex with importin 7 to import histone H1 (Mattaj and Englmeier, 1998; Weis, 2002). A growing list of cargoes, including a number of viral proteins such as HIV-1 Rev and Tat, and HTLV Rex, as well as ribosomal proteins L23a and L5, the transcriptional activators CREB, Jun and Fos, and cyclin B1 are imported into the nucleus by direct interaction with importin β (see Cingolani *et al.*, 2002; Macara, 2001). These proteins contain so-called non-classical NLSs (ncNLSs) which, although somewhat basic in character, can be much longer than cNLSs and do not conform to a clear consensus. Resolution of the crystal structure of an N-terminal fragment of importin β1 bound to the ncNLS of parathyroid hormone-related protein (PTHrP) shows that the ncNLS binds in an extended conformation to a site that overlaps the RanGTP binding site and is distinct from the site to which importin α binds (Cingolani *et al.*, 2002). Another type of well-characterized NLS is the M9 sequence of hnRNPA1, a 38-aa stretch enriched in aromatic residues and glycine. The M9 sequence, responsible for rapid shuttling of hnRNPA1, binds directly to the carrier transportin1/Kap104, a transport receptor related to importin β that imports mRNA-binding proteins and ribosomal proteins (Macara, 2001).

The best-characterized nuclear export sequence, referred to as NES, is the short leucine-rich motif recognized by Crm1 (LxxxLxxLxL, where x is variable in number), although other NESs that do not conform to this consensus have been identified and Crm1 can recognize large NESs (e.g. the 150 aa NES in snurportin). This prototype NES was first characterized in protein kinase A inhibitor protein (PKI) and HIV Rev, and has since been identified in a myriad of cellular and viral proteins (Fischer *et al.*, 1995; Macara, 2001; Wen *et al.*, 1995; Whittaker and Helenius, 1998). The leucine residues are essential for recognition as an NES, although some of these can be replaced by other hydrophobic residues. A hallmark of Crm1-dependent nuclear export is its inhibition by the antifungal antibiotic Leptomycin B (LMB) in all organisms tested save *S. cerevisiae*, which is resistant because it contains a threonine in place of the conserved cysteine modified by LMB. A recent study identified biochemically the calcium-binding ER chaperone calreticulin (CRT) as an export receptor, both for proteins containing a leucine-rich NES and, *via* a Crm1-independent pathway, for diverse members of the nuclear receptor superfamily of transcription factors, in which CRT was shown to interact with the DNA-binding domain (Black *et al.*, 2001). This finding contradicts many studies reporting a complete dependence on Crm1 for the export of leucine-rich NES-containing proteins, and to date neither nuclear nor cytoplasmic pools of CRT have been identified *in vivo*. However, in this context it is interesting that calcium depletion from the ER lumen has been reported to correlate with inhibition of signal-mediated import and passive diffusion into the nucleus, and was proposed to be a potential cellular response to stress (Greber and Gerace, 1995).

In addition to Ran interactions with import and export complexes, nucleo-cytoplasmic transport can be regulated at the levels of the cargo and the nuclear transport machinery itself (Kehlenbach and Gerace, 2000; Komeili and O'Shea, 2001; Macara, 2001). There are many examples of phosphorylation or de-phosphorylation affecting the nucleo-cytoplasmic distribution of transcription factors or other proteins by masking or unmasking an NLS or NES directly, or indirectly by influencing cargo binding to another protein (e.g. members of the 14-3-3 family). For example, phosphorylation of Pho4 allows it to be recognized by Msn5 for export and inhibits its binding to the importin Pse1. The cell cycle regulator Cdc25, the transcription factor NFAT and the histone deacetylase HDAC4 have all been shown to shuttle due to the presence of both NES and NLS sequences, and each, when phosphorylated, has been reported to accumulate in the cytoplasm as a result of binding 14-3-3 proteins (Wang and Yang, 2001). For Cdc25, 14-3-3 association has been reported to mask its NLS, but association of 14-3-3 with HDAC4 has been reported to both inhibit its nuclear import and stimulate its nuclear export. Acetylation of lysine residues within the NLS of CIITA has been reported to stimulate the nuclear accumulation of this transcriptional trans-activator (Spilianakis et al., 2000). The formation of a complex can also be required to create a functional NLS or NES, as reported for the capsid proteins of SV40 for nuclear import, or the yeast transcription factor yAP1 in nuclear export (Macara, 2001; Whittaker and Helenius, 1998). Inhibition of nuclear import by phosphorylation of insoluble components of the nuclear transport machinery, perhaps nucleoporins, has been suggested as a mechanism to globally control cellular activity (Kehlenbach and Gerace, 2000).

2.2 Plant cells: The current picture

The breathtaking advance in our understanding of nucleo-cytoplasmic transport in vertebrates and yeast in the past 10 years is in large measure the result of the development of in vitro permeabilized cell nuclear import and, more recently, export assays in which transport could be reconstituted biochemically so that factors could be identified in an unbiased manner and tested for function (Adam et al., 1990; Holaska and Paschal, 1998; Kehlenbach et al., 1998). Several studies have begun to explore this process in plant cells, but progress has been slowed by the lack of such a reconstitutable in vitro transport system. Thus, the current picture of plant nucleo-cytoplasmic transport is based mainly on the identification of NLS and NES motifs and of plant homologues to known animal and yeast factors, combined with in vitro studies of the interaction of α importin homologues with NLSs. While the picture is far from complete, the basic transport machinery appears to be conserved. Some distinctions that have been noted in terms of import may represent what we now view as a spectrum of variation in signal-dependent nuclear import pathways.

A few α importin homologues have been identified in Arabidopsis and rice, and two importin β homologues have been cloned from rice (Yamamoto and Deng, 1999). Ran homologues have been cloned from Arabidopsis, tomato, tobacco and Vicia faba, and the tomato and tobacco homologues were shown to be functional by their ability to suppress a *S. pombe ts pim1* (RCCl) mutation (Ach and Gruissem, 1994; Merkle *et al.*, 1994). The *Arabidopsis* homologue of Crm1 (*AtXPO1*) has been cloned, based on homology to human Crm1, and *Arabidopsis* homologues of RanBP1 and RanBP2 have been identified in a yeast two-hybrid screen for proteins interacting with *Arabidopsis* RanGTP (Haasen *et al.*, 1999; Haizel *et al.*, 1997). Recently, the *Arabidopsis* orthologue of exportin-t (termed *PAUSED* or *PSD*) was identified independently and cloned by two groups based on the pleiotropic effects of *PSD* null alleles on *Arabdiopsis* development, which included delays in leaf formations and the transition from vegetative to floral development (Hunter *et al.*, 2003; Li and Chen, 2003). The *Arabidopsis* orthologue of exportin 5 has also been recently identified (*HASTY* or *HST*) based on the effects of *hst* loss of function mutants on phase change and morphogenesis in *Arabidpsis* (Bollman *et al.*, 2003). Notably, *PSD* null alleles and *psd/hst* double mutants, although displaying obvious developmental defects, are viable, suggesting the existence of one or more additional tRNA export pathways in *Arabidopsis*. The plant Ran, importin β, Crm1, and RanBP homologues have not been shown to function in a plant nucleo-cytoplasmic transport assay, although the rice importin β homologues have been examined in a heterologous permeabilized HeLa cell assay, and nuclear export of a green fluorescent protein (GFP)-NES reporter construct in tobacco protoplasts was shown to be LMB- sensitive (Jiang *et al.*, 1998b; Haasen *et al.*, 1999). What has been reported is the identification of functional plant NLS and NES motifs, and several features of plant importin-α-dependent nuclear import and NES-mediated nuclear export.

Several approaches have identified functional NLSs in plant cellular proteins and in proteins encoded by some plant pathogenic viruses or bacteria, where the latter deliver proteins to the plant cell nucleus (e.g. *Agrobacterium tumefaciens* VirD2 and VirE2 proteins, which target the bacterial tumour-inducing DNA into the plant cell nucleus for integration, and *Xanthomonas campestris* nuclear-targeted avirulence factor AvrBs3, a crucial pathogenicity determinant) (Carrington *et al.*, 1991; Citovsky *et al.*, 1992; Hicks and Raikhel, 1995; Howard *et al.*, 1992; Qin *et al.*, 1998; Leclerc *et al.*, 1999; Sanderfoot *et al.*, 1996; Szurek *et al.*, 2002). These have included mutational analysis of the endogenous protein, which identified clusters of basic amino acids required for function, and the ability to redirect a reporter protein, usually β-glucuronidase (GUS) or GFP, from the cytoplasm to the nucleus when appropriate expression constructs were transiently expressed in biolistically bombarded onion epidermal cells or in cultured tobacco protoplasts, or used to generate transgenic plants (most commonly tobacco or *Arabidopsis*). Most of the NLSs have been classified as monopartite or bipartite cNLSs and can function in mammalian, insect or yeast cells, just as the SV40 large T and nucleoplasmin NLSs have been shown to function in plant cells (Hicks and Raikhel, 1995; Raikhel, 1992; Sanderfoot and Lazarowitz, 1995). However, the NLS consensus motifs are sufficiently vague that we suggest some plant 'bipartite' NLSs (e.g. the tobacco etch potyvirus NIa protein and maize transcription factor O2) could as well be viewed as ncNLSs of the type found in proteins like PTHrP and ribosomal protein L23, which directly bind to importin β. Our studies also suggest that the SqLCV coat protein (CP) may contain such a ncNLS (*Table 1*) (Qin *et al.*, 1998). In addition, the maize transcription factor R contains a cNLS (termed 'C') that resembles the yeast Mat α2 NLS (*Table 1*), which is reported not to function in mammalian cells. Differences have been noted for interactions of this R NLS-C with different plant α importin homologues (see below) (Lanford *et al.*, 1990).

Table 1. Examples of plant NLS and NES targeting signals[a]

cNLS monopartite[b]	B_4, $P(B_3x)$, $Pxx(B_3x)$, $B_3(H/P)$
R NLS M (maize)	MSE**RKRR**EKL
NSP (SqLCV)	VKTVP**NRTRTY**ILK
cNLS bipartite[c]	$BBx_{10}(B_3x_2)$
O2 NLS B (maize)	**RKRK**ESNRESA**RRSRYRK**
Tga-1a (tobacco)	**RR**LAQN**REAARKSRLRKK**
VirD2	**KR**PREDDDGEPSE**RKRER**
Vir E2 NSE1	**KL**RPEDRYIQTE**KYGRR**
Vir E2 NSE2	**KT**KYGSDTEU**KLKSK**
NSP (SqLCV)	**KR**SYGAARGD**DRRR**
Mat a2-like	KIPIK
R NLS C	MISEAL**RKAIGKR**
NcNLS[d]	————
NIa (potyvirus TEV)	**KKN**QKHKLKM-32aa-**KRK**
CP (SqLCV)	**KL**KNYTNSVMFWLVRD**RRP**-33aa- VMHRFYAKVTGGQYASNEQALV**RR**
NES (leu-rich)	$Lx_{2,3}(F/I/L/V/M)X_{1,2,3}LX(I/V/L)$
NSP (SqLCV)	LEKDTLLIDL
AtRanBP1a	LLEKLTV

[a]Residues in bold have been shown to be essential for function
[b]SV40 large T: **PKKKRK**
[c]Nucleoplasmin: **KR**PAATKKAGQA**KKKK**LD
[d]PTHrP: RYLTQETNKVETYKEQPLKTPGKKKKGKP
L23a: HSHKKKKIRTSPTFRRPKTLRLRRQPKYPRKSAPRRNY

Two groups have documented protein nuclear export in plant cells mediated by a leucine-rich NES, deduced to be Crm1-dependent based on its LMB sensitivity (Haasen *et al.*, 1999; Ward and Lazarowitz, 1999). Based on our finding that the SqLCV movement protein NSP is a nuclear shuttle protein that is trapped in the cytoplasm by the second viral movement protein MP, we developed a transient expression assay in tobacco protoplasts that allowed us to analyse NSP–GUS fusion proteins and thereby identify the NES in NSP as a leucine-rich sequence resembling that found in other rapidly shuttling nuclear proteins such as PKI and HIV-1 Rev. The NSP NES, as well as its monopartite and bipartite cNLSs (*Table 1*), are all essential for viral infectivity, and each can function in insect cells (Sanderfoot and Lazarowitz, 1995; Sanderfoot *et al.*, 1996; Ward and Lazarowitz, 1999). We further showed that the *Xenopus* TFIIIa NES could functionally substitute for the NSP NES in nuclear export in tobacco protoplasts and partially restore virus infectivity (Ward and Lazarowitz, 1999). In a similar approach that used GFP-fusion proteins, a leucine-rich NES was identified in an *Arabidopsis* homologue of RanBP1, and the

HIV-1 Rev NES was shown to function in tobacco protoplasts *via* an LMB-sensitive pathway (Haasen *et al.*, 1999).

There is evidence for an importin α/β-dependent nuclear import pathway in plants. A rice α importin homologue can assemble into a complex with mouse importin β and, with mouse importin β, mediate nuclear import in a permeabilized HeLa cell assay. In this same assay, rice importin β also mediated nuclear translocation (Jiang *et al.*, 1998a, b). However, as in yeast and vertebrates, there are variant pathways. Detailed analysis of the *Arabidopsis* α importin homologue AtIMPα show it to be more like mammalian β, rather than α, importin. In a co-immune precipitation assay, AtIMPα can specifically bind the maize R Mat α2-like and monopartite cNLSs and the O2 'bipartite' NLS in the absence of importin β (Smith *et al.*, 1997). That this is high-affinity importin-β-independent binding was shown in a quantitative solid phase binding assay in which AtIMPα, but not mouse importin α, bound the SV40 large T cNLS and a *Xenopus* bipartite cNLS, as well as the yeast GAL4 ncNLS, with high affinity. In this same assay, mouse importin β also directly bound the GAL4 ncNLS and, like AtIMPα, the maize O2 'bipartite' NLS (Hubner *et al.*, 1999). Similar to β, and in contrast to α importin from yeast and mammalian cells, a fraction of AtIMPα is found concentrated at the NE by immuno-staining of tobacco protoplasts and, in the absence of added mouse importin β, as well as Ran and NTF2, it can mediate binding of an appropriate NLS-containing fluorescent import substrate to the NE in a reconstituted rat hepatoma *in vitro* import assay (Hubner *et al.*, 1999; Smith *et al.*, 1997). Consistent with these results, AtIMPα alone could mediate nuclear import in the rat hepatoma *in vitro* assay of substrates with the large T antigen or *Xenopus* bipartite cNLS to levels comparable to those achieved with the mouse importin α/β complex. Thus, an importin-β-independent import pathway may exist in *Arabidopsis*, although it remains to be seen whether AtIMPα also mediates importin-β-dependent nuclear import in plants. AtIMPα can bind both yeast and mouse β importins, consistent with the importin β binding domain (IBB) of AtIMPα being highly similar (69% and 77%) to that of mouse and yeast α importins. Intriguingly, other regions of AtIMPα, including the first ARM (armadillo) repeat adjacent to the IBB, do not show high conservation. Thus, relevant to AtIMPα cargo-binding properties are recent structural analyses that show while the N-terminus of mammalian importin α, which contains the IBB, is not structured in solution, a short stretch of basic residues contacts the major NLS-binding pocket (ARM repeats 2–4) and may function as an auto-inhibitor of cargo loading. As in animals, there are multiple α-importin isoforms in plants, including *Arabidopsis*. The six mammalian α importin isoforms recognize the SV40 monopartite NLS with similar affinity, but differ in their affinities for other NLSs and, in contrast to AtIMPα, a rice importin α only binds the monopartite and bipartite NLSs *in vitro* (Jiang *et al.*, 1998a; Macara, 2001).

Two groups, both using cultured tobacco and maize cells, have reported NLS-specific, saturable nuclear import in permeabilized protoplasts from higher plants (Hicks *et al.*, 1996; Merkle *et al.*, 1996). In both studies, protoplasts were evacuolated by centrifugation through Percoll (the large plant vacuoles, responsible for protoplast fragility, can occupy 50–90% of the cell volume) to obtain cells more amenable to manipulation. As plant cell membranes contain very low amounts of cholesterol, cells could not be permeabilized with digitonin or streptolysin. Rather, cells were permeabilized by a controlled decrease in the external osmoticum (mannitol) or by brief treatment with 0.01% Triton X-100. For mammalian cells, digitonin releases α and β importins, Ran, and NTF2 as soluble factors, which requires that these, as well as GTP and an ATP-regenerating system, be exogenously supplied as purified factors or in the form of cytosol. In contrast, the permeabilized plant protoplasts retained all components needed for nuclear import, including GTP and ATP. Only a suitable import substrate had to be supplied (Hicks *et al.*, 1996; Holaska and Paschal, 1998; Kehlenbach *et al.*, 1998; Merkle *et al.*, 1996). In both plant cell assays, the need for energy in the form of GTP, and thus a role for Ran, was inferred based on partial inhibition of nuclear import by non-hydrolysable guanine nucleotide analogues. However, ATP-γ-S did not inhibit import, although an indirect nuclear import assay using parsley protoplasts did indicate an ATP requirement (Harter *et al.*, 1994). While the tight NE association of a fraction of AtIMPα might be relevant here, these differences in plant *versus* mammalian *in vitro* nuclear import may in large part reflect the different methods used to permeabilize cells. Indeed, in addition to the non-selective aspect of Triton X-100 treatment (0.1–1.0% Triton X-100 will severely inhibit import or destroy the cells) (Hicks *et al.*, 1996; Merkle *et al.*, 1996; Smith *et al.*, 1997), it remains unclear how lowering the osmoticum permeabilizes plant protoplasts. In the latter case, vesiculation of the plasma membrane and a rough appearance of the cell surface were noted, which may be factors in the entry of substrates and antibodies (Hicks *et al.*, 1996). However, one difference from mammalian cells, which cannot be explained by the different treatments, is in the temperature dependence: nuclear import was temperature dependent in plant cells, but was slowed at 4°C, rather than blocked as reported for mammalian cells. It has been suggested that this difference might reflect plant adaptation to the low temperatures which they often encounter in nature (Hicks *et al.*, 1996).

Important an advance as these plant *in vitro* assays are, they are limited by the fact that essential cytoplasmic factors cannot be depleted. The inability to reconstitute nuclear import, as well as export, is a major obstacle to continued progress in this field, for it means that essential plant factors and regulators cannot be identified in an unbiased manner and that these, as well as plant proteins identified based on homology to known yeast and vertebrate components, cannot be directly tested for function in nuclear transport, except in heterologous animal cell- or yeast-based assays. To address this limitation, other approaches to developing *in vitro* plant nuclear transport assays or investigating nucleo-cytoplasmic transport in plants cells are needed. Our detailed characterization of the nuclear shuttling of NSP and its entrapment by movement protein (MP) provides the tools to develop such alternative approaches.

3. Plant virus movement proteins

3.1 *Geminiviruses encode a nuclear shuttle protein*

The bipartite geminiviruses such as squash leaf curl virus (SqLCV) and the closely related cabbage leaf curl virus (CLCV) (Hill *et al.*, 1998), with their single strand (ss) DNA genomes, encode two movement proteins, a nuclear shuttle protein (NSP) and a cell-to-cell movement protein (MP[2]). Investigation of SqLCV established that these two proteins interact in a co-operative and regulated manner to move the viral genome from its site of replication in the nucleus to the cytoplasm and into adjacent plant cells

---

[2] NSP: <u>N</u>uclear <u>S</u>huttle <u>P</u>rotein, formerly BR1. MPB: cell-to-cell <u>M</u>ovement <u>P</u>rotein, formerly BL1.

(colour *Plate 5a*) (Pascal *et al.*, 1994; Sanderfoot and Lazarowitz, 1995; Sanderfoot *et al.*, 1996; Ward and Lazarowitz, 1999; Ward *et al.*, 1997). NSP is a ssDNA-binding protein that shuttles newly replicated viral genomes between the nucleus and the cytoplasm (Pascal *et al.*, 1994; Sanderfoot *et al.*, 1996; Ward and Lazarowitz, 1999). MP, associated with unique endoplasmic reticulum (ER)-derived tubules that cross the walls of infected plant cells, traps these NSP–genome complexes in the cytoplasm and redirects them to and across the plant cell wall (Noueiry *et al.*, 1994; Sanderfoot and Lazarowitz, 1995; Ward *et al.*, 1997). In the adjacent cell, NSP–genome complexes are released, and NSP targets the viral genome to the nucleus to initiate new rounds of infection. Three independent lines of evidence indicate that the MP-associated tubules are derived from the ER. First, these unique tubules, found only in infected plant cells, are specifically labelled with anti-BiP antisera. MP also co-fractionates with ER-membrane containing fractions from SqLCV-infected plants (Ward *et al.*, 1997). Finally, SqLCV MP, when transiently expressed in tobacco protoplasts, specifically co-localizes with BiP to cortical ER (Lazarowitz and Beachy, 1999). This ER association is significant since the core structure of plasmodesmata, the desmotubule, is initially derived from cortical ER that is 'trapped' at the cell plate during cell division as the plant cell wall forms (Hepler, 1982; Hepler *et al.*, 1990; Overall *et al.*, 1982). We have proposed that SqLCV, which infects and moves in immature dividing phloem cells (Lazarowitz *et al.*, 1998), usurps the cellular machinery that normally forms plasmodesmata, with MP creating virus-specific channels by targeting to cortical ER regions that will develop into desmotubules. How MP, which lacks an N-terminal signal sequence, specifically targets to and associates with cortical ER, and whether the MP-associated tubules, which deeply penetrate the cytoplasm (Ward *et al.*, 1997), extend back to the nuclear envelope, are important unanswered questions.

For SqLCV, the co-operative interaction of NSP and MP has been visualized by indirect immunofluorescent staining and confocal laser scanning microscopy (CLSM) using a transient expression assay in plant protoplasts. With this assay and the ability to determine the effects of specific mutations on virus infectivity, we identified the functional domains within NSP and MP, and established that the mechanism of nuclear shuttling is highly conserved in plants, as well as in animal cells and yeast (Sanderfoot and Lazarowitz, 1995; Sanderfoot *et al.*, 1996; Ward and Lazarowitz, 1999). These studies demonstrated that SqLCV NSP is a typical rapidly shuttling nuclear protein that contains two cNLSs and an essential leucine-rich NES of the type found in rapidly shuttling nuclear proteins such as HIV Rev and TFIIIa (*Table 1*). When expressed individually, NSP targets to the nucleus and MP to the cortical region (cortical ER) of tobacco protoplasts. In contrast, when both proteins are co-expressed in the same cell, MP redirects NSP from the nucleus to the cortical ER (Sanderfoot and Lazarowitz, 1995; M. Carvalho and S. Lazarowitz, unpublished observations). Substitution of alanine residues for leucines in the NSP NES (NSP$^{L189A/L190A/L193A}$) prevents NSP from exiting the nucleus to be trapped in the cytoplasm by MP and eliminates SqLCV infectivity (Ward and Lazarowitz, 1999). Insertion of the *Xenopus laevis* TfIIIa NES into NSP$^{L189A/L190A/L193A}$ fully restores the ability of NSP to exit from the nucleus, as is evident by its ability to now interact with MP in tobacco protoplasts, and partially restores virus infectivity (colour *Plate 5b*) (Ward and Lazarowitz, 1999). As predicted by the model for SqLCV movement (colour *Plate 5a*), the interaction of NSP and MP is highly regulated. Following its interaction with MP, NSP is released and now targets to the perinuclear region of the cell, which indicates that NSP has been post-translationally modified as a result of this interaction with MP. The nature of this post-translational change remains to be defined. Mutational studies suggest that it may involve an alteration in the phosphorylation state of NSP (Sanderfoot and Lazarowitz, 1996; Sanderfoot *et al.*, 1996). In addition, SqLCV coat protein (CP), although not essential for cell-to-cell movement or systemic infection, synergistically aids NSP in the nuclear phase of virus movement by sequestering replicated progeny ssDNA genomes away from the replication pool to make them available for nuclear export by NSP (Qin *et al.*, 1998; Sanderfoot *et al.*, 1996).

To identify the host cell transport and signalling pathways that are required for and regulate NSP function, and thereby begin to investigate the regulation of nuclear shuttling in plant cells, we have used the classic yeast two-hybrid screen to identify *Arabidopsis thaliana* proteins that interact with NSP encoded by CLCV, for which *Arabidopsis* is a host. This assay could potentially identify host proteins involved in the nuclear import and export of NSP, its interactions with MP, or its binding to or release of the viral genome. One NSP-interacting protein we identified is AtNSI (*A.* thaliana NSP Interactor), a novel *Arabidopsis* acetyltransferase that is highly conserved in plants (McGarry *et al.*, 2003).

3.2 *AtNSI, a novel* Arabidopsis *acetyltransferase that interacts with NSP*

We have shown that AtNSI, which directly interacts with NSP in an *in vitro* GST pulldown assay, is a novel *Arabidopsis* nuclear acetyltransferase involved in regulating NSP-mediated nuclear export of the viral ssDNA genome (McGarry *et al.*, 2003). Based on cloning the full-length cDNA by 5'- and 3'-RACE, *AtNSI* is a single copy gene on *Arabidopsis* chromosome 1 predicted to encode a novel 258-aa protein. It is highly conserved among divergent plant species from monocots such as rice to dicots such as tomato and *Arabidopsis*, but not found in yeast, *Drosophila*, *C. elegans* or humans. We determined the subcellular localization of AtNSI using a polyclonal rabbit antisera generated against bacterial-expressed AtNSI (McGarry *et al.*, 2003). To do this, we expressed AtNSI from a potato virus X (PVX) vector and detected it in nuclei by indirect immunofluorescent staining and CLSM of methacrylate-embedded sections from infected tobacco leaves (colour *Plate 6*). We used this approach because we could not detect AtNSI in *Arabidopsis* extracts by immunoblotting or by indirect immunofluorescent staining of methacrylated-embedded *Arabidopsis* tissues (McGarry *et al.*, 2003). This inability to detect AtNSI in *Arabidopsis* suggests that its expression may be tightly regulated quantitatively as well as developmentally.

Semi-quantitative RT-PCR analyses showed that the pattern of *AtNSI* transcription in *Arabidopsis* was consistent with the path of geminivirus infection. *AtNSI* was most highly expressed in cauline leaves and was expressed at somewhat lower (~40%) levels in stems, siliques, inflorescences and rosette leaves. It was expressed at only very low levels in roots (McGarry *et al.*, 2003). Consistent with this pattern of expression, bipartite geminiviruses such as CLCV infect leaf tissue and move through the phloem, but do not invade developing embryos or roots. To show that AtNSI does play a role *in vivo* in bipartite geminivirus infection, we tested the ability of CLCV to infect transgenic *Arabidopsis* lines that over-expressed AtNSI from the strong cauliflower mosaic virus 35S promoter (McGarry *et al.*, 2003). CLCV infectivity is enhanced from 20% to 30% in these AtNSI-over-expressing *Arabidopsis* lines (*Figure 1a*). In addition,

AtNSI stably interacts with SqLCV NSP, as well as with CLCV NSP, in the yeast two-hybrid assay (McGarry *et al.*, 2003). This further supports a general role for AtNSI in geminivirus infection.

The C-terminal region of AtNSI (aa 175–244) is predicted to contain an acetyltransferase domain belonging to the GNAT (GCN5-like *N*-acetyltransferase) family of acetyltransferases. However, it is distantly related to characterized GNAT family members or to acetyltransferases belonging to the MYST family (McGarry *et al.*, 2003). AtNSI lacks conserved structural features found in these acetyltransferases, most of which are transcriptional co-activators, including the bromodomain, which anchors nuclear histone acetyltranferases and other co-activators onto active chromatin (Dhalluin *et al.*, 1999; Jacobson *et al.*, 2000; Owen *et al.*, 2000). To show that AtNSI is an active acetyltransferase, GST-AtNSI or HIS$_6$-AtNSI fusion proteins, expressed in and purified from *E. coli*, were tested for their ability to acetylate

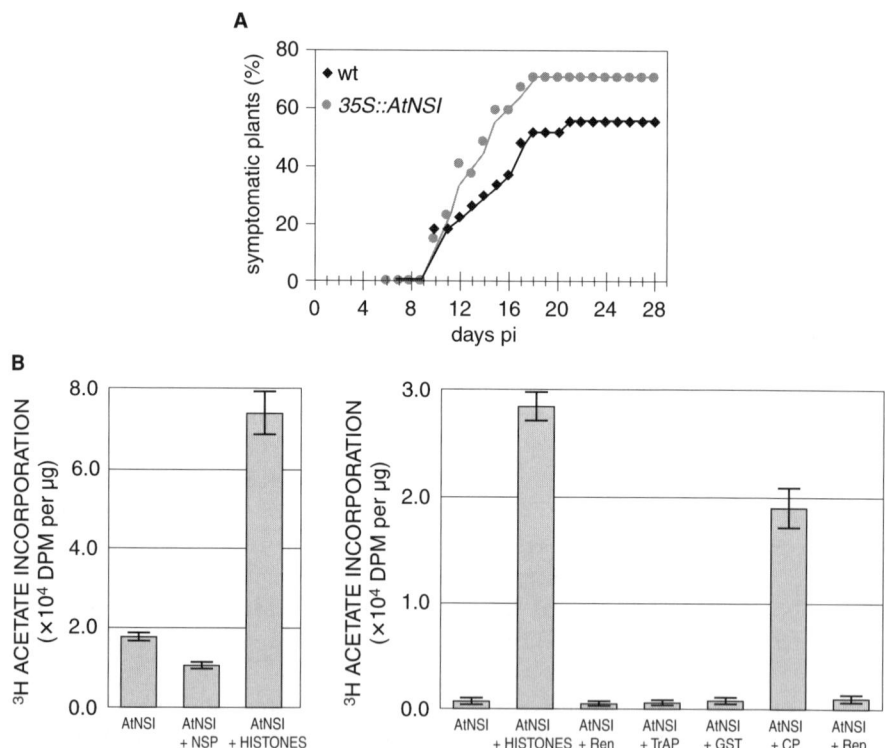

Figure 1. *(A) Over expression of AtNSI enhances CLCV infection efficiency in* Arabidopsis. *Infectivity assay of CLCV on wild type (wt)* Arabidopsis *plants or a transgenic* Arabdiopsis *T3 line that over expresses AtNSI (35S::AtNSI). Shown are the numbers of plants displaying systemic symptoms at different days postinoculation (pi). (B) AtNSI acetylates histones and CLCV CP in vitro, but does not acetylate NSP. 300 ng of purified AtNSI was incubated with* ^{3}*H-acetyl-CoA alone, or in the presence of 1 mg of calf thymus histones or purified CLCV HIS$_6$-NSP, GST-Ren, GST-TrAP, GST, HIS$_6$-CP or HIS$_6$-Rep, as indicated. Use of AtNSI cleaved from a GST-AtNSI fusion with thrombin (left panel) vs. HIS$_6$-AtNSI (right panel) accounts for different relative levels of autoacetylation ('AtNSI') resulting from internal cleavages introduced by thrombin. Adapted from McGarry et al. (2003), The Plant Cell, © ASPB Press, with permission.*

histones *in vitro* when incubated with [³H]-acetyl CoA. AtNSI acetylates both commercial calf thymus histone preparations and purified *Drosophila* histones with a specific activity comparable to that of the co-activator p300 from *Drosophila* (*Figure 1b*). However, when the *in vitro* acetylated histones are analysed by SDS-PAGE and auto-radiography, AtNSI predominantly acetylates H3 and H2A. This is in contrast to p300, which mainly acetylates histones H3 and H4 (McGarry *et al.*, 2003).

Given the differences between AtNSI and characterized transcriptional co-activators in terms of their structures and histone substrate specificities, we tested whether AtNSI would act as a co-activator *in vitro* when tested in a chromatin assembly and transcription assay (McGarry *et al.*, 2003). The transcriptional activator p300, when tested in this assay with reporter plasmids that contain the adenovirus E4 core promoter fused to either the oestrogen responsive element or Gal4-binding site, enhances transcriptional activation by the oestrogen receptor or a Gal4–VP16 fusion protein, respectively, by ~2-fold (Kraus and Kadonaga, 1999). In contrast, AtNSI, tested over a range of concentrations, does not boost the activity of these sequence-specific transcriptional activators using this same assay (McGarry *et al.*, 2003). Thus, AtNSI does not act as a transcriptional co-activator *in vitro*.

To further ascertain the potential role of AtNSI in the nuclear aspects of geminivirus replication and movement, we tested the five nuclear-targeted geminivirus-encoded proteins — NSP, CP, the viral transcriptional transactivator TrAP, the replication initiation protein Rep, and the replication enhancer Ren — as possible substrates for acetylation by AtNSI. NSP was a potential substrate given its interaction with AtNSI. However, CP was also considered a likely substrate given its ability to sequester replicated progeny genomes and make them available for nuclear export by NSP (Qin *et al.*, 1998). Of these five CLCV nuclear-targeted proteins, AtNSI specifically acetylated CP *in vitro* (*Figure 1b*) (McGarry *et al.*, 2003). We therefore tested whether CP could stably interact with AtNSI. In contrast to the stable interaction of NSP and AtNSI, CLCV CP, when tested in a yeast two-hybrid assay, did not stably interact with AtNSI. Nor did CLCV NSP and CLCV CP stably interact (McGarry *et al.*, 2003).

3.3 *A model for AtNSI–NSP interactions in geminivirus infection*

Our studies suggest that AtNSI regulates the nuclear export of the geminivirus ssDNA genome and potentially other non-transcriptional nuclear events in plant cells. NSP and MP co-operatively act to move newly replicated viral genomes from the nucleus through the cytoplasm and across the cell wall. The interactions of NSP with the viral ssDNA genome and MP must be regulated to assure that progeny genomes are directed into the cell-to-cell movement pathway at the right time, and not prior to transcription and replication of the incoming viral DNA. Previous studies showed that geminivirus CP binds replicated viral ssDNA without encapsidating it to prevent the viral ssDNA from re-entering the replication pool and thus make it available for NSP-mediated nuclear export (Ingham *et al.*, 1995; Qin *et al.*, 1998). Our identification of AtNSI as a nuclear acetyltransferase that specifically interacts with NSP and our finding that it can acetylate CLCV CP *in vitro* suggests that protein acetylation is important in regulating these nuclear events to facilitate virus movement.

We propose that AtNSI acetylation of CLCV CP allows NSP to displace CP and bind progeny genomes to export these from the nucleus. Following uncoating of the incoming viral genome, the geminivirus ssDNA is converted by host nuclear enzymes into double-stranded DNA templates for transcription and for rolling circle replication to generate progeny ssDNA genomes (Lazarowitz, 2001). Given the role of viral CP in making unencapsidated viral ssDNAs available for NSP binding and export (Qin et al., 1998), it was proposed that NSP, prior to the accumulation of sufficiently high levels of CP to assemble virus particles, would co-operatively bind to the newly synthesized progeny genomes to displace CP and export the viral ssDNA to the cytoplasm. This predicted that there were at least two pools of CP in the nucleus. Given our findings that NSP is not acetylated by AtNSI and that AtNSI does not stably interact with CP, we propose that NSP, bound to newly replicated viral ssDNA, recruits AtNSI into a ternary complex with the genome-bound CP to acetylate CP and disrupt its binding to the viral ssDNA. In this manner, NSP would displace CP as NSP itself co-operatively bound the viral genome to export it from the nucleus. Consistent with this model, is our finding that CLCV infection is enhanced on transgenic Arabidopsis lines that over-express AtNSI (McGarry et al., 2003). Over-expression of AtNSI would not be expected to affect CP binding as such to viral ssDNA since in the absence of NSP AtNSI is not recruited into a complex to acetylate CP. However, it would be expected to accelerate the release of CP from the ssDNA, and thus export of the CLCV genome and virus movement, once NSP recruits AtNSI into the ternary complex with CP bound to the viral genome.

4. Concluding remarks

Nuclear acetyltransferases play a key role in chromatin remodelling during eukaryotic transcriptional activation and elongation. However, beyond this they have been implicated in DNA replication, DNA recombination, and DNA repair and apoptosis (Burke et al., 2001; Iizuka and Stillman, 1999; Ikura et al., 2000; McMurry and Krangel, 2000). In particular, p300/CBP and PCAF, through their action on non-histone target proteins, have been proposed to be key integrators in affecting the stability of protein complexes to modulate transcription and a potential range of biological processes, including cell proliferation and differentiation. In this context, it is not surprising that several cellular and viral proteins have recently been found to modulate and possibly retarget the acetyltransferase activities of p300/CBP or PCAF (reviewed in Hottiger and Nabel, 2000). The majority of eukaryotic DNA viruses require host cell DNA-synthesizing enzymes for their replication. These viruses ensure that these host enzymes are available, and also inhibit the apoptotic response to unscheduled DNA synthesis, by inducing cells to enter S phase or maintaining cells in a dedifferentiated state (DiMaio and Coen, 2001). To do this, SV40, HPV-16, an oncogenic human papilloma virus and adenoviruses all encode proteins that bind and inactivate Rb (and the related p107 and p130 proteins) and p53 (Flint and Shenk, 1997). Some of these same viral-encoded proteins also appear to target and modulate the activities of p300 and CBP. Adenovirus E1A, long known to bind Rb to lead to release of E2F family transcription factors and induction of S phase, has been reported to recruit p300 into a ternary complex with Rb that leads to increased acetylation of Rb, although the functional consequences of this remain to be defined (Chan et al., 2001). SV40 large T-antigen, which targets both Rb and p53, and HPV-16 E6, which inactivates p53, bind both p300 and CBP to competitively disrupt their interactions with p53 and other cellular targets (Eckner *et al.*, 1996; Patel *et al.*, 1999). Although the consequences of E1A binding to p300 and CBP may be more complex, E1A can directly inhibit the histone acetyltransferase activities of p300 and PCAF to repress p300-dependent transcription and PCAF-dependent histone modifications, and to block acetylation of p53 (Chakravarti *et al.*, 1999; Hamamori *et al.*, 1999; Vries *et al.*, 2001; Yang *et al.*, 1996).

Given the distinct features and properties of AtNSI when compared to known transcriptional co-activators such as GCN5 and p300/CPB, and that AtNSI does not act as a transcriptional co-activator when tested *in vitro*, what might be the function of AtNSI in plant cells? Relevant to this question is our finding that AtNSI is unique to plants and highly conserved among divergent plant species (McGarry *et al.*, 2003). Thus, if AtNSI has a role in chromosome replication, this is likely to involve aspects unique to this process in plant cells. Alternatively, AtNSI could have a role in regulating unique aspects of plant cell growth and differentiation, acetylating proteins that induce cellular differentiation similar to targets of p300 or TIP60. A number of geminiviruses have been reported to infect differentiated plant cells, and for these geminiviruses their Rep or Rep-related proteins have been reported to interact with plant proteins related to mammalian Rb (Kong *et al.*, 2000; Xie *et al.*, 1995). This has led to the suggestion that these viral proteins, analogous to those encoded by mammalian DNA viruses, may act to induce plant cells to enter S phase. In contrast to these geminiviruses, SqLCV replication and infection are limited to undifferentiated phloem cells, and differences in viral Rep do not appear to account for this phloem limitation (Kong *et al.*, 2000; Lazarowitz *et al.*, 1998). Consistent with our inability to detect AtNSI in *Arabidopsis* extracts or *AtNSI* transcripts by *in situ* hybridization, our recent analysis of transgenic *Arabidopsis* lines expressing an *AtNSI* promoter–GUS (*uidA*) fusion indicates that *AtNSI* expression is developmentally regulated and restricted to immature tissues in the developing plant, including veins in developing leaves (M. Carvalho and S. Lazarowitz, unpublished observations). Thus, as proposed for interactions of the MYST family acetyltransferase HBO1 with the replication protein MCM2 and the origin recognition protein ORC1 (Burke *et al.*, 2001), we suggest that an additional function of NSP binding to AtNSI may be to down-regulate AtNSI acetyltransferase activity to maintain infected phloem or mesophyll cells in a dedifferentiated state and thereby favour viral replication. Analysis of *Arabidopsis AtNSI* mutants will examine this proposed role of AtNSI in growth and differentiation.

Beyond these specific considerations of NSP–AtNSI interactions is the more general utility of NSP for identifying plant cell components of the nuclear import/export apparatus and regulators of nucleo-cytoplasmic transport in plant cells. These might be missed using approaches based on similarity to components identified in animal cells or yeast. The completed *Arabidopsis* genome sequence is predicted to encode eight α importins, three β importins (none are cloned), and three exportins (*AtXPO1*, and a putative second homologue and exportin t). Given the heterogeneity among members of the importin β superfamily, new members are difficult to identify in the databases. There also is no way to identify and test tissue-specific or plant-specific regulators of nuclear transport, which current evidence suggests do exist. For example, regulated nucleo-cytoplasmic transport is implicated in light- and gibberellin-induced developmental responses in plants (Igarashi *et al.*, 2001; Yamamoto and Deng, 1999). In addition, inspection of the plant databases suggests that several animal NE inner membrane and nucleoplasmic proteins are lacking in plants — lamins, lamin B receptor, lamina-associated polypeptide-1 and 2, emerin, MAN1, otefin and nurim —— nor are homologues of Nup358 as such, or RCC1 evident (Meier, 2001; Rose and Meier, 2001). The tight regulation of NSP nuclear shuttling in the presence or absence of MP has been used to develop an *in vivo* cell-based assay for investigating nuclear shuttling in plant cells (Ward and Lazarowitz, 1999), and can now be used to develop an *in vitro* plant cell nuclear import and export assay along the lines of the mammalian permeabilized cell assays. Given that NSP, with its dominant NLSs, is nuclear in the absence of MP and cytoplasmic in its presence (Sanderfoot and Lazarowitz, 1995; Sanderfoot *et al.*, 1996), such an assay could present some advantages over the two current mammalian *in vitro* nuclear export assays, which are based on PKI (which diffuses into the nucleus) exporting a pre-loaded reporter protein, or the use of cells that stably express NFAT and the manipulation NFAT DNA-binding and phosphorylation states (Holaska and Paschal, 1998; Kehlenbach *et al.*, 1998). This *in vitro* assay, combined with our NSP-MP cell-based *in vivo* assay and interactive screens to identify NSP-interacting plant proteins, will be an important advance in our ability to identify unique components of the plant nuclear transport apparatus and investigate the regulation of nucleo-cytoplasmic transport in plants cells. Using such approaches, continued investigation of AtNSI may point to a role in regulating plant cell nuclear export beyond its specific regulation of NSP-mediated viral genome export.

References

Ach, R. and Gruissem, W. (1994) A small nuclear GTP-binding protein from tomato suppresses a *Schizosaccahromyces pombe* cell cycle mutant. *Proc. Natl Acad. Sci. USA* **91**: 5863–5867.

Adam, S., Marr, R. and Gerace, L. (1990) Nuclear protein import in permeabilized mammalian cells requires soluble cytoplasmic factors. *J. Cell. Biol.* **111**: 807–816.

Atabekov, J. and Taliansky, M. (1990) Expression of a plant virus-coded transport function by different virus genomes. *Adv. Virus Res.* **38**: 201–248.

Black, B., Holaska, J., Rastinejad, F. and Paschal, B. (2001) DNA binding domains in diverse nuclear receptors function as nuclear export signals. *Curr. Biol.* **11**: 1749–1758.

Bollman, K.M., Aukerman, M.J., Park, M.-Y., Hunter, C., Berardini, T.Z. and Poethig, R.S. (2003) HASTY, the *Arabdiopsis* ortholog of exportin 5/MSN5, regulates phase change and morphogenesis. *Development* **130**: 1493–1504.

Burke, T., Cook, J., Asano, M. and Nevins, J. (2001) Replication factors MCM2 and ORC1 interact with the histone acetyltransferase HBO1. *J. Biol. Chem.* **276**: 15397–15408.

Carrington, J., Freed, D. and Leinicke, A.J. (1991) Bipartite signal sequence mediates nuclear translocation of the plant potyviral NIa protein. *Plant Cell* **3**: 953–962.

Carrington, J., Kasschau, K., Mahajan, S. and Schaad, M. (1996) Cell-to-cell and long-distance transport of viruses in plants. *Plant Cell* **8**: 1669–1681.

Chakravarti, D., Ogryzko, V., Kao, H.Y., Nash, A., Chen, H., Nakatani, Y. and Evans, R. (1999) A viral mechanism for inhibition of p300 and PCAF acetyltransferase activity. *Cell* **96**: 393–403.

Chan, H., Krstic-Demonacos, M., Smith, L., Demonacos, C. and La Thangue, N. (2001) Acetylation control of the retinoblastoma tumour-suppressor protein. *Nat. Proc. Natl Acad. Sci. USA Cell Biol.* **3**: 667–674.

Cingolani, G., Bednenko, J., Gillespie, M. and Gerace, L. (2002) Molecular basis for the recognition of a nonclassical nuclear localization signal by importin b. *Molec. Cell* **10**: 1345–1353.

Citovsky, V., Zupan, J., Warnick, D. and Zambryski, P. (1992) Nuclear localization of the VirE2 protein in plants cells. *Science* **256**: 1802–1805.

Dawson, W. and Hilf, M. (1992) Host-range determinants of plant viruses. *Ann. Rev. Plant Physiol. Plant Mol. Biol.* **43**: 527–555.

Dhalluin, C., Carlson, J., Zeng, L., He, C., Aggarwal, A. and Zhou, M. (1999) Structure and ligand of a histone acetyltransferase bromodomain. *Nature* **399**: 491–496.

DiMaio, D. and Coen. (2001) In: Knipe, D.M. and Howley, P.M. (eds) *Fields' Virology*, pp. 119–132. Philadelphia: Lippincott, Williams and Wilkins.

Dingwall, C., Sharnick, S. and Laskey, R. (1982) A polypeptide domain that specifies migration of nucleoplasmin into the nucleus. *Cell* **30**: 449–458.

Eckner, R., Ludlow, J., Lill, N., Oldread, E., Arany, Z., Modjtahedi, N., DeCaprio, J., Livingston, D. and Morgan, J. (1996) Association of p300 and CBP with simian virus 40 large T antigen. *Molec. Cell. Biol.* **16**: 3454–3464.

Fischer, U., Huber, J., Boelens, W., Mattaj, I. and Luhrmann, R. (1995) The HIV-1 Rev activation domain is a nuclear export signal that accesses an export pathway used by specific cellular RNAs. *Cell* **82**: 475–483.

Flint, J. and Shenk, T. (1997) Viral transactivating proteins. *Ann. Rev. Genet.* **31**: 177–212.

Greber, U. and Gerace, L. (1995) Depletion of calcium from the lumen of endoplasmic reticulum reversibly inhibits passive diffusion and signal-mediated transport into the nucleus. *J. Cell. Biol.* **128**: 5–14.

Haasen, D., Kohler, C., Neuhaus, G. and Merkle, T. (1999) Nuclear export of proteins in plants: AtXPO1 is the export receptor for leucine-rich nuclear export signals in *Arabidopsis thaliana. Plant J.* **20**: 695–705.

Haizel, T., Merkle, T., Pay, A., Fejes, E. and Nagy, F. (1997) Characterization of proteins that interact with the GTP-bound form of the regulatory GTPase Ran in Arabidopsis. *Plant J.* **11**: 93–103.

Hamamori, Y., Sartorelli, V., Ogryzko, V., Puri, P., Wu, H., Wang, J., Nakatani, Y. and Kedes, L. (1999) Regulation of histone acetyltransferases p300 and PCAF by the bHLH protein twist and adenoviral oncoprotein E1A. *Cell* **96**: 405–413.

Harter, K., Kircher, S., Frohnmeyer, H., Krenz, M., Nagy, F. and Schäfer, E. (1994) Light-regulated modrificaiton and nuclear translocation of cytosolic G-box binding factors in parsley. *Plant Cell* **6**: 545–559.

Heese-Peck, A. and Raikhel, N. (1998) The nuclear pore complex. *Plant Mol. Biol.* **38**: 145–162.

Hepler, P. (1982) Endoplasmic reticulum in the formation of the cell plate and plasmodesmata. *Protoplasma* **111**: 121–133.

Hepler, P., Palevitz, B., Lancelle, S., McCauley, M. and Lichtscheidl, I. (1990) Cortical endoplasmic reticulum in plants. *J. Cell. Sci.* **96**: 355–373.

Hetzer, M., Gruss, O. and Mattaj, I. (2002) The Ran GTPase as a marker of chromosome position in spindle formation and nuclear envelope assembly. *Nat. Cell. Biol.* **4**: E177–184.

Hicks, G. and Raikhel, N. (1995). Protein import into the nucleus: an integrated view. *Ann. Rev. Cell. Dev. Biol.* **11**: 155–188.

Hicks, G., Smith, H., Lobreaux, S. and Raikhel, N. (1996) Nuclear import in permeabilized protoplasts from higher plants has unique features. *Plant Cell* **8**: 1337–1352.

Hill, J., Strandberg, J., Hiebert, E. and Lazarowitz, S. (1998) Asymmetric infectivity of pseudorecombinants of cabbage leaf curl virus and squash leaf curl virus: implications for bipartite geminivirus evolution and movement. *Virology* **250**: 283–292.

Holaska, J. and Paschal, B. (1998) A cytosolic activity distinct from crm1 mediates nuclear export of protein kinase inhibitor in permeabilized cells. *Proc. Natl Acad. Sci. USA* **95**: 14739–14744.

Hottiger, M. and Nabel, G. (2000) Viral replication and the coactivators p300 and CBP. *Trends Microbiol.* **8**: 560–565.

Howard, E., Zupan, J., Citovsky, V. and Zambryski, P. (1992) The VirD2 protein of *A. tumefaciens* contains a C-terminal bipartite nuclear localization signal: implications for nuclear uptake of DNA in plant cells. *Cell* **68**: 109–118.

Hubner, S., Smith, H., Hu, W., Chan, C., Rihs, H., Paschal, B., Raikhel, N. and Jans, D. (1999) Plant importin a binds nuclear localization sequences with high affinity and can mediate nuclear import independent of importin b. *J. Biol. Chem.* **274**: 22610–22617.

Hunter, C.A., Aukerman, M.J., Sun, H., Fokina, M. and Poethig, R.S. (2003) PAUSED encodes the *Arabidopsis* exportin-t ortholog. *Plant Physiol.* **132**: 2135–2143.

Igarashi, D., Ishida, S., Fukazawa, J. and Takahashi, Y. (2001) 14-3-3 proteins regulate intra-cellular localization of the bZIP transcriptional activator RSG. *Plant Cell* **13**: 2483–2497.

Iizuka, M. and Stillman, B. (1999) Histone acetyltransferase HBO1 interacts with the ORC1 subunit of the human initiator protein. *J. Biol. Chem.* **274**: 23027–23034.

Ikura, T., Ogryzko, V., Grigoriev, M., Groisman, R., Wang, J., Horikoshi, M., Scully, R., Qin, J. and Nakatani, Y. (2000) Involvement of the TIP60 histone acetylase complex in DNA repair and apoptosis. *Cell* **102**: 463–473.

Ingham, D., Pascal, E. and Lazarowitz, S. (1995) Both geminivirus movement proteins define viral host range, but only BL1 determines viral pathogenicity. *Virology* **207**: 191–204.

Jacobson, R., Ladurner, A., King, D. and Tjian, R. (2000) Structure and function of a human TAF(II)250 double bromodomain module. *Science* **288**: 1422–1425.

Jiang, C., Imamoto, N., Matsuki, R., Yoneda, Y. and Yamamoto, N. (1998a) Functional charac-terization of a plant importin a homologue. Nuclear localization signal (NLS)-selective binding and mediation of nuclear import of NLS proteins *in vitro. J. Biol. Chem.* **273**: 24083–24087.

Jiang, C., Imamoto, N., Matsuki, R., Yoneda, Y. and Yamamoto, N. (1998b) *In vitro* character-ization of rice importin beta1: molecular interaction with nuclear transport factors and mediation of nuclear protein import. *FEBS Lett.* **437**: 127–130.

Kalderon, D., Roberts, G., Richardson, W. and Smith, A. (1984) A short amino acid sequence able to specify nuclear location. *Cell* **39**: 499–509.

Kehlenbach, R., and Gerace, L. (2000) Phosphorylation of the nuclear transport machinery down-regulates nuclear protein import in vitro. *J. Biol. Chem.* **275**: 17848–17856.

Kehlenbach, R., Dickmanns, A. and Gerace, L. (1998) Nucleo-cytoplasmic shuttling factors including Ran and CRM1 mediate nuclear export of NFAT in vitro. *J. Cell. Biol.* **141**: 863–874.

Komeili, A. and O'Shea, E. (2001) New perspectives on nuclear transport. *Ann. Rev. Genet.* **35**: 341–364.

Kong, L., Orozco, B., Roe, J., *et al.* (2000) A geminivirus replication protein interacts with the retinoblastoma protein through a novel domain to determine symptoms and tissue specificity of infection in plants. *EMBO J.* **19**: 3485–3495.

Kraus, W. and Kadonaga, J. (1999) Ligand- and cofactor-regulated transcription with chromatin templates. In: Picard, D. (ed.) *Steroid/Nuclear Receptor Superfamily: A Practical Approach*, pp. 167–189. Oxford: Oxford University Press.

Kuersten, S., Ohno, M. and Mattaj, I. (2001) Nucleo-cytoplasmic transport: Ran, beta and beyond. *Trends Cell. Biol.* **11**: 497–503.

Kunzler, M. and Hurt, E. (2001) Targeting of Ran: variation on a common theme? *J. Cell. Sci.* **114**: 3233–3241.

Lanford, R., Feldherr, C., White, R., Dunham, R. and Kanda, P. (1990) Comparison of diverse transport signals in synthetic peptide-induced nuclear transport. *Exp. Cell. Res.* **186**: 32–38.

Lazarowitz, S. (1999) Probing plant cell structure and function with viral movement proteins. *Curr. Opin. Plant Biol.* **2**: 332–338.

Lazarowitz, S. (2001) Plant viruses. In: Howley, P.M. (ed.) *Fundamental Virology*, pp. 377–442. Philadelphia: Lippincott, Williams, and Wilkins.

Lazarowitz, S. and Beachy, R. (1999) Viral movement proteins as probes for investigating intra-cellular and intercellular trafficking in plants. *Plant Cell* **11**: 535–548.

Lazarowitz, S., Ward, B., Sanderfoot, A. and Laukaitis, C. (1998) Intercellular and intracellular trafficking: what we can learn from geminivirus movement. In: Morelli, G., Lo Schiavo, F., Last, R.L. and Raikhel, N.V. (eds) *Cellular Integration of Signalling Pathways in Plant Development*, pp. 275–288. Heidelberg: Springer-Verlag.

Leclerc, D., Chapdelaine, Y. and Hohn, T. (1999) Nuclear targeting of the cauliflower mosaic virus coat protein. *J. Virol.* **73**: 553–560.

Li, J. and Chen, X. (2003) PAUSED, a putative exportin-t, acts pleiotrophically in *Arabdiopsis* development but is dispensable for viability. *Plant Physiol.* **132**: 1913–1924.

Macara, I. (2001) Transport into and out of the nucleus. *Microbiol. Mol. Biol. Rev.* **65**: 570–594.

Mattaj, I. and Englmeier, L. (1998) Nucleo-cytoplasmic transport: the soluble phase. *Ann. Rev. Biochem.* **67**: 265–306.

McGarry, R., Barron, Y., Carvalho, M., Hill, J., Gold, D., Cheung, E., Kraus, W. and Lazarowitz, S. (2003) A novel Arabidopsis acetyltransferase interacts with the geminivirus movement protein NSP. *Plant Cell* **15**: 1605–1618.

McMurry, M. and Krangel, M. (2000) A role for histone acetylation in the developmental regulation of V(D)J recombination. *Science* **287**: 495–498.

Meier, I. (2001) The plant nuclear envelope. *Cell. Mol. Life Sci.* **58**: 1774–1780.

Merkle, T., Leclerc, D., Marshallsay, C. and Nagy, F. (1996) A plant in vitro system for the nuclear import of proteins. *Plant J.* **10**: 1177–1186.

Merkle, T., Haizel, T., Matsumoto, T., Harter, K., Dallman, G. and Nagy, F. (1994) Phenotype of the fission yeast cell cycle regulatory mutant pim1-46 is suppressed by a tobacco cDNA encoding a small, Ran-like GTP-binding protein. *Plant J.* **6**: 555–565.

Noueiry, A., Lucas, W. and Gilbertson, R. (1994) 2 proteins of a plant DNA virus coordinate nuclear and plasmodesmal transport. *Cell* **76**: 925–932.

Oparka, K., Boevink, P. and SantaCruz, S. (1996) Studying the movement of plant viruses using green fluorescent protein. *Trends Plant Sci.* **1**: 412–418.

Ossareh-Nazari, B., Gwizdek, C. and Dargemont, C. (2001) Protein export from the nucleus. *Traffic* **2**: 684–689.

Overall, R., Wolf, J. and Gunning, B. (1982) Intercellular communication in *Azolla* roots: I. Ultrastructure of Plasmodesmata. *Protoplasma* **111**: 134–150.

Owen, D., Ornaghi, P., Yang, J., Lowe, N., Evans, P., Ballario, P., Neuhaus, D., Filetici, P. and Travers, A. (2000) The structural basis for the recognition of acetylated histone H4 by the bromodomain of histone acetyltransferase Gcn5p. *EMBO J.* **19**: 6141–6149.

Pascal, E., Sanderfoot, A., Ward, B., Medville, R., Turgeon, R. and Lazarowitz, S. (1994) The geminivirus BR1 movement protein binds single-stranded DNA and localizes to the cell nucleus. *Plant Cell* **6**: 995–1006.

Patel, D., Huang, S., Baglia, L. and McCance, D. (1999) The E6 protein of human papillomavirus type 16 binds to and inhibits co-activation by CBP and p300. *EMBO J.* **18**: 5061–5072.

Qin, S., Ward, B. and Lazarowitz, S. (1998) The bipartite geminivirus coat protein aids BR1 function in viral movement by affecting the accumulation of viral single-stranded DNA. *J. Virol.* **72**: 9247–9256.

Raikhel, N. (1992) Nuclear targeting in plants. *Plant Physiol.* **100**: 1627–1632.

Rose, A. and Meier, I. (2001) A domain unique to plant RanGAP is responsible for its targeting to the plant nuclear rim. *Proc. Natl Acad. Sci. USA* **98**: 15377–15382.

Sanderfoot, A. and Lazarowitz, S. (1995) Cooperation in viral movement: the geminivirus BL1 movement protein interacts with BR1 and redirects it from the nucleus to the cell periphery. *Plant Cell* **7**: 1185–1194.

Sanderfoot, A., Ingham, D. and Lazarowitz, S. (1996) A viral movement protein as a nuclear shuttle: The geminivirus BR1 movement protein contains domains essential for interaction with BL1 and nuclear localization. *Plant Physiol.* **110**: 23–33.

Sanderfoot, A.A. and Lazarowitz, S. (1996) Getting it together in plant virus movement: cooperative interactions between bipartite geminivirus movement proteins. *Trends Cell Biol.* **6**: 353–358.

Sazer, S. and Dasso, M. (2000) The ran decathlon: multiple roles of Ran. *J. Cell. Sci.* **113** (**Pt 7**): 1111–1118.

Smith, H., Hicks, G. and Raikhel, N. (1997) Importin alpha from Arabidopsis thaliana is a nuclear import receptor that recognizes three classes of import signals. *Plant Physiol.* **114**: 411–417.

Spilianakis, C., Papamatheakis, J. and Kretsovali, A. (2000) Acetylation by PCAF enhances CIITA nuclear accumulation and transactivation of major histocompatibility complex class II genes. *Mol. Cell. Biol.* **20:** 8489–8498.

Strom, A. and Weis, K. (2001) Importin-beta-like nuclear transport receptors. *Genome Biol.* **2:** REVIEWS3008.

Szurek, B., Rossier, O., Hause, G. and Bonas, U. (2002) Type III-dependent translocation of the *Xanthomonas* AvrBs3 protein into the plant cell. *Mol. Microbiol.* **46:** 13–23.

Tzfira, T., Rhee, Y., Chen, M., Kunik, T. and Citovsky, V. (2000) Nucleic acid transport in plant–microbe interactions: the molecules that walk through the walls. *Ann. Rev. Microbiol.* **54:** 187–219.

Vries, R., Prudenziati, M., Zwartjes, C., Verlaan, M., Kalkhoven, E. and Zantema, A. (2001) A specific lysine in c-jun is required for transcriptional repression by E1A and is acetylated by p300. *EMBO J.* **20:** 6095–6103.

Wang, A. and Yang, X. (2001) Histone deacetylase 4 possesses intrinsic nuclear import and export signals. *Mol. Cell. Biol.* **21:** 5992–6005.

Ward, B. and Lazarowitz, S. (1999) Nuclear export in plants: Use of geminivirus movement proteins for an *in vivo* cell based export assay. *Plant Cell* **11:** 1267–1276.

Ward, B., Medville, R., Lazarowitz, S. and Turgeon, R. (1997) The geminivirus BL1 movement protein is associated with endoplasmic reticulum-derived tubules in developing phloem cells. *J. Virol.* **71:** 3726–3733.

Weis, K. (2002) Nucleo-cytoplasmic transport: cargo trafficking across the border. *Curr. Opin. Cell. Biol.* **14:** 328–335.

Wen, W., Meinkoth, J., Tsien, R. and Taylor, S. (1995) Identification of a signal for rapid export of proteins from the nucleus. *Cell* **82:** 463–473.

Whittaker, G. and Helenius, A. (1998) Nuclear import and export of viruses and viral genomes. *Virology* **46:** 1–23.

Wiese, C., Wilde, A., Moore, M., Adam, S., Merdes, A. and Zheng, Y. (2001) Role of importin-b in coupling Ran to downstream targets in microtubule assembly. *Science* **291:** 653–656.

Xie, Q., Suarezlopez, P. and Gutierrez, C. (1995) Identification and analysis of a retinoblastoma binding motif in the replication protein of a plant DNA virus – requirement for efficient viral-DNA replication. *EMBO J.* **14:** 4073–4082.

Yamamoto, N. and Deng, X. (1999) Protein nucleo-cytoplasmic transport and its light regulation in plants. *Genes Cells* **4:** 489–500.

Yang, X., Ogryzko, V., Nishikawa, J., Howard, B. and Nakatani, Y. (1996) A p300/CBP-associated factor that competes with the adenoviral oncoprotein E1A. *Nature* **382:** 319–324.

Zhang, C. and Clarke, P. (2000) Chromatin-independent nuclear envelope assembly induced by Ran GTPase in *Xenopus* egg extracts. *Science* **288:** 1429–1432.

Zhang, C. and Clarke, P. (2001) Roles of Ran-GTP and Ran-GDP in precursor vesicle recruitment and fusion during nuclear envelope assembly in a human cell-free system. *Curr. Biol.* **11:** 208–212.

10

Dynamics of nuclear lamina assembly and disassembly

Jos L.V. Broers and Frans C.S. Ramaekers

1. Introduction

Since the discovery of nuclear lamins some 20 years ago, functions in nuclear stability and chromatin organization have been suggested for these intermediate filament-type proteins. While initially the nuclear lamina was considered to be a rigid structure, supporting the nuclear membrane, the use of GFP-tagged lamins in studying nuclear lamina behaviour provides us with a growing insight that the lamina is highly flexible (Broers *et al.*, 1999).

The lamins provide structural support to the nuclear envelope through a meshwork of filaments that is located at the inner layer of the nuclear membrane, forming the lamina (Gruenbaum *et al.*, 2000; Moir *et al.*, 1995). Two main types of lamins can be discerned, A-type and B-type lamins. B-type lamins, represented by the lamin B1 and lamin B2 gene products, are to a large extent ubiquitously expressed in animal cells (see, however, Broers *et al.*, 1997). In contrast, the expression of A-type lamins is low or absent in cells with a low degree of differentiation and/or in highly proliferating cells (Broers *et al.*, 1997; Röber *et al.*, 1989). The lamin A/C (LMNA) gene encodes at least four different A-type lamin proteins, lamin A, lamin AΔ10, lamin C and lamin C2, resulting from alternative splicing (Furukawa and Kondo, 1998; Lin and Worman, 1993; Machiels *et al.*, 1996). Lamins are also found in the nucleoplasm, distinct from the lamina, where they assemble into a number of structures, such as distinct foci (Bridger *et al.*, 1993; Kennedy *et al.*, 2000; Moir *et al.*, 1994), and intranuclear and transnuclear channels (*Figure 1;* Broers *et al.*, 1999; Fricker *et al.*, 1997). In addition, lamins can be found in a nucleoplasmic veil-like structure, which is found throughout the non-nucleolar region of the nucleoplasm (*Figure 1;* Broers *et al.*, 1999; Liu *et al.*, 2000; Moir *et al.*, 2000). Live cell observations with GFP-tagged lamins indicate that only part of these nucleoplasmic lamins is integrated into stable structures.

The nuclear lamins and lamin-associated proteins contain chromatin-binding domains (Glass *et al.*, 1993; Taniura *et al.*, 1995), suggesting that lamins may also be involved in the organization of chromatin (Moir *et al.*, 1995; Wilson *et al.*, 2001), and

The Nuclear Envelope, edited by D.E. Evans, C. Hutchison & J.A. Bryant.
© 2004 Garland Science/BIOS Scientific Publishers

Lamin Adel10-GFP   Lamin B1-GFP

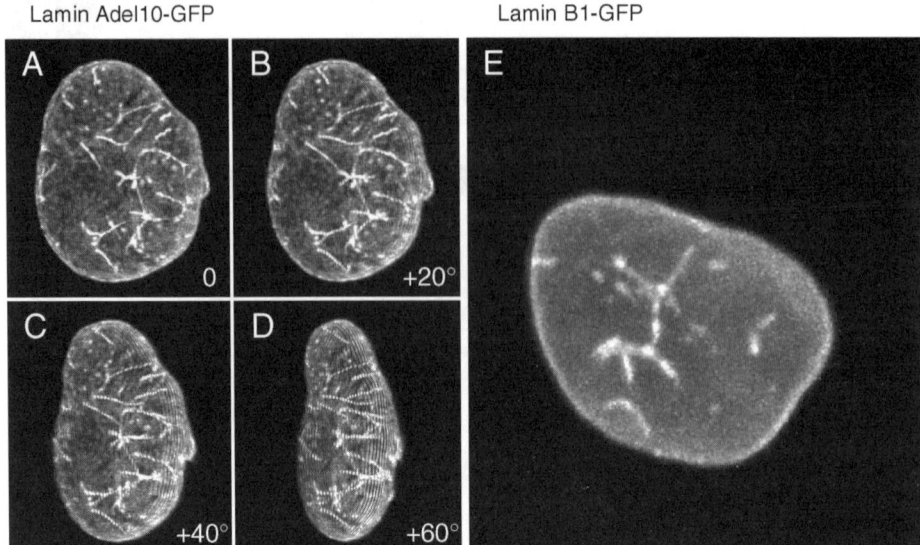

Figure 1. (A–D) *Projection at different angles (0, 20 40 and 60 degrees rotation) of a 3D-reconstructed view of a lamin AΔ10-GFP transfected cell. Note that next to the nuclear lamina, prominent nucleoplasmic tubule-like structures are seen. In addition, a diffuse nucleoplasmic fluorescence (intranuclear veil) can be observed. (E) Z-projection of a lamin B1-GFP transfected cell. Note that fluorescence patterns in the lamina and nucleoplasm for lamin B1 are very similar to lamin AΔ10-GFP.*

possibly also in the regulation of gene expression (Cohen *et al.*, 2001; Holaska *et al.*, 2002; Jagatheesan *et al.*, 1999). To perform these functions, lamins are believed to exhibit a dynamic behaviour rather than forming only a rigid structure.

This chapter will discuss the dynamic behaviour of nuclear lamins in interphase cells. In addition we will discuss the role and behaviour of lamins during mitosis and apoptosis.

2.   Lamina dynamics in interphase cells

While the most obvious localization of lamins during interphase is at the nuclear periphery, a growing number of studies suggest that in interphase cells lamins can be found in nucleoplasmic areas as well. Based on the most recent insights, three levels of lamin organization can be distinguished: (i) lamins associated with the nuclear membrane; (ii) lamins organized into intranuclear tubules and aggregates; and (iii) lamins visible as veil-like structures. We will discuss each of the organizational levels and discuss their (potential) functions.

2.1 *Dynamics of nuclear membrane-associated lamins*

The most prominent concentration of lamins is seen at the nucleoplasmic site of the nuclear periphery, were lamins assemble into a thick meshwork of intermediate filaments, called the nuclear lamina (Aebi *et al.*, 1986). It is unclear until now whether the typical reticulate lamina organisation as seen in *Xenopus* oocytes (Aebi *et al.*, 1986) is also the organization type in other animal cells. Lamina organization at the molecular level is just one of many questions on lamins and their function which are to be elucidated. Based on their suggested rigidity, ubiquitous expression in all animal species and their localization, lamins have long been suggested to provide structural support to the nuclear membrane, and to play an important role in nuclear membrane reassembly after mitosis (see below). Next to lamins, the main lamina-associated proteins include several LAP2 isoforms, the lamin B receptor (LBR), nurim, the MAN antigen, otefin and emerin (Dechat et al., 2000; Gruenbaum et al., 2000). Recently, several new inner nuclear membrane proteins were discovered. These include the ring-finger binding protein (RFBP), luma, and most importantly, the group of nesprin/myne-1 proteins (Mislow et al., 2002; for a review see Mounkes et al., 2003). The interaction between these lamina-associated proteins, lamins, and the underlying chromatin are potentially of great importance for several nuclear functions. However, the exact role of these interactions in cellular function remains unknown. Direct or indirect association of both A- and B-type lamins with (hetero-) chromatin has been suggested, amongst others based on *in vitro* binding studies (Glass et al., 1993; Ludérus et al., 1992; Taniura et al., 1995). The importance of the presence of A-type lamins for heterochromatin organization has recently been shown in cells lacking LMNA gene expression, which display severely disorganized heterochromatin at the nuclear periphery (Sullivan et al., 1999).

The dynamics of the nuclear lamina during interphase have been investigated using fluorescence bleaching techniques. These techniques make use of bleaching of fluorochrome molecules upon exposure to high-power light, such as a laser beam used for confocal scanning microscopy. After bleaching, bleached molecules will become replaced by unaffected fluorescent molecules, provided that in these systems freely diffusing fluorescent molecules are available. Using chimeric proteins consisting of green fluorescent protein (GFP) and, in this case, lamins, it is possible to study the dynamics of these proteins by examining their behaviour upon bleaching. Speed of recovery from photobleaching in a bleached area can be examined (fluorescence recovery after photobleaching, FRAP), or the amount of fluorescence lost after (repetitive) bleaching in a neighbouring region outside of the bleached area can be measured (fluorescence loss in photobleaching, FLIP).

Using these techniques it was deduced that lamins organized into the lamina show a very low turnover. We have demonstrated by FRAP that after bleaching of GFP-tagged A-type lamins in the lamina no recovery of fluorescent signal was observed within 90 minutes after photobleaching (Broers et al., 1999). FLIP experiments showed, however, that not all lamins localized at or close to the lamina are immobile. *Figure 2a–c* shows the loss of fluorescence in the lamina of CHO-K1 cells transfected with either lamin A or lamin C, both tagged with GFP. From this figure it is obvious that the main fraction of lamin A-GFP in the lamina cannot be bleached by a repetitive bleaching regimen elsewhere within the nucleus (*Figure 2a*). In contrast, after bleaching of lamin C-GFP, several cells show a considerable decrease of the fluorescent signal in the lamina (compare *Figures 2b, c*). In some cells more than 50% of the lamin C signal is lost (*Figure 2d*). These results confirm earlier biochemical studies which showed that lamin C is more soluble than lamin A- or B-type lamins in interphase cells (Gerace and Blobel, 1982), and that lamin C rather than lamin A or lamin B can be partially solubilized upon high-salt treatment (Dagenais et al., 1990). FRAP studies on lamin B1 dynamics showed that bleached lamin B1 remained immobile even 45 hours after bleaching, since in non-dividing cells no noticeable recovery of GFP-fluorescence had occurred within this time period (Daigle *et al.*, 2001). These findings are in agreement with biochemical extraction studies which show that in interphase cells, B-type lamins are highly insoluble and resistant to detergents, RNase and DNase treatment (Gerace and Blobel, 1982). The functional implication of the more dynamic behaviour of lamin C within the nuclear lamina is unclear at the moment. Lamin C may act as a vehicle for attaching different regions of (hetero-)chromatin to the nuclear membrane, in order to inactivate gene expression. Alternatively, lamin C may shuttle between the lamina and the intranuclear lamin pool in response to replication and/or transcription regulation (see below).

2.2 *Dynamics of lamins in intranuclear foci*

Immunohistochemical staining of lamin proteins often reveals bright intranuclear dots, next to a perinuclear rim. These dots vary in size and number, depending on the fixation method, cellular state and antibody used. The presence of these intranuclear lamin aggregates has long been considered a fixation artefact. Yet there is growing evidence that intranuclear lamin foci are native structures, possibly associated with initiation of replication sites (Kennedy *et al.*, 2000; Spann *et al.*, 1997; see, however, Dimitrova and Berezney, 2002) or transcriptional complexes (Jagatheesan *et al.*, 1999). Two different types of intranuclear A-type lamin foci should, however, be distinguished. Firstly, it has been described that after *de novo* synthesis lamin C (and not lamin A) accumulates into intranuclear stores and from there gradually becomes integrated into the nuclear lamina. This can be observed after micro-injection of lamin C (Pugh *et al.*, 1997), but also shortly after transfection of GFP-labelled lamin C in cells (not shown). Similarly, lamin A precursor protein (prelamin A) can be found in intranuclear aggregates prior to processing and transport into the lamina (Sasseville and Raymond, 1995; Sinensky *et al.*, 1994).

Secondly, A- and B-type lamin foci can be found in a variety of interphase cells, which cannot directly be attributed to storage of pre-processed lamins. Over-expression of (transfected) lamin C results in intranuclear aggregation in cell cultures (see e.g. *Figure 2b*), but intranuclear foci can also be found in untransfected cells. While Bridger *et al.* (1993) reported the presence of A-type lamin foci in dermal fibroblasts during G1 phase of the cell cycle after a special fixation procedure, Moir *et al.* (1994), showed that in particular lamin B is associated with DNA replication foci in S-phase cells. A more recent study indicated that A-type lamins are present in foci of DNA replication surrounding the nucleolus, which contain replication proteins such as p150 and PCNA (Kennedy *et al.*, 2000). These foci are established in early G1 phase and also contain members of the pRb family. Later, in S phase, DNA replication sites distribute to regions located throughout the nucleus. As cells progress through S phase, the association of A-type lamins with replication foci and pRb family members is lost.

These authors suggest that the onset of DNA synthesis is co-ordinately regulated at a small number of perinucleolar sites that are selected in early G1 phase. Strikingly, the number of replication foci is much lower in primary cell cultures than in established cell lines. Importantly, they also explain why other studies were unable to show co-localization of BrdU-labelled regions of the nucleus (indicating DNA replication)

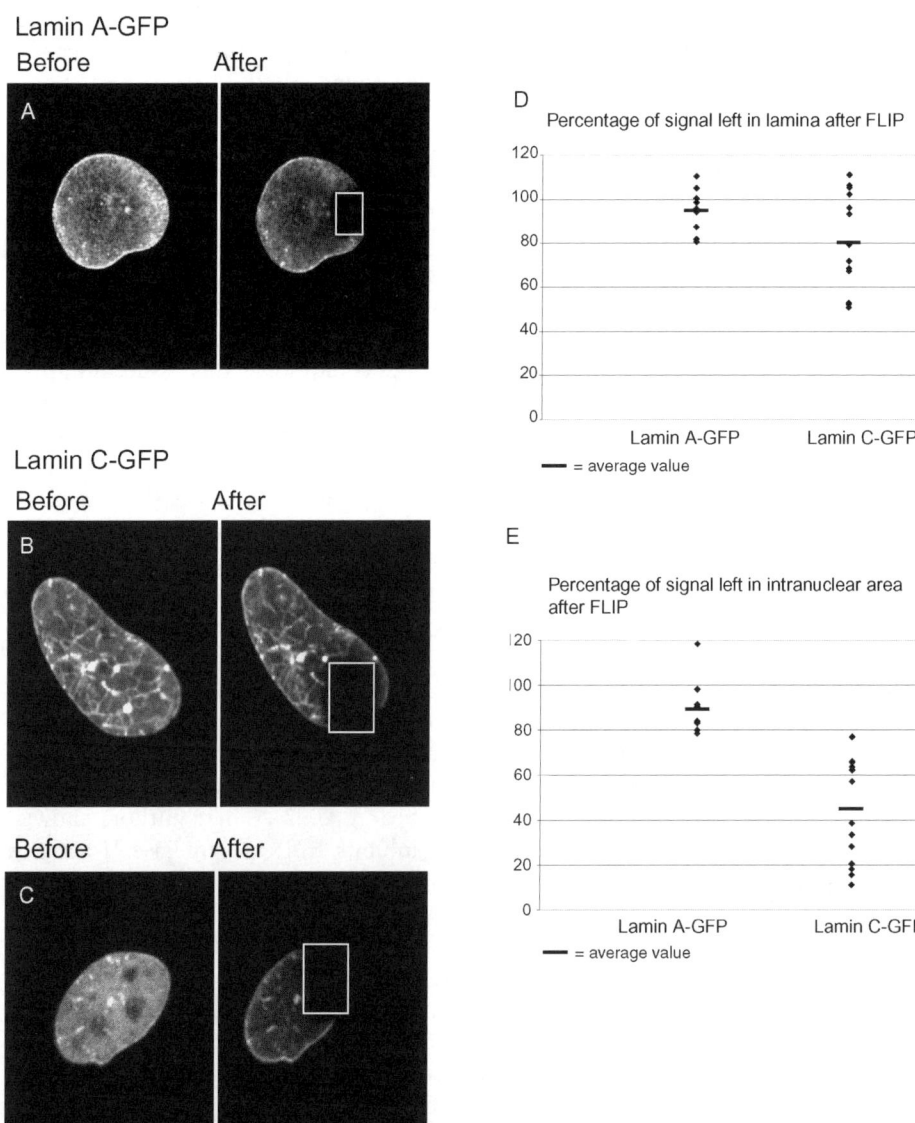

Figure 2. *Bleaching of fluorescent signal (FLIP) in A-type lamin-GFP transfected cells. Living cells were repetitively bleached (15 times 10 bleaching scans with an interval of 15 seconds between bleaching rounds). (A) Signal before and after bleaching in a lamin A-GFP transfected cells. Note that outside of the bleached area (rectangle) almost no fluorescent signal is lost. (B, C) Signal before and after bleaching in two different lamin C-GFP transfected cells. Note that in (B) most of the fluorescence is retained outside of the bleached area, while in another cell (C) a considerable amount of nucleoplasmic lamin C-GFP signal is lost. (D, E) Calculated percentages of lamin A and lamin C-GFP retained after photobleaching for individual cells. Note that the lamin A-GFP signal is strongly retained both in lamina (average fraction retained is 0.95 ±0.09) and intranuclear areas (average 0.90 ±0.13), while for lamin C-GFP a larger loss is seen with much more variation between cells (lamina 0.80 ±0.2, intranuclear areas 0.45 ±0.24).*

with intranuclear lamin localization. They show that, depending on the DNA denaturation method used (HCl vs. DNAse I), nuclear protein structures can become disrupted and dispersed throughout the nucleus (Kennedy *et al.*, 2000).

Studies with mutant lamins suggest that normal lamina assembly is required to establish DNA replication centres (Ellis *et al.*, 1997) and that lamins are essential for the elongation phase of DNA synthesis (Spann *et al.*, 1997).

A recent study, however, argues against a direct role for the retinoblastoma or nuclear lamin proteins in mammalian DNA synthesis under normal physiological conditions. These authors were unable to detect spatial coincidence between the early firing replicons (based on BrdU labelling) and nuclear lamin proteins, the retinoblastoma protein or the nucleolus in primary human and rodent cells (Dimitrova and Berezney, 2002). Possibly, bleaching studies could solve this issue, although currently most live-cell imaging systems do not meet the requirements to answer this question. Prolonged imaging of at least one complete cell cycle should be performed with the possibility of generating fast z-stacks and the capability of bleaching a specific region of the nucleus. Standard confocal microscopy does not meet these requirements, since it is difficult to image cells for more than a few hours without inducing prominent photobleaching. In addition, the process of bleaching is highly toxic to cells. Other systems, such as the spinning (Nipkow) disk confocal microscope system are less toxic to cells, but unfortunately do not allow bleaching of specific areas. Finally, two-photon laser scanning microscopy yields an intrinsic lower (z-) resolution than conventional confocal microscopy and in our hands also causes significant bleaching, if imaging requires excitation with relatively high laser power.

Next to the association of lamins with replication foci, other investigators have reported on the concentration of lamins in nuclear areas with increased RNA polymerase II activity, indicative of transcription (Spann, 2002). These authors showed that disruption of normal lamin organization inhibits RNA polymerase II activity, suggesting that lamins are involved in the synthesis of RNA probably by acting as a scaffold upon which the transcription factors required for RNA polymerase II activation are organized. Jagatheesan *et al.* (1999) and Muralikrishna *et al.* (2001), showed a potential role for A-type lamins in the RNA splicing process. They found the presence of intranuclear A-type lamin foci, which associate with RNA splicing speckles in C2C12 myoblasts and myotubes. Lamin speckles were observed in dividing myoblasts but disappeared early during the course of differentiation in postmitotic myocytes, and were absent in myotubes and muscle fibres. These results suggest that muscle cell differentiation is accompanied by regulated rearrangements in the organization of the A-type lamins (Jagatheesan *et al.*, 1999; Muralikrishna *et al.*, 2001).

2.3 *Lamins in nuclear tubules*

It is likely that the intranuclear lamin foci seen with immunofluorescence are similar to the intranuclear and transnuclear channels observed after microinjection (Fricker *et al.*, 1997) or vital imaging with GFP-lamin-transfected cells (Broers *et al.*, 1999). These intranuclear channels are visible both after transfection with GFP-tagged A-type lamins or with GFP-lamin B1 (*Figures 1* and colour *Plate 7*). Comparison of the distribution patterns of GFP-lamin fluorescence to those obtained after antibody recognition of individual lamins in the same cell shows that the intranuclear tubules seen with GFP(-lamin) fluorescence were often visible by immunofluorescence microscopy as irregular nuclear concentrations, similar to the above-mentioned intranuclear foci (Broers *et al.*, 1999). A possible explanation for this discrepancy is the lack of penetration of antibodies deep into the chromatin-dense, histone-rich nuclear areas. The number of intranuclear channels is highly variable, ranging from zero to a few dozen. In our studies, no correlation with the cell cycle state and channels was observed. Moreover, we found no correlation with level of lamin (over-)expression and the presence of these tubules. Most of these tubules contain membrane lipids as well as nuclear pore complex proteins. It is suggested that these tubules could serve as transport channels between different cytoplasmic regions (Fricker *et al.*, 1997). Vital imaging indicates that these channels can persist for a prolonged period of time, and appear to be rather stable, but flexible, similar to the lamins present in the nuclear lamina as seen in three-dimensional imaging in time (for movie, see http://molcelb2.unimaas.nl/mcb/cytoskeleton.html). Bleaching studies showed that fluorescent GFP-tagged lamin channels are stable with a very low turnover of fluorescent molecules, similar to lamins in the nuclear lamina (Broers *et al.*, 1999).

2.4 *Dynamics of nucleoplasmic lamin*

Ultrastructural investigations suggest the presence of a dispersed intermediate filament lamin network throughout the nucleus (Hozák *et al.*, 1995). The existence of a more finely dispersed, veil-like nucleoplasmic lamin network was suggested after the use of different bleaching techniques (Broers *et al.*, 1999; Moir *et al.*, 2000). FLIP experiments show that a considerable amount of the intranuclear lamins, visible as a diffuse fluorescence pattern in GFP-lamin-transfected cells, cannot be bleached by repetitive bleaching in a nearby area of the nucleus. This implies that these lamins are protected from bleaching as a result of retainment in a particular compartment of the nucleus. Strikingly a large intercellular variation in fluorescence retainment is observed after lamin C-GFP transfection. (*Figure 2e*), which is more pronounced than in lamin A-GFP-transfected cells. Interaction with other intranuclear structures, including (temporary) chromatin association or binding to nuclear histone proteins, known to show *in vitro* interaction with lamins (Taniura *et al.*, 1995), seems an obvious explanation for this phenomenon. The role of this fine network in cellular processes is unclear so far. Bleaching studies on cell lines transfected with GFP- lamins, in which mutations similar to those seen in Emery–Dreifuss muscle dystrophy patients were induced, showed that these mutated lamins do not incorporate properly into a nucleoplasmic veil (J.L.V. Broers, *et al.* (to be published)).

3. Lamin dynamics during mitosis

The most dramatic changes in lamina architecture occur during the process of cell division. At the transition from prophase to prometaphase the nuclear membrane and the lamina disassemble. Lamins are targets for phosphorylation by $p32^{cdc2}$ kinase (Peter *et al.*, 1990; Ward and Kirschner, 1990). As a result of this hyperphosphorylation the lamina polymers disassemble and the lamina proteins become dispersed throughout the cytoplasm. Initially it was assumed that A- and B-type lamins show different responses to phosphorylation. A-type lamins were suggested to become solubilized and disperse completely into the cytoplasm, while B-type lamin particles remained associated with nuclear membrane structures (Nigg, 1992). This view has recently been questioned, and evidence is accumulating that both A- and B-type are solubilized at the onset of mitosis (Beaudouin et al., 2002; Daigle et al., 2001).

For a long time it has been thought that phosphorylation of lamins, together with other proteins, marks the onset of nuclear envelope breakdown. However, recently it has been shown that at the end of prophase microtubules bind to the nuclear membrane via dynein and tear away membrane fragments from the nucleus. As a result the nuclear envelope becomes partially disrupted, allowing kinases to enter the nucleus and to phosphorylate lamin molecules, which then become solubilized (Beaudouin et al., 2002).

Vital imaging of A-type lamin-GFP-transfected cells (Broers et al., 1999; Moir et al., 2000) showed that after metaphase lamina reassembly of all three A-type lamins does not commence until after the separation of the cytoplasm of the daughter cells (cytokinesis). The majority of all three A-type lamin molecules do not move towards the newly formed nucleus until cytokinesis is completed (Broers et al., 1999; and see video images at http://molcelb2.unimaas.nl/mcb/cytoskeleton.html#GFP). At that stage the A-type lamins surround the chromatin very rapidly, since in our studies no GFP signal is visible in the cytoplasm surrounding the chromosomes within 3 minutes after initiation of lamin-GFP condensation. These findings correlate well with earlier immunocytochemical studies that indicate that reassembly of the lamina starts during later stages of telophase and the beginning of cytokinesis, while the bulk of lamins repolymerise into the lamina when the daughter chromosomes decondense (Chaudhary and Courvalin, 1993; Figure 3a). From this figure it is clear that GFP fluorescence (left panel of each stage) does not completely correspond with simultaneous immunocytochemical detection of lamins. In immunofluorescence less staining is seen in areas with dense chromatin, while, in contrast, more fluorescence is detected in cytoplasmic regions. Our studies cannot exclude that some of the lamin molecules are already concentrated at parts of the chromosome surface at late anaphase as suggested previously (Foisner, 1997; Yang et al., 1997). However, it is clear that the majority of A-type lamin molecules only reassemble after cytokinesis.

Some contradictory data exist about the reassembly of B-type lamins, in particular lamin B1, after mitosis. Recent studies with GFP-tagged human lamin B1 in mitotic cells have shown that this lamin begins associating with the peripheral regions of chromosomes during late anaphase to mid-telophase suggesting that lamin B1 polymerization is required for both chromatin decondensation and the binding of nuclear membrane precursors during the early stages of normal nuclear envelope assembly (Lopez-Soler et al., 2001; Moir et al., 2000). However, other studies found that concentration of lamin B1 around chromatin could only be detected in late telophase/early cytokinesis, a stage when chromatin is already sealed by a pore-containing membrane (Daigle et al., 2001). Figure 3b shows the video-recording of reassembly of lamin B1-GFP into the nuclear lamina after mitosis. From this it is evident that lamin B1-GFP is solubilized at metaphase, and is diffusely dispersed throughout the cytoplasm, with no binding to the endoplasmic reticulum, as seen for LBR (Daigle et al., 2001). Starting at telophase, lamin B1 seems to (re-)associate with membrane particles which, however, do not yet surround the chromosomes. Only at late telophase/cytokinesis does lamin B1-GFP reassemble into a nuclear membrane structure (see video material at http://molcelb2.unimaas.nl/mcb/cytoskeleton.html). This clearly suggests that in the

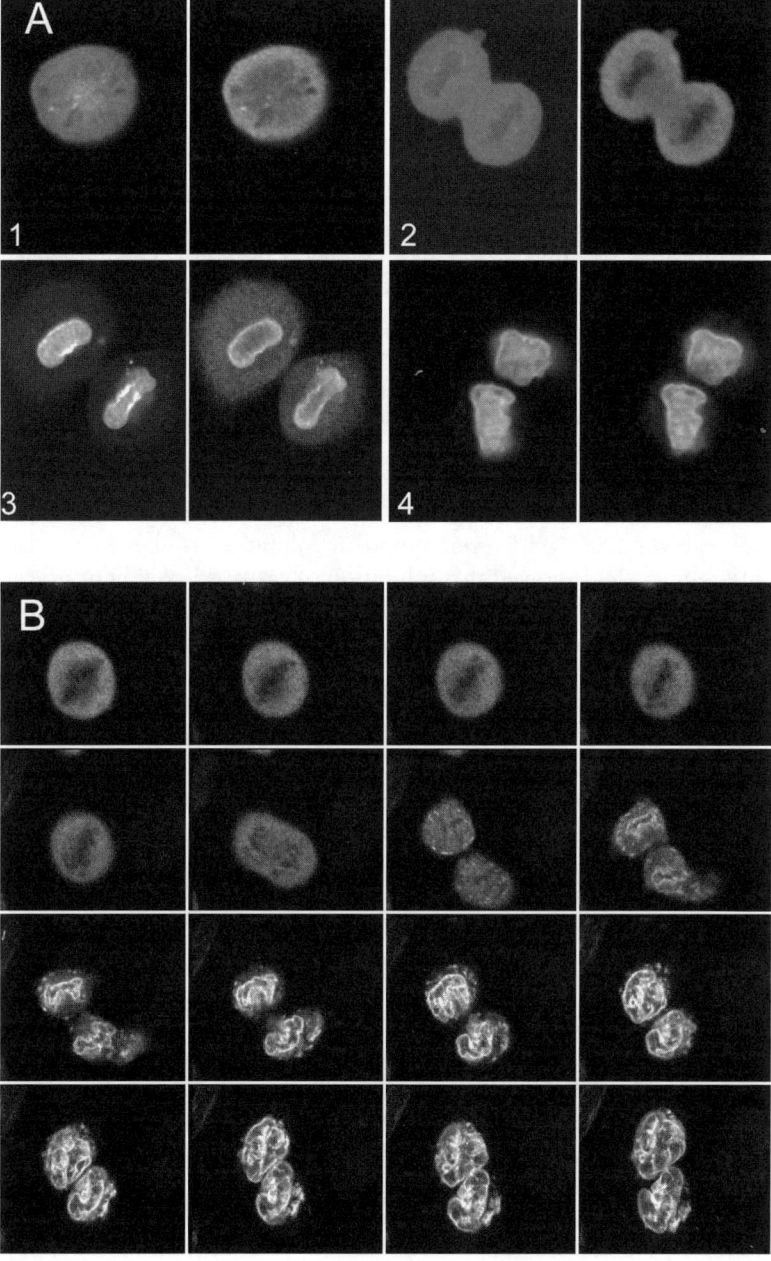

Figure 3. *(A) Double fluorescence recordings of lamin A-GFP (left picture of each panel) and immunofluorescent signal after staining with the lamin A antibody 133A2 (right picture), at different mitotic stages: (1) Metaphase; (2) Telophase; (3) Telophase/cytokinesis; (4) Early G1 phase after cytokinesis. Note that while the compartments of the cell containing both GFP signal and lamin A antibody reactivity are similar, the intensity levels are clearly different. (B) Montage from movie of lamin B1-GFP transfected cell recorded during mitosis. Time interval between the images shown is 5 minutes. The complete movie can be observed at site http://molcelb2.unimaas.nl/mcb/ cytoskeleton.html. Note that lamin-B1 reassembly occurs mainly after cytokinesis.*

cell-system we have used (stably transfected CHO-K1 cells) reorganization of lamin B1 into the nuclear membrane is a late event in mitosis as suggested by Daigle *et al.* (2001).

4. Dynamics of the lamina during apoptosis

During the process of programmed cell death (apoptosis) proteolytic enzymes, called caspases, degrade the lamina networks. Caspase 6 has been identified as the major protease responsible for lamin A degradation (Slee *et al.*, 2000). The A-type lamins are cleaved at their conserved VEID site, which is located in the non-helical linker region L12 at position 227–230. This region contains chromatin-binding sites (Glass *et al.*, 1993) and is also involved in the formation of the lamina structure (McKeon, 1991). It has been assumed that caspase 6 is also responsible for B-type lamin cleavage at their conserved VEVD site at position 227–230. However, a recent study using cell-free extracts immunodepleted of either caspase 3 or 6 showed that proteolysis of lamin B was unaffected by the removal of caspase 6. This suggests that lamin B degradation during the execution phase of apoptosis could also be achieved through caspase 3 activity (Slee *et al.*, 2000). However, apoptosis inhibition studies in our laboratory using a different method showed that inhibition of caspase 3 could not prevent lamin B cleavage, indicating that caspase 6 and not only caspase 3 is responsible for lamin B cleavage.

Although it has been shown that lamins are substrates for caspases, little is known about the steps that govern the proteolysis of the higher-order structures of the nuclear lamina. Lamina degradation is likely to be important in terminating cell function by degrading the structural support of a cell's nucleus, since in cells transfected with lamins with a mutated cleavage site nuclear degradation during apoptosis was largely delayed (Rao *et al.*, 1996). However, blocking caspase 6 activity, which prevents lamina breakdown, alone is insufficient to prevent chromatin condensation (colour *Plate 7*). We performed a comprehensive study on the dynamics and the consequences of lamin cleavage during apoptosis in living cells, in order to obtain a better understanding of the rapid and co-ordinated breakdown of the complex structure of the nucleus (Broers *et al.*, 2002). Induction of apoptosis by staurosporin in A-type lamin-GFP-transfected cells revealed a loss of the GFP-tagged lamina structure at the periphery of the nucleus, with a simultaneous translocation of the GFP to the nucleoplasm and cytoplasm of the cell. Very rapidly after A-type lamin relocalization (within a few minutes) chromatin condensation became visible. Immunoblotting of apoptotic cells revealed that only part of the A-type lamin proteins had been cleaved, while the majority of these proteins were apparently only solubilized.

Oberhammer *et al.* (1994) reported a small increase in soluble depolymerized intact lamin monomers before the major onset of DNA condensation in apoptosis. In contrast to the A-type GFP-tagged lamins, no lamin B1-GFP signal was observed in the cytoplasm in early apoptotic cells. Furthermore, a lamina-like structure at the periphery of the nucleus could still be observed in cells several hours after the morphological onset of DNA condensation (colour *Plate 7*). FLIP studies in living cells supported this finding, since repetitive bleaching did not cause complete bleaching of lamin B1-GFP signal as was the case for lamin C-GFP. The observation that lamin B1 fragments remain visible as a nuclear structure during apoptosis is in accordance with findings of Buendia *et al.* (1999), who showed that during apoptosis, the majority of the proteolytic fragments of lamin B2 remain associated with an insoluble structure, whereas the majority of the proteolytic fragments of Nup153, a component of the nuclear pore complexes, is released into the cytoplasmic compartment. Differences in extractability with different fixations might account for different findings in the literature. For instance, Kihlmark *et al.* (2001) did not distinguish between formaldehyde and ethanol/acetone-fixed cells.

The differences in translocation behaviour between GFP-tagged lamin B1 and A-type lamins during the apoptotic cascade, as observed in our study, most likely reflect differences in molecular organization between the lamin subtypes at three different levels.

Firstly, A- and B-type lamins differ in the processing of their C-terminus. After lamin processing, B-type lamins contain a hydrophobic isoprene tail, whereas the A-type lamins lack this anchoring site for the nuclear membrane. It has previously been reported that the CaaX motif in this tail domain is necessary for the efficient integration of lamin B into an already formed lamina, since lamin B CaaX- mutants showed reduced targeting to the lamina, and appeared to be soluble and not associated with membranes at mitosis. This indicates that the modifications at the CaaX motif are responsible for the association of lamin B with nuclear membranes (Mical and Monteiro, 1998). It has been previously shown that nuclear membranes still surround chromatin fragments formed during apoptosis (Kerr, 1971; Lazebnik *et al.*, 1993; Oberhammer *et al.*, 1993). Apparently the lateral or head-to-tail assembly of lamin B1 molecules (Stuurman *et al.*, 1996) is sufficient to ensure that both the amino terminal and carboxy-terminal fragments of lamin B1 remain attached to this nuclear membrane, even after apoptotic lamin cleavage.

Secondly, A- and B-type lamins interact with different lamina-associated proteins, which could cause differences in translocation during apoptosis. Of these, LAP2α associates specifically with A-type lamins in intranuclear regions (Dechat *et al.*, 2000). The C-terminus of LAP2α binds directly to residues 319–566 in A-type lamins, which include the C-terminus of the rod and the entire tail (Furukawa and Kondo, 1998). During apoptosis LAP2α is cleaved in a caspase-dependent manner. The apoptotic C-terminal fragment of LAP2α remains associated with a residual framework upon extraction using detergent/salt buffers, whereas the N-terminal fragment is extracted from intranuclear structures (Gotzmann *et al.*, 2000).

LAP2β binds specifically to a region within coil 1B of B-type lamins (Furukawa and Kondo, 1998) and to chromatin, but has low affinities for A-type lamins (Foisner and Gerace, 1993; Hutchison *et al.*, 2001). In contrast to LAP2α, LAP2β is integrated into the inner nuclear membrane by its C-terminal domain. It has been reported that LAP2β is cleaved during apoptosis, probably by the activity of caspase 3 (Buendia *et al.*, 1999). Furthermore, it was shown that cleavage of both LAP2β and lamin B2 started at about the same time point in the apoptotic process.

LBR (also called p58) is an integral protein of the inner nuclear membrane that interacts with B-type lamins (Duband-Goulet *et al.*, 1998; Stuurman *et al.*, 1998; Ye and Worman, 1994), but not with A-type lamins (see however Mical and Monteiro, 1998). It has been reported that the amino-terminal domain of LBR is specifically cleaved at a late stage of apoptosis, subsequent to the cleavage of lamin B (Duband-Goulet *et al.*, 1998). Another study failed to observe any proteolytic apoptotic products of LBR even after long periods of apoptosis induction, whereas lamin B2 had already been proteolysed (Buendia *et al.*, 1999).

Interactions between LAP2β and LBR, which are cleaved at late stages of apoptosis, with the relatively early cleaved lamin B1 can cause the apoptotic lamin B1 fragments to remain attached to the remnants of the nuclear membrane. Specific fragments of proteolysed lamins might remain in place during apoptosis and continue to impart structural support to the nuclear membrane, a property that might be important in the formation of apoptotic bodies.

Thirdly, phosphorylation of lamins may play a role in apoptosis. It has been reported that efficient lamina disassembly during apoptosis in ara-C incubated HL60 cells requires both lamin hyperphosphorylation by protein kinase C-δ and caspase-mediated proteolysis (Brodie and Blumberg, 2003; Cross *et al.*, 2000). If phosphorylation is important for lamin solubilization, one would not expect solubilization of A-type lamins in the presence of staurosporin, a well-known kinase inhibitor, which we have used to induce apoptosis (Broers *et al.*, 2002). Apparently, other mechanisms play a role in lamin degradation.

5. Summary

What can be concluded about lamin dynamics?

a. While the nuclear lamina forms a tight network of proteins, individual lamina members, such as the lamin C proteins, are only partially bound to the lamina.
b. A prominent pool of nucleoplasmic lamins exists in most cells, which interacts with intranuclear structures (DNA? Histones? Replication and/or transcription complexes?) in a dynamic fashion.
c. During mitosis lamins do not play a key role in the initial reformation of the nuclear envelope. However, they are important for the correct functioning of the nucleus immediately after mitosis.

Many questions remain unanswered, although nuclear lamins have been studied at different levels for over 20 years. We have only just begun to understand their crucial role in several cellular processes. Some of the important questions that still remain are:

a. The function of intranuclear lamin foci. Are these native nuclear structures (nuclear channels/tubules) and do they play a role in replication and transcription?
b. The function of the nucleoplasmic veil of lamins. Are these molecules only temporarily bound to intranuclear structures as a result of affinity to chromosomes or nuclear proteins, or is there a functional interaction with these molecules?
c. The extent to which lamin molecules add to nuclear membrane organisation. Do these molecules play a key role in keeping the membrane intact and the nucleus functional, or are they only one of many supportive component of the nuclear membrane?

Acknowledgements

The authors which to thank Helma Kuijpers, Jorike Endert, Barbie Machiels, Nancy Bronnenberg and all others with whom we have been working on lamin function for the last decade. The vital imaging studies were financially supported by a grant from the Netherlands Organisation for Scientific Research (NWO grant 901-28-134).

References

Aebi, U., Cohn, J., Buhle, L. and Gerace, L. (1986) The nuclear lamina is a meshwork of intermediate filaments. *Nature* **323**: 560–564.

Beaudouin, J., Gerlich, D., Daigle, N., Eils, R. and Ellenberg, J. (2002) Nuclear envelope breakdown proceeds by microtubule-induced tearing of the lamina. *Cell* **108**: 83–96.

Bridger, J.M., Kill, I., O'Farrell, M. and Hutchison, C.J. (1993) Internal lamin structure within G1 nuclei of human dermal fibroblasts. *J. Cell Sci.* **104**: 297–306.

Brodie, C. and Blumberg, P.M. (2003) Regulation of cell apoptosis by protein kinase c delta. *Apoptosis* **8**: 19–27.

Broers, J.L., Machiels, B.M., Kuijpers, H.J., Smedts, F., van den Kieboom, R., Raymond, Y. and Ramaekers, F.C. (1997) A- and B-type lamins are differentially expressed in normal human tissues. *Histochem. Cell Biol.* **107**: 505–517.

Broers, J.L.V., Machiels, B.M., van Eys, G.J.J., Kuijpers, H.J.H., Manders, E.M.M., van Driel, R. and Ramaekers, F.C.S. (1999) Dynamics of the nuclear lamina as monitored by GFP-tagged A-type lamins. *J. Cell Sci.* **112**: 3463–3475.

Broers, J.L., Bronnenberg, N.M., Kuijpers, H.J., Schutte, B., Hutchison, C.J. and Ramaekers, F.C. (2002) Partial cleavage of A-type lamins concurs with their total disintegration from the nuclear lamina during apoptosis. *Eur. J. Cell Biol.* **81**: 677–691.

Buendia, B., Santa-Maria, A. and Courvalin, J.C. (1999) Caspase-dependent proteolysis of integral and peripheral proteins of nuclear membranes and nuclear pore complex proteins during apoptosis. *J. Cell Sci.* **112**: 1743–1753.

Chaudhary, N. and Courvalin, J.-C. (1993) Stepwise reassembly of the nuclear envelope at the end of mitosis. *J. Cell Biol.* **122**: 295–306.

Cohen, M., Lee, K.K., Wilson, K.L. and Gruenbaum, Y. (2001) Transcriptional repression, apoptosis, human disease and the functional evolution of the nuclear lamina. *Trends Biochem. Sci.* **26**: 41–47.

Cross, T., Griffiths, G., Deacon, E., Sallis, R., Gough, M., Watters, D. and Lord, J.M. (2000) PKC-delta is an apoptotic lamin kinase. *Oncogene* **19**: 2331–2337.

Dagenais, A., LeMyre, A. and Bibor-Hardy, V. (1990) Differential transport and integration into the nuclear lamina for lamins A, B, and C. *Biochem. Cell Biol.* **68**: 827–831.

Daigle, N., Beaudouin, J., Hartnell, L., Imreh, G., Hallberg, E., Lippincott-Schwartz, J. and Ellenberg, J. (2001) Nuclear pore complexes form immobile networks and have a very low turnover in live mammalian cells. *J. Cell Biol.* **154**: 71–84.

Dechat, T., Vlcek, S. and Foisner, R. (2000) Lamina-associated polypeptide 2 isoforms and related proteins in cell cycle-dependent nuclear structure dynamics. *J. Struct. Biol.* **129**: 335–345.

Dimitrova, D.S. and Berezney, R. (2002) The spatio-temporal organization of DNA replication sites is identical in primary, immortalized and transformed mammalian cells. *J. Cell Sci.* **115**: 4037–4051.

Duband-Goulet, I., Courvalin, J.C. and Buendia, B. (1998) LBR, a chromatin and lamin binding protein from the inner nuclear membrane, is proteolyzed at late stages of apoptosis. *J. Cell Sci.* **111**: 1441–1451.

Ellis, D.J., Jenkins, H., Whitfield, W.G. and Hutchison, C.J. (1997) GST-lamin fusion proteins act as dominant negative mutants in Xenopus egg extract and reveal the function of the lamina in DNA replication. *J. Cell Sci.* **110**: 2507–2518.

Foisner, R. (1997) Dynamic organisation of intermediate filaments and associated proteins during the cell cycle. *Bioessays* **19**: 297–305.

Foisner, R. and Gerace, L. (1993) Integral membrane proteins of the nuclear envelope interact with lamins and chromosomes, and binding is modulated by mitotic phosphorylation. *Cell* **73**: 1267–1279.

Fricker, M., Hollinshead, M., White, N. and Vaux, D. (1997) Interphase nuclei of many mammalian cell types contain deep, dynamic, tubular membrane bound invaginations of the nuclear envelope. *J. Cell Biol.* **136**: 531–544.

Furukawa, K. and Kondo, T. (1998) Identification of the lamina-associated-polypeptide-2-binding domain of B-type lamin. *Eur. J. Biochem.* **251**: 729–733.

Gerace, L. and Blobel, G. (1982) Nuclear lamina and the structural organization of the nuclear envelope. *Cold Spring Harb. Symp. Quant. Biol.* **46**: 967–978.

Glass, C.A., Glass, J.R., Taniura, H., Hasel, K.W., Blevitt, J.M. and Gerace, L. (1993) The alfa-helical rod domain of human lamins A and C contains a chromatin binding site. *EMBO J.* **12**: 4413–4424.

Gotzmann, J., Vlcek, S. and Foisner, R. (2000) Caspase-mediated cleavage of the chromosome-binding domain of lamina-associated polypeptide 2 alpha. *J. Cell Sci.* **113**: 3769–3780.

Gruenbaum, Y., Wilson, K.L., Harel, A., Goldberg, M. and Cohen, M. (2000) Nuclear lamins – Structural proteins with fundamental functions. *J. Struct. Biol.* **129**: 313–323.

Holaska, J.M., Wilson, K.L. and Mansharamani, M. (2002) The nuclear envelope, lamins and nuclear assembly. *Curr. Opin. Cell Biol.* **14**: 357–364.

Hozák, P., Sasseville, M.-J., Raymond, Y. and Cook, P.R. (1995) Lamin proteins form an internal nucleoskeleton as well as a peripheral lamina in human cells. *J. Cell Sci.* **108**: 635–644.

Hutchison, C.J., Alvarez-Reyes, M. and Vaughan, O.A. (2001) Lamins in disease: why do ubiquitously expressed nuclear envelope proteins give rise to tissue-specific disease phenotypes? *J. Cell Sci.* **114**: 9–19.

Jagatheesan, G., Thanumalayan, S., Muralikrishna, B., Rangaraj, N., Karande, A.A. and Parnaik, V.K. (1999) Colocalization of intranuclear lamin foci with RNA splicing factors. *J. Cell Sci.* **112**: 4651–4661.

Kennedy, B.K., Barbie, D.A., Classon, M., Dyson, N. and Harlow, E. (2000) Nuclear organization of DNA replication in primary mammalian cells. *Genes Dev.* **14**: 2855–2868.

Kerr, J.F. (1971) Shrinkage necrosis: a distinct mode of cellular death. *J. Pathol.* **105**: 13–20.

Kihlmark, M., Imreh, G. and Hallberg, E. (2001) Sequential degradation of proteins from the nuclear envelope during apoptosis. *J. Cell Sci.* **114**: 3643–3653.

Lazebnik, Y.A., Cole, S., Cooke, C.A., Nelson, W.G. and Earnshaw, W.C. (1993) Nuclear events of apoptosis in vitro in cell-free mitotic extracts: a model system for analysis of the active phase of apoptosis. *J. Cell Biol.* **123**: 7–22.

Lin, F. and Worman, H.J. (1993) Structural organization of the human gene encoding nuclear lamin A and nuclear lamin C. *J. Biol. Chem.* **268**: 16321–16326.

Liu, J., Ben-Shahar, T.R., Riemer, D., Treinin, M., Spann, P., Weber, K., Fire, A. and Gruenbaum, Y. (2000) Essential roles for *Caenorhabditis elegans* lamin gene in nuclear organization, cell cycle progression, and spatial organization of nuclear pore complexes. *Mol. Biol. Cell.* **11**: 3937–3947.

Lopez-Soler, R.I., Moir, R.D., Spann, T.P., Stick, R. and Goldman, R.D. (2001) A role for nuclear lamins in nuclear envelope assembly. *J. Cell Biol.* **154**: 61–70.

Ludérus, M.E.E., de Graaf, A., Mattia, E., den Blaauwen, J.L., Grande, M.A., de Jong, L. and van Driel, R. (1992) Binding of matrix attachment regions to lamin B1. *Cell* **70**: 949–959.

Machiels, B.M., Zorenc, A.H.G., Endert, J.M., Kuijpers, H.J.H., van Eys, G.J.J.M., Ramaekers, F.C.S. and Broers, J.L.V. (1996) An alternative splicing product of the lamin A/C gene lacks exon 10. *J. Biol. Chem.* **271**: 9249–9253.

McKeon, F. (1991) Nuclear lamin proteins: domains required for nuclear targeting, assembly, and cell-cycle-regulated dynamics. *Curr. Opin. Cell Biol.* **3**: 82–86.

Mical, T.I. and Monteiro, M.J. (1998) The role of sequences unique to nuclear intermediate filaments in the targeting and assembly of human lamin B: evidence for lack of interaction of lamin B with its putative receptor. *J. Cell Sci.* **111**: 3471–3485.

Mislow, J., Holaska, J., Kim, M., Lee, K., Segura-Totten, M., Wilson, K. and McNally, E. (2002) Nesprin-1alpha self-associates and binds directly to emerin and lamin A in vitro. *FEBS Lett.* **525**: 135.

Moir, R.D., Montag-Lowy, M. and Goldman, R.D. (1994) Dynamic properties of nuclear lamins: lamin B is associated with sites of DNA replication. *J. Cell Biol.* **125**: 1201–1212.

Moir, R.D., Spann, T.P. and Goldman, R.D. (1995) The dynamic properties and possible functions of nuclear lamins. *Int. Rev. Cytol.* **162B**: 141–182.

Moir, R.D., Yoon, M., Khuon, S. and Goldman, R.D. (2000) Nuclear lamins A and B1:
different pathways of assembly during nuclear envelope formation in living cells. *J. Cell Biol.*
151: 1155–1168.

Mounkes, L., Kozlov, S., Burke, B. and Stewart, C.L. (2003) The laminopathies: nuclear
structure meets disease. *Curr. Opin. Genet. Dev.* **13**: 223–230.

Muralikrishna, B., Dhawan, J., Rangaraj, N. and Parnaik, V.K. (2001) Distinct changes in intranuclear lamin A/C organization during myoblast differentiation. *J. Cell Sci.* **114**: 4001–4011.

Nigg, E.A. (1992) Assembly and cell cycle dynamics of the nuclear lamina. *Semin. Cell Biol.*
3: 245–253.

Oberhammer, F., Fritsch, G., Schmied, M., Pavelka, M., Printz, D., Purchio, T., Lassmann, H.
and Schulte-Hermann, R. (1993) Condensation of the chromatin at the membrane of an apoptotic nucleus is not associated with activation of an endonuclease. *J. Cell Sci.* **104**: 317–326.

Oberhammer, F.A., Hochegger, K., Froschl, G., Tiefenbacher, R. and Pavelka, M. (1994)
Chromatin condensation during apoptosis is accompanied by degradation of lamin A+B,
without enhanced activation of cdc2 kinase. *J. Cell Biol.* **126**: 827–837.

Peter, M., Nakagawa, J., Dorée, M., Labbé, J.C. and Nigg, E.A. (1990) In vitro disassembly of
the nuclear lamina and M phase-specific phosphorylation of lamins by cdc2 kinase. *Cell*
61: 591–602.

Pugh, G.E., Coates, P.J., Lane, E.B., Raymond, Y. and Quinlan, R.A. (1997) Distinct nuclear
assembly pathways for lamins A and C lead to their increase during quiescence in Swiss 3T3
cells. *J. Cell Sci.* **110**: 2483–2493.

Rao, L., Perez, D. and White, E. (1996) Lamin proteolysis facilitates nuclear events during
apoptosis. *J. Cell Biol.* **135**: 1441–1455.

Röber, R.A., Weber, K. and Osborn, M. (1989) Differential timing of nuclear lamin A/C
expression in the various organs of the mouse embryo and the young animal: a developmental
study. *Development* **105**: 365–378.

Sasseville, A.M.J. and Raymond, Y. (1995) Lamin A precursor is localized to intranuclear foci.
J. Cell Sci. **108**: 273–285.

Sinensky, M., Fantle, K., Trujillo, M., McLain, T., Kupfer, A. and Dalton, M. (1994) The
processing pathway of prelamin A. *J. Cell Sci.* **107**: 61–67.

Slee, E.A., Adrain, C. and Martin, S.J. (2000) Executioner caspases-3, -6 and -7 perform distinct,
non-redundant, roles during the demolition phase of apoptosis. *J. Biol. Chem.* **276**: 7320–7326.

Spann, T.P., Moir, R.D., Goldman, A.E., Stick, R. and Goldman, R.D. (1997) Disruption of
nuclear lamin organization alters the distribution of replication factors and inhibits DNA
synthesis. *J. Cell Biol.* **136**: 1201–1212.

Spann, T.P., Goldman, A.E., Wang, C., Huang, S. and Goldman, R.D. (2002) Alteration of
nuclear lamin organization inhibits RNA polymerase II-dependent transcription. *J. Cell Biol.*
156: 603–608.

Stuurman, N., Sasse, B. and Fisher, P.A. (1996) Intermediate filament protein polymerization:
molecular analysis of Drosophila nuclear lamin head-to-tail binding. *J. Struct. Biol.* **117**: 1–15.

Stuurman, N., Heins, S. and Aebi, U. (1998) Nuclear lamins: their structure, assembly, and
interactions. *J. Struct. Biol.* **122**: 42–66.

Sullivan, T., Escalante-Alcalde, D., Bhatt, H., Anver, M., Bhat, N., Nagashima, K., Stewart,
C.L. and Burke, B. (1999) Loss of A-type lamin expression compromises nuclear envelope
integrity leading to muscular dystrophy. *J. Cell Biol.* **147**: 913–920.

Taniura, H., Glass, C. and Gerace, L. (1995) A chromatin binding site in the tail domain of
nuclear lamins that interacts with core histones. *J. Cell Biol.* **131**: 33–44.

Ward, G.E. and Kirschner, M.W. (1990) Identification of cell-cycle regulated phosphorylation
sites on nuclear lamin C. *Cell* **61**: 561–577.

Wilson, K.L., Zastrow, M.S. and Lee, K.K. (2001) Lamins and disease: insights into nuclear infrastructure. *Cell* **104**: 647–650.

Yang, L., Guan, T. and Gerace, L. (1997) Integral membrane proteins of the nuclear envelope are dispersed throughout the endoplasmic reticulum during mitosis. *J. Cell Biol.* **137**: 1199–1210.

Ye, Q. and Worman, H.J. (1994) Primary structure analysis of lamin B and DNA binding of human LBR, an integral protein of the nuclear envelope inner membrane. *J. Biol. Chem.* **269**: 11306–11311.

Spatial and temporal control of nuclear envelope assembly by Ran GTPase

Paul R. Clarke and Chuanmao Zhang

1. Introduction

1.1 *Ran GTPase and its regulators*

Ran, a member of the Ras superfamily of small GTPases, is highly conserved in all eukaryotic cells from the 'primitive' *Giardia lamblia* to yeasts, plants and animals. The biological activity of Ran is determined both by its localization and its guanine-nucleotide bound state (Clarke and Zhang, 2001). During interphase, Ran is concentrated in the nucleus by an active import mechanism involving NTF-2. Like other GTPases, Ran exists in GTP- and GDP-bound states that interact differently with its regulators and effectors (Kunzler and Hurt, 2001). The intrinsic GTPase activity of Ran is very low, but it is greatly stimulated by the interaction of a GTPase-activating protein (RanGAP1) located at the cytoplasmic face of the nuclear pore and in the cytoplasm. Ran GTPase activity is further stimulated by interaction with the Ran-GTP binding domains of Ran binding protein 1 (RanBP1), a mobile but predominantly cytoplasmic protein, and Ran binding protein 2 (RanBP2/Nup358), a component of the cytoplasmic filaments of the nuclear pore complex (NPC). These activities ensure that the relatively low concentration of Ran in the cytoplasm is converted to the GDP-bound state. By contrast, a high concentration of Ran-GTP is generated in the nucleus by RCC1 (Kalab *et al.*, 2002). Generation of Ran-GTP in the nucleus may also be promoted by the nucleotide-destabilizing factor Mog1, which may act together with Ran-GTP binding proteins to dissociate GDP from Ran and reloading with GTP (Nicolás *et al.*, 2001).

1.2 *Ran determines the direction of nucleocytoplasmic transport*

The compartmentalization of Ran, RCC1 and Mog1 to the nucleoplasm, while RanGAP, RanBP1and RanBP2 are kept cytoplasmic, is thought to produce a high

The Nuclear Envelope, edited by D.E. Evans, C. Hutchison & J.A. Bryant.
© 2004 Garland Science/BIOS Scientific Publishers

concentration of Ran-GTP in the nucleus and a much lower concentration of Ran-GDP in the cytoplasm (Görlich and Kutay, 1999; Kalab *et al.*, 2002; Nachury and Weis, 1999; Smith *et al.*, 2002). Experimentally this model has been verified in *Xenopus* egg extracts using fluorescent reporter systems based on the Ran-binding domains of RanBP1 or importin-β (Kalab *et al.*, 2002). Boundary of the nucleus by the nuclear envelope (NE) produces a homogenously high concentration of Ran-GTP in the nucleoplasm and a steep concentration gradient across the nuclear pore. One function of this differential concentration of Ran-GTP is to control the directionality of nucleocytoplasmic transport by controlling the assembly and disassembly of complexes formed between transported cargoes and proteins of the importin/exportin family that act as receptors for targeting sequences on the cargo (Görlich and Kutay, 1999). In the nucleus, interaction of Ran-GTP with the exportin Crm1 causes the assembly of complexes with proteins containing a leucine-rich nuclear export signal (NES) and their export from the nucleus through nuclear pores. In the cytoplasm, import complexes are formed between importins and karyophilic cargo. Proteins carrying a lysine-rich nuclear localization signal (NLS) interact with importin-β directly or through an adaptor, importin-α. Other importin family members play more specialized roles in the transport of specific proteins that have distinct signal sequences. After translocation through the nuclear pore, import complexes are dissociated in the nucleoplasm by Ran-GTP, which binds importin-β and ejects the cargo. During translocation through the pore, transport cargoes may interact with nuclear pore complex (NPC) proteins (nucleoporins) through importins in transient interactions controlled by Ran (Stewart *et al.*, 2001).

1.3 *Role of Ran in mitotic spindle assembly*

Disruption of Ran or its interacting proteins suggested a requirement in cell cycle progression in genetically amenable yeasts, as well as in cultured mammalian cells (Sazer, 1996). However, it was difficult in these systems to distinguish secondary effects that might have been due to defects in nuclear transport. In 1999, several groups, utilizing *Xenopus* egg extracts as a model cell-free system suitable for biochemical analysis, demonstrated a distinct role for Ran in microtubule aster and spindle assembly (Carazo-Salas *et al.*, 1999; Kalab *et al.*, 1999; Ohba *et al.*, 1999; Wilde and Zheng, 1999; Zhang *et al.*, 1999). Ran is present in *Xenopus* egg extracts at a high concentration (1–2 μM) and judging by its interaction with specific binding proteins, is predominantly GDP-bound (Hughes *et al.*, 1998; Kalab *et al.*, 2002). When the concentration of Ran-GTP was increased by adding the exchange factor RCC1 or Ran mutants that are deficient in GTPase activity and are thereby stabilized in the GTP-bound state were used, microtubule assembly was promoted throughout the extract, resulting in ectopic asters containing typical centrosome-associated proteins (Carazo-Salas *et al.*, 1999; Kalab *et al.*, 1999; Ohba *et al.*, 1999; Wilde and Zheng, 1999; Zhang *et al.*, 1999). With incubation, these asters formed spindle-like structures in the absence of chromatin or centrioles, albeit smaller than proper spindles formed following addition of sperm heads (Carazo-Salas *et al.*, 1999; Kalab *et al.*, 1999).

Chromatin has a positional effect on spindle formation by decreasing the catastrophe rate and increasing the rescue frequency of dynamic microtubules, thereby promoting the elongation of spindle microtubules specifically towards the chromatin.

Generation of Ran-GTP by RCC1 in the locality of chromatin during mitosis could thereby account for the stabilizing influence of chromatin on spindles (Dasso, 2001). In mitotic *Xenopus* egg extracts, Ran-GTP not only stabilizes microtubule elongation, but also regulates motor proteins involved in spindle dynamics (Carazo-Salas *et al.*, 2001; Wilde *et al.*, 2001). The effect of Ran-GTP is mediated through a similar biochemical mechanism to nuclear transport: in mitotic extracts, Ran-GTP dissociates factors such as NuMA and TPX2 involved in nuclear protein import from inhibitory complexes with importins (Nachury *et al.*, 2001). Whether Ran-GTP also acts through stabilizing complexes with export factors such as Crm1 is unclear.

The marking of chromatin by local generation of Ran-GTP may be particularly important in a system such as the *Xenopus* early embryo in which cells are very large and the orientation of spindle assembly is not constrained. In *Xenopus* egg extracts, a cloud of Ran-GTP is generated around chromatin in mitosis, whereas Ran dispersed in the extract is predominantly GDP-bound (Kalab *et al.*, 2002). Interestingly, the amount of RCC1 bound to chromosomes increases during mitosis in *Xenopus* egg extracts and RCC1 plays a role in a spindle checkpoint controlling cell cycle progression (Arnaoutov and Dasso, 2003).

In somatic cells, the localization of Ran and its regulators during mitosis may be complex. The gradient of Ran-GTP between chromatin and the rest of the cell is likely to be more gradual than during interphase when the nucleus is enclosed by the NE. Ran appears to be dispersed in mitotic mammalian cells and is not concentrated on chromosomes (Ren *et al.*, 1993; Zhang *et al.*, 1999). However, localization studies with mutants indicate that there is a dynamic interaction of Ran with mitotic chromosomes (Moore *et al.*, 2002). By contrast, RCC1 is predominantly localized to mitotic chromosomes in mammalian cells (Moore *et al.*, 2002), although this interaction is highly dynamic (Li *et al.*, 2003). RanGAP1 modified by SUMO-1 and RanBP2 is present on the mitotic spindle, suggesting that hydrolysis of GTP by Ran may be localized there (Joseph *et al.*, 2002; Matunis *et al.*, 1996). Recent results show that RanBP2 plays a critical role in mitotic progression (Salina *et al.*, 2003). RanGAP is required for spindle assembly and chromosome positioning in the nematode worm *Caenorhabditis elegans*, consistent with a requirement for GTP hydrolysis by Ran (Bamba *et al.*, 2002; Gönczy *et al.*, 2000). Targeting of Ran or Ran-interacting proteins to the spindle microtubules, centrosomes or kinetochore regions of chromosomes (Bamba *et al.*, 2002; Keryer *et al.*, 2003; Zhang *et al.*, 1999) may permit localized activity of Ran.

2. Control of nuclear envelope assembly by Ran

2.1 *Initial evidence from* Xenopus *egg extracts*

Xenopus egg extracts provide a valuable model system to study the assembly of the vertebrate nucleus *in vitro*. Nuclear assembly is usually initiated by the addition of sperm chromatin, which first undergoes decondensation, a process that involves the exchange of basic proteins for histones mediated by nucleoplasmin. Subsequently, membrane vesicles bind to chromatin and fuse to form a double membrane, NPCs are assembled and nuclear growth occurs. The mechanism of NE assembly is not well understood, but vesicle fusion is inhibited by GTPγS, suggesting the involvement of a GTPase, or by N-ethylmaleimide (NEM), a reagent that reacts with thiol groups on proteins (Macaulay and Forbes, 1996). Vesicle fusion may be a prerequisite for NPC assembly, since GTPγS or NEM prevent NPC formation, while NPC formation can be uncoupled from membrane formation, since the metal cation chelator BAPTA results in the formation of a NE without NPCs (Goldberg et al., 1997; Macaulay and Forbes, 1996). The initial recruitment of vesicles to chromatin may involve binding between chromatin and integral membrane proteins that become constituents of the inner membrane, but the mechanism is not understood (Vasu and Forbes, 2001).

In this system, Ran is required for nuclear assembly from sperm chromatin and the establishment of DNA replication, a process that is disrupted by dominant Ran mutants that are deficient in GTP binding or GTP hydrolysis (Dasso et al., 1994; Hughes et al., 1998; Kornbluth et al., 1994; Zhang et al., 1999). Depletion of RCC1, the guanine nucleotide exchange factor for Ran, or addition of excess RanBP1, which opposes RCC1, also prevent proper nuclear assembly (Dasso et al., 1992; Nicolás et al., 1997; Pu and Dasso, 1997). RanBP1 addition only disrupts nuclear assembly when added early in the assembly reaction and does not directly inhibit nuclear protein import (Nicolás et al., 1997). Using field emission in-lens scanning electron microscopy (FEISEM) in collaboration with M.W. Goldberg and T.D. Allen (Paterson Institute for Cancer Research, Manchester, UK), we found that excess RanBP1 resulted in defective NEs that were highly convoluted. Together, these results suggested that both generation of Ran-GTP from Ran-GDP by RCC1 and GTP hydrolysis by Ran are required during an early stage of nuclear assembly (Nicolás et al., 1997).

2.2 Ran induces nuclear envelope formation and nuclear pore complex assembly around chromatin

To examine the possible role of Ran in NE assembly, we supplemented extracts with high concentrations (around 10 μM) of Ran mutants to disrupt the normal Ran-GDP/GTP balance. RanQ69L, locked in the GTP-bound form, had a dramatic effect on NE assembly, inhibiting the fusion of membrane vesicles on the surface of chromatin that remained highly compacted (Zhang et al., 2002a). By contrast, RanT24N, which fails to bind GTP and inhibits RCC1, caused the formation of small, rounded up nuclei with highly convoluted NEs very similar to those formed when RanBP1 was added. These results suggested that both generation of Ran-GTP by RCC1 and GTP hydrolysis by Ran play roles in NE assembly. However, addition of high concentrations of wild-type Ran-GDP, the predominant exogenous form in the extracts, dramatically induced the formation of NEs and the density of NPCs, as well as promoting chromatin decondensation, even in extracts centrifuged at high speed (200 000 g) to remove most membrane vesicles. However, in these experiments manipulation of the Ran system caused significant changes in the morphology of chromatin, so it remained possible that the primary effect of Ran was on chromatin structure rather than a direct role in vesicle attachment and fusion.

In a parallel series of experiments, Hetzer et al. (2000) addressed the possible role of Ran in NE assembly using a novel assay in which chromatin was first decondensed by treatment with a heat-stable fraction of extract containing nucleoplasm, then vesicles labelled with two different coloured lipophilic dyes were incubated with the decondensed chromatin with the addition of a soluble extract fraction. Membrane formation was assayed by the mixing of the two colours as fusion between the two vesicle populations occurred. Hetzer et al. (2000) found that addition of RanQ69L or RanT24N inhibited vesicle fusion in this assay. When extracts were depleted of RCC1 using RanT24N, exogenous RCC1 or Ran-GTP, but not Ran-GDP, were able to overcome the inhibition of vesicle fusion around chromatin assembled on DNA, indicating that generation of Ran-GTP by RCC1 is required. However, Ran bound with the non-hydrolysable GTP analogue GTPγS failed to rescue RCC1 depletion, indicating that GTP hydrolysis by Ran is also required. Immunodepletion experiments also showed that Ran itself is essential for membrane fusion and nucleoporin incorporation.

An important point in interpreting the results from both series of experiments is to note that addition of exogenous Ran proteins may affect both the overall level of Ran-GTP in the extracts and its normal localization. Addition of RanT24N (or excess RanBP1) is likely to disrupt the generation of Ran-GTP that normally occurs specifically only around chromatin, whereas RanQ69L will result in widespread elevation of Ran-GTP throughout the extract. Thus, the apparent inhibitory effects of Ran-Q69L on NE assembly around chromatin could indicate a requirement for GTP hydrolysis by Ran or could be due to the formation of complexes distal to chromatin that sequester necessary components.

2.3 *Interaction of Ran and RCC1 with chromatin*

During nuclear assembly in *Xenopus* egg extracts, endogenous Ran associates rapidly with demembranated sperm chromatin following its addition (Zhang *et al.*, 1999). The recruitment of Ran to chromatin occurs before envelope assembly and is not dependent on membranes, but does correlate with chromatin decondensation when the protein composition of chromatin changes and there is structural rearrangement. Ran in either the GTP- or GDP-bound form interacts directly with isolated chromatin through core histones (Bilbao-Cortes *et al.*, 2002). However, in egg extracts, Ran-GDP is recruited specifically and promotes the binding of RCC1 (Zhang *et al.*, 2002a). RCC1 interacts with chromatin in egg extracts through core histones H2A and H2B (Nemergut *et al.*, 2001). We have found that the association of RCC1 with chromatin is regulated by Ran, being inhibited by excess Ran-GTP and promoted by Ran-GDP (Zhang *et al.*, 2002a). The recruitment of RCC1 to chromatin may provide a specialized mechanism to kick-start nuclear assembly during the first cell cycle following fertilization of the egg.

In higher vertebrate cells, however, RCC1 is localized predominantly to chromatin throughout the cell cycle (Moore *et al.*, 2002). Nevertheless, relocalization of Ran to chromosomes during telophase (Zhang *et al.*, 1999) and the consequent increase in its local concentration may play an important role during exit from mitosis, providing a spatial cue for changes in microtubule dynamics and the initiation of NE assembly. The temporal regulation of this process is likely to be coupled to changes in mitotic protein kinase activity during progression through cell divisions. However, the regulation of the interaction of Ran and RCC1 with chromatin in cells is poorly understood at present.

2.4 *Nuclear envelope assembly around Ran beads*

Using *Xenopus* egg extracts, we tested the hypothesis that concentration of Ran might play a direct role in NE formation by immobilizing and concentrating recombinant Ran proteins on inert Sepharose beads. In extracts lacking DNA or chromatin, beads coated with wild-type Ran rapidly accumulated membrane vesicles that fused to form a continuous lipid layer that incorporated nucleoporins (Zhang and Clarke, 2000). In electron micrographs, NPCs were apparent crossing a double membrane, indicating that a complete NE was assembled (Zhang and Clarke, 2000; Zhang et al., 2002a). The envelopes were functional, since a fluorescent dextran that is too large to diffuse across the nuclear pores was excluded, whereas a karyophilic protein containing an NLS was concentrated within the beads. In other words, simply concentrating Ran on the surface of beads was sufficient to induce membrane vesicle binding and fusion, as well as the assembly of NPCs and the subsequent initiation of nucleocytoplasmic transport. In contrast to wild-type Ran, beads coated with the mutants RanT24N or RanQ69L failed to make intact NEs, indicating that both loading of Ran with GTP and GTP hydrolysis are required for envelope assembly around Ran beads. Immobilization of RCC1 onto beads was insufficient to induce NE formation, illustrating the importance of the high local concentration of Ran.

The ability of Ran concentrated on the surface of beads to induce the formation of NEs is not restricted to *Xenopus* egg extracts, since extracts prepared from mitotic human cells also work (Zhang and Clarke, 2001). Using human cell extracts, we showed that if beads coated with Ran-GDP are used, then RCC1 is required to generate Ran-GTP, whereas beads coated with Ran-GTP do not require RCC1. Conversely, if RanGAP is inhibited with an antibody, then vesicle binding and fusion are reduced, showing that GTP hydrolysis on Ran is involved. We have also induced membrane formation around Ran-beads in insect and yeast cell extracts, strongly suggesting that this mechanism is conserved between eukaryotes (authors' unpublished data).

This work showed that concentration of Ran is sufficient to induce complete NE assembly and NPC assembly in the absence of chromatin. This is reminiscent of studies in which stacks of NE-like membranes containing pore complexes, called annulate lamellae, can be formed in *Xenopus* egg extracts (Dabauvalle et al., 1993). Indeed, addition of exogenous wild-type Ran induces annulate lamellae formation in this system (*Figure 1*; see also Harel et al., 2003; Walther et al., 2003) showing that increased concentrations of Ran without immobilization induces assembly of NPCs into membranes. Together, these results suggest the hypothesis that induction of NE assembly at the end of mitosis specifically around chromatin is driven by the relocalization of Ran to chromatin and the subsequent local generation of Ran-GTP at a higher local concentration than during mitosis.

2.5 *Roles for importin-β in nuclear envelope assembly*

How does Ran control NE assembly? To address this question, we have used an assay in which *Xenopus* egg extracts were depleted of Ran-binding proteins (ΔRanBP extracts) using RanQ69L (Zhang et al., 2002b). ΔRanBP extracts were deficient in ability to promote membrane vesicle recruitment and fusion to form continuous membranes around Ran beads, but NE assembly was restored by addition of 5 μM importin-β, a concentration similar to that of the endogenous protein in non-depleted extracts. By contrast, other related import factors were ineffective and a supraphysiological concentration of importin-β (50 μM) inhibited membrane formation around Ran beads. A truncated protein (importin-β^{1-409}) that lacks importin-α binding activity

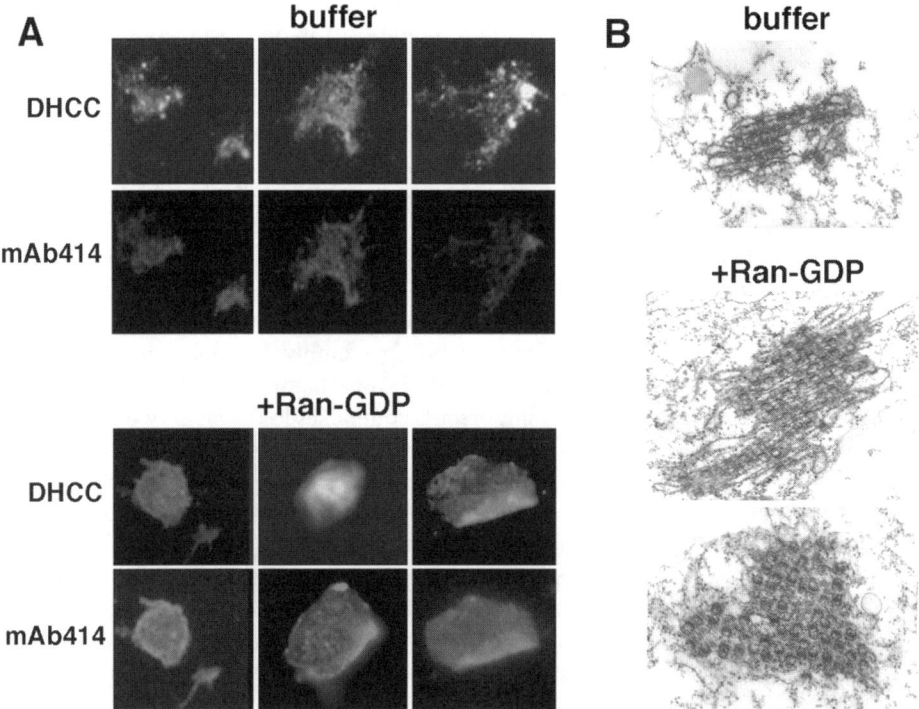

Figure 1. Ran induces nuclear envelope formation and nuclear pore complex assembly in the absence of chromatin. In interphase Xenopus egg extracts, excess wild-type Ran (10 μM) induces the formation of membrane stacks containing nuclear pore complexes (annulate lamellae) in the absence of DNA or chromatin. (A) Occasional membrane aggregations stained by a lipophilic dye (DHCC) are formed in extracts during incubation (upper panel), but these are irregular and stain only weakly with an antibody (mAB414) that recognises a subset of nucleoporins. Addition of 10 μM Ran, preloaded with GDP induces more regular structures that incorporate nucleoporins (lower panel). (B) Using transmission electron microscopy, the structures formed following the addition of Ran can be seen to consist of regular stacks of membranes (transverse view) which are studded with nuclear pores complexes (apical view), identical to annulate lamellae.

(Kutay *et al.*, 1997) also restored NE assembly around Ran-beads in ΔRanBP extracts, indicating that importin-β does not function in NE assembly by interaction through importin-α with karyophilic proteins carrying Lys-rich NLS motifs. By contrast, importin-α[45-462], which lacks the Ran-binding region was not functional. These results suggested that importin-β acts as an adaptor that recruits target proteins to Ran to initiate NE formation.

In addition to transported cargoes and Ran, importin-β interacts directly with protein components of the nuclear pore (nucleoporins) containing FxFG (Phe-x-Phe-Gly, where x is usually Ser, Gly or Ala) repeats. An importin-β mutant in which Ile[178] is changed to Asp (I178D), shows decreased affinity of importin-β for FxFG nucleoporins but not nucleoporins containing GLFG (Gly-Leu-Phe-Gly) repeats or transport cargoes (Bayliss *et al.*, 2000) This mutation inhibited the ability of importin-β to promote NE assembly around Ran beads, suggesting that importin-β acts by recruiting certain FxFG nucleoporins to Ran. Treatment of importin-β with N-ethylmaleimide (NEM), which reacts with cysteine residues on importin-β and inactivates it (Chi and Adam, 1997), abolished the ability to induce NE formation around beads. Importin-β may therefore account, at least in part, for the NEM-sensitivity of NE assembly. These results indicate that importin-β plays a role during NE assembly induced by Ran through interaction with certain nucleoporins.

Importins are also required for NE assembly in invertebrates (Timinszky *et al.*, 2002), consistent with this mechanism being conserved. However, recent work from Walther *et al.* (2003) and Harel *et al.* (2003) has questioned a positive role for importin-β during NE assembly around chromatin in *Xenopus* egg extracts. These views may perhaps be reconciled in a model (*Figure 2*) in which importin-β controls NE assembly distal from chromatin by inhibiting vesicle fusion and NPC assembly through interacting with specific components such as those of the Nup107–160 subcomplex. These components may be released from suppression by importin-β complexes in the vicinity of chromatin by locally generated Ran-GTP. An excess of importin-β overcomes the effect of Ran-GTP and inhibits NE and NPC assembly. Nevertheless, in the absence of importin-β, NE assembly cannot be directed specifically around chromatin by Ran. Importin-β is therefore essential for physiological NE assembly.

Ran is likely to play multiple roles during nuclear envelope assembly: in the recruitment of vesicles to chromatin, their fusion to form a double membrane, and in the assembly of NPCs. Interestingly, Harel *et al.* (2003) showed that an excess of importin-β did not inhibit vesicle recruitment to chromatin, whereas vesicle fusion and NPC assembly were blocked. High concentrations of RanQ69L promoted membrane formation, but did not overcome the inhibitory effect of excess importin-β on NPC assembly. Thus, Ran and importin-β may act through multiple types of inter-action to control complete NE formation.

It is not clear how the interactions between Ran, importin-β and target proteins are changed at the end of mitosis. In part, the switch is likely to be determined by a change in phosphorylation status of specific components controlled by mitotic protein kinases. In addition, the increased concentration of Ran on chromatin at the end of mitosis may favour lower affinity interactions than those occurring during mitosis when Ran is dispersed.

2.6 *Is the function of Ran in nuclear envelope assembly conserved in all eukaryotes?*

The high degree of similarity of components of the Ran system amongst eukaryotes suggests that control of NE formation by Ran is likely to have been conserved during evolution. In the nematode *Caenorhabditis elegans*, inhibition of Ran, RanGAP, RanBP1 or importin expression by RNAi treatment of embryos disrupts spindle structure and prevents proper nuclear assembly following mitosis (Askjaer *et al.*, 2002; Gönczy *et al.*, 2000). Ran localizes to kinetochores during metaphase/anaphase in *C. elegans*, relocalizing to the periphery of the reforming nucleus at telophase. Depletion of Ran by RNAi treatment prevents NE assembly, as well as causing defects in mitotic spindle positioning (Bamba *et al.*, 2002). In *Drosophila*, injection of a dominant mutant of importin-β blocks NE assembly in cleavage embryos (Timinszky *et al.*, 2002).

Does the Ran system also play a role in NE integrity in species such as yeasts that have a closed mitosis in which the NE is not disassembled? In the fission yeast

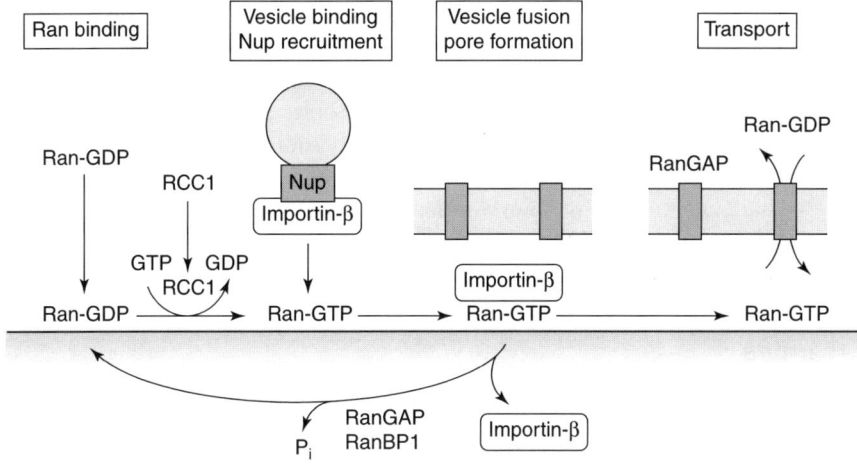

Figure 2. *A model for the role of Ran in NE assembly. Ran-GDP binds to the surface of chromatin and recruits RCC1, which generates Ran-GTP. Ran-GTP recruits vesicles and soluble nucleoporins through interaction with importin-β releasing the targets from importin-β and thereby inducing vesicle fusion to form an intact membrane and NPC assembly. Ran-GTP/ importin-β complexes are recycled by GTP-hydrolysis promoted by RanGAP. Nuclear envelope formation encloses the chromatin and allows a high concentration of Ran-GTP to accumulate on the nucleoplasmic side, initiating nuclear transport through the nuclear pores.*

Schizosaccharomyces pombe, a temperature-sensitive mutation of the RCC1 homologue Pim1p produces fragmentation of the NE at exit from mitosis, although the NE normally remains intact during mitosis in this species (Demeter *et al.*, 1995). Generation of Ran-GTP by Pim1p may be required at this point in the cell cycle because additional NE needs to be formed during nuclear division. In *Saccharomyces cerevisiae*, Ran plays an essential role in NPC assembly (Ryan *et al.*, 2003). It can be envisaged that the expansion of NE membranes during cell division in yeast and the formation of the NE at the end of an open mitosis are closely related processes that are likely to both involve the fusion of dispersed vesicles and/or the feeding in of new membranes to the NE via the ER, as well as insertion of NPCs. Ran is likely to play critical roles in these processes in all eukaryotes.

3. Summary

Using evidence derived primarily from studies using *Xenopus* egg extracts, a model for the role of Ran in multiple stages during NE assembly can be proposed (*Figure 2*). Ran is concentrated on chromatin prior to NE assembly and recruits RCC1 that generates Ran-GTP locally. Recruitment of RCC1 to chromatin may be a specialized mechanism to initiate NE assembly following fertilization of the egg, whereas in somatic cells, RCC1 may be present on chromatin throughout mitosis. Ran-GTP recruits vesicles to the surface of chromatin, and promotes vesicle fusion to form the double membrane of the NE. Ran-GTP may recruit membrane vesicles to chromatin through binding to integral membrane proteins through importin-β. A transient complex would be formed between Ran-GTP, importin-β and the target protein, which would be released locally to promote assembly of a precursor complex. GTP hydrolysis by Ran would release importin-β, but may also play a role in vesicle fusion. Ran-GTP also promotes NPC assembly by releasing nucleoporins such as Nup107 from inhibitory complexes with importin-β.

In vertebrate cells undergoing mitosis, the majority of Ran molecules are excluded from the chromosomes and dispersed into the cytoplasm. Relocalization of Ran to chromatin at the end of mitosis may co-ordinate the initiation of NE assembly with disassembly of the mitotic spindle. The function of Ran in this transition is likely to be coupled to changes in the activity of cyclin-dependent protein kinases and other activities that control the progression of the cell cycle. Thus, changes in the localization of Ran and its regulators provide temporal and spatial control of NE assembly at the end of mitosis.

References

Arnaoutov, A. and Dasso, M. (2003) The Ran GTPase regulates kinetochore function. *Dev. Cell* **5**: 99–111.

Askjaer, P., Galy, V., Hannak, E. and Mattaj, I.W. (2002) Ran GTPase cycle and importins alpha and beta are essential for spindle formation and nuclear envelope assembly in living *Caenorhabditis elegans* embryos. *Mol. Biol. Cell* **13**: 4355–4370.

Bamba, C., Bobinnec, Y., Fukuda, M. and Nishida, E. (2002) The GTPase Ran regulates chromosome positioning and nuclear envelope assembly in vivo. *Curr. Biol.* **12**: 503–507.

Bayliss, R., Littlewood, T. and Stewart, M. (2000) Structural basis for the interaction between FxFG nucleoporin repeats and importin-beta in nuclear trafficking. *Cell* **102**: 99–108.

Bilbao-Cortes, D., Hetzer, M., Langst, G., Becker, P.B. and Mattaj, I.W. (2002) Ran binds to chromatin by two distinct mechanisms. *Curr. Biol.* **12**: 1151–1156.

Carazo-Salas, R., Guarguaglini, G., Gruss, O.J., Segref, A., Karsenti, E. and Mattaj, I.W. (1999) Generation of GTP-bound Ran by RCC1 is required for chromatin-induced mitotic spindle formation. *Nature* **400**: 178–181.

Carazo-Salas, R.E., Gruss, O.L. and Karsenti, E. (2001) Ran-GTP coordinates regulation of microtubule nucleation and dynamics during mitotic-spindle assembly. *Nature Cell Biol.* **3**: 228–234.

Chi, N.C. and Adam, S.A. (1997) Functional domains in nuclear import factor p97 for binding the nuclear localization sequence receptor and the nuclear pore. *Mol. Biol. Cell* **8**: 945–956.

Clarke, P.R. and Zhang, C. (2001) Ran GTPase: a master regulator of nuclear structure and function during the eukaryotic cell division cycle? *Trends Cell Biol.* **11**: 366–371.

Dabauvalle, M.C., Loos, K., Merkert, H. and Scheer, U. (1993) Spontaneous assembly of pore complex-containing membranes (annulate lamellae) in *Xenopus* egg extract in the absence of chromatin. *J. Cell Biol.* **112**: 1073–1082.

Dasso, M. (2001) Running on Ran: Nuclear transport and the mitotic spindle. *Cell* **104**: 321–324.

Dasso, M., Nishitani, H., Kornbluth, S., Nishimoto, T. and Newport, J.W. (1992) RCC1, a regulator of mitosis, is essential for DNA replication. *Mol. Cell. Biol.* **12**: 3337–3345.

Dasso, M., Seki, T., Azuma, Y., Ohba, T. and Nishimoto, T. (1994) A mutant form of the Ran/TC4 protein disrupts nuclear function in *Xenopus* laevis egg extracts by inhibiting the RCC1 protein, a regulator of chromosome condensation. *EMBO J.* **13**: 5732–5744.

Demeter, J., Morphew, M. and Sazer, S. (1995) A mutation in the RCC1-related protein pim1 results in nuclear envelope fragmentation in fission yeast. *Proc. Natl Acad. Sci. USA* **92**: 1436–1440.

Goldberg, M.W., Wiese, C., Allen, T.D. and Wilson, K.L. (1997) Dimples, pores, star-rings, and thin rings on growing nuclear envelopes: evidence for structural intermediates in nuclear pore complex assembly. *J. Cell Sci.* **110**: 409–420.

Gönczy, P., Echeverri, C., Oegema, K., *et al.* (2000) Functional genomic analysis of cell division in *C. elegans* using RNAi of genes on chromosome III. *Nature* **408**: 331–336.

Görlich, D. and Kutay, U. (1999) Transport between the cell nucleus and the cytoplasm. *Ann. Rev. Cell Dev. Biol.* **15**: 607–660.

Harel, A., Chan, R.C., Lachish-Zalait, A., Zimmerman, E., Elbaum, M. and Forbes, D.J. (2003) Importin beta negatively regulates nuclear membrane fusion and NPC assembly. *Mol. Biol. Cell* **14**: 4387–4396

Hetzer, M., Bilbao-Cortés, D., Walter, T.C., Gruss, O.J. and Mattaj, I.W. (2000) GTP hydrolysis by Ran is required for nuclear envelope assembly. *Mol. Cell* **5**: 1013–1024.

Hughes, M., Zhang, C., Avis, J.M., Hutchison, C.J. and Clarke, P.R. (1998) The role of Ran GTPase in nuclear assembly and DNA replication: characterisation of the effects of Ran mutants. *J. Cell Sci.* **111**: 3017–3026.

Joseph, J., Tan, S.H., Karpova, T.S., McNally, J.G. and Dasso, M. (2002) SUMO-1 targets RanGAP1 to kinetochores and mitotic spindles. *J. Cell Biol.* **156**: 595–602.

Kalab, P., Pu, R.T. and Dasso, M. (1999) The Ran GTPase regulates mitotic spindle assembly. *Curr. Biol.* **9**: 481–484.

Kalab, P., Weis, K. and Heald, R. (2002) Visualization of a Ran-GTP gradient in interphase and mitotic *Xenopus* egg extracts. *Science* **295**: 2452–2456.

Keryer, G., Di Fiore, B., Celati, C., Lechtreck, K.F., Mogensen, M., Delouvee, A., Lavia, P., Bornens, M. and Tassin, A.M. (2003) Part of Ran is associated with AKAP450 at the centrosome: involvement in microtubule-organizing activity. *Mol. Biol. Cell* **14**: 4260–4271.

Kornbluth, S., Dasso, M. and Newport, J. (1994) Evidence for a dual role for TC4 protein in regulating nuclear structure and cell cycle progression. *J. Cell Biol.* **125**: 705–719.

Kunzler, M. and Hurt, E. (2001) Targeting of Ran: variation on a common theme? *J. Cell Sci.* **114**: 3233–3241.

Kutay, U., Izaurralde, E., Bischoff, F.R., Mattaj, I.W. and Gorlich, D. (1997) Dominant-negative mutants of importin-beta block multiple pathways of import and export through the nuclear pore complex. *EMBO J.* **16**: 1153–1163.

Li, H.Y., Wirtz, D. and Zheng, Y. (2003) A mechanism of coupling RCC1 mobility to RanGTP production on the chromatin in vivo. *J. Cell Biol.* **160**: 635–644.

Macaulay, C. and Forbes, D.J. (1996) Assembly of the nuclear pore: biochemically distinct steps revealed with NEM, GTP gamma S, and BAPTA. *J. Cell Biol.* **132**: 5–20.

Matunis, M.J., Coutavas, E. and Blobel, G. (1996) A novel ubiquitin-like modification modulates the partitioning of the Ran-GTPase-activating protein RanGAP1 between the cytosol and the nuclear pore complex. *J. Cell Biol.* **135**: 1457–1470.

Moore, W.J., Zhang, C. and Clarke, P.R. (2002) Targetting of RCC1 to chromosomes is required for proper mitotic spindle assembly in human cells. *Curr. Biol.* **12**: 1442–1447.

Nachury, M.V. and Weis, K. (1999) The direction of transport through the nuclear pore can be inverted. *Proc. Natl Acad. Sci. USA* **96**: 9622–9627.

Nachury, M.V., Maresca, T.J., Salmon, W.C., Waterman-Storer, C.M., Heald, R. and Weiss, K. (2001) Importin β is a mitotic target of the small GTPase Ran in spindle assembly. *Cell* **104**: 95–106.

Nemergut, M.E., Mizzen, C.A., Stukenberg, T., Allis, C.D. and Macara, I.G. (2001) Chromatin docking and exchange activity enhancement of RCC1 by histones H2A and H2B. *Science* **292**: 1540–1543.

Nicolás, F.J., Moore, W.J., Zhang, C. and Clarke, P.R. (2001) XMog1, a nuclear Ran-binding protein in *Xenopus*, is a functional homologue of *Schizosaccharomyces pombe* Mog1p that co-operates with RanBP1 to control generation of Ran-GTP. *J. Cell Sci.* **114**: 3013–3023.

Nicolás, F.J., Zhang, C., Hughes, M., Goldberg, M.W., Watton, S.J. and Clarke, P.R. (1997) *Xenopus* Ran-binding protein 1: molecular interactions and effects on nuclear assembly in *Xenopus* egg extracts. *J. Cell Sci.* **110**: 3019–3030.

Ohba, T., Nakamura, M., Nishitani, H. and Nishimoto, T. (1999) Self-organization of microtubule asters induced in *Xenopus* egg extracts by GTP-bound Ran. *Science* **284**: 1356–1358.

Pu, R.T. and Dasso, M. (1997) The balance of RanBP1 and RCC1 is critical for nuclear assembly and nuclear transport. *Mol. Biol. Cell* **8**: 1955–1970.

Ren, M., Drivas, G., D'Eustachio, P. and Rush, M.G. (1993) Ran/TC4: a small nuclear GTP-binding protein that regulates DNA synthesis. *J. Cell Biol.* **120**: 313–323.

Ryan, K.J., McCaffery, J.M. and Wente, S.R. (2003) The Ran GTPase cycle is required for yeast nuclear pore complex assembly. *J. Cell Biol.* **160**: 1041–1053.

Salina, D., Enarson, P., Rattner, J.B. and Burke, B. (2003) Nup358 integrates nuclear envelope breakdown with kinetochore assembly. *J. Cell Biol.* **162**: 991–1001.

Sazer, S. (1996) The search for the primary function of the Ran GTPase continues. *Trends Cell Biol.* **6**: 81–85.

Smith, A.E., Slepchenko, B.M., Schaff, J.C., Loew, L.M. and Macara, I.G. (2002) Systems analysis of Ran transport. *Science* **295**: 488–491.

Stewart, M., Baker, R.P., Bayliss, R., Clayton, L., Grant, R.P., Littlewood, T. and Matsuura, Y. (2001) Molecular mechanism of translocation through nuclear pore complexes during nuclear protein import. *FEBS Lett.* **498**: 145–149.

Timinszky, G., Tirian, L., Nagy, F.T., Toth, G., Perczel, A., Kiss-Laszlo, Z., Boros, I., Clarke, P.R. and Szabad, J. (2002) The importin-beta P446L dominant-negative mutant protein loses RanGTP binding ability and blocks the formation of intact nuclear envelope. *J. Cell Sci.* **115**: 1675–1687.

Vasu, S.K. and Forbes, D.J. (2001) Nuclear pores and nuclear assembly. *Curr. Opin. Cell Biol.* **13**: 363–375.

Walther, T.C., Askjaer, P., Gentzel, M., Habermann, A., Griffiths, G., Wilm, M., Mattaj, I.W. and Hetzer, M. (2003) RanGTP mediates nuclear pore complex assembly. *Nature* **424**: 689–694.

Wilde, A., Lizarraga, S.B., Zhang, L., Wiese, C., Gliksman, N.R., Walczak, C.E. and Zheng, Y. (2001) Ran stimulates spindle assembly by altering microtubule dynamics and the balance of motor activities. *Nature Cell Biol.* **3**: 221–227.

Wilde, A. and Zheng, Y. (1999) Stimulation of microtubule aster formation and spindle assembly by the small GTPase Ran. *Science* **284**: 1362–1365.

Zhang, C. and Clarke, P.R. (2000) Chromatin-independent nuclear envelope assembly induced by Ran GTPase in *Xenopus* egg extracts. *Science* **288**: 1429–1432.

Zhang, C. and Clarke, P.R. (2001) Roles of Ran-GTP and Ran-GDP in precursor vesicle recruitment and fusion during nuclear envelope assembly in a human cell-free system. *Curr. Biol.* **11**: 208–212.

Zhang, C., Hughes, M. and Clarke, P.R. (1999) Ran-GTP stabilises microtubule asters and inhibits nuclear assembly in *Xenopus* egg extracts. *J. Cell Sci.* **112**: 2453–2461.

Zhang, C., Goldberg, M.W., Moore, W.J., Allen, T.D. and Clarke, P.R. (2002a) Concentration of Ran on chromatin induces decondensation, nuclear envelope formation and nuclear pore complex assembly. *Eur. J. Cell Biol.* **81**: 623–633.

Zhang, C., Hutchins, J.R., Muhlhausser, P., Kutay, U. and Clarke, P.R. (2002b) Role of importin-beta in the control of nuclear envelope assembly by Ran. *Curr. Biol.* **12**: 498–502.

12

Nuclear envelope dynamics during mitosis

Brian Burke, Melissa Crisp and Davide Salina

1. Introduction

The acquisition of organelles and the evolution of compartmental boundaries have provided eukaryotes with unique mechanisms with which to modulate multiple cellular activities. This is particularly evident in the sequestration of chromosomes within a nuclear envelope, which effectively separates the nuclear processes of gene transcription and replication from translation. Clearly, control of molecular traffic across the nuclear envelope is a prerequisite for normal eukaryotic cell function. This enhanced regulatory potential in eukaryotes, however, comes at a cost of considerably complicating the mechanics of cell division. Progression through mitosis requires that chromosomes within the cell nucleus engage with microtubules (MTs) of the mitotic spindle. In organisms such as yeast, the spindle poles are embedded in the nuclear envelope (NE) and spindle MTs form within the nucleus. This is a 'closed' mitosis. In higher cells, however, the mitotic spindle is a cytoplasmic structure, and consequently, in order for mitotic chromosomes to align at the spindle equator, the NE must be either partially or completely dispersed.

2. The nuclear envelope

The most prominent features of the NE are a pair of inner and outer nuclear membranes (INM and ONM, *Figure 1*) (Gant and Wilson, 1997; Gerace and Burke, 1988; see Chapter 1). While the ONM displays frequent connections with the ER and features numerous ribosomes, the INM contains a unique set of membrane proteins, is ribosome-free and maintains close contacts with chromatin (Wilson, 2000). Regardless of their biochemical differences, the INM and ONM are joined in regions where they are spanned by nuclear pore complexes (NPCs), the channels that mediate trafficking between the nucleus and cytoplasm. In this way, the INM, ONM and ER form a single continuous membrane system. Metazoans contain an additional NE structure, the nuclear lamina (Gerace and Burke, 1988). In mammalian somatic cells

---

The Nuclear Envelope, edited by D.E. Evans, C. Hutchison & J.A. Bryant.
© 2004 Garland Science/BIOS Scientific Publishers

this appears as a thin (~20 nm) protein meshwork lining the INM and which maintains interactions with both chromatin and INM-specific proteins. The lamina is composed primarily of the A- and B-type lamin family of intermediate filament proteins and plays an essential role in the maintenance of both NE integrity and nuclear organization (Gruenbaum *et al.*, 2000; Wilson *et al.*, 2001).

During mitosis in mammalian somatic cells, the nuclear lamina and NPCs are disassembled. At the same time, nuclear membrane components are dispersed throughout the cell (Ellenberg *et al.*, 1997; Ostlund *et al.*, 1999; Yang *et al.*, 1997). Disassembly of the lamina and NPCs occurs in response to phosphorylation of both lamina and NPC subunits (Gerace and Blobel, 1980; Heald and McKeon, 1990; Macaulay *et al.*, 1995). While the majority of these components become distributed throughout the mitotic cytoplasm, certain NPC proteins (nucleoporins or Nups) and associated molecules, including Rae1, Nup107 and Nup133 become preferentially associated with kinetochores (Babu *et al.*, 2003; Belgareh *et al.*, 2001; Wang *et al.*, 2001). Another nucleoporin, Nup358, which is a component of the short (100 nm) filaments that extend from the cytoplasmic face of the NPC during interphase, relocates to both spindle microtubules and kinetochores (Joseph *et al.*, 2002). This relocation of Nup358 occurs in association with Ran GTPase activating protein 1 (RanGAP1), a molecule with which Nup358 also interacts during interphase. Conversely, certain mitotic checkpoint proteins, such as

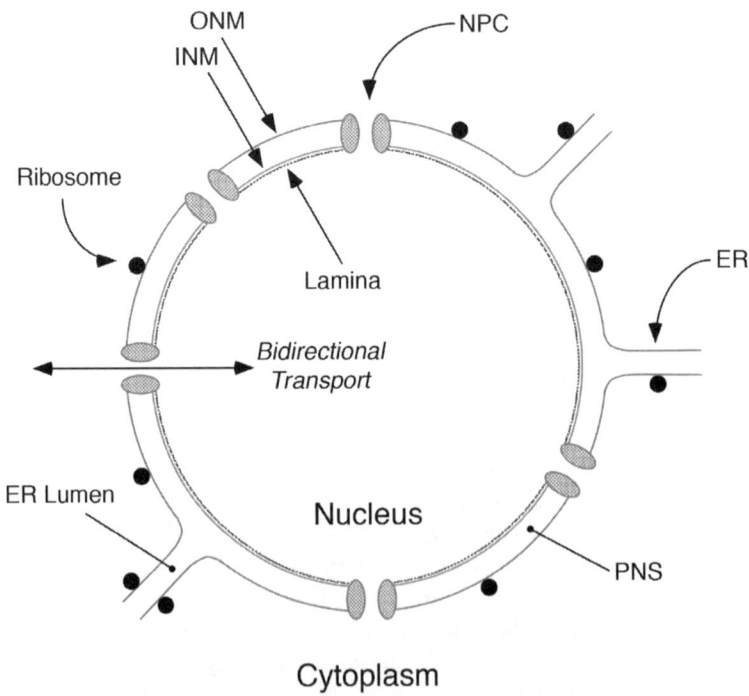

Figure 1. *Organization of the nuclear envelope. The inner and outer nuclear membranes (INM and ONM) are joined where they are spanned by nuclear pore complexes (NPCs). The ONM is also continuous with the endoplasmic reticulum. Similarly the perinuclear space (PNS) is continuous with the lumen of the ER.*

Mad1, Mad2 and Mps1 that are kinetochore-associated during mitosis are found associated with NPCs during interphase (Campbell *et al.*, 2001; Liu *et al.*, 2003). In yeast,
this localization is mediated by Nup53p, part of a larger complex of NPC proteins that
includes Nup157p and Nup170p (Iouk *et al.*, 2002). Remarkably, yeast strains deficient
in Mad1p exhibit a reduced rate of nuclear protein import as well as decreased stability
of the Nup53p complex (Iouk *et al.*, 2002). The implication is that there is a functional
relationship between the mitotic apparatus and the nuclear envelope.

3. Nuclear envelope breakdown

Prophase in higher cells is defined by condensation of chromatin and initiation of
events leading to NE breakdown. Nuclear envelope breakdown involves the disassembly and dispersal of all major NE structural components (Lee *et al.*, 2000),
including the nuclear lamina (Gerace and Blobel, 1980; Heald and McKeon, 1990;
Peter *et al.*, 1990; Pfaller and Newport, 1995; Stick *et al.*, 1988; Ward and Kirschner,
1990), NPCs (Macaulay *et al.*, 1995; Snow *et al.*, 1987) and membranes. The disruption
of the nuclear membranes marks the end of prophase, and at this time, integral
proteins of the INM and NPCs are lost from the nuclear periphery and become
distributed throughout the cell (Chaudhary and Courvalin, 1993; Ellenberg *et al.*,
1997; Yang *et al.*, 1997). By mid-prometaphase the NE has largely dispersed and the
nuclear contents are released into the cytoplasm.

The mechanism of nuclear membrane breakdown has, in recent years, been a focus
of considerable attention. Subcellular fractionation, and studies on nuclear disassembly and reassembly in *Xenopus* egg extracts suggest that dividing cells contain
unique populations of NE-derived vesicles (Newport and Spann, 1987; Vigers and
Lohka, 1991). These findings provide a basis for models in which nuclear membrane
breakdown is accomplished by a process of vesiculation. Other studies, primarily in
mammalian systems, however, suggest that nuclear envelope breakdown results in the
intermingling of ER and INM components (Ellenberg *et al.*, 1997; Yang *et al.*, 1997).
Indeed, ultrastructural analyses in several mammalian cell types, including thyroid
epithelia (Zeligs and Wollman, 1979), PtK2 cells (Roos, 1973) and Hela cells (Robbins
and Gonatas, 1964) consistently reveal the detachment of membrane cisternae, often
described as ER-like, from the nuclear periphery, without extensive vesiculation. While
these two views of nuclear membrane breakdown and dispersal are clearly poles apart,
the notion of intermixing of nuclear membrane components with bulk ER during
prophase can still be reconciled with data supporting the vesicular model (Collas and
Courvalin, 2000). For instance, if nuclear membrane components were to enter or to
form microdomains within the ER, then subcellular fractionation would be anticipated
to yield populations of microsomal vesicles enriched for NE components. Indeed
certain nuclear membrane components, including members of the LAP2 family of
INM proteins, appear not to be homogeneously distributed within the peripheral ER
of cells exiting mitosis. Instead, local cytoplasmic concentrations of these proteins are
frequently observed that persist into early G1 (D. Salina, M. Crisp and B. Burke,
unpublished observations). The existence of these relatively long-lived foci is
consistent with the notion that NE proteins do in fact form microdomains within the
mitotic ER.

4. The role of dynein

In the absence of mechanisms inducing vesiculation in mammalian somatic cells, the question arises as to what processes actually promote the rupturing of the nuclear membranes. There is now a consensus emerging that at least in certain cell types cytoplasmic dynein plays a central role in facilitating nuclear membrane dispersal (Beaudouin et al., 2002; Salina et al., 2002). Dynein, a microtubule minus-end directed motor protein associates with the cytoplasmic face of the NE during early mitotic prophase. The action of dynein during NE breakdown is two-fold. Firstly, NE-associated dynein has been shown to stabilize the growing ends of astral microtubules in the vicinity of the NE (Piehl and Cassimeris, 2003). Secondly, dynein mediates the withdrawal of NE components towards the separating centrosomes (to which the microtubule minus ends are anchored) (Salina et al., 2002). This results in the application of tension to regions of the nuclear membranes that are distal to the centrosomes (Beaudouin et al., 2002). It is in these distal regions that significant breaches in the nuclear membranes are first observed. Although yet to be conclusively demonstrated, it is likely that membrane fenestrae that result from the disassembly of nuclear pore complexes, form the epicentre for the initial rupturing of the nuclear membranes. In this way, cytoplasmic dynein functions to literally tear open the nuclear membranes (Beaudouin et al., 2002; Salina et al., 2002).

The role that dynein is thought to play in nuclear envelope breakdown is summarized in *Figure 2*. In this model, the presence of dynein on the prophase NE provides a unifying mechanism for observations from a number of laboratories, in particular, the microtubule-dependent deformation of the NE during prophase and the concentration of NE components around the centrosomes (Georgatos et al., 1997; Robbins and Gonatas, 1964; Roos, 1973; Stafstrom and Staehelin, 1984; Waterman-Storer et al., 1993; Zeligs and Wollman, 1979) following nuclear envelope breakdown. It must be stressed however, that dynein and microtubules only play a facilitative role in nuclear envelope breakdown. Interference with dynein function or microtubule integrity will merely delay, but not prevent, release of chromosomes into the cytoplasm. Furthermore, in certain cell types, for instance starfish oocytes, nuclear envelope breakdown occurs via a dynein-independent mechanism (Lenart et al., 2003). As in mammalian systems, however, breaches in the nuclear membranes still appear to originate at disassembled NPCs (Lenart et al., 2003; Terasaki et al., 2001).

The role of dynein in the dispersal of the nuclear membranes raises the question of how dynein associates with the cytoplasmic face of the NE during mitotic prophase. It is known that dynein binding to a variety of organellar membranes (including, in all likelihood, the NE) is mediated in part by a large multiprotein assembly, the dynactin complex (Allan and Schroer, 1999). While the dynein/dynactin binding partner on the NE has yet to be identified it must possess certain characteristics. For instance, it must be enriched in the ONM *versus* the peripheral ER, and furthermore, it must be able to transmit force to the INM and nuclear lamina. Such features have led to suggestions that NPCs might actually contain the dynein binding site. Certainly they are concentrated within the NE, are exposed to the cytoplasm and are anchored to the nuclear lamina. However, immunolocalization studies indicate that dynein is not specifically associated with NPCs during mitotic prophase, rather it localizes to regions of the ONM that lie between NPCs (D. Salina and B. Burke, unpublished observations). The implication is that there are ONM-specific membrane proteins that are somehow

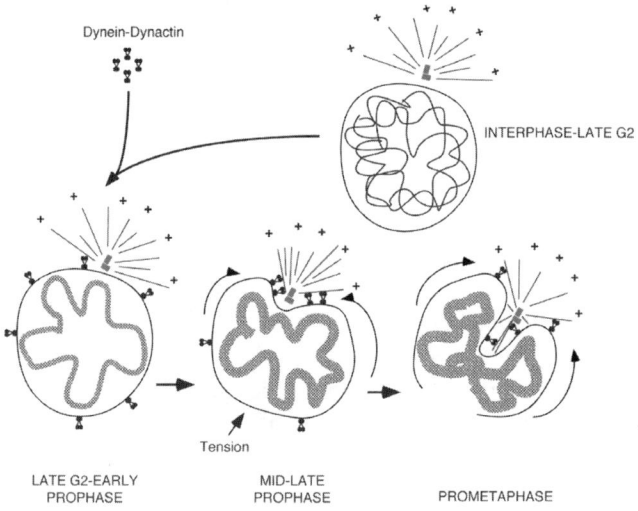

Figure 2. Recruitment of cytoplasmic dynein to the nuclear envelope at the end of G2 leads to withdrawal of NE components towards the minus ends of microtubules anchored at the centrosomes. This results in increased tension across the nuclear membranes and ultimately membrane tearing. For clarity, only one of the two centrosomes is shown.

linked to the INM and lamina perhaps via protein complexes that span the perinuclear space. The most promising candidates for dynein/dynactin binding partners are currently represented by members of the nesprin family of nuclear membrane proteins (Zhang *et al.*, 2001).

5. The kinetochore connection

While NPCs appear not to play an active role in dynein-mediated nuclear membrane disruption, certain NPC proteins nevertheless have an essential function in mitotic progression. In the case of one of these, Nup358, this mitotic function has been revealed using RNA interference approaches to specifically deplete cells of this protein. Nup358 is also known as Ran binding protein 2 (RanBP2) and is the major component of the 100-nm filaments that extend from the cytoplasmic face of NPCs (Walther *et al.*, 2002; Yokoyama *et al.*, 1995). The Nup358 molecule contains multiple phenylalanine-glycine (FG) repeats that form binding sites for transport receptors of the importin/karyopherin-α family (Delphin *et al.*, 1997). All available evidence suggests that Nup358 is a docking site for NLS-bearing import susbstrates in association with their cognate receptors. Such a function provides a basis for observations that NLS-bearing molecules concentrate at the cytoplasmic filaments prior to translocation through the NPC (Panté and Aebi, 1996; Richardson *et al.*, 1988). Nup358 also binds RanGAP1, the activating protein for Ran, a member of the Ras superfamily of small GTPases. Ran is an important regulator of nucleocytoplasmic transport (Saitoh *et al.*, 1997) and defines the directionality of the transport process. Only RanGAP1 molecules that have been modified with the small ubiquitin-like protein SUMO associate with Nup358 at the nuclear periphery (Mahajan *et al.*, 1997; Matunis *et al.*, 1998;

see Chapter 6). Intriguingly, Nup358 itself has SUMO E3 ligase activity (Pichler *et al.*, 2002). The implication here is that Nup358 functions not just as a docking protein but also as a modulator of the cytoplasmic levels of Ran-GDP versus Ran-GTP. It comes as no small surprise therefore that Nup358 is apparently redundant with respect to the transport process. Nuclei assembled *in vitro* in *Xenopus* egg extracts depleted of Nup358 are transport competent, although the NPCs within these nuclei lack cytoplasmic filaments (Walther *et al.*, 2002). Similarly, depletion of Nup358 in HeLa cells by means of RNA interference has little effect on the qualitative import of proteins into the nucleus. Such redundancy in nucleoporin function has previously been extensively documented in yeast.

While siRNA-mediated depletion of Nup358 in HeLa cells has little obvious consequence on nuclear protein import, its effects on mitotic progression are dramatic (Salina *et al.*, 2003). Depletion of this nucleoporin in HeLa cells results in the emergence of two unusual cell populations: prometaphase cells in which there is a failure in chromosome congression (*Figure 3*) and interphase cells containing multiple micronuclei. The most reasonable explanation for the appearance of these cell populations is a failure of spindle microtubules to capture chromosomes followed eventually by decay of the spindle assembly checkpoint. The outcome of this would be mitotic exit and NE reformation around dispersed chromosomes or groups of chromosomes. In this way, defective prometaphase cells would represent the precursors of the multinucleate cells. That this sequence of events does indeed occur is suggested by increasing numbers of multinucleate telophase and early G1 cells evident in Nup358-depleted HeLa cell cultures. In many anaphase and telophase cells following 4 days of Nup358 siRNA treatment, the presence of lagging chromosomes and chromatin strands spanning the intracellular bridge is yet further indication of chromosome congression failure.

In mitotic cells Nup358 is associated with both spindle microtubules and kinetochores, findings that complement a number of recent studies that have revealed a direct role for components of the nucleocytoplasmic transport system in mitotic spindle assembly. Members of the importin/karyopherin family as well as Ran are implicated in this process. The role of the importin α/β heterodimer in spindle assembly is exactly analogous to its role in nucleocytoplasmic transport (Hetzer *et al.*, 2002). During mitosis, importin α/β sequesters factors such as TPX2 are required in spindle formation. In the presence of RanGTP, whose concentration is highest in the vicinity of the mitotic chromosomes, importin α/β releases TPX2, leading to microtubule nucleation. In this way both Ran and importins provide important spatial cues for mitotic spindle assembly. A role for Nup358 in spindle formation and chromatid segregation has been highlighted in several recent studies. In *Caenorhabditis elegans* early embryos (Askjaer *et al.*, 2002), an RNA interference approach has revealed that depletion of Nup358/RanBP2 leads to inhibition of chromosome congression associated with aberrant spindle morphology. A very similar situation has been described in HeLa cells. In both experimental systems, while asters do form, chromosome capture is at best inefficient and only aberrant spindles are formed. Identical effects have also been observed in *C. elegans* embryos that have been depleted of CENP-A, a centromere-specific histone that is essential for normal kinetochore formation (Oegema *et al.*, 2001). These latter observations confirm an essential role for kinetochores in the establishment of normal spindle morphology. They further indicate that

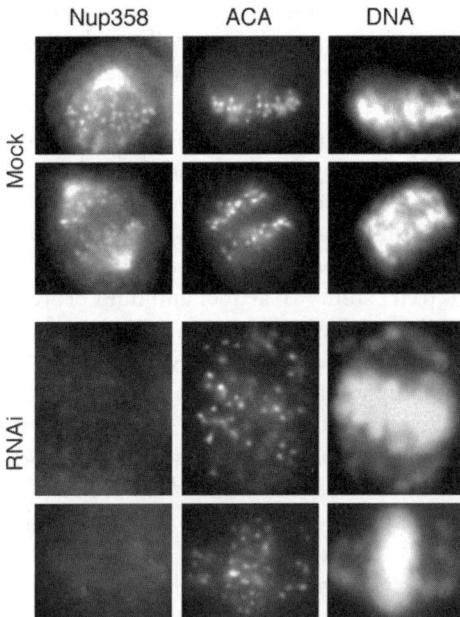

Figure 3. *Depletion of cells of Nup358 leads to a failure in chromosome congression and prometaphase arrest. This is most clearly demonstrated by the random distribution of centromeres (revealed by labelling with the anti-centromere autoantiserum, ACA) in the RNAi-treated cells. Mock treated cells on the other hand, exhibit centromeres associated with the mitotic apparatus. In addition to anti-Nup358 and ACA antibodies, the cells are also labelled with DNA-specific Hoechst dye to reveal the mitotic chromosomes.*

the effects of Nup358 depletion could be acting at the level either of kinetochore or spindle microtubule function. Given that Nup358 is associated with both kinetochores and spindle microtubules in mitotic cells, defects in either or both of these structures could potentially give rise to the types of aberrations in chromatid segregation that have been described both in *C. elegans* embryos and in HeLa cells.

There is now a growing body of information that Nup358 is at least required for normal kinetochore formation and assembly. EM studies of Nup358-depleted prometaphase cells, reveal highly abnormal kinetochore morphologies that feature partial or complete loss of the characteristic trilaminar plate-like structure as well as incomplete condensation of subjacent centromeric heterochromatin. These types of abnormal kinetochore morphologies have previously been reported following premature chromatin condensation in cell fusion experiments (Rattner and Wang, 1992), or after exposure of cells to caffeine (Brinkley *et al.*, 1988). Significantly, certain kinetochore components, including the checkpoint proteins, Mad1, Mad2 and Zw10, are mislocalized in prometaphase cells depleted of Nup358. Studies by Chan *et al.* (2000) employing a strategy based on antibody microinjection have demonstrated quite convincingly that interference with one of these, Zw10, leads to bypass of the spindle assembly checkpoint, appearance of lagging chromatids and aneuploidy. In other studies, Yao *et al.* (2000) have shown that depletion of CENP-E in mammalian cells gives rise to a spectrum of abnormalities that is virtually identical to those observed following Nup358 depletion in HeLa cells. Indeed CENP-E is one of the kinetochore proteins that is mislocalized in cells depleted of Nup358. Very similar defects have also been reported in studies involving depletion of several other kineto-chore proteins including Hec1, hMis12 and *Drosophila* Mast/Orbit (Goshima *et al.*, 2003; Maiato *et al.*, 2002; Martin-Lluesma *et al.*, 2002).

While depletion of Nup358 clearly perturbs kinetochore structure, preventing microtubule capture and leading to defective chromosome congression, it is not yet clear how Nup358 actually functions at the kinetochore. The most likely role for Nup358 might be related to its ability to attract and bind proteins of the Ran system. Indeed, RanGap1 and SUMO-I have both been found at the kinetochore (Joseph *et al.*, 2002). Given that SUMO-modified RanGAP1 binds Nup358 at the NPC, it is tempting to imagine that Nup358 might perform a similar function at the kineto-chore during mitosis. In fact, a population of Ran is actually found at the kinetochore in mitotic cells. In addition, Ran's nucleotide exchange factor, RCC1 also remains chromatin-associated during mitosis (Moore *et al.*, 2002). In *C. elegans*, depletion of RCC1 by RNA interference produces effects similar, although less severe, to those observed following Nup358 depletion (Askjaer *et al.*, 2002). Since Ran has been shown to be essential for kinetochore–microtubule interaction these various Ran system components could well function to modulate the cycling of proteins on and off the kinetochore. Interference with one branch of the Ran system, i.e. depletion of Nup358, might then result in the structural and compositional defects reported by Salina *et al.* (2003). Such a view is lent considerable support by the finding that Ran does indeed have a critical role in kinetochore function (Arnaoutov and Dasso, 2003). A second, although not exclusive possibility is that since Nup358 is a SUMO ligase (Pichler *et al.*, 2002), this activity might be required for proper kinetochore organi-zation and function. Support for this view is provided by the intriguing finding that SUMO-1 can act as a suppressor of certain CENP-C mutations in vertebrate cells (Fukagawa *et al.*, 2001).

Nup358 clearly has an important role in the recruitment of several kinetochore proteins, including those involved in the spindle assembly checkpoint. However, Nup358 does not itself provide a checkpoint function. Rather, the spindle assembly checkpoint remains substantially intact in cells depleted of Nup358. Such cells display only a relatively slow escape from the mitotic arrest that is associated with defective chromosome congression. Escape from mitosis in the absence of Nup358 can, however, be accelerated by co-depletion of the authentic checkpoint protein Mad1 (Salina *et al.*, 2003). One implication of these results is that the various displaced checkpoint components may remain functional at other cytoplasmic sites in cells depleted of Nup358 leading to the conclusion that kinetochore association is not absolutely required for checkpoint complex function.

Still unresolved is the issue of why NPC or NE components should play any role at all in mitotic progression. It has been suggested, however, that the reciprocal relationship between the NPC and the mitotic spindle (and associated structures), represented by the cycling of proteins between the two structures, might provide a fail-safe signal that defines the interphase versus mitotic status of the cell (Joseph *et al.*, 2002). Applying this argument to the kinetochore, this structure could only become functional once NE breakdown and disassembly of NPCs has commenced, permitting the relocation of some NPC components, Nup358 for instance, and leading to recruitment of other kinetochore proteins. In this way an orderly and stepwise progression of mitotic events is ensured.

The relationship between the NPC (or at least the NE) and the kinetochore may have its roots in the evolutionary history of the nucleus. Certain primitive cell types that undergo a closed mitosis, have well-differentiated kinetochores that remain closely associated with the nuclear surface of the NE (Kubai, 1975; Ris, 1975). In the dinoflagellate *Trichonympha agilis*, spindle microtubules, which are exclusively cytoplasmic, make contact not with the kinetochore itself, but with the patch of nuclear membrane that overlies the kinetochore (Kubai, 1975). In this way, chromosome segregation, although still driven by the mitotic spindle, is actually mediated by components of the nuclear envelope. It is possible that this mechanism has been conserved in organisms that have evolved an open mitosis such that disassembled NE components still maintain their ancient role. It is also intriguing that certain spindle assembly checkpoint proteins, notably Mad1 and Mad2, that are functional at the kinetochore, localize to the NPC during interphase. It appears that the NPC or NE functions as a storage site for these proteins. Is it possible, then, that proteins such as Nup358 originally evolved as kinetochore components? Only later, by virtue of sequestration at the NPC might they have been co-opted to have a function in nucleocytoplasmic transport. Such a notion is provided some support by the findings of Wozniak and colleagues (Iouk *et al.*, 2002) that at least in yeast, Mad1 and Mad2 appear to contribute to the functionality of the NPC. Clearly there may be important evolutionary lessons to be learned from studies on mitotic versus interphase roles of NE proteins.

References

Allan, V.J. and Schroer, T.A. (1999) Membrane motors. *Curr. Opin. Cell Biol.* 11: 476–482.

Arnaoutov, A., and Dasso, M. (2003) The Ran GTPase regulates kinetochore function. *Dev. Cell.* 5: 99–111.

Askjaer, P., Galy, V., Hannak, E. and Mattaj, I.W. (2002) Ran GTPase cycle and importins alpha and beta are essential for spindle formation and nuclear envelope assembly in living *Caenorhabditis elegans* embryos. *Mol. Biol. Cell.* 13: 4355–4370.

Babu, J.R., Jeganathan, K.B., Baker, D.J., Wu, X., Kang-Decker, N. and van Deursen, J.M. (2003) Rae1 is an essential mitotic checkpoint regulator that cooperates with Bub3 to prevent chromosome missegregation. *J. Cell. Biol.* 160: 341–353.

Beaudouin, J., Gerlich, D., Daigle, N., Eils, R. and Ellenberg, J. (2002) Nuclear envelope breakdown proceeds by microtubule-induced tearing of the lamina. *Cell* 108: 83–96.

Belgareh, N., Rabut, G., Bai, S.W., *et al.* (2001) An evolutionarily conserved NPC subcomplex, which redistributes in part to kinetochores in mammalian cells. *J. Cell. Biol.* 154: 1147–1160.

Brinkley, B.R., Zinkowski, R.P., Mollon, W.L., Davis, F.M., Pisegna, M.A., Pershouse, M. and Rao, P.N. (1988) Movement and segregation of kinetochores experimentally detached from mammalian chromosomes. *Nature* 336: 251–254.

Campbell, M.S., Chan, G.K. and Yen, T.J. (2001) Mitotic checkpoint proteins HsMAD1 and HsMAD2 are associated with nuclear pore complexes in interphase. *J. Cell Sci.* 114: 953–963.

Chan, G.K., Jablonski, S.A., Starr, D.A., Goldberg, M.L. and Yen, T.J. (2000) Human Zw10 and ROD are mitotic checkpoint proteins that bind to kinetochores. *Nat. Cell Biol.* 2: 944–947.

Chaudhary, N. and Courvalin, J.-C. (1993) Stepwise reassembly of the nuclear envelope at the end of mitosis. *J. Cell Biol.* 122: 295–306.

Collas, I. and Courvalin, J.C. (2000) Sorting nuclear membrane proteins at mitosis. *Trends Cell. Biol.* 10: 5–8.

Delphin, C., Guan, T., Melchior, F. and Gerace, L. (1997) RanGTP targets p97 to RanBP2, a filamentous protein localized at the cytoplasmic periphery of the nuclear pore complex. *Mol. Biol. Cell.* **8**: 2379–2390.

Ellenberg, J., Siggia, E.D., Moreira, J.E., Smith, C.L., Presley, J.F., Worman, H.J. and Lippincott-Schwartz, J. (1997) Nuclear membrane dynamics and reassembly in living cells: targeting of an inner nuclear membrane protein in interphase and mitosis. *J. Cell Biol.* **138**: 1193–1206.

Fukagawa, T., Regnier, V. and Ikemura, T. (2001) Creation and characterization of temperature-sensitive CENP-C mutants in vertebrate cells. *Nucleic Acids Res.* **29**: 3796–3803.

Gant, T.M. and Wilson, K.L. (1997) Nuclear assembly. *Ann. Rev. Cell Dev. Biol.* **13**: 669–695.

Georgatos, S.D., Pyrpasopoulou, D. and Theodoropoulos, P.A. (1997) Nuclear envelope breakdown in mammalian cells involves stepwise lamina disassembly and microtubule-driven deformation of the nuclear membranes. *J. Cell Sci.* **110**: 2129–2140.

Gerace, L. and Blobel, G. (1980) The nuclear envelope lamina is reversibly depolymerized during mitosis. *Cell* **19**: 277–287.

Gerace, L. and Burke, B. (1988) Functional organization of the nuclear envelope. *Ann. Rev. Cell Biol.* **4**: 335–374.

Goshima, G., Kiyomitsu, T., Yoda, K. and Yanagida, M. (2003) Human centromere chromatin protein hMis12, essential for equal segregation, is independent of CENP-A loading pathway. *J. Cell Biol.* **160**: 25–39.

Gruenbaum, Y., Wilson, K.L., Harel, A., Goldberg, M. and Cohen, M. (2000) Review: nuclear lamins – structural proteins with fundamental functions. *J. Struct. Biol.* **129**: 313–323.

Heald, R. and McKeon, F. (1990) Mutations of phosphorylation sites in lamin A that prevent nuclear lamina disassembly in mitosis. *Cell* **61**: 579–589.

Hetzer, M., Gruss, O.J. and Mattaj, I.W. (2002) The Ran GTPase as a marker of chromosome position in spindle formation and nuclear envelope assembly. *Nat. Cell Biol.* **4**: E177–184.

Iouk, T., Kerscher, O. Scott, R.J., Basrai, M.A. and Wozniak, R.W. (2002) The yeast nuclear pore complex functionally interacts with components of the spindle assembly checkpoint. *J. Cell Biol.* **159**: 807–819.

Joseph, J., Tan, S.H., Karpova, T.S., McNally, J.G. and Dasso, M. (2002) SUMO-1 targets RanGAP1 to kinetochores and mitotic spindles. *J. Cell Biol.* **156**: 595–602.

Kubai, D.F. (1975) The evolution of the mitotic spindle. *Int. Rev. Cytol.* **43**: 167–227.

Lee, K.K., Gruenbaum, T., Spann, P., Liu, J. and Wilson, K.L. (2000) *C. elegans* nuclear envelope proteins emerin, MAN1, lamin, and nucleoporins reveal unique timing of nuclear envelope breakdown during mitosis [In Process Citation]. *Mol. Biol. Cell.* **11**: 3089–3099.

Lenart, P., Rabut, G., Daigle, N., Hand, A.R., Terasaki, M. and Ellenberg, J. (2003) Nuclear envelope breakdown in starfish oocytes proceeds by partial NPC disassembly followed by a rapidly spreading fenestration of nuclear membranes. *J. Cell Biol.* **160**: 1055–1068.

Liu, S.T., Chan, G.K., Hittle, J.C., Fujii, G., Lees, E. and Yen, T.J. (2003) Human MPS1 kinase is required for mitotic arrest induced by the loss of CENP-E from kinetochores. *Mol. Biol. Cell.* **14**: 1638–1651.

Macaulay, C., Meier, E. and Forbes, D.J. (1995) Differential mitotic phosphorylation of proteins of the nuclear pore complex. *J. Biol. Chem.* **270**: 254–262.

Mahajan, R., Delphin, C., Guan, T., Gerace, L. and Melchior, F. (1997) A small ubiquitin-related polypeptide involved in targeting RanGAP1 to nuclear pore complex protein RanBP2. *Cell* **88**: 97–107.

Maiato, H., Sampaio, P., Lemos, C.L., Findlay, J., Carmena, M., Earnshaw, W.C. and Sunkel, C.E. (2002) MAST/Orbit has a role in microtubule-kinetochore attachment and is essential for chromosome alignment and maintenance of spindle bipolarity. *J. Cell Biol.* **157**: 749–760.

Martin-Lluesma, S., Stucke, V.M. and Nigg, E.A. (2002) Role of Hec1 in spindle checkpoint signaling and kinetochore recruitment of Mad1/Mad2. *Science* **297**: 2267–2270.

Matunis, M.J., Wu, J. and Blobel, G. (1998) SUMO-1 modification and its role in targeting the Ran GTPase-activating protein, RanGAP1, to the nuclear pore complex. *J. Cell Biol.* **140**: 499–509.

Moore, W., Zhang, C. and Clarke, P.R. (2002) Targeting of RCC1 to chromosomes is required for proper mitotic spindle assembly in human cells. *Curr. Biol.* **12**: 1442–1447.

Newport, J. and Spann, T. (1987) Disassembly of the nucleus in mitotic extracts: membrane vesicularization, lamin disassembly, and chromosome condensation are independent processes. *Cell* **48**: 219–230.

Oegema, K., Desai, A., Rybina, S., Kirkham, M. and Hyman, A.A. (2001) Functional analysis of kinetochore assembly in *Caenorhabditis elegans. J. Cell Biol.* **153**: 1209–1226.

Ostlund, C., Ellenberg, J., Hallberg, E., Lippincott-Schwartz, J. and Worman, H.J. (1999) Intracellular trafficking of emerin, the Emery-Dreifuss muscular dystrophy protein. *J. Cell Sci.* **112**: 1709–1719.

Panté, N. and Aebi, U. (1996) Sequential binding of import ligands to distinct nucleopore regions during their nuclear import. *Science* **273**: 1729–1732.

Peter, M., Nakagawa, J., Dorée, M., Labbé, J.C. and Nigg, E.A. (1990) In vitro disassembly of the nuclear lamina and M phase-specific phosphorylation of lamins by cdc2 kinase. *Cell* **61**: 591-602.

Pfaller, R. and Newport, J.W. (1995) Assembly/disassembly of the nuclear envelope membrane. Characterization of the membrane-chromatin interaction using partially purified regulatory enzymes. *J. Biol. Chem.* **270**: 19066–19072.

Pichler, A., Gast, A., Seeler, J.S., Dejean, A. and Melchior, F. (2002) The nucleoporin RanBP2 has SUMO1 E3 ligase activity. *Cell* **108**: 109–120.

Piehl, M. and Cassimeris, L. (2003) Organization and dynamics of growing microtubule plus ends during early mitosis. *Mol. Biol. Cell.* **14**: 916–925.

Rattner, J.B. and Wang, T. (1992) Kinetochore formation and behaviour following premature chromosome condensation. *J. Cell Sci.* **103**(Pt 4): 1039–1045.

Richardson, W.D., Mills, A.D., Dilworth, S.M., Laskey, R.A. and Dingwall, C. (1988) Nuclear protein migration involves two steps: rapid binding at the nuclear envelope followed by slower translocation through the nuclear pores. *Cell* **52**: 655–664.

Ris, H. (1975) Primitive mitotic mechanisms. *Biosystems* **7**: 298–301.

Robbins, E. and Gonatas, N.K. (1964) The ultrastructure of a mammalian cell during the mitotic cycle. *J. Cell Biol.* **21**: 429–463.

Roos, U.-P. (1973) Light and electron microscopy of rat kangaroo cells in mitosis. l. Formation and breakdown of the mitotic apparatus. *Chromosoma* **40**: 43–82.

Saitoh, H., Pu, R., Cavenagh, M. and Dasso, M. (1997) RanBP2 associates with Ubc9p and a modified form of RanGAP1. *Proc. Natl Acad. Sci. USA* **94**: 3736–3741.

Salina, D., Bodoor, K., Eckley, D.M., Schroer, T.A., Rattner, J.B. and Burke, B. (2002) Cytoplasmic dynein as a facilitator of nuclear envelope breakdown. *Cell* **108**: 97–107.

Salina, D., Enarson, P., Rattner, J.B. and Burke, B. (2003) Nup358 integrates nuclear envelope breakdown with kinetochore assembly. *J. Cell Biol.* **162**: 991–1001.

Snow, C.M., Senior, A. and Gerace, L. (1987) Monoclonal antibodies identify a group of nuclear pore complex glycoproteins. *J. Cell Biol.* **104**: 1143–1156.

Stafstrom, J.P. and Staehelin, L.A. (1984) Dynamics of the nuclear envelope and of nuclear pore complexes during mitosis in the Drosophila embryo. *Eur. J. Cell Biol.* **34**: 179–189.

Stick, R., Angres, B., Lehner, C.F. and Nigg, E.A. (1988) The fates of chicken nuclear lamin proteins during mitosis: Evidence for a reversible distribution of lamin B2 between the inner nuclear membrane and elements of the endoplasmic reticulum. *J. Cell Biol.* **107**: 397–406.

Terasaki, M., Campagnola, P., Rolls, M.M., Stein, P.A., Ellenberg, J., Hinkle, B. and Slepchenko, B. (2001) A new model for nuclear envelope breakdown. *Mol. Biol. Cell* **12**: 503–510.

Vigers, G.P.A. and Lohka, M.J. (1991) A distinct vesicle population targets membranes and pore complexes to the nuclear envelope in Xenopus eggs. *J. Cell Biol.* **112**: 545–556.

Walther, T.C., Pickersgill, H.C., Cordes, V.C., Goldberg, M.W., Allen, T.D., Mattaj, I.W. and Fornerod, M. (2002) The cytoplasmic filaments of the nuclear pore complex are dispensable for selective nuclear protein import. *J. Cell Biol.* **158**: 63–77.

Wang, X., Babu, J.R., Harden, J.M., Jablonski, S.A., Gazi, M.H., Lingle, W.L., de Groen, P.C., Yen, T.J. and van Deursen, J.M. (2001) The mitotic checkpoint protein hBUB3 and the mRNA export factor hRAE1 interact with GLE2p-binding sequence (GLEBS)-containing proteins. *J. Biol. Chem.* **276:** 26559–26567.

Ward, G.E. and Kirschner, M.W. (1990) Identification of cell-cycle regulated phosphorylation sites on nuclear lamin C. *Cell* **61:** 561–577.

Waterman-Storer, C.M., Sanger, J.W. and Sanger, J.M. (1993) Dynamics of organelles in the mitotic spindles of living cells: membrane and microtubule interactions. *Cell. Motil. Cytoskeleton.* **26:** 19–39.

Wilson, K.L. (2000) The nuclear envelope, muscular dystrophy and gene expression. *Trends Cell. Biol.* **10:** 125–129.

Wilson, K.L., Zastrow, M.S. and Lee, K.K. (2001) Lamins and disease: insights into nuclear infrastructure. *Cell* **104:** 647–650.

Yang, L., Guan, T. and Gerace, L. (1997) Integral membrane proteins of the nuclear envelope are dispersed throughout the endoplamic reticulum during mitosis. *J. Cell Biol.* **137:** 1199–1210.

Yao, X., Abrieu, A., Zheng, Y., Sullivan, K.F. and Cleveland, D.W. (2000) CENP-E forms a link between attachment of spindle microtubules to kinetochores and the mitotic checkpoint. *Nat. Cell Biol.* **2:** 484–491.

Yokoyama, N., Hayashi, N., Seki, T., *et al.* (1995) A giant nucleopore protein that binds Ran/TC4. *Nature* **376:** 184–188.

Zeligs, J.D. and Wollman, S.H. (1979) Mitosis in rat thyroid epithelial cells in vivo: 1. ultrastructural changes in cytoplasmic organelles during the mitotic cycle. *J. Ultrastruct. Res.* **66:** 53–77.

Zhang, Q., Skepper, J.N., Yang, F., Davies, J.D., Hegyi, L., Roberts, R.G., Weissberg, P.L., Ellis, J.A. and Shanahan, C.M. (2001) Nesprins: a novel family of spectrin-repeat-containing proteins that localize to the nuclear membrane in multiple tissues. *J. Cell Sci.* **114:** 4485–4498.

13

Nuclear dynamics in higher plants

David W. Galbraith

1. Introduction

1.1 *What do we know about nuclear structure?*

The nucleus is recognized as one of the characteristic features of eukaryotic cells. It comprises the location of the bulk of the cellular genome separated during interphase from the cytoplasm by a pair of bilamellar nuclear membranes. Most of the ultrastructural details of the nucleus have been defined using microscopy, particularly transmission and scanning electron microscopy. These techniques have revealed a great deal of information about nuclear structure, and provide the basis of our conventional conceptual image of the nucleus as an oblate spheroid bounded by paired bilamellar membranes containing grommet-like nuclear pores linking nucleoplasm and cytoplasm. Empirically, employing extraction with non-ionic detergents under high salt conditions, the nucleus can also be divided into two structural components, chromatin and the nuclear matrix. Based on studies on metazoans, the latter comprises nuclear lamina, the nuclear pore complexes, RNP proteins, and nucleoli, and is thought to function as a nucleoskeletal framework within the context of which the various mechanical and enzymatic nuclear activities occur.

The reliance on conventional light and electron microscopy for analysis of nuclear structure has, until recently, under-emphasized the dynamic nature of the nucleus. This is a consequence of the shortcomings of these types of microscopy. Under the light microscope, the interphase nucleus in living cells appears rather featureless. With the exception of the nucleolus, visible using DIC optics, the nucleoplasm seems structurally homogeneous. Its component macromolecules are not intrinsically pigmented or fluorescent and therefore do not provide natural contrast at visible wavelengths. This problem can be resolved using fixed tissues and cells, and work involving immunolabelling, hybridization with fluorescent nucleic acid probes, and direct incorporation of fluorescent precursors, has led to the identification of a high degree of spatial organization within the nucleoplasm (for recent reviews, see Cremer and Cremer, 2001; Janicki and Spector, 2003) strongly implicated in the

---

The Nuclear Envelope, edited by D.E. Evans, C. Hutchison & J.A. Bryant.
© 2004 Garland Science/BIOS Scientific Publishers

regulation of gene expression, amongst other things. Nonetheless, even under these conditions, the resolving power of light microscopy ultimately limits the spatial information that can be retrieved. Electron microscopy, in contrast to light microscopy, provides resolving power down to the level of individual biological molecules. However, almost all imaging is done using fixed tissues, which also in many cases are embedded and sectioned. This restricts the recovery of kinetic and dynamic information, and can, due to sampling considerations, miss the identification of various aspects of spatial heterogeneity and sample non-uniformity (De Solorzano *et al.*, 2001). It is well known that fixation has the potential to introduce artifacts, whether at the electron or light microscope level.

1.2 *A definition of nuclear dynamics*

For the purposes of this review, I have chosen to focus on: (i) the characterization of the three-dimensional shape of the interphase nucleus; (ii) measurement of alterations over time of this shape; (iii) the analysis of nuclear movement within interphase cells; and (iv) the analysis of the dynamics of the nucleoplasm. In all cases, I primarily employ plants as the experimental system, and provide speculations as to what might be the functional significance of these changes in nuclear dynamics.

1.3 *How can we measure nuclear dynamics?*

The most convenient and appropriate methods for analysis of dynamic changes in living organisms are those that provide high-resolution spatial imaging with high contrast whilst maintaining the viable state of the organism (Gerlich and Ellenberg, 2003). Confocal laser scanning microscopy of organisms transgenically expressing fluorescent protein (FP) markers has recently emerged as a powerful means for obtaining information about many aspects of cellular dynamics (Lippincott-Schwartz and Patterson, 2003; Lippincott-Schwartz *et al.*, 2003; Tsien, 1998, 2003). The FP class of proteins, of which the green fluorescent protein (GFP) of *Aequorea victoria* is the archetype, achieve fluorescence through covalent intramolecular reaction of their amino acid side chains, a process which requires no other cofactors or enzymes (Tsien, 1998). This means that expression of the FP coding sequence within the target organism is all that is needed for production of fluorescence. Further, the three-dimensional GFP structure has been found to accommodate both N- and C-terminal translational fusions, such that it can be targeted to a wide variety of specific subcellular locations (Tsien, 1998). Finally, expression of FPs in general appears non-toxic, with the usual caveat that excessive production of any protein can be toxic, and that specific FP–gene fusions may, unusually, have toxic (dominant negative) effects due to interference with quaternary structures required for cellular function. A variety of FPs now are available, along with a wide variety of sequence variants of GFP, which provide a palette of distinct spectral colors for use in subcellular imaging. These include variants that are optimized for expression in different organisms, as well as variants that possess kinetic properties such as time-dependent alterations in spectral characteristics, and the ability to undergo photoactivation (Lippincott-Schwartz and Patterson, 2003; Lippincott-Schwartz *et al.*, 2003).

2. The structure of plant nuclei as revealed using GFP targeting

Nuclear targeting of GFP can be readily achieved by fusing the GFP coding sequence to proteins containing effective nuclear targeting signals (see, for example, Grebenok *et al.*, 1997a, 1997b). A further requirement is that the fusion protein be larger than the passive size exclusion limit of the nuclear pores (about 40 kDa), such that the fusion protein is retained within the nucleus after import. In early work, we satisfied this criterion using a small nuclear localization sequence (NLS) from an orphan *N. tabacum* transcription factor fused to the complete coding sequences of β-glucuronidase (GUS) and GFP (Grebenok *et al.*, 1997a, 1997b). GFP and GUS are both functional in the fusion protein, and we found that GFP was targeted to the nuclei of transgenic tobacco and *Arabidopsis*, using Agrobacterium-mediated transformation. Other workers have employed this construction for nuclear targeting of GFP within monocots (Collings *et al.*, 2000) and lower plants (Benzanilla *et al.*, 2003). Recently, we have found that a number of *Arabidopsis* nuclear proteins can be employed as GFP fusions to confer nuclear targeting and accumulation of green fluorescence, including histones of the HI, H2A, H2B, and H3 families, and members of the MBD (methyl-CpG-binding domain containing), CRD (chromodomain/chromo shadow containing), and NFD (nucleosome/chromatin assembly factor group D) families of chromatin-associated proteins (Zhang and Galbraith, unpublished). Transgenic expression can be done with constitutive promoters (such as the cauliflower mosaic virus (CaMV) 35S promoter), or regulated (tissue-specific or other) promoters (Chytilova *et al.*, 1999). For plants that are recalcitrant to transformation, transient transfection is readily achieved via particle bombardment (Collings *et al.*, 2000).

In transgenic *Arabidopsis thaliana* plants, ntGFP expression driven by constitutive promoters highlights most of the nuclei within the different plant tissues and organs. The fluorescence can be conveniently detected and imaged using confocal microscopy. In most cases, transgenic expression of ntGFP appears innocuous, although some depression of growth is seen for plants expressing histone–GFP fusions (Zhang, C. and Galbraith, D. unpublished). The nuclei of somatic cells are characteristically endoreduplicated (Galbraith *et al.*, 1991), and different sizes of fluorescent nuclei can be distinguished corresponding to the 2C, 4C, 8C and 16C nuclei that make up the majority of the population. Different nuclear structures are observed within the different somatic cells. For cells that are roughly isodiametric rectangular oblongs or similarly shaped cylinders, the nuclei are oblate spheroids. During cell division, the nuclei become spherical prior to prometaphase. For cells that are grossly anisometric (for example, root hairs, which are cylinders of diameter ~20 μm and lengths of ~ 400 μm), the nuclei are highly pleiomorphic. Many exhibit elongated, spindle-shaped forms, which change both plastically and elastically over time.

In onion, transient expression of GFP has revealed deep invaginations of the nuclear membrane, which can result in complete fenestration of the nucleus (Collings *et al.*, 2000). Similar invaginations have been described for mammalian cells (Fricker *et al.*, 1997). More recently, an extensive intranuclear reticular network has been described in human hepatoma (SKHep1) cells that is implicated in calcium signalling (Echevarria *et al.*, 2003). Its contiguity with the ER is implied by labelling with the ER-Tracker dye, and by the presence of calreticulin within the same nuclear reticulum. This reticulum is also labelled by fluorescent calcium indicators, and FRAP measurements are consistent with the enclosure of the dyes within the reticulum. The type II InsP$_3$ receptor is also found on this nucleoplasmic reticulum, and uncaging of InsP$_3$ in very close proximity to the nucleoplasmic reticulum instigated increases in Ca^{2+} spreading from that location. Potential downstream effects of nuclear and cytoplasmic changes in Ca^{2+} levels are implied by the observation that separate uncaging of Ca^{2+} within the nucleus and within the cytoplasm selectively alters the distributions of nuclear and cytosolic forms of GFP protein kinase C–γ fusions. This work strongly suggests the existence of spatial separation of calcium signalling pathways within the cell.

The functional continuity of reticula between the nucleus and cytoplasm may explain differential retention of nuclear-targeted GFP fusions following tissue homogenization. We have observed that chimeric GFP–beta glucuronidase fusions targeted to the nucleoplasm using a monopartite nuclear localization signal (NLS) are rapidly lost from the nucleus after gentle tissue disruption. In contrast, GFP fusions to chromatin-associated proteins remain tightly associated with the nucleus, in some cases remaining associated with chromosomes during mitosis (Zhang, C. and Galbraith, D. unpublished). Mechanical stresses on the nuclear membrane accompanying disruption of nucleo-cytoplasmic reticula, particularly in the absence of strengthening provided by lamins (as is apparently the case in plants), might lead to losses of GFP proteins that are accumulated within the soluble components of nucleoplasm.

3. Nuclear movement

Intracellular movement is an obvious feature of somatic cell nuclei within transgenic *Arabidopsis* plants. This movement is exaggerated within elongated cells, such as root hairs, and is spatially regulated during root hair development (Chytilova *et al.*, 2000, 2001; Ketelaar *et al.*, 2002; Van Bruaene *et al.*, 2003). In these cell types, movement occurs over long distances (tens to hundreds of microns), is energy dependent, and is abolished by treatment with drugs directed against actin, such as latrunculin B, cytochalasins, and N-ethyl maleimide. In contrast, microtubules appear not to be involved in movement, since treatments with any of taxol, oryzalin, vinblastine and colchicine have no effects on nuclear movement (Chytilova *et al.*, 2000). Movement involves divalent cations, as would be expected for ATP-dependent processes (Sliwinska *et al.*, 2002). Surprisingly, we found that treatment with the myosin-ATPase-specific inhibitor 2,3-butanedione monoxime (BDM) had no effect on nuclear movement. The effectiveness of BDM against *Arabidopsis* myosin-ATPase has been demonstrated in studies of peroxisomal movement (Jedd and Chua, 2002). Either different myosin isoforms, differing in BDM sensitivity, are involved in nuclear and peroxisomal movement, or nuclear movement does not involve myosin.

The involvement of actin networks in nuclear movement and positioning is well established for *C. elegans*, *Drosophila* and yeast (Starr and Han, 2003). In the nematode worm, syncytial nuclear positioning is achieved by UNC-84, which localizes to the nuclear envelope, and ANC-1, an enormous protein linking UNC-64 to cytoplasmic actin (Starr and Han, 2002). Proteins related to ANC-1 have been found in mammals and there is some evidence that these can bind lamins and other components of the nuclear lamin and therefore may have an intranuclear as well as extranuclear function (Starr and Han, 2003).

4. Nuclear fragmentation

One surprising feature of the nuclei of anisometric cells is their ability to fragment and reassemble (*Figure 1*; Chytilova *et al.*, 2000, 2001). This feature is readily observed in root hair cells. Fragmentation can produce subnuclear structures that are unequal in size, but at the same time, it does not compromise the integrity of the nuclear membrane. It does not appear to be pathological, since the cells continue active cytoplasmic streaming, nor is it akin to apoptotic changes, since nuclear fragmentation is freely reversible. Although the different subnuclear structures can move in different directions, they always appear connected by thread-like structures. It is therefore likely that very thin tubules of nuclear membrane retain topological continuity between the fragmented products, although we cannot rule out the possibility that these products are indeed separate entities that retain the ability to reassemble.

The ability of root hair nuclei to fragment non-mitotically is reminiscent of behaviour of the macronucleus during cell division in ciliates (Raikov, 1996). Macronuclear division does not involve chromosomal condensation and is not mediated by a microtubule spindle. It occurs by a stretching and pinching process that can be asymmetric, and further segregation of macronuclei during cytokinesis involves nuclear movement. The macronucleus is also endoreduplicated, a feature shared with root hair nuclei, which may be relevant to the shared behaviour. This shared behaviour evidently could be an evolutionary relic that might cast light on the possible origins of obvious differences existing between the structures of the nuclei of plants, yeast and multicellular animals.

5. Dynamic movement of chromosomes and chromatin within the nucleus

Genomic integration of *lac* operator arrays in eukaryotic cells combined with transgenic expression of GFP–lac repressor protein fusions provides a means for directly measuring, via fluorescence microscopy, the ability of chromosomes and chromatin to move within the context of the nucleus (Gasser, 2002). Kato and Lam (2003) have directly examined the dynamics of interphase chromatin in plants, through use of a similar tagging method involving dexamethasone-inducible GFP-LacI expression (Kato and Lam, 2001). Induction results in the intranuclear appearance of fluorescent spots at the sites of *lac* operator insertion. Measurement of the relative positions of the different spots over time can be used to calculate the chromatin mobility and to ascertain whether or not this mobility is constrained. The confinement area was found to be smaller in diploid guard cell nuclei than in endoreduplicated pavement cells. In the latter, correlation of the between-cell variation in spot number with c-value suggests approximately half of the chromatids remain cohered following endoreduplication.

In general, these measurements suggest the physiochemical properties of plant chromatin are similar to those of other organisms, the chromatin being capable of constrained diffusional movement within the nucleus (Chubb *et al.*, 2002; Heun *et al.*, 2001; Ishii *et al.*, 2002; Marshall *et al.*, 1997). It also implies that plant chromatin is specifically tethered within the nucleus. In other organisms, chromatin constraints are increased closer to the nuclear periphery and in association with nucleoli (Chubb *et al.*,

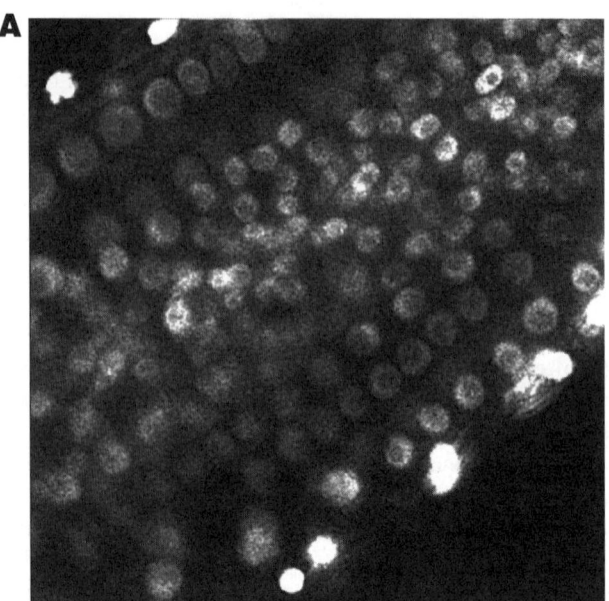

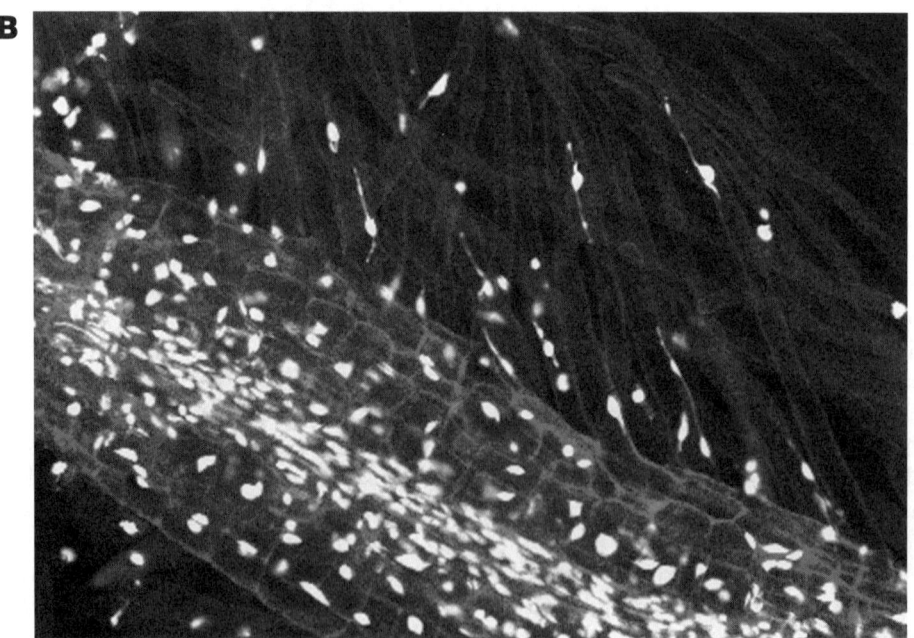

Figure 1. Polymorphic nuclei within Arabidopsis thaliana roots, as highlighted by nuclear targeted GFP. (A) Nuclei within the body of the root are mostly spherical. Magnification, ×720 (from Figure 6 of Chytilova et al., 2000. Reprinted from Molecular Biology of the Cell (Mol. Biol. Cell 2000 11: 2733–2741) with permission by The American Society for Cell Biology). (B) In this example, within the root hairs the majority of the nuclei are non-spherical, and exhibit grossly distorted and elongated shapes. Magnification, ×410.

2002). Mechanistically, chromatin can be tethered by lamin-mediated attachment to the nuclear envelope (Lopez-Soler *et al.*, 2001), and to nuclear pores (Galy *et al.*, 2000). The absence of lamins from plants (see below) eliminates this potential means for chromatin constraint, and implies the existence of other nuclear anchoring mechanisms. The potential importance of interchromosomal interactions in epigenetic mechanisms of gene regulation, its dependence on the degree to which chromatin is able to diffuse, and its relationship to endoreduplication has been noted (Kato and Lam, 2003).

6. What mediates the nuclear shape changes?

In animals, a feature of the inner nuclear membrane is the presence of the nuclear lamina, a fibrous network of protein connected to the membrane via the lamin B receptor (Gruenbaum *et al.*, 2003). The nuclear lamina is thought to provide mechanical strength to the overall structure of the nucleus, which can be directly measured using micropressure devices. In plants and yeast, sequences homologous to the nuclear lamins have not been found (Gindullis *et al.*, 2002; Goldman *et al.*, 2002), nor have homologues been found to the N-terminal region of the Lamin B Receptor (LBR; Irons *et al.*, 2003). It has been hypothesized that non-metazoans have evolved other nuclear envelope proteins as functional replacements for the lamins. It is clear that plant nuclear structure has maintained some functional similarities to those of metazoans. For example, nuclear matrices can be extracted from plant nuclei using conditions similar to those employed for animals (Calikowski *et al.*, 2003; Hall *et al.*, 1991).

Lamins comprise a subset both of the family of intermediate filament (IF) proteins and of the larger grouping of coiled-coil proteins. Lamins certainly play a central role in defining nuclear shape in metazoans (Goldman *et al.*, 2002). They are typically located within a band of variable thickness located immediately underneath the inner nuclear membrane and linked to it via specific interactions with membrane proteins including the LBR. Lamins can also extend into the nucleoplasm where they are in a considerably dynamic state (Broers *et al.*, 1999). Extraction of metazoan cells with non-ionic detergents, which solubilize most cytoplasmic and nuclear proteins but leave IFs intact, does not affect nuclear shape, and the lamins continue to occupy a position immediately underneath the inner nuclear membrane (Capco and Penman, 1983). Further evidence of the role of lamins in nuclear structure comes from studies of *in vitro* nuclear assembly, depletion of lamins resulting in the production of small, fragile nuclei (Newport *et al.*, 1990). Studies of knockout mice, and of the various human laminopathies resulting from mutations in the three lamin genes, have led to the suggestion that muscle dystrophies might be a consequence of deformation or disruption of fragile nuclei (Sullivan *et al.*, 1999). The absence of lamins in plants may reflect the fact that plants and yeast, having sturdy cell walls, lack an evolutionary need to provide structural protection for the nucleus.

In that lamins are also known to be involved in a variety of critical cellular processes, including nuclear disassembly/reassembly during cell division, DNA synthesis, transcription regulation, and interactions with chromatin, it appears most important to uncover the identity of plant and yeast genes whose products might replace the functions of lamins. In plants, one method of gene discovery has taken advantage of the observation that lamins form a subset of the coiled-coil proteins, that one such long coiled-coil protein (MFP1) is localized to the nuclear periphery to tobacco cells (Gindullis and Meier, 1999; Meier *et al.*, 1996), and that a small, plant-specific protein (MAF1) interacts with this protein and is localized to the nuclear envelope (Gindullis *et al.*, 1999). A small family of *Arabidopsis* genes encoding long coiled-coil proteins was identified using the two-hybrid interaction assay with MAF1 as the bait, followed by BLAST searches (Gindullis *et al.*, 2002). It will be of interest to determine whether the protein products of these genes are localized to the nuclear membrane, and whether knockouts display altered nuclear morphology or altered pleomorphism within root hair cells.

Similarly to the situation for lamins, no plant homologues have been found for several well-characterized mammalian nuclear envelope proteins implicated in binding of interphase chromatin to the inner nuclear membrane (Meier, 2001). It has been suggested that evolution of an 'open' mitosis may have occurred separately for plants and metazoa, with a corresponding lack of homologies between plant and animal inner membrane proteins involved with membrane dissociation and reassembly during mitosis (Meier, 2001). Otherwise, the inner nuclear membranes of plants and metazoa do share functional commonalities; for example, fusions between the LBR and GFP expressed in plants are primarily targeted to the inner nuclear membrane (Irons *et al.*, 2003; see Chapter 15). In this work, the first 238 amino acids of LBR were fused to GFP. Such a fusion protein, containing the nucleoplasmic N terminal domain and one transmembrane domain of LBR, is localized primarily to the nuclear envelope and secondarily to the ER of mammalian cells (Ellenberg *et al.*, 1997; Irons *et al.*, 2003). A similar situation is seen in transgenic plant cell lines, a consequence perhaps of saturation of the nuclear envelope to which the protein is targeted. Examination of the time course of mitosis suggests the LBR–GFP fusion disperses by late metaphase from the nuclear envelope into the ER, but fluorescent tubular membrane processes subsequently appear, and as mitosis proceeds, move to the spindle poles and finally encircle the daughter nuclei. It would be interesting to compare, using FRAP, the mobility of the targeted protein *in situ* for both plants and animals, both during interphase and at various stages of cell division. A restriction of mobility of GFP–LBR within the plant NE would suggest specific interactions with other macromolecules. The interacting partners might then be identified through two-hybrid analysis.

An alternative approach to the discovery of nuclear cytoskeletal proteins has involved the recent use of high-throughput proteomic methods (Calikowski *et al.*, 2003). This is based on the observation that methods for the isolation of animal nuclear matrices can also be applied to plants. A total of 365 protein spots were resolved by 2D-PAGE and 36 identified by mass spectrometry following 1D SDS-PAGE. It will be interesting to determine the phenotypes of plants carrying null mutations of these genes and to discover those proteins with which these proteins interact.

A further aspect of potential interest to nuclear movement and fragmentation is the observation of an involvement of actomyosin motors in transcription (Pestic-Dragovich *et al.*, 2000). This stems from the observation of a myosin I isoform in the nucleus which colocalizes with RNA polymerase II. This colocalization is abolished by treatment with α-amanitin and decreased by treatment with actinomycin D. It is suggested that nuclear myosin I might therefore provide a power source for transcription. It is now accepted that actin is present in the nucleus (Pederson and Aebi,

2002), but probably not in an extended filamentous form (Shumaker et al., 2003) although short filaments cannot be ruled out. In specific cases, actin appears to specifically associated with mRNA during transcription, and then accompanies the mRNA through the nuclear pores to the polyribosomes (Percipalle et al., 2001). Protein molecules capable of linking actin to lamins have also been described (Shumaker et al., 2003).

The possibility that plant nuclear movement is a consequence of the operation of a transcriptionally coupled actomyosin motor appears unlikely given that treatment of transgenic plants expressing nuclear-targeted GFP with either α-amanitin or actinomycin D had no effect on nuclear movement (Galbraith et al., unpublished). Further studies are required to determine whether actinomycin D treatment might affect the degree of nuclear fragmentation.

7. Prospects and future directions

A number of unanswered questions remain concerning the regulation of nuclear shape, the occurrence of nuclear fragmentation, and the observation of nuclear movement. These can roughly be grouped into questions of why and how. The question as to why a nucleus is generally an oblate spheroid, yet in some cases can become highly polymorphic with cytoplasmic fenestrations and invaginations, touches on fundamental aspects of physics, chemistry and thermodynamics, subjects which are neglected by many practicing biologists. A spherical structure minimizes its surface area to volume ratio. If an object of given volume is required to produce a flux of components that exceeds the carrying capacity of the surface area that these components are required to cross, one solution is for the object to deviate from a spherical shape. This can only be achieved if the amounts of the individual components of the surface area also co-ordinately increase, and leads to the prediction that changes in gene expression associated with cells containing polymorphic nuclei should include up-regulation of genes encoding components of the nuclear membrane. Dissecting out the contributions of these genes would be difficult, one possibility being the prior identification of cellular systems within which nuclear polymorphism might naturally occur or be experimentally manipulated. In any case, solving these problems leads naturally to the mechanism underlying the observation.

The question as to why nuclei might fragment is also presumably based on physicochemical issues, but also reflects the important observation that such types of extreme topological perturbation can occur within normal cells. This is important in considering the structural components of the nucleus, and as to how structural integrity is maintained within this compartment.

The observation of long distance, active nuclear movement within plant cells raises the question as to whether such movement is universally observed within the cells of different species and within organisms of different kingdoms, the associated question as to whether this movement is obligatorily found within grossly anisotropic cells, and the general question as to whether the movement reflects a mechanistic purpose or is an unimportant consequence of the operation of existing mechanisms. Further research will be required to answer these questions, but the general applicability of the methods described above should facilitate this research.

Acknowledgements

This work was supported in part by funding from the U.S.D.A. N.R.I., and from the N.S.F. Plant Genome Program. I thank Changqing Zhang for providing the image comprising *Figure 1b*, and Georgina Lambert for editorial assistance with this manuscript.

References

Bezanilla, M., Pan, A. and Quatrano, R.S. (2003) RNA interference in the moss *Physcomitrella patens*. *Plant Physiol.* **133**: 470–474.

Broers, J.L.V., Machiels, B.M., van Eys, G.J.J.M., Kuijpers, H.J.H., Manders, E.M.M., van Driel, R. and Ramaekers, F.C.S. (1999) Dynamics of the nuclear lamina as monitored by GFP tagged A-type lamins. *J. Cell Sci.* **112**: 3463–3475.

Calikowski, T.T., Meulia, T. and Meier, I. (2003) A proteomic study of the *Arabidopsis* nuclear matrix. *J. Cell. Biochem.* **90**: 361–378.

Capco, D.G. and Penman, S. (1983) Mitotic architecture of the cell: the filament networks of the nucleus and cytoplasm. *J. Cell Biol.* **96**: 896–906.

Chubb, J.R., Boyle, S., Perry, P. and Bickmore, W.A. (2002) Chromatin motion is constrained by association with nuclear compartments in human cells. *Curr. Biol.* **12**: 439–445.

Chytilova, E., Macas, J. and Galbraith, D.W. (1999) Green fluorescent protein targeted to the nucleus, a transgenic phenotype useful for studies in plant biology. *Ann. Bot.* **83**: 645–654.

Chytilova, E., Macas, J., Sliwinska, E., Rafelski, S., Lambert, G. and Galbraith, D.W. (2000) Nuclear dynamics in *Arabidopsis thaliana*. *Molec. Biol. Cell* **11**: 2733–2741.

Chytilova, E., Macas, J., Sliwinska E., Rafelski, S., Lambert, G. and Galbraith, D.W. (2001) Nuclear dynamics in *Arabidopsis thaliana*. In: Ludin, B. (ed.) *GFP in Motion, volume II*. CD-ROM, published by Trends in Cell Biology, Elsevier Press.

Collings, D.A., Carter, C.N., Rink, J.C., Scott, A.C., Wyatt, S.E. and Allen, N.S. (2000) Plant nuclei can contain extensive grooves and invaginations. *Plant Cell.* **12**: 2425–2439.

Cremer, T. and Cremer, C. (2001) Chromosome territories, nuclear architecture and gene regulation in mammalian cells. *Nature Rev. Genet.* **2**: 292–301.

De Solorzano, C.O., Malladi, R., Lelievre, S.A. and Lockett, S.J. (2001) Segmentation of nuclei and cells using membrane related protein markers. *J. Microscopy* **201**: 404–415.

Echevarria, W., Leite, M.F., Guerra, M.T., Zipfel, W.R. and Nathanson, M.H. (2003) Regulation of calcium signals in the nucleus by a nucleoplasmic reticulum. *Nature Cell Biol.* **5**: 440–446.

Ellenberg, J., Siggia, E.D., Moriera, J.E., Smith, C.L., Presley, J.F., Worman, H.J. and Lippincott-Schwartz, J. (1997) Nuclear membrane dynamics and reassembly in living cells: Targeting of an inner nuclear membrane protein in interphase and mitosis. *J. Cell Biol.* **138**: 1193–1206.

Fricker, M., Hollinshead, M., White, N. and Vaux, D. (1997) Interphase nuclei of many mammalian cell types contain deep, dynamic, tubular membrane-bound invaginations of the nuclear envelope. *J. Cell Biol.* **136**: 531–544.

Galbraith, D.W., Harkins, K.R. and Knapp, S. (1991) Systemic endopolyploidy in *Arabidopsis thaliana*. *Plant Physiol.* **96**: 985–989.

Galy, V., Olivo-Marin, J.C., Scherthan, H., Doye, V., Rascalou, N. and Nehrbass, U. (2000) Nuclear pore complexes in the organization of silent telomeric chromatin. *Nature* **403**: 108–112.

Gasser, S.M. (2002) Visualizing chromatin dynamics in interphase nuclei. *Science* **296**: 1412–1416.

Gerlich, D. and Ellenberg, J. (2003) 4D imaging to assay complex dynamics in live specimens. *Nature Rev. Molec. Cell Biol.* **4(suppl.)**: SS14–SS19.

Gindullis, F. and Meier, I. (1999) Matrix attachment region binding protein MFP1 is located in discrete domains at the nuclear envelope. *Plant Cell.* **11**: 1117–1128.

Gindullis, F., Peffer, N.J. and Meier, I. (1999) MAF1, a novel plant protein interacting with matrix attachment region binding protein MPF1, is located at the nuclear envelope. *Plant Cell.* **11**: 1755–1768.

Gindullis, F., Rose, A., Patel, S. and Meier, I. (2002) Four signature motifs define the first class of structurally related large coiled-coil proteins in plants. *BMC Genomics* **3**: 9.

Goldman, R.D., Gruenbaum, Y., Moir, R.D., Shumaker, D.K. and Spann, T.P. (2002) Nuclear lamins: building blocks of nuclear architecture. *Genes Dev.* **16**: 533–547.

Grebenok, R.J., Pierson, E.A., Lambert, G.M., Gong, F.-C., Afonso, C.L., Haldeman-Cahill, R., Carrington, J.C. and Galbraith, D.W. (1997a) Green-fluorescent protein fusions for efficient characterization of nuclear localization signals. *Plant J.* **11**: 573–586.

Grebenok, R.J., Lambert, G.M. and Galbraith, D.W. (1997b) Characterization of the targeted nuclear accumulation of GFP within the cells of transgenic plants. *Plant J.* **12**: 685–696.

Gruenbaum, Y., Goldman, R.D., Meyuhas, R., Mills, E., Margalit, A., Fridkin, A., Dayani, Y., Prokocimer, M. and Enosh, A. (2003) The nuclear lamina and its functions in the nucleus. *Intl Rev. Cytol.* **226**: 1–62.

Hall, G., Jr., Allen, G.C., Loer, D.S., Thompson, W.F. and Spiker, S. (1991) Nuclear scaffolds and scaffold attachment regions in higher plants. *Proc. Natl Acad. Sci. USA* **88**: 9320–9324.

Heun, P., Laroche, T., Shimada, K., Ferrer, P. and Gasser, S.M. (2001) Chromosome dynamics in the yeast interphase nucleus. *Science* **294**: 2181–2186.

Irons, S.L., Evans, D.E. and Brandizzi, F. (2003) The first 238 amino acids of the human lamin B receptor are targeted to the nuclear envelope in plants. *J. Exp. Bot.* **54**: 943–950.

Ishii, K., Arib, G., Lin, C., van Houwe, G. and Laemmli, U.K. (2002) Chromatin boundaries in budding yeast: the nuclear pore connection. *Cell* **109**: 551–562.

Janicki, S.M. and Spector, D.L. (2003) Nuclear choreography: interpretations from living cells. *Curr. Opin. Cell Biol.* **15**: 149–157.

Jedd, G. and Chua, N.H. (2002) Visualization of peroxisomes in living plant cells reveals acto-myosin-dependent cytoplasmic streaming and peroxisome budding. *Plant Cell Physiol.* **43**: 384–392.

Kato, N. and Lam, E. (2001) Detection of chromosomes tagged with green fluorescent protein in live *Arabidopsis thaliana* plants. *Genome Biology* **2**: research0045.1 – research 0045.10.

Kato, N. and Lam, E. (2003) Chromatin of endoreduplicated pavement cells has greater range of movement than that of diploid guard cells in *Arabidopsis thaliana*. *J. Cell Sci.* **116**: 2195–2201.

Ketelaar, T., Faivre-Moskalenko, C., Esseling, J.J., de Ruijter, N.C.A., Grierson, C.S., Dogterom, M. and Emons, A.M.C. (2002) Positioning of nuclei in *Arabidopsis* root hairs: an actin-regulated process of tip growth. *Plant Cell.* **14**: 2941–2955.

Lippincott-Schwartz, J. and Patterson, G.H. (2003) Development and use of fluorescent protein markers in living cells. *Science* **300**: 87–91.

Lippincott-Schwartz, J., Altan-Bonnet, N. and Patterson, G.H. (2003) Photobleaching and photoactivation: following protein dynamics in living cells. *Nature Rev. Molec. Cell Biol.* **4(suppl.)**: SS7–SS14.

Lopez-Soler, R.I., Moir, R.D., Spann, T.P., Stick, R. and Goldman, R.D. (2001) A role for nuclear lamins in nuclear envelope assembly. *J. Cell Biol.* **154**: 61–70.

Marshall, W.F., Straight, A., Marko, J.F., Swedlow, J., Dernburg, A., Belmont, A., Murray, A.W., Agard, D.A. and Sedat, J.W. (1997) Interphase chromosomes undergo constrained diffusional motion in living cells. *Curr. Biol.* **7**: 930–939.

Meier, I. (2001) The plant nuclear envelope. *Cell. Molec. Life Sci.* **58**: 1774–1780.

Meier, I., Phelan, T., Gruissem, W., Spiker, S. and Schneider, D. (1996) MFP1, a novel plant filament-like protein with affinity for matrix attachment region DNA. *Plant Cell.* **8**: 105–115.

Newport, J.W., Wilson, K.L. and Dunphy, W.G. (1990) A lamin-independent pathway for nuclear envelope assembly. *J. Cell Biol.* **111**: 2247–2259.

Pederson, T. and Aebi, U. (2002) Actin in the nucleus: what form and what for? *J. Struct. Biol.* **16**: 533–547.

Percipalle, P., Zhao, J., Pope, B., Weeds, A., Lindberg, U. and Daneholt, B. (2001) Actin bound to the heterogeneous nuclear ribonucleoprotein hrp36 is associated with Balbiani ring mRNA from the gene to the polysomes. *J. Cell Biol.* **153**: 229–236.

Pestic-Dragovich, L., Stojiljkovic, L., Philimonenko, A.A., Nowak, G., Ke, Y., Settlage, R.E., Shabanowitz, J., Hunt, D.F., Hozak, P. and de Lanerolle, P. (2000) A myosin I isoform in the nucleus. *Science* **290**: 337–341.

Raikov, I.B. (1996). Nuclei of ciliates. In: Hausmann, K. and Bradbury, P.C. (eds) *Ciliates: Cells as Organisms*, pp. 221–242. Stuttgart, Germany: Gustav Fischer.

Schumaker, D.K., Kuczmarski, E.R. and Goldman, R.D. (2003) The nucleoskeleton: lamins and actin are major players in essential nuclear functions. *Curr. Opin. Cell Biol.* **15**: 358–366.

Sliwinska, E., Lambert, G.M. and Galbraith, D.W. (2002) Factors affecting nuclear dynamics and Green Fluorescent Protein targeting to the nucleus in *Arabidopsis thaliana* roots. *Plant Sci.* **163**: 425–430.

Starr, D.A. and Han, M. (2002) Role of ANC-1 in tethering nuclei to the actin cytoskeleton. *Science* **298**: 406–409.

Starr, D.A. and Han, M. (2003) ANChors away: an actin based mechanism of nuclear positioning. *J. Cell Sci.* **116**: 211–216.

Sullivan, T., Escalante-Alcalde, D., Bhatt, H., Anver, M., Bhat, N., Nagashima, K., Stewart, C.L. and Burke, B. (1999) Loss of A-type lamin expression compromises nuclear envelope integrity leading to muscular dystrophy. *J. Cell Biol.* **147**: 913–920.

Tsien, R. (1998) The green fluorescent protein. *Ann. Rev. Biochem.* **67**: 509–544.

Tsien, R. (2003) Imagining imaging's future. *Nature Rev. Molec. Cell Biol.* **4(suppl.)**: SS16–SS21.

Van Bruaene, N., Joss, G., Thas, O. and Van Oostveldt, P. (2003) Four-dimensional imaging and computer-assisted track analysis of nuclear migration in root hairs of *Arabidopsis thaliana*. *J. Microscop.* **211**: 167–178.

14

The nuclear envelope in the plant cell cycle

David E. Evans, Sarah L. Irons, Mekdes H. Debela and
Federica Brandizzi

1. Introduction

Plants, like all other eukaryotes, have a nuclear envelope (NE) which separates the nucleoplasm and genetic material from the cytoplasm. In electron micrographs, the plant nuclear envelope resembles that of cells of other kingdoms. It is a double-membrane structure, is perforated by nuclear pores and it is linked to surrounding endoplasmic reticulum (ER). Unlike yeast and most other unicells, but in common with most eukaryotes, plants have an open cell division and the NE breaks down in mitosis. When this occurs, the disrupted NE membrane appears to integrate with ER; however, the membranes of the plant mitotic apparatus are complex. The context of the plant NE varies greatly from cell to cell; in many cells, it is appressed between a large vacuole and the plasma membrane and surrounded by very little cytoplasm. In other cells – especially meristematic cells, where cell division is taking place, it is much more prominent and the nucleus is central and surrounded by cytoplasm.

In this chapter, we will present current understanding of the dynamic behaviour of the plant NE during cell division, particularly emphasizing recent data using *in vivo* markers for NE constituents. Other chapters of this volume describe the NE in animal and yeast cells in detail. While there are many similarities between the nuclear envelopes of plants and other higher eukaryotes, there are a number of differences of functional significance, which relate both to the NE itself and to the structures with which it is associated. Some of these differences, which will be dealt with in detail later in the chapter, are highlighted below:

- Higher plants undergo an open cell division in which nuclear envelope breakdown precedes chromosome segregation. However, the mechanism for NE breakdown and reformation differs from the other eukaryotes in which an open mitosis occurs.

The Nuclear Envelope, edited by D.E. Evans, C. Hutchison & J.A. Bryant.
© 2004 Garland Science/BIOS Scientific Publishers

- Plant NE is associated with the cytoskeleton and nucleoskeleton, but plant nuclei lack sequence homologues of the nuclear lamins.
- Plants lack centrosomes (microtubule organizing centres, MTOCs, or spindle pole bodies) with the entire nuclear envelope able to act as an MTOC.
- The position of the plane of cell division is crucial to plant development, as the presence of a cell wall prevents movement of newly formed cells. Establishing the plane of division involves the cytoskeleton and the membranes of the mitotic apparatus.
- The membranes of the mitotic apparatus are essential to cell wall synthesis. While it is assumed that membrane from the NE becomes part of the pool of ER in the mitotic apparatus, comparatively little is known about its traffic and targeting during NE breakdown and reformation.

2. An overview of plant cell division

Plant cell division normally occurs in cells of the meristem, which are less vacuolate, have thin cell walls and in which the nucleus is prominent. It commences with an increase in cell mass (G1 phase), followed by S phase (DNA synthesis) and a delay (G2 phase) before mitosis occurs (M phase). The whole process from onset of G1 phase to the end of M phase takes typically 12–24 h. During G2, a process unique to plants is observed; microtubules that had been uniformly distributed around the cell cortex become concentrated into a ring just below the plasma membrane. This ring, the preprophase band, predicts the plane at which cell division will occur and the new cell wall be formed (*Figure 1*). At the end of G2, the mitotic spindle forms outside the NE while chromosome condensation occurs. The microtubules of the spindle gradually align from pole to pole. The NE then disrupts, the preprophase band of microtubules breaks down and the microtubules contact the chromosomes. Microtubules (the kinetochore microtubules, running to the spindle poles) then attach to the kinetochores of the chromosomes and oscillatory movements of the chromosomes ensue until they are aligned at the metaphase plate. At the onset of anaphase, chromosomes separate as the kinetochore microtubules shorten until they become concentrated at the spindle poles. Polar microtubules (from spindle pole to spindle pole) become more visible and a rapid increase in the number of microtubules in the mid region becomes evident, along a line predicted by the preprophase band. These phragmoplast microtubules will be involved in the deposition of the new cell wall.

3. A survey of key elements of the endomembrane system involved in the mitotic apparatus

3.1 *Nuclear envelope, endoplasmic reticulum and phragmoplast formation*

The plant nuclear envelope may be regarded as a specialized region of the ER, with a distinct protein composition (Staehelin, 1997). The outer NE (ONE) is a functional continuum with perinuclear ER and is capable of protein translocation as rough ER. The degree of continuity between ONE and ER is clearly important in maintaining the protein composition of the ONE as a distinct domain. Rapid-freeze electron microscopy suggests that the connections are constricted and 25–30 nm in diameter (Craig and Staehelin, 1988) and it has been suggested (Staehelin, 1997) that they may restrict protein movement from ER to ONE.

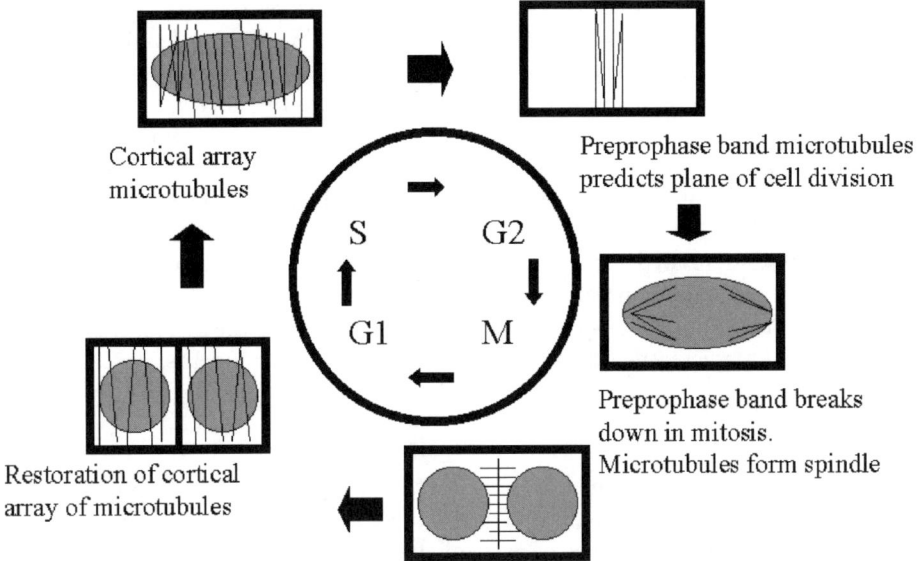

Figure 1. *Plant cell microtubules behave dynamically during cell division. The initial cortical array of microtubules condenses to a central band, the preprophase band predicting the plane of division. In prophase, the preprophase band breaks down and microtubules form the spindle. After division, microtubules are active in forming the phragmoplast which forms the new cell plate. This contacts the existing cell walls at the point where the preprophase band was located. Finally, the cortical array of microtubules is restored.*

The inner NE (INE) is only connected to the ONE via nuclear pore complexes (NPCs) and INE proteins can only diffuse into it through these structures. Therefore, the INE represents a distinct membrane domain with unique characteristics when compared to the ER membranes.

The higher plant NE breaks down in prophase and reforms at telophase. In interphase it is intact and linked to ER surrounding the nucleus. ER is also present at the cell cortex (cortical ER) and in transvacuolar strands. In prophase, the NE is disrupted and masses of ER membranes are already visible at the spindle poles. By metaphase, the spindle is surrounded by tubular and cisternal ER and the NE is indistinguishable from it. In anaphase, the spindle is almost entirely surrounded by membrane, with strands of membrane present between the chromosomes. Sheets of cisternal ER surround the spindle while tubular ER invaginates into it and aligns with the spindle microtubules. This tubular ER appears to form channels through which the daughter chromatids are pulled whilst simultaneously extending towards the pole and along the plane of phragmoplast formation as the spindle apparatus pulls apart (Hawes *et al.*, 1981; Hepler, 1980). During telophase and cytokinesis, a complex network of ER tubules interweaves the developing cell plate and stays connected to the daughter nuclei by a system of ER tubules (*Figure 2*). High-voltage electron microscopy has shown that this tubular ER forms a three-dimensional network at the plane of the phragmoplast which appears to predict the site of phragmoplast formation (Hawes *et al.*, 1981). This forms the new cell plate that will develop into the cell wall dividing the two daughter cells. Its location was predicted by location of the preprophase band of microtubules. The cell plate grows from the centre outwards until it contacts the adjacent cell walls. Its growth requires a large amount of vesicle traffic and this derives from the ER in the mitotic apparatus (MA). During this process the cortical ER network of the cell remains relatively unaffected and reformation of the NE around the daughter nuclei is now clearly visible.

The processes described above clearly require highly organized traffic and assembly of membrane and membrane components throughout mitosis and cytokinesis. This must be both temporally and spatially organized and involves the assembly and disassembly of membrane interacting with a variety of cellular components.

3.2 *Nuclear pore complexes*

Nuclear envelopes contain nuclear pores situated where inner and outer nuclear membranes join. The pore is composed of about 30 proteins that form the nuclear pore complex (NPC) (Chapter 7). The pore channel is approximately 10 nm in diameter and 100 nm long (Pante and Aebi, 1994). Small molecules, between 20 and 70 kDa, can

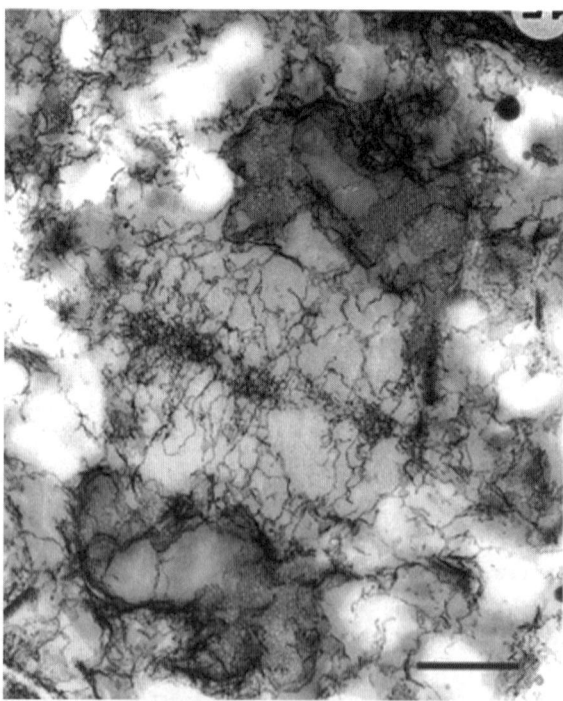

Figure 2. *Telophase nuclei of maize. Tubular ER has aggregated around the developing cell plate (phragmoplast) and is connected by long strands of ER to the developing telophase NE. Reproduced with permission from Planta, Low and high voltage electron microscopy of mitosis and cytokinesis in maize roots. Hawes et al., Vol. 152, 397–402, figure 4, (1981) © Springer-Verlag. Bar = 2 μm.*

passively diffuse through the pore (Bustamante *et al.*, 1995) while larger molecules require nuclear localization sequences for transport. Their translocation is driven by ATP and GTP hydrolysis and can require chaperones, such as Hsp70 (Dingwall and Laskey, 1992), and other accessory factors (Melchior and Gerace, 1995).

The pore complexes are attached to the nuclear lamina, and hence are not dependent on the phospholipid membrane for maintaining their structure (Lyman and Gerace, 2001). The use of nucleoporins fused to green fluorescent protein (GFP) has allowed the visualization of NPC dynamics *in vivo* in animal cells. Daigle *et al.* (2001) produced nucleoporin GFP fusions of POM121, an integral membrane protein localized to the ring spoke region, and Nup153 a peripheral membrane protein found in the nuclear fibrils (forming part of the basket like structural portion of the NPC). These fluorescent chimeras demonstrated different dynamics through the cell cycle, suggesting that the NPC has a core set of proteins with low turnover such as POM121 and other proteins such as Nup153 that show rapidly cycling (Daigle *et al.*, 2001).

The presence of NPCs in plant cells has been known for some time, as demonstrated by freeze fracture of tobacco cells. However, as yet no plant nuclear pore proteins have been characterized (Meier, 2001). Early electron microscopy studies in plants (de la Torre *et al.*, 1979) showed a growth in NE from G1 to G2 and a concomitant increase in pore number from G1 to mid S phase. Presence of nuclear pores persisting in membranes of the MA has been suggested in some, but not all, EM studies (e.g. Hawes *et al.*, 1981 and references therein).

3. Traffic of components of the plant NE and of other membranes during cell division

What happens to the components of the NE when plant cells divide? Microscopy suggests that the NE breaks down to join the ER of the mitotic apparatus and reforms from this pool; but is this also true of individual proteins of the NE and do they all behave similarly? To date, a number of studies have followed the localization of NE and NE-associated proteins either *in vivo* or *in vitro* during cell division. The proteins or probes used are summarized in *Table 1*. It is evident from this table that there is a real paucity of NE markers available for study in plants and the generation of further NE markers remains a major research goal.

3.1 *LCA1-antigen during mitosis in tomato root cells*

LCA1 was identified (Ewing *et al.*, 1990) as a homologue of mammalian SR/ER type Ca ATPase (SERCA) in tomato. When an antipeptide antibody raised to the C-terminus of LCA1 was used to immuno-localize the antigen in tomato roots, it was found to localize to the NE with remarkably high specificity. This was shown both at the light microscope level (*Figure 3*) and by silver-enhanced electron microscopy (*Figure 4*). While the absence of a suitable probe precluded an *in vivo* study, we were able to describe the localization of the antigen by electron and light microscopy of fixed and stained material (Downie *et al.*, 1998).

Confocal microscopy in combination with immuno-electron microscopy of cells in metaphase revealed antigen distributed with the membranes associated with the spindle poles (*Figures 3 and 4*). Later, the antigen was associated with the membrane

Table 1. Proteins of the NE and associated membranes used in localization studies

Probe	Description	Reference
MFP1	81kDa coiled coil matrix binding protein predominantly located in the plant nuclear matrix, but also NE; observed as small punctate structures at nuclear rim. Shows structural similarities to mysosin, tropomyosin and intermediate filament binding proteins	Meier *et al.*, 1996, Gindullis *et al.*, 1999; Samaniego *et al.*, 2001
MAF1	Small ser/thr rich MFP1-binding partner. Localized to the nuclear envelope and nuclear matrix	Gindullis *et al.*, 1999
LCA-1 antigen	Antigen localized to the nuclear envelope of tomato by an anti-peptide antibody raised to the C-terminus of a plant SR/ER Ca ATPase homologue in tomato	Downie *et al.*, 1998
LBR	Chimeric expression of the N-terminal 238 amino acids of the mammalian lamin B receptor in tobacco. Localized to the nuclear envelope	Irons *et al.*, 2003
p80 antigen	Antigen detected by an antibody to calf thymus centrosomes. Located at nuclear periphery in prophase and telophase onwards.	Schmit *et al.*, 1994
NMCP1	Coiled coil protein located at nuclear periphery in interphase and at spindle poles after NE breakdown	Masuda *et al.*, 1997

distributed around the daughter chromatids before being present in the reforming NE of the daughter cells in telophase. Comparison of these data with those shown in *Figure 2* makes it clear that this antigen (which is likely to be an integral membrane protein) (Downie *et al.*, 1998) becomes distributed with specific regions of the membranes of the mitotic apparatus and is not uniformly dispersed. If the antibody is specifically recognizing the Ca-ATPase LCA, this may suggest that it is associated with a Ca^{2+}-signalling pool in the membranes of the mitotic apparatus that is involved in the regulation of events in mitosis. More significantly, it suggests that the NE constituents are targeted to specific ER domains in the mitotic apparatus.

3.2 *Mammalian LBR expressed in plants*

Our recent work has revealed that when a GFP fusion of the N-terminal 238 amino acids of the mammalian lamin B receptor (LBR) containing the N-terminus and first transmembrane domain is expressed in plants, it localizes with a high degree of specificity to the nuclear envelope (Irons *et al.*, 2003; colour *Plates 8 and 9*). The LBR is a constitutively expressed integral membrane protein found in the inner NE in animal, but not plant, cells (Worman *et al.*, 1990). It is a 58-kDa protein with a large globular N terminal nucleoplasmic domain that is hydrophilic and rich in basic amino acids, and a hydrophobic C terminus consisting of eight transmembrane segments with high homology to C-14 sterol reductases in plants (Schrick *et al.*, 2000) and animals

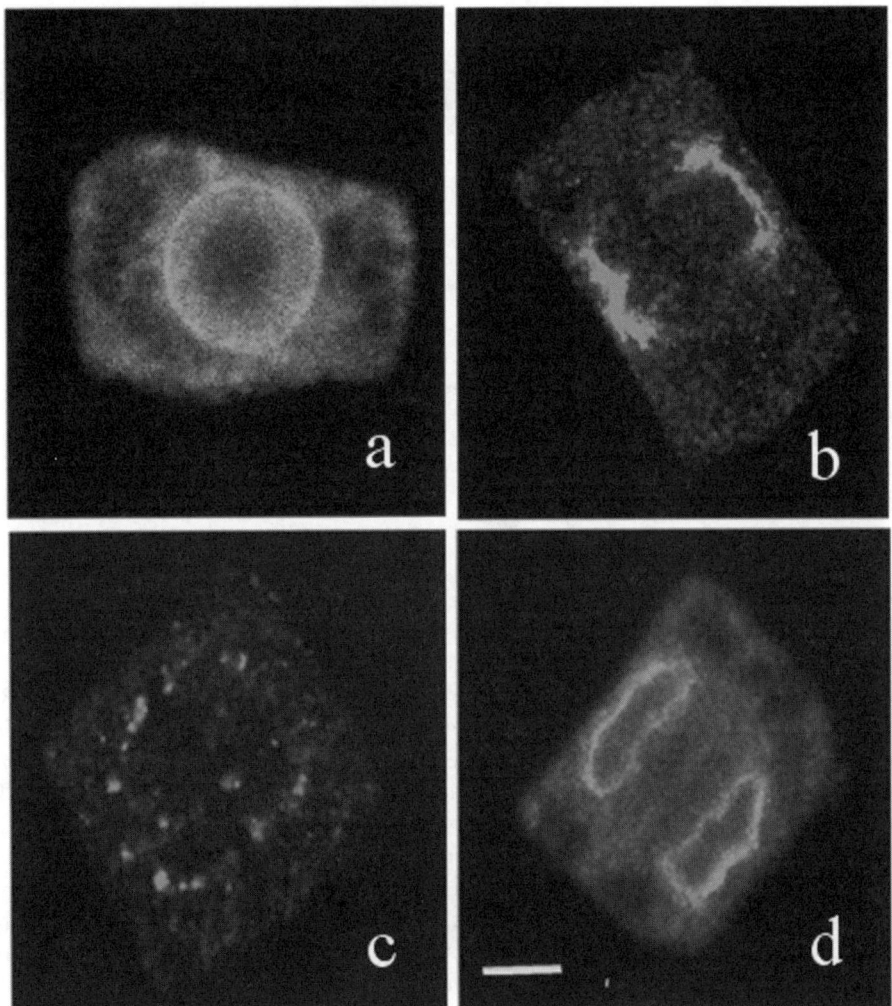

Figure 3. Light micrographs of fixed and permeabilized tomato root cells stained with LCA-1 antibody at various stages of the cell cycle. (a) Interphase; (b) metaphase (note membranes aggregated to the poles; (c) anaphase (note staining distributed throughout mitotic apparatus); (d) telophase (note limited labelling of cell plate with strong labelling of daughter nuclear envelopes). Bar = 2.5 μm. Reprinted from FEBS letters, Vol. 429(1), Downie et al., A calcium pump at the higher plant nuclear envelope? pp 44–48 (1998) with permission from Elsevier.

(Holmer *et al.*, 1998). The N-terminus of LBR binds chromatin (Pyrpasopoulou *et al.*, 1996; Ye and Worman, 1994), and this may be of importance in nuclear membrane reassembly and in localizing LBR to the inner NE. The amino terminal domain including the first transmembrane domain is responsible for targeting LBR to the nuclear membrane and contains a bipartite type nuclear signal sequence (Smith and Blobel, 1993; Soullam and Worman, 1993, 1995).

Ellenberg *et al.* (1997) used the N terminal 238 amino acids of the LBR fused to EGFP in animal cells (COS-7) to observe its activity during mitosis. In this study, LBR-EGFP was targeted to NE membranes at interphase. Two populations of

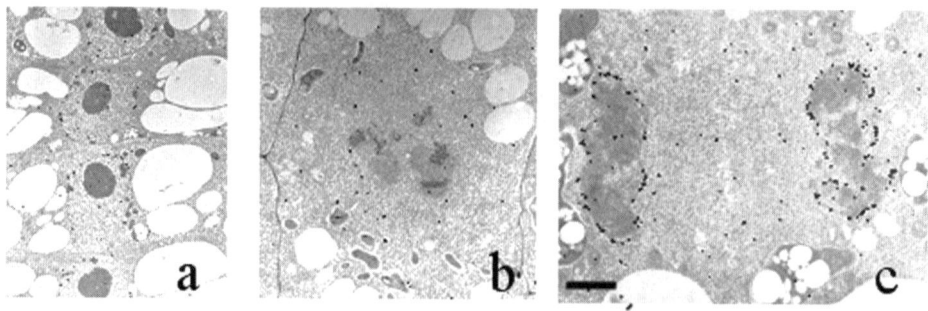

Figure 4. Electron micrographs of thin sections of tomato root cells stained with LCA 1 antibody and silver enhanced. Black deposits indicate the original location of the antigen. (a) Interphase cells. (b) Anaphase. Note stain distributed throughout the mitotic apparatus. (c) Telophase. Note the staining of the daughter nuclear envelopes and limited staining of the cell plate. Compare with Figures 1 and 2. Reprinted from FEBS letters, Vol. 429(1), Downie et al., A calcium pump at the higher plant nuclear envelope? pp 44–48 (1998) with permission from Elsevier. Bar = 1 μm.

LBR-EGFP were observed in the cells in this phase; an immobilized NE population, possibly immobilized through binding to lamins or chromatin, and a small pool of LBR-EGFP that was freely diffusible in the ER. During mitosis the LBR-EGFP of the NE became highly mobile and dispersed to the ER. When nuclear membrane reformation was recorded using time-lapse confocal imaging, a redistribution of LBR-EGFP from diffusely distributed ER to membranes tightly associated with chromatin was observed, followed by a subsequent expansion to the spherical NE. When the construct was expressed at high levels it clearly localized with DNA, producing NE invaginations.

Confocal microscopy of the amino-terminal 238 amino acids of LBR in an LBR GFP$_5$ chimera in plants revealed that it associated with membranes as an integral component of the NE (Irons *et al.*, 2003). At interphase, specific NE labelling was present with only limited labelling of ER (Colour *Plates 8 and 9*). When cells from a population of a stably expressing tobacco suspension culture were observed in real time following NE breakdown during division (*Figure 5*), LBR EGFP$_5$ was observed first to concentrate in the mass of ER membrane at the poles. Subsequently, as labelling at the poles diminished slightly, it was also apparent in tubular membrane structures between the metaphase plate and the poles. As the cell entered telophase, labelling was apparent around the daughter nuclei (*Figure 5*, 2069s) before strong labelling of the membranes of the phragmoplast, perpendicular to the poles, was also apparent. At this stage, the outlines of the reforming NEs were evident, with labelling of the phragmoplast and some diffuse labelling of membranes between the nuclei and the phragmoplast. Finally, the labelling reverted to being specific to the nuclear envelope.

A similar study of chimeric expression of LBR in yeast (Smith and Blobel, 1993) also showed its location at the inner nuclear envelope, together with its presence in membrane stacks associated with the NE. These stacks were associated with overexpressed LBR. In the study of Ellenberg *et al.* (1997) LBR EGFP fluorescence, which had been immobile in the interphase NE, was observed within interconnected tubular membrane of the mitotic apparatus. Finally, it became associated with chromatin and within a 2–3 minute period the fluorescent ER membrane appeared to wrap itself

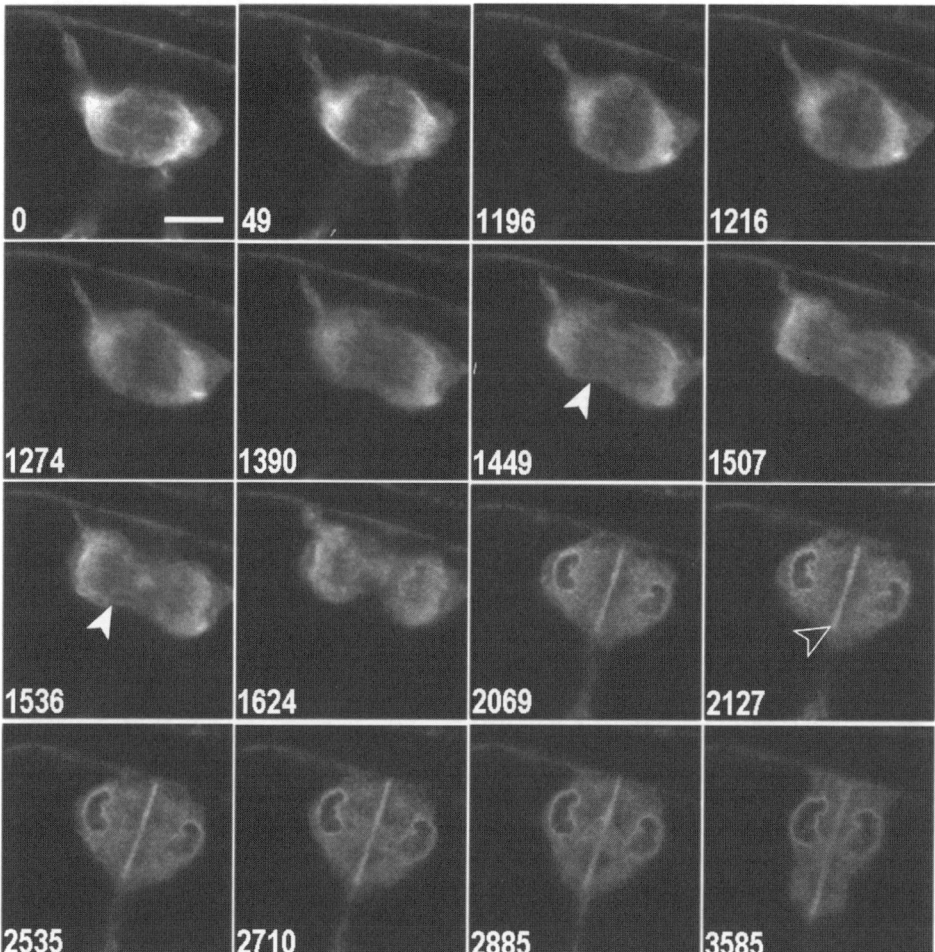

Figure 5. *LBR time course. Numbers refer to time, in seconds, elapsed from the start of images. Cells in late metaphase expressing LBR-GFP5 show fluorescence distributed through the ER membranes (time 0 ±1216 s). Tubular membraneous structures form through the mitotic apparatus (arrow, time1274–1507 s). As division progresses the membranes move towards opposite poles as the chromosomes separate (time 0–1536 s). The ER membranes encircle the newly formed daughter nuclei (1624 s). The NE begins to form around each nucleus (2069 s). The phragmoplast (marked with an empty arrow), which is the basis for the cell wall formation in dividing the cells, forms between the nuclei, this grows across the cell as more wall is assembled (2069 ±3585 s). Scale bar 20 mm. Reproduced from Irons et al. (2003). The first 238 amino acids of the human lamina B receptor are targeted to the nuclear envelope in plants. Journal of Experimental Botany, Vol. 54, with permission of the society for Experimental Biology.*

around the chromatin forming an intact NE. The NE then expanded to its full size over the next 30–80 minutes. Clearly this process showed strong similarities to the events observed in tobacco cells, with the exception of the membranes of the phragmoplast, a structure unique to the plant system.

Comparison of the distribution of the LBR with the distribution of ER during plant cell division (e.g. *Figure 2*) suggests that it migrates into all the ER of the mitotic apparatus and is not targeted to specific domains. This is further suggested by the fact that over-expression of LBR GFP$_5$ results in ER, as well as NE, labelling and that LBR becomes associated with the phragmoplast, a site of vigorous ER membrane activity. If, in common with mammalian LBR-GFP, the protein is retained in the INE by chromatin binding, such mobility into the ER when chromatin binding is lost should not be surprising.

3.3 *Traffic of other plant NE proteins in cell division*

Early studies of the plant NE in cell division involved immunostaining fixed material. As centrosomes are not present, the surface of the plant NE acts as a microtubule organizing centre (MTOC). Using a monoclonal antibody to calf thymus centrosomes, Schmit *et al.* (1994) followed the behaviour of a putative component of plant MTOCs through the cell cycle. During interphase the antigen was associated with the nuclear periphery; following NE breakdown, it migrated to the centromere region of the chromosomes and remained there until telophase, when it was observed to associate with the newly forming daughter nuclear envelopes. The antigen (p80; see *Table 1*) is likely to be a peripheral, rather than an integral, component of the NE and its mobility is therefore likely to be governed by microtubule binding rather than membrane traffic. Another coiled coil protein, NMCP1, identified in carrot by Masuda *et al.* (1997) was localized to the nuclear rim during interphase. At NE breakdown, it migrates to the spindle poles (Masuda *et al.*, 1999). Its function is unknown, but it is likely to be a component of the nuclear matrix. Recently, significant progress in the dynamics of nuclear envelope traffic has been made using fluorescent probes *in vivo*. Dixit and Cyr (2002) made use of a fluorescent chimera of a predominantly Golgi-resident protein (N-acetylglucosaminyl transferase) as a marker for NE breakdown in tobacco. While this construct gave sufficient labelling of the NE to permit the events of its breakdown to be followed, its lack of specificity means that conclusions on the traffic of NE-resident proteins could not be drawn, although it was possible to use it to obtain accurate sequencing of events during NE break down.

Two other NE-associated proteins have also been investigated; matrix attachment region-binding protein (MFP1) and its binding partner matrix attachment region-binding protein-associated factor (MAF1). MFP1 is a coiled coil protein identified in tomato, with similarity to myosin, tropomyosin and intermediate filament proteins and with two putative transmembrane domains and a binding region for attachment to DNA (Meier *et al.*, 1996). It is proposed that MFP1 is a key protein attaching chromatin to the NE (Meier, 2000). In tomato, it is associated with the NE in discrete regions; labelling gives a speckled appearance to the nuclear rim (Gindullis and Meier, 1999). The only study of MFP1 through the cell cycle at an ultrastructural level was conducted in onion (Samaniego *et al.*, 2001). In this species, two putative MFP1 homologues were identified and a nucleoplasmic location was most strongly suggested, as concentrations of the protein associated with replication factories. These were sometimes closely associated with the inner NE. MAF1 is also associated with the NE and the nuclear matrix (Gindullis *et al.*, 1999); its behaviour in the cell cycle has yet to be studied.

4. Targeting to the plant NE

To enter the inner NE (INE), proteins synthesized in the ER must move through the outer NE and a nuclear pore. Once in the INE, they may be retained by binding to components of the nucleoplasm (Smith and Blobel, 1993; Soullam and Worman, 1993, 1995).

Retention of the lamin B receptor at the mammalian inner nuclear envelope is believed to be due both to its interaction with B-type lamins and with chromatin and two binding sites have been proposed. The N-terminal 60 amino acids of LBR bind to B-type lamins (Holmer *et al.*, 1998) while this region and another domain in the N-terminus (residues 54–89) bind to chromatin (Takano *et al.*, 2002). The latter region, the SR repeat region, was found to be sufficient for INM localization (Meyer and Radsak, 2000; Meyer *et al.*, 2002; Takano *et al.*, 2002). The phosphorylation of this RS repeat region has been associated with chromatin binding (Takano *et al.*, 2002) which would immobilize the protein in the INE. This explains the immobility of LBR EGFP in the interphase NE, revealed by photobleaching/recovery studies when, in contrast, it is mobile in the ER of the mitotic apparatus (Ellenberg *et al.*, 1997). The LBR–chromatin complex in mammalian cells includes heterochromatin protein 1 (HP1) and histones H3 and H4 forming a large complex anchoring NE to the nucleoskeleton (Polioudaki *et al.*, 2001; Ye and Worman, 1994). Genome sequencing has not revealed the presence of plant lamin homologues (Meier, 2001; see Chapter 5); however biochemical and ultrastructural evidence suggests the presence of proteins fulfilling similar functions (Beven *et al.*, 1991; Minguez and Delaespina, 1993). Pea nuclear fractions contain proteins that immuno-label with mammalian anti-lamin B and anti-intermediate filament antibodies, thus demonstrating the presence of similar epitopes in plants (McNulty and Saunders, 1992).

In the absence of lamin homologues, therefore, it is likely that the chimeric LBR LBRGFP$_5$ construct is retained in the NE by its binding to chromatin through the RS repeat region. Experiments in our laboratory suggest that the construct does not accumulate in the NE when this region is mutated, but accumulates either in the ER or in the nucleoplasm (authors unpublished data) depending on the mutation.

The targeting of LBR, an animal protein without a plant homologue, to the plant NE indicates at least some common mechanisms exist in NE targeting in plants and animals. Similar targeting of a chimeric construct of LBR-GFP has also been observed in yeast.

5. Mechanisms for the breakdown and reformation of the plant NE and for protein traffic during this process

There are two theories as to the fate of the nuclear membrane during the cell cycle; vesiculation or ER absorption (see Buendia *et al.*, 2001 for a review). The vesiculation theory was suggested by experiments with *Xenopus* cell-free extracts that showed that the nuclear membranes disrupt to form vesicles which appear to reform around chromatin (Vigers and Lohka, 1991). *In vivo* studies however strongly suggest the alternative – that the NE is absorbed into the ER. Use of fluorescently labelled NE proteins (e.g. LBR, described in detail elsewhere in this chapter) showed continuity of labelling within mitotic ER membranes and subsequent NE reformation around daughter nuclei *in vivo* in animal cells (Ellenberg *et al.*, 1997).

Two co-ordinated processes have been suggested to initiate NE breakdown. The first is protein phosphorylation or dephosphorylation, implicated in NE disassembly during mitosis in animal cells (Foisner and Gerace, 1993). A change in phosphorylation state dissociates inner NE proteins from the nucleoskeleton. In animal cells, lamins depolymerise when phosphorylated by p34^{cdc2} kinase, the β form of protein kinase C and mitogen-activated protein kinase (MAPK) (Goldberg *et al.*, 1999). LAP2 binding of lamin B and chromatin is disrupted by phosphorylation specific to mitosis and the lamin B receptor is known to undergo phosphorylation by kinases such as p34^{cdc2} kinase and SR protein specific kinase (SRPK) (Gerace and Foisner, 1994; Takano *et al.*, 2002). Such protein phosphorylation changes allow reversible dissociation of inner nuclear membrane proteins and their ligands, effectively removing their anchorage to the nuclear structure and so allowing movement into the ER membranes.

The second, related, event is the initiation of membrane breakdown. Here, recent studies in mammalian cells indicate that the first stage of NE breakdown is caused by microtubules tearing the lamina (Beaudouin and Daigle, 2002). In an elegant study, these authors showed that early spindle microtubules caused folds and invaginations of the NE up to 1 hour prior to NE breakdown causing tension in the underlying lamina. As the NE loses its anchorage to the lamina (due to altered phosphorylation) this is followed by tearing of the NE at the point where maximal tension is occurring. The tear then grows rapidly until breakdown of the NE is complete. Dynein and dynactin, concentrated on the NE, are part of the force-generating system (Salina *et al.*, 2002).

Electron microscopy of early plant prophase nuclei reveals areas of disrupted NE with other areas remaining intact (e.g. Hawes *et al.*, 1981). Using a dual-expressing tobacco cell line, in which microtubules were visualized using GFP-tubulin and NE visualized with a red fluorescent protein (RFP) N-acetylglucosaminyl transferase 1 construct, Dixit and Cyr (2002) obtained indirect evidence for the involvement of microtubules in plant NE breakdown. In this system, NE breakdown precedes the disappearance of the preprophase band microtubules. Its breakdown was accompanied by a ruffling of the NE and occurred in the region of the nucleus closest to the preprophase band. Conversely, breakdown of the preprophase band was most rapid in the region closest to the disrupted NE. These results are entirely consistent with plant NE breakdown being initiated by tearing and involving the attachment of microtubules to the surface of the NE. An added dimension in plants is the fact that microtubules and other cytoskeletal elements must be correctly orientated to establish the plane of cell division and the correct positioning of the new cell wall.

Reformation of the NE occurs in telophase when membranes containing NE proteins from the ER are seen to associate with the chromatin of newly forming daughter nuclei. Lamin and chromatin binding are central to this process. Recently, a role for GTP hydrolysis by the small GTP binding protein Ran and the presence of RanGDP have been shown to be required for NE vesicles to associate with chromatin (Hetzer *et al.*, 2000; Zhang and Clarke, 2000). Nuclear-envelope-like structures, capable of transport and containing nuclear pore complexes will form around beads coated with Ran (Zhang and Clarke, 2000). Plant Ran has been identified (Ach and Gruissem, 1994; Merkle *et al.*, 1994) and a plant RAN GTPase activating protein (RanGAP) has also been described. Plant RanGAP differs from mammalian in that it possesses an N-terminal domain (termed the WPP domain) that is necessary and sufficient for NE binding (Rose and Meier, 2001). This 110 amino-acid domain contains tryptophan and proline residues and is highly conserved in plant RanGAPs and also in the nuclear-envelope-associated protein MAF1 (see above). The domain is not found in yeast or animals. MAF1 binds at the NE to the filament-like protein

MFP1 (which is believed to link chromatin with the NE; (Gindullis and Meier, 1999; and see above).

Thus, the association of INE with chromatin during the formation of the plant NE involves processes that initiate or permit its binding to chromatin. The current hypothesis developed for non-plant systems is that Ran acts as a marker for the position of chromatin at the end of mitosis, with high levels of RanGDP promoting NE formation (Zhang and Clarke, 2000). The details of this hypothesis, including the role of nucleotide exchange and GTP hydrolysis are presented elsewhere in this volume (Chapter 11). It seems likely that Ran performs similar functions in the reformation of the NE in plants; however, the proteins with which Ran, chromatin and the NE interact are different. Given that RanGAP and MAF1 show homologous sequences and that MAF1 interacts with MFP1 (Meier, 2000) suggests that MAF1 and MFP1 are components of this system in plants, with MAF1 binding to MFP1 blocking RanGAP binding. This in turn would prevent RanGAP associating with nuclear vesicles and prevent premature NE reformation.

6. Unanswered questions and future prospects

The plant nuclear envelope shows some remarkable similarities to animal NE during the cell cycle. However, it does this using a group of distinct proteins not found in animals; there is strong functional homology without an equivalent protein sequence homology. It is therefore all the more fascinating, though perhaps not surprising, that a chimeric construct of a mammalian protein, the LBR should localize to the NE in plants and behave, at least in terms of traffic and targeting, in an identical manner in plants as in animals. The NE localization of LBR in plants indicates that plants and animals have at least one common targeting and retention mechanism for NE proteins. Identifying those targeting and retention mechanisms involves mutation of the regions of LBR known to interact with components of the mammalian nucleoskeleton to retain the protein in the NE. To understand the system fully, however, it is essential that the *in vivo* binding partners of the retention mechanism are identified. Using LBR as bait in two-hybrid or other assays for binding partners might well prove fruitful in identifying them.

A major 'knowledge gap' also exists about the protein composition, traffic and assembly of nuclear pore complexes in plants. Recent approaches to the identification of components of nuclear pore complexes have included the use of GFP-labelled expression libraries. Identification and labelling plant NPC components will permit comparative studies of their behaviour and assembly in mitosis; their use as NE markers would also be beneficial to understanding the behaviour of NE components.

Possession of good fluorescent markers for the NE is a major asset for research and provides the opportunity to follow the behaviour of a number of NE markers during the cell cycle. LBR is useful, because it is reasonably specific; however, contrasting the location of LCA during the cell cycle with LBR suggests that not all NE proteins behave identically. While LCA appears to remain closely associated with membrane at the spindle poles and at the periphery of the MA before specifically labelling the NE in telophase, LBR appears to enter the general pool of ER in the mitotic apparatus and, as well as re-associating with the NE, also remains in ER membranes of the phragmoplast. There is a great need for other, membrane-intrinsic, fluorescent NE markers to further study NE protein traffic during the cell cycle.

References

Ach, R.A. and Gruissem, W. (1994) A small nuclear GTP-binding protein from tomato suppresses a *Schizosaccharomyces-pombe* cell-cycle mutant. *Proc. Natl Acad. Sci. USA* **91**: 5863–5867.

Beaudouin, J. and Daigle, N. (2002) Dynamics of nuclear cell membrane revealed in vivo. *M S Med. Sci.* **18**: 41–43.

Beven, A., Guan, Y.H., Peart, J., Cooper, C. and Shaw, P. (1991) Monoclonal-antibodies to plant nuclear matrix reveal intermediate filament-related components within the nucleus. *J. Cell Sci.* **98**: 293–302.

Buendia, B., Courvalin, J.C. and Collas, P. (2001) Dynamics of the nuclear envelope at mitosis and during apoptosis. *Cell. Mol. Life Sc.* **58**: 1781–1789.

Bustamante, J.O., Liepins, A., Prendergast, R.A., Hanover, J.A. and Oberleithner, H. (1995) Patch-clamp and atomic-force microscopy demonstrate TATA-binding protein (TBP) interactions with the nuclear-pore complex. *J. Memb. Biol.* **146**: 263–272.

Craig, S. and Staehelin, L.A. (1988) High-pressure freezing of intact plant-tissues – evaluation and characterization of novel features of the endoplasmic-reticulum and associated membrane systems. *Euro. J. Cell Biol.* **46**: 80–93.

Daigle, N., Beaudouin, J., Hartnell, L., Imreh, G., Hallberg, E., Lippincott-Schwartz, J. and Ellenberg, J. (2001) Nuclear pore complexes form immobile networks and have a very low turnover in live mammalian cells. *J. Cell Biol.* **154**: 71–84.

de la Torre, C., Sacristan-Garate, A. and Navarette, M.H. (1979) Dynamics of the nuclear envelope during the cell cycle in plants. *Cytobios* **24**: 25–31.

Dingwall, C. and Laskey, R. (1992) The nuclear membrane. *Science* **258**: 942–947.

Dixit, R. and Cyr, R.J. (2002) Spatio-temporal relationship between nuclear-envelope breakdown and preprophase band disappearance in cultured tobacco cells. *Protoplasma* **219**: 116–121.

Downie, L., Priddle, J., Hawes, C. and Evans, D.E. (1998) A calcium pump at the higher plant nuclear envelope? *Febs Lett.* **429**: 44–48.

Ellenberg, J., Siggia, E.D., Moreira, J.E., Smith, C.L., Presley, J.F., Worman, H.J. and Lippincott-Schwartz, J. (1997) Nuclear membrane dynamics and reassembly in living cells: Targeting of an inner nuclear membrane protein in interphase and mitosis. *J. Cell Biol.* **138**: 1193–1206.

Ewing, N.N., Wimmers, L.E., Meyer, D.J., Chetelat, R.T. and Bennett, A.B. (1990) Molecular-cloning of tomato plasma-membrane H$^+$-ATPase. *Plant Physiol.* **94**: 1874–1881.

Foisner, R. and Gerace, L. (1993) Integral membrane-proteins of the nuclear-envelope interact with lamins and chromosomes, and binding is modulated by mitotic phosphorylation. *Cell* **73**: 1267–1279.

Gerace, L. and Foisner, R. (1994) Integral membrane proteins and dynamics of the nuclear envelope. *Trends Cell Biol.* **4**: 127–131.

Gindullis, F. and Meier, I. (1999) Matrix attachment region binding protein MFP1 is localized in discrete domains at the nuclear envelope. *Plant Cell* **11**: 1117–1128.

Gindullis, P., Peffer, N.J. and Meier, I. (1999) MAF1, a novel plant protein interacting with matrix attachment region binding protein MFP1, is located at the nuclear envelope. *Plant Cell* **11**: 1755–1767.

Goldberg, M., Harel, A. and Gruenbaum, Y. (1999) The nuclear lamina: Molecular organization and interaction with chromatin. *Crit. Rev. Eukary. Gene Exp.* **9**: 285–293.

Hawes, C.R., Juniper, B.E. and Horne, J.C. (1981) Low and high-voltage electron-microscopy of mitosis and cytokinesis in maize roots. *Planta* **152**: 397–407.

Hepler, P.K. (1980) Membranes in the mitotic apparatus of barley cells. *J. Cell Biol.* **86**: 490–499.

Hetzer, M., Bilbao-Cortes, D., Walther, T.C., Gruss, O.J. and Mattaj, I.W. (2000) GTP hydrolysis by Ran is required for nuclear envelope assembly. *Mol. Cell* **5**: 1013–1024.

Holmer, L., Pezhman, A. and Worman, H.J. (1998) The human lamin B receptor/sterol reductase multigene family. *Genomics* **54**: 469–476.

Irons, S.L, Evans, D.E. and Brandizzi, F. (2003) The first 238 amino acids of the human lamin B receptor are targeted to the nuclear envelope in plants. *J. Exp. Bot.* **54**: 1–8.

Lyman, S.K. and Gerace, L. (2001) Nuclear pore complexes: dynamics in unexpected places. *J. Cell Biol.* **154**: 17–20.

Masuda, K., Haruyama, S. and Fujino, K. (1999) Assembly and disassembly of the peripheral architecture of the plant cell nucleus during mitosis. *Planta* **210**: 165–167.

Masuda, K., Xu, Z.J., Takahashi, S., Ito, A., Ono, M., Nomura, K. and Inoue, M. (1997) Peripheral framework of carrot cell nucleus contains a novel protein predicted to exhibit a long alpha-helical domain. *Exp. Cell Res.* **232**: 173–181.

McNulty, A.K. and Saunders, M.J. (1992) Purification and immunological detection of pea nuclear intermediate filaments – evidence for plant nuclear lamins. *J. Cell Sci.* **103**: 407–414.

Meier, I. (2000) A novel link between Ran signal transduction and nuclear envelope proteins in plants. *Plant Physiol.* **124**: 1507–1510.

Meier, I. (2001) The plant nuclear envelope. *Cell. Mol. Life Sci.* **58**: 1774–1780.

Meier, I., Phelan, T., Gruissem, W., Spiker, S. and Schneider, D. (1996) MFP1, a novel plant filamnet-like protein with affinity for matrix attachment region DNA. *Plant Cell* **8**: 2105–2115.

Melchior, F. and Gerace, L. (1995) Mechanisms of nuclear-protein import. *Curr. Opin. Cell Biol.* **7**: 310–318.

Merkle, T., Haizel, T., Matsumoto, T., Harter, K., Dailmann, G. and Nagy, F. (1994) Phenotype of the fission yeast cell cycle regulatory mutant *pim1-46* is suppressed by a tobacco cDNA encoding a small, Ran-like GTP-binding protein. *Plant J.* **6**: 555–565.

Meyer, G., Gicklhorn, D., Strive, T., Radsak, K. and Eickmann, M. (2002) A three-residue signal confers localization of a reporter protein in the inner nuclear membrane. *Biochem. Biophys. Res. Commun.* **291**: 966–971.

Meyer, G.A. and Radsak, K.D. (2000) Identification of a novel signal sequence that targets transmembrane proteins to the nuclear envelope inner membrane. *J. Biol. Chem.* **275**: 3857–3866.

Minguez, A. and Delaespina, S.M.D. (1993) Immunological characterization of lamins in the nuclear matrix of onion cells. *J. Cell Sci.* **106**: 431–439.

Pante, N. and Aebi, U. (1994) Toward the molecular details of the nuclear-pore complex. *J. Struct. Biol.* **113**: 179–189.

Polioudaki, H., Kourmouli, N., Drosou, V., Bakou, A., Theodoropoulos, P.A., Singh, P.B., Giannakouros, T. and Georgatos, S.D. (2001) Histones H3/H4 form a tight complex with the inner nuclear membrane protein LBR and heterochromatin protein 1. *EMBO Rep.* **2**: 920–925.

Pyrpasopoulou, A., Meier, J., Maison, C., Simos, G. and Georgatos, S.D. (1996) The lamin B receptor (LBR) provides essential chromatin docking sites at the nuclear envelope. *EMBO J.* **15**: 7108–7119.

Rose, A. and Meier, I. (2001) A domain unique to plant RanGAP is responsible for its targeting to the plant nuclear rim. *Proc. Natl Acad. Sci. USA* **98**: 15377–15382.

Salina, D., Bodoor, K., Eckley, D.M., Schroer, T.A., Rattner, J.B. and Burke, B. (2002) Cytoplasmic dynein as a facilitator of nuclear envelope breakdown, *Cell* **108**: 97–107.

Samaniego, R., Yu, W.D., Meier, I. and de la Espina, S.M.D. (2001) Characterisation and high-resolution distribution of a matrix attachment region-binding protein (MFP1) in proliferating cells of onion. *Planta* **212**: 535–546.

Schmit, A.C., Stoppin, V., Chevrier, V., Job, D. and Lambert, A.M. (1994) Cell-cycle dependent distribution of a centrosomal antigen at the perinuclear mtoc or at the kinetochores of higher-plant cells. *Chromosoma* **103**: 343–351.

Schrick, K., Mayer, U., Horrichs, A., Kuhnt, C., Bellini, C., Dangl, J., Schmidt, J. and Jurgens, G. (2000) FACKEL is a sterol C-14 reductase required for organized cell division and expansion in Arabidopsis embryogenesis. *Genes Develop.* **14**: 1471–1484.

Smith, S. and Blobel, G. (1993) The 1st membrane spanning region of the lamin-b receptor is sufficient for sorting to the inner nuclear-membrane. *J. Cell Biol.* **120**: 631–637.

Soullam, B. and Worman, H.J. (1993) The amino-terminal domain of the lamin-b receptor is a nuclear-envelope targeting signal. *J. Cell Biol.* **120:** 1093–1100.

Soullam, B. and Worman, H.J. (1995) Signals and structural features involved in integral membrane-protein targeting to the inner nuclear-membrane. *J. Cell Biol.* **130:** 15–27.

Staehelin, L.A. (1997) The plant ER: A dynamic organelle composed of a large number of discrete functional domains. *Plant J.* **11:** 1151–1165.

Takano, M., Takeuchi, M., Ito, H., Furukawa, K., Sugimoto, K., Omata, S. and Horigome, T. (2002) The binding of lamin B receptor to chromatin is regulated by phosphorylation in the RS region. *Euro. J. Biochem.* **269:** 943–953.

Vigers, G.P.A. and Lohka, M.J. (1991) A distinct vesicle population targets membranes and pore complexes to the nuclear-envelope in *xenopus* eggs. *J. Cell Biol.* **112:** 545–556.

Worman, H.J., Evans, C.D. and Blobel, G. (1990) The lamin B receptor of the nuclear-envelope inner membrane – a polytopic protein with 8 potential transmembrane domains. *J. Cell Biol.* **111:** 1535–1542.

Ye, Q. and Worman, H.J. (1994) Primary structure-analysis and lamin-b and DNA-binding of human LBR, an integral protein of the nuclear-envelope inner membrane. *J. Biol. Chem.* **269:** 11306–11311.

Zhang, C.M. and Clarke, P.R. (2000) Chromatin-independent nuclear envelope assembly induced by Ran GTPase in Xenopus egg extracts. *Science* **288:** 1429–1432.

15

Signalling to the nucleus via A-kinase anchoring proteins

Philippe Collas, Sandra B. Martins and Helga B. Landsverk

1. Summary

The cell nucleus is a highly dynamic organelle whose function and structure during the cell cycle is tightly controlled. A number of signals triggered by external stimuli or intracellular clocks are relayed to the nucleus by protein kinases and phosphatases. Specificity of action of kinases and phosphatases can be achieved by their recruitment into multiprotein complexes targeted to discrete subcellular or subnuclear loci. One class of molecules targeting signalling units within single complexes are A-kinase anchoring proteins or AKAPs. AKAPs not only target enzymes to their substrate but may also regulate enzyme activity. This chapter highlights the role of nuclear AKAPs in relaying and modulating protein kinase and phosphatase signals to the nucleus or chromosomes.

2. Introduction

Multiple hormones signal through common second messengers to activate the same protein kinase and phosphatase cascades but elicit distinct intracellular responses. Many kinases and phosphatases have broad substrate specificities and can be used in various combinations to achieve distinct biological responses. Mechanisms must exist to properly guide the repertoire of enzymes into specific signalling pathways. One form of regulation of these various effects is the intracellular compartmentalization of the kinases and phosphatases involved to specific sites of action. This is achieved by the recruitment of signalling molecules into multiprotein signalling networks or the activation of silent enzymes already positioned in the vicinity of their substrate (Pawson and Scott, 1997).

The cell nucleus is highly responsive to environmental cues relayed by cytoplasmic signals. Nuclear responses to stimulation or inhibition of cell growth, differentiation, transformation or apoptosis are as broad as modulation of DNA replication, gene activity, transcription, mRNA translation, protein processing and alterations in

---

The Nuclear Envelope, edited by D.E. Evans, C. Hutchison & J.A. Bryant.
© 2004 Garland Science/BIOS Scientific Publishers

nuclear architecture manifested by nuclear disassembly at mitosis or by the reformation and growth of the nucleus upon re-entry into interphase. Signalling to the
nucleus must therefore be spatially and temporally accurate. Signalling to the nucleus
is mediated by several classes of molecules, including ions, metabolites, nucleotides
and nucleotide-binding proteins, cyclic nucleotides (such as cAMP), protein kinases,
protein phosphatases and possibly structural proteins.

Compartmentalization of protein kinase and phosphatase signalling at the level of
the nuclear envelope and intranuclear structures such as the nuclear matrix or chromatin is essential for specificity of action of these molecules on DNA replication, gene
expression, chromatin structure and nuclear envelope composition and dynamics.

Paradigms of compartmentalized protein kinases and phosphatases important for
the regulation of nuclear structure and function are the cAMP-dependent protein
kinase (PKA) and protein phosphatase 1 (PP1). Both are broad-specificity enzymes
with ubiquitous patterns of expression (Bollen, 2001; Bollen and Beullens, 2002;
Cohen, 2002; Tasken *et al.*, 1997). A large body of evidence indicates that substrate
specificity of both enzyme classes is mediated by subcellular localization. This chapter
focuses on a specific class of protein kinase and phosphatase adaptors, termed
A-kinase anchoring proteins (AKAPs). We review the role of AKAPs of the nucleus on
the regulation of nuclear structure and function and cell cycle progression. The
function of HA95, a recently identified nuclear protein related to AKAP95, on nuclear
dynamics is also discussed.

3.   Intracellular targeting of cAMP signalling by A-kinase anchoring proteins

3.1 *cAMP signalling via PKA*

The biological effects of cAMP are mediated by PKA. Type-I and type-II PKA holoenzymes consist of two catalytic and two regulatory (RI or RII, respectively) subunits
(Tasken *et al.*, 1997). Specificity of PKA is largely determined by the structure and
properties of the R subunits, whereas the catalytic subunits exhibit similar kinetics
features and substrate specificities. PKA is activated by binding of two cAMP molecules to each R subunit followed by release of the two catalytic subunits from the
R-cAMP complex. The activated catalytic subunits phosphorylate specific substrates,
an event that enhances hormonal responses by altering the activity of key enzymes and
structural proteins of the cytoplasm and the nucleus. Free catalytic subunits also
translocate to the nucleus where they have a role in the activation of cAMP-responsive
genes (Riabowol *et al.*, 1988).

3.2 *Specifying cAMP signalling by A-kinase anchoring proteins (AKAPs)*

The specificity of cellular and nuclear responses to cAMP results from interactions of
the RII subunit dimer of PKA with AKAPs in various cellular compartments. A variety
of techniques, including PKA-RII overlays, have been used to identify over 30 AKAPs,
some of which are members of gene families with several splice variants. AKAPs are
functionally defined by their binding to the RI and/or RII subunit dimers of the PKA
holoenzymes. Classically, RI or RII binding occurs via a well-characterized amphipathic

helix in which the hydrophobic side chains constitute the main binding determinants (Edwards and Scott, 2000; Vijayaraghavan *et al.*, 1999); however, exceptions have been identified (Diviani *et al.*, 2000).

The targeting domain of AKAPs determines the localization of the PKA-mediated cAMP signal at discrete subcellular compartments including the nucleus (*Figure 1a, b*). Targeting is mediated by protein–protein interactions to structural components or protein–lipid interactions with membranes. Consequently, AKAPs have been identified in many cellular compartments, including plasma membrane, Golgi, endoplasmic or sarcoplasmic reticulum, vesicles, mitochondria, microtubules, centrosome, nuclear envelope and nuclear matrix/chromatin fractions (Edwards and Scott, 2000; Smith and Scott, 2002). Targeting may also be dictated by the alternative splicing of AKAP genes (Edwards and Scott, 2000).

An interesting property of AKAPs is their ability to simultaneously anchor other protein kinases such as protein kinase C or glycogen synthase kinase-3, protein phosphatases including PP1, PP2A or PP2B (Colledge and Scott, 1999; Edwards and Scott, 2000; Smith and Scott, 2002) and phosphodiesterases (PDEs) (Tasken *et al.*, 2001) (*Figure 1b*). As such, AKAPs have emerged as scaffolding proteins that can integrate, and perhaps regulate, multiple signalling pathways.

4. Regulation of chromosome dynamics and nuclear function by AKAP95

4.1 *Mitotic chromosome condensation* in vitro *and in somatic cells*

Condensation of chromosomes at mitosis requires the highly conserved 13S condensin complex (Hirano, 2000). In humans, the condensin complex consists of two structural maintenance of chromosomes (SMC) subunits (hCAP-C and hCAP-E) and three regulatory non-SMC subunits (hCAP-D2/CNAP1, hCAP-G and hCAP-H) (Kimura *et al.*, 2001; Schmiesing *et al.*, 1998, 2000). The 13S condensin complex

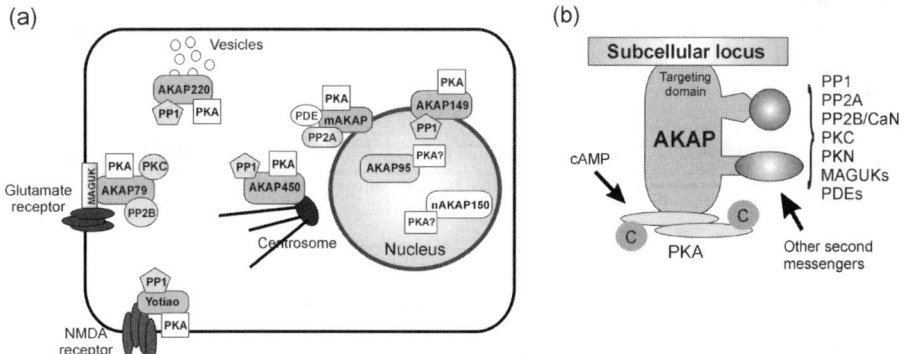

Figure 1. *AKAPs as adaptor proteins for multiple signalling molecules. (a) Subcellular localization of selected AKAPs, including all known AKAPs of the cell nucleus, and associated signalling molecules. AKAP149 and mAKAP are associated with the nuclear envelope whereas AKAP95 and nAKAP150 localize primarily in the nuclear matrix. AKAP149 is also found in the endoplasmic reticulum and mitochondria (not shown) (b) A generic AKAP.*

isolated from *Xenopus* egg extracts displays ATPase activity *in vitro*, which introduces positive supercoils into DNA and thereby probably assists in chromosome condensation (Kimura and Hirano, 1997; Kimura *et al.*, 1999). Condensin recruitment to chromatin at mitosis correlates with mitotic phosphorylation of the non-SMC subunits (Hirano *et al.*, 1997).

Another important factor for mitotic chromosome condensation is AKAP95. AKAP95 is a 95-kDa protein that harbours two zinc fingers (designated ZF1 and ZF2) in its COOH-terminal half upstream of the RII-binding domain (Coghlan *et al.*, 1994; Eide *et al.*, 1998) (*Figure 2*). In interphase, AKAP95 is localized exclusively in the nucleus and associates with the nuclear matrix but does not anchor RIIα (Collas *et al.*, 1999) (*Figure 3a*). During mitosis, AKAP95 redistributes from the nuclear matrix to the condensing chromosomes as the nuclear envelope breaks down (Collas *et al.*, 1999; Steen *et al.*, 2000a). AKAP95 binds chromatin and has been proposed to act as a targeting molecule for the condensin complex once nuclear envelope breakdown has taken place (Steen *et al.*, 2000a) (*Figure 3a*). AKAP95-mediated recruitment of the condensin complex is required for chromosome condensation and the amount of condensin (as judged by immunological analyses of hCAP-D2, a non-SMC component of the condensin complex) correlates with the extent of chromosome condensation (Steen *et al.*, 2000a). Subsequently, AKAP95 anchors RIIα onto or near the metaphase plate. Targeting of RIIα from the centrosome-Golgi area (Carlson *et al.*, 2001) to chromatin at mitosis and binding to AKAP95 require phosphorylation of RIIα on Thr54 by cyclin-dependent kinase 1 (Landsverk *et al.*, 2001) (*Figure 3b*). Notably, PKA binding to AKAP95 is dispensable for the chromosome condensation activity of AKAP95. However, maintenance of chromosomes in a condensed form throughout mitosis requires PKA activity and binding and blocking of PKA activity or disruption of anchoring leads to premature chromatin decondensation (Collas *et al.*, 1999).

We have identified determinants of RII binding, chromatin binding, condensin targeting and chromosome condensation activities of AKAP95 (Eide *et al.*, 2002) (*Figure 2*). Binding of AKAP95 to chromatin requires the zinc finger ZF1. A disruption in ZF1 that abolishes zinc chelation abrogates chromosome condensation but not condensin recruitment; however ZF2 is required for condensin targeting. Thus, condensin recruitment to chromatin is not sufficient to promote

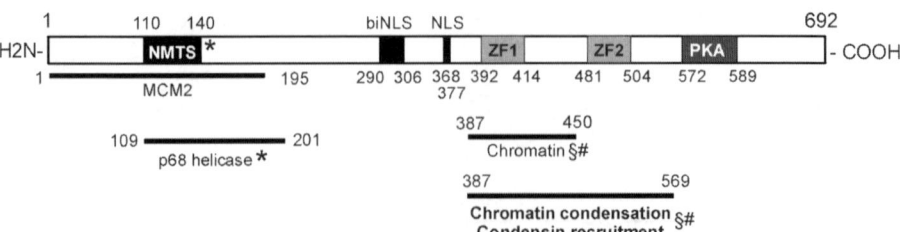

Figure 2. *Interaction and functional domains of AKAP95. Identified binding domains and functions (in bold) of human AKAP95. Numbers refer to amino acid positions. NMTS, nuclear matrix targeting sequence; biNLS, bipartite nuclear localization signal; NLS, nuclear localization signal; ZF, zinc finger; PKA, RII-binding domain. *(Akileswaran et al., 2001); § (Eide et al., 2002); #(Bomar et al., 2002).*

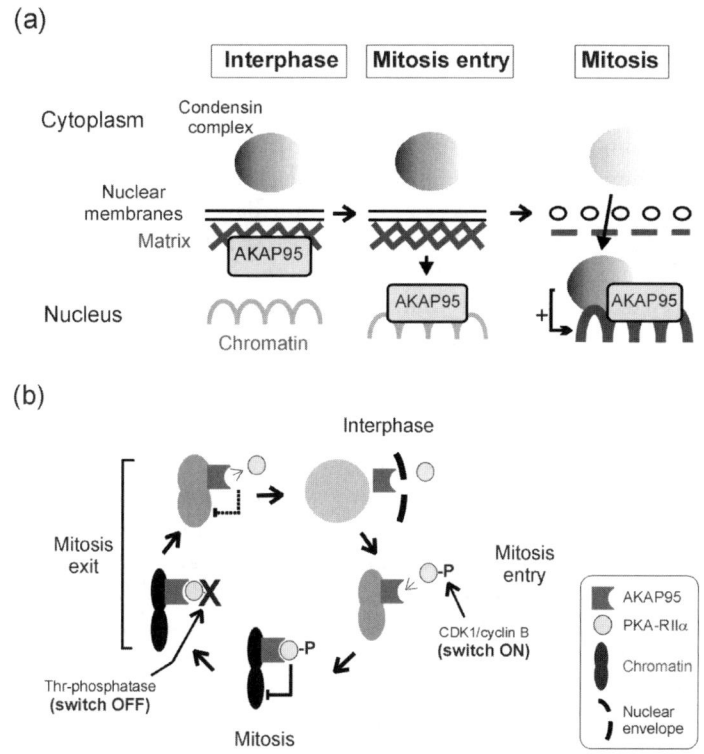

Figure 3. Roles of AKAP95 on chromosome structure during the cell cycle. (a) AKAP95-mediated recruitment of the condensin complex to chromatin at mitosis. In interphase, condensins are primarily cytoplasmic, whereas AKAP95 resides in the nuclear matrix. At mitosis entry, AKAP95 is released from the matrix and associates with chromatin. As nuclear membranes break down and the nuclear matrix solubilizes at mitosis, AKAP95 targets condensins to chromatin to elicit chromosome condensation. (b) Model for how reversible AKAP95–PKA interaction (mediated by phosphorylation of the RIIα subunit of PKA) controls chromatin structure during mitosis. During interphase, AKAP95 and PKA-RIIα localize to distinct compartments, separated by the nuclear envelope. At mitosis entry, AKAP95 associates with condensing chromatin. RIIα is phosphorylated by the cyclin-dependent kinase 1. RIIα phosphorylation turns on a molecular switch promoting RIIa anchoring to AKAP95 and maintenance of condensed chromosomes during mitosis. Although anchoring of RIIα to AKAP95 has been demonstrated, anchoring of the catalytic subunit of PKA is only suggested. Throughout mitosis, anchoring of phosphorylated RIIα to chromatin-bound AKAP95 is required to prevent premature decondensation of chromosomes. At mitosis exit, dephosphorylation of RIIα by a threonine phosphatase induces RIIα dissociation from chromatin-bound AKAP95. This relieves inhibition of chromatin decondensation, allowing chromosome decondensation as the nuclear envelope reassembles.

chromosome condensation. Furthermore, AKAP95 interacts with *Xenopus* XCAP-H *in vitro* and *in vivo* but not with human hCAP-D2 (Eide *et al.*, 2002), indicating that AKAP95 can bind directly with at least one subunit of the condensin complex. Therefore, overlapping but distinct domains of AKAP95 are involved in condensin recruitment and chromosome condensation activities and these appear to be independent of RII anchoring.

4.2 Involvement of AKAP95 in maternal and paternal chromosome condensation in mouse zygotes

The behaviour of chromosomes during first embryonic mitosis relates to their parental origin. Fertilization is marked by extensive remodelling of the highly compacted, protamine-containing sperm chromatin into a large male pronucleus. The oocyte, arrested in metaphase II in most mammals, resumes meiosis and assembles a usually more compact female pronucleus (Bouniol-Baly et al., 1997). In the mouse, the pronuclei migrate towards each other and at mitosis maternal and paternal chromosomes condense into a single metaphase plate. Yet, both chromosome sets remain topologically separated up to the four-cell stage (Mayer et al., 2000). Condensed maternal chromosomes appear shorter and more spiralized than paternal chromosomes (Donahue, 1972) but the significance of this finding is unclear. Male and female pronuclei also respond differently to premature chromosome condensation, with maternal chromosomes condensing faster than paternal chromosomes (Ciemerych and Czolowska, 1993). These observations suggest a different structural organization of maternal and paternal chromosomes and distinct mechanisms regulating the dynamics of each parental chromosome complement.

The differential regulation of condensation of maternal and paternal chromosomes in mitotic mouse zygotes has been proposed to be mediated by AKAP95 (Bomar et al., 2002). AKAP95 is synthesized upon oocyte activation, targeted to the female pronucleus and specifically associates with maternal chromosomes at mitosis (Bomar et al., 2002). Curiously, the AKAP95 mRNA is restricted to the vicinity of the meiotic spindle in metaphase II oocytes, perhaps providing a mechanism for rapid and specific uptake by the female pronucleus after oocyte activation. In vivo displacement of endogenous AKAP95 in female pronuclei by microinjection of competitor peptides and rescue experiments show that AKPA95 is required for the recruitment of the condensin complex to, and condensation of, maternal chromosomes. In contrast, AKAP95 is dispensable for condensin recruitment to, and condensation of, paternal chromosomes. Thus at first embryonic mitosis, paternal chromosomes target condensins and condense independently of AKAP95, whereas maternal chromosomes require AKAP95 for condensin recruitment and condensation (Bomar et al., 2002). This suggests a concept whereby condensation of chromosomes in gametes, zygotes and somatic cells might involve related but distinct mechanisms.

4.3 Interphase functions of AKAP95

Association of AKAP95 with the nuclear matrix is determined by a 30 amino acid domain in the NH2-terminus of the protein (Akileswaran et al., 2001). A yeast two-hybrid screen verified by in vitro binding and in vivo co-localization, identified p68 RNA helicase as a binding partner of AKAP95 (Akileswaran et al., 2001). Interestingly, the AKAP95-related protein, HA95 (see below) also binds an RNA helicase (RHA). Although RHA and p68 RNA helicase share little sequence similarity, both proteins display identical activities and belong to the same family of DEAD/H box helicases. Thus, multiple AKAP95/HA95–RNA helicase complexes might exist in mammalian cells (Akileswaran et al., 2001). The significance of these interactions remains speculative. Nevertheless, RHA mediates association of cAMP-response element-binding protein (CREB)-binding protein (CBP) with RNA polymerase II (Nakajima et al.,

1997) and p68 RNA helicase also associates with CBP (Endoh *et al.*, 1999). Akileswaran and colleagues (Akileswaran *et al.*, 2001) proposed that AKAP95 (and/or HA95) might provide a docking site for PKA and target the kinase for its role in regulating CBP/CREB transcription initiation complexes.

The c-Myc-binding protein, AMY-1, was recently found to bind AKAP95 in the nucleus whereas it associates with S-AKAP94/AKAP149 in the cytoplasm (Furusawa *et al.*, 2002) (see below). *In vitro*, AKAP95, AMY-1 and RII can form a ternary complex. However, presence of AMY-1 in the complex prevents binding of the catalytic subunit of PKA to the AKAP-RII complex and leads to a suppression of AKAP-associated PKA activity (Furusawa *et al.*, 2002). This makes AMY-1 a novel potential modulator of PKA activity.

AKAP95 emerges as a protein with multiple functions. Besides its per-definition function in RII anchoring, AKAP95 may play a role in the assembly of hormonally responsive transcriptional complexes through interactions with RNA helicases and the AMY-1–c-Myc complex. Relocation of AKAP95 from the nuclear matrix to chromatin at mitosis and loading of the condensin complex to chromosomes are necessary for chromosome condensation. Subsequent targeting of RII to AKAP95 ensures that chromosomes remain condensed throughout mitosis (substrates of PKA involved in this function at mitosis, however, remain to be identified). Chromosome decondensation at mitosis exit or *in vitro* correlates with the release of RII from chromatin-bound AKAP95 (Landsverk *et al.*, 2001) and, presumably, reassociation of the AKAP with the nuclear matrix. Identification of novel binding partners of AKAP95 in the nucleus is likely to provide new insights on AKAP95 function.

5. Integration of cAMP and Ca^{2+} signalling at the cardiomyocyte nuclear envelope by mAKAP

Muscle AKAP (mAKAP) is a 255-kDa scaffolding protein expressed in myocytes, skeletal muscle and brain. mAKAP assembles a signalling unit at the nuclear envelope and the sarcoplasmic reticulum of cardiomyocytes and at intercalated discs in adult rat heart tissue (Dodge *et al.*, 2001; Kapiloff *et al.*, 1999; Marx *et al.*, 2000). Two independent mAKAP regions containing spectrin-like repeat motifs are sufficient for targeting to the nuclear envelope (Kapiloff *et al.*, 1999). Displacement studies have shown that targeting of mAKAP can be saturated (Kapiloff *et al.*, 1999). These findings are consistent with the idea that targeting involves interaction with another protein expressed at limiting concentrations at the nuclear envelope (Kapiloff *et al.*, 1999).

mAKAP is a part of a multicomponent signalling complex at the nuclear envelope, which includes PKA, the ryanodine receptor (RyR), a phosphodiesterase (the PDE4D3 isoform) and the protein phosphatase PP2A (Dodge *et al.*, 2001; Kapiloff *et al.*, 2001) (*Figure 4*). The RyR is a Ca^{2+} channel that is also found in the sarcoplasmic reticulum where it is involved in striated muscle excitation–contraction (Fill and Copello, 2002). At the nuclear envelope, the RyR may be involved in controlling nucleoplasmic Ca^{2+} concentrations (Kapiloff *et al.*, 2001). The RyR is a substrate for PKA and phosphorylation of the RyR by PKA has been reported to increase Ca^{2+} conductance of the RyR (Fill and Copello, 2002) (*Figure 4*). It is possible that mAKAP-associated PP2A turns off the cAMP signal by dephosphorylating the RyR (Kapiloff *et al.*, 2001) in a manner similar to the antagonistic actions of PKA and PP2A on L-type calcium channels (Davare *et al.*, 2000). PP2A has also been shown to be involved in cAMP-stimulated dephosphorylation (Feschenko *et al.*, 2002) in a PKA-independent manner. The mAKAP-associated PDE is involved in attenuating the cAMP signal by hydrolysing cAMP, thus bringing cAMP back to resting levels (Houslay and Adams, 2003). In addition, PDE4D3 is a substrate for PKA and as PKA in the mAKAP complex becomes activated by an increase in cAMP concentrations, PDE4D3 activity increases, resulting in a negative feedback (Dodge *et al.*, 2001). It is thus possible that cAMP stimulates both activation – through PKA (Kapiloff *et al.*, 2001) – and inactivation, through PDE4D3 and PP2A (Dodge *et al.*, 2001; Kapiloff *et al.*, 2001), of the RyR. The mAKAP Ca^{2+} complex might function in the integration of these responses. Finally, the RyR is regulated by Ca^{2+} (Fill and Copello, 2002), therefore the PKA–mAKAP complex may also permit integration of cAMP and Ca^{2+} signals at the cardiomyocyte nuclear envelope (Kapiloff *et al.*, 2001).

Assembly of the mAKAP complex in the perinuclear region is induced by hypertrophic stimuli in rat neonatal ventriculocytes and is thought to be associated with cellular differentiation and development of a ventricular hypertrophic phenotype (Kapiloff *et al.*, 1999). Induction of mAKAP expression also results in the redistribution of RII to the nuclear envelope (Kapiloff *et al.*, 1999). This is of interest because PKA phosphorylation induces cAMP-responsive genes involved in propagation in cardiac hypertrophy (Zimmer, 1997) and concurrent anchoring of PDE4D3 might serve to establish a negative feedback loop (Dodge *et al.*, 2001).

6. Targeting of PP1 to the nuclear envelope by AKAP149: implications on cell cycle progression

6.1 *AKAP149*

AKAP149 belongs to a family of splice variant proteins that share a 525 amino acid NH_2-terminus but vary in their COOH-terminal region (Chen *et al.*, 1997; Huang *et al.*, 1997; Lin *et al.*, 1995). Additional NH_2-terminal splice variants in the mouse

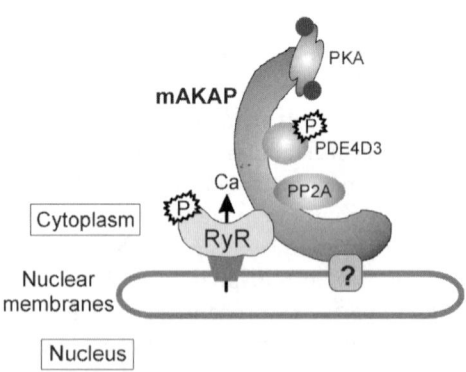

Figure 4. *Model of association of the mAKAP complex with the ryanodine receptor at the nuclear envelope. mAKAP associates with PKA, PDE4D3, PP2A and the RyR at the nuclear envelope of cardiomyocytes. PKA can phosphorylate RyR and PDE4D3, promoting an increase in Ca^{2+} ion channel activity. PP2A is proposed to reverse RyR and PDE4D3 phosphorylation. Redrawn from Kapiloff et al. (2001) J Cell Sci 2001 Vol 114: 3167–3176 with permission from The Company of Biologists Ltd.*

proteins (termed N0 and N1) have been shown to be involved in targeting the AKAP to the mitochondrial membrane and to the endoplasmic reticulum-nuclear envelope membrane system, respectively (Huang *et al.*, 1999). Whether such NH_2-terminal isoforms exist in humans is currently unknown. Human AKAP149 (Trendelenburg *et al.*, 1996) (*Figure 5a*) is an integral membrane protein found in mitochondria as well as the endoplasmic reticulum–nuclear envelope network (Steen *et al.*, 2000b). AKAP149 binds PKA-RII *in vitro* (Steen *et al.*, 2000b) and co-immunoprecipitates PKA and PKA activity from nuclear envelopes (authors' unpublished data). AKAP149 contains a consensus RV*X*F motif that binds PP1 (Steen *et al.*, 2000b). Finally, AKAP149 contains a single KH motif and a TUDOR domain (*Figure 5a*), suggestive of an RNA-binding and/or dimer-ization property. The mouse homologue of AKAP149, AKAP121, tethers PKA to the outer mitochondrial membrane (Feliciello *et al.*, 1998). AKAP121 has been shown to bind specific 3'UTR sequences of the mRNA encoding Mn superoxide dismutase (MnSOD), a mitochondrial protein (Ginsberg *et al.*, 2003). RNA binding requires a structural motif in the 3'UTR and is stimulated by cAMP (Ginsberg *et al.*, 2003). Over-expressed AKAP121 induces translocation of MnSOD mRNA from the cytosol to mitochondria and leads to an increase in mitochondrial MnSOD (Ginsberg *et al.*, 2003). The KH motif of AKAP149 may have a similar role or it might be involved in other RNA-binding functions.

6.2 *Targeting of PP1 to the nuclear envelope by AKAP149*

AKAP149 and PP1 are associated at the nuclear envelope in a cell cycle-dependent manner. At the end of mitosis, AKAP149 recruits a fraction of chromatin-bound PP1 to the reforming nuclear membranes (Steen and Collas, 2001; Steen *et al.*, 2000b). This process is inhibited by competitor peptides harbouring the PP1-binding RV*X*F motif of AKAP149 (*Figure 5b*). Mistargeting of PP1 to the nuclear membranes inhibits the assembly of B-type lamins into a nuclear lamina (Steen and Collas, 2001), presumably as a result of absence of dephosphorylation of the lamins by PP1 (Thompson *et al.*, 1997). Moreover, addition of the specific PP1 inhibitor, Inhibitor-2, during nuclear reassembly *in vitro* does not alter the nuclear envelope targeting of PP1, but it inhibits nuclear lamina assembly indicating that, in addition to proper targeting, PP1 activity is required (Steen *et al.*, 2003). Failure to assemble B-type lamins into the nuclear envelope is followed by rapid apoptosis and extensive proteolytic degradation of nuclear envelope markers (Steen and Collas, 2001). As lamin B knockouts in *Caenorhabditis elegans* are embryonic lethal (Liu *et al.*, 2000), this supports the view that proper assembly of B-type lamins is essential for cell survival. Nevertheless, it cannot at present be excluded that RV*X*F motif-containing peptides sequester PP1 and prevent it from dephosphorylating additional targets that are directly responsible for inducing apoptosis (Steen and Collas, 2001).

PP1 remains associated with AKAP149 throughout G1 phase and is released from the AKAP (and the nuclear envelope) upon S phase entry (Steen *et al.*, 2003) (*Figure 5b*). Association of PP1 with its regulatory subunits has been shown to involve, in some instances, phosphorylation of Ser residues near the RV*X*F motif (Beullens *et al.*, 1999; Liu and Brautigan, 2000; McAvoy *et al.*, 1999) (*Figure 5b*). Dissociation of AKAP149 and PP1 at the nuclear envelope correlates with phosphorylation of AKAP149 (Steen *et al.*, 2003).

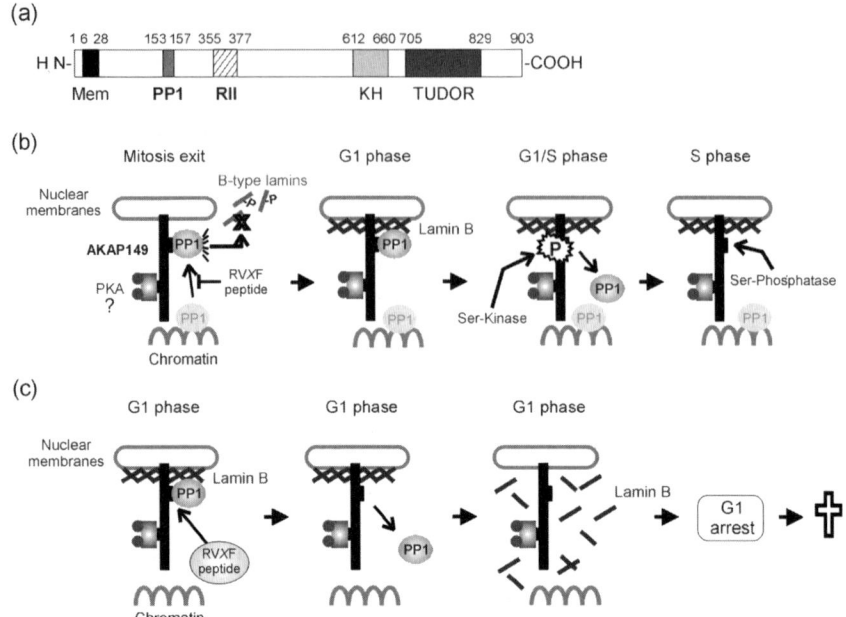

Figure 5. *Features and models of AKAP149 function. (a) Identified functional and binding domains of human AKAP149. Mem, membrane-association domain. PP1 and RII refer to motifs binding to PP1 and RII, respectively. KH and TUDOR domains are shown. Numbers indicate amino acid positions. (b) Model for the role of AKAP149 in nuclear lamina assembly. At the end of mitosis, membranes harbouring AKAP149 are targeted to chromatin. Whether AKAP149 binds PKA at this stage is uncertain. AKAP149 recruits a fraction of chromatin-bound PP1, a process inhibited by peptides containing the PP1-binding RVXF motif of AKAP149. AKAP149 acts as a B-type lamin specifier and thereby stimulates the phosphatase activity of PP1 towards the phosphorylated, depolymerized, B-type lamins. PP1-mediated lamin dephosphorylation induces the lamins to polymerize and assemble into a nuclear lamina. Only B-type lamins are represented. A-type lamins assemble through a different pathway and whether PP1 also elicits A-type lamin dephosphorylation is unknown. PP1 remains associated with AKAP149 throughout G1. At the G1/S phase transition, serine-phosphorylation of AKAP149 coincides with release of PP1 from the AKAP. To where released PP1 is retargeted is not known. During S phase, a Ser-phosphatase dephosphorylates AKAP149. This, however, is not sufficient to promote PP1 retargeting to the AKAP, which only takes place upon exit from the next mitosis. (c) Association of PP1 with nuclear membrane-bound AKAP149 during G1 is essential for maintenance of nuclear integrity. Introduction of AKAP149-derived RVXF peptides into G1-phase nuclei in vitro or in vivo causes dissociation of PP1 from the AKAP. As a result, both A- and B-type lamins become phosphorylated and solubilize into the nuclear interior. Only B-type lamins are shown. The cells arrest with a G1 morphology for several hours and eventually undergo apoptosis.*

6.3 *AKAP149 is a B-type lamin-specifying subunit of PP1*

We recently found that AKAP149 is not only a novel anchoring protein for PP1 but also a substrate-dependent regulator of associated PP1 activity (Steen *et al.*, 2003). *In vitro*, AKAP149 stimulates the phosphatase activity of associated PP1 towards immunoprecipitated B-type lamins, whereas it inhibits PP1 activity towards an irrelevant substrate such as phosphorylase *a* (Steen *et al.*, 2003). This indicates that AKAP149 acts as a B-type lamin specifying subunit of PP1.

6.4 *AKAP149–PP1 association in G1 phase: control of nuclear integrity*

Selective displacement of PP1 from AKAP149 by intranuclear microinjection of RV*X*F-containing AKAP149 peptides during G1 phase in living cells leads to a G1 arrest (*Figure 5c*). This argues that the association of PP1 with AKAP149 is required for cell cycle progression into S phase. The G1 arrest phenotype is maintained for up to ~8 h before the cells enter apoptosis (Steen *et al.*, 2003) (*Figure 5c*). Furthermore, lamin distribution is dramatically affected in the G1-arrested cells. Both A- and B-type lamins are displaced from their perinuclear localization, become phosphorylated and are solubilized in the nucleoplasm (Steen *et al.*, 2003) (*Figure 5c*). These observations suggest that AKAP149-anchored PP1 continuously contributes to dephosphorylating B-type lamins during G1 and that this is necessary to maintain the lamins in a poly-merized form. Anchoring of PP1 to AKAP149 seems to be essential for this function, as overloading G1 cells containing the inhibitor peptide with excess active PP1 cannot rescue the phenotype. It is therefore likely that accurate anchoring of PP1 to specific subnuclear loci is important for the proper modulation of PP1 activity towards its nuclear G1-phase substrates.

6.5 *Anchoring of AMY-1 to S-AKAP84/AKAP149*

AMY-1 is ubiquitously expressed in human tissues, is mostly located in the cytoplasm, but translocates to the nucleus during S phase (Taira *et al.*, 1998). It has been found to competitively bind *in vitro* and *in vivo* to the RII binding region of AKAP149 or of AKAP95 (see above) (Furusawa *et al.*, 2001, 2002; Petersen *et al.*, 1994). The presence of AMY-1 in a ternary complex with S-AKAP84 and RII inhibits PKA activity in the complex by preventing binding of its catalytic subunit. In this way, AMY-1 may serve to modulate PKA activity (Furusawa *et al.*, 2002). AAT-1 is an AMY-1-binding protein specifically expressed in testis. It binds to the NH_2-terminal region of S-AKAP84/ AKAP149, within which the PP1 binding RV*X*F motif is located, suggesting that AAT-1 and PP1 might bind S-AKAP84/AKAP149 competitively. AAT-1 is a PKA substrate and weakly stimulates PKA activity when it is present in a quaternary complex with AMY-1, S-AKAP84/AKAP149 and PKA (Yukitake *et al.*, 2002). It would be inter-esting to determine whether nuclear envelope-associated AKAP149 also binds AMY-1 and/or AAT-1. One might speculate that AMY-1 preferentially binds AKAP149 or AKAP95 during various stages of S-phase in the nucleus.

7. Developmentally regulated nAKAP150

nAKAP150 is a 150-kDa AKAP immunologically unrelated to AKAP149 or AKAP150. nAKAP150 is abundant in the nuclear matrix of pre-cartilage nuclei but is essentially absent from nuclei of differentiated chondrocytes (Zhang *et al.*, 1996). Interestingly, cartilage differentiation is accompanied by a marked reduction in the level of nuclear PKA-RII. This suggests that PKA type II is imported into the mesenchymal cell nucleus prior to chondrogenesis and this process may rely on the activity of nAKAP150 (Zhang *et al.*, 1996). Thus, developmental regulation of nAKAP150 could represent a way of concentrating the PKA holoenzyme inside the nucleus during the very stages at which cAMP signalling elicits chondrocyte-specific gene expression (Kosher *et al.*, 1979).

8. A novel nuclear envelope–chromatin interaction: involvement in DNA replication

8.1 *Nuclear envelope–chromatin interactions*

Perhaps the most interesting feature of the nuclear envelope is its involvement in the mediation of several key nuclear functions by directly interacting with chromatin. The inner nuclear membrane (INM) harbours integral proteins that interact with the nuclear lamina and chromosomes. For example, the lamin B receptor (LBR) interacts directly with DNA (Worman *et al.*, 1988, 1990), histones H3/H4 (Polioudaki *et al.*, 2001) and heterochromatin protein HP1 (Ye and Worman, 1996; Ye *et al.*, 1997). Interestingly, Lys9 methylation of histone H3 tails provides a specific recognition site for proteins (including HP1) containing a chromodomain, which is generally required for heterochromatin formation and gene repression (Jenuwein and Allis, 2001). This suggests that the nuclear envelope might be involved in the formation and/or anchoring of peripheral heterochromatin. Another integral protein of the INM, lamina-associated polypeptide (LAP)2β, binds chromatin in a phosphorylation-dependent manner (Foisner and Gerace, 1993; Furukawa *et al.*, 1995). LAP2β contains a LEM domain shared with emerin and MAN1, two other INM integral proteins (Lin *et al.*, 2000). The LEM domain interacts with the DNA-bridging protein, barrier-to-autointegration factor (BAF) (Furukawa, 1999). At least seven LAP2 isoforms are produced from alternative mRNA splicing in mammals (Berger *et al.*, 1996). LAP2 proteins share an NH$_2$-terminal 187 amino acid domain which binds DNA (Cai *et al.*, 2001) and contains the LEM motif (Furukawa, 1999; Shumaker *et al.*, 2001) (*Figure 6*). LAP2β also interacts with the germ cell less (GCL) transcription regulator and mediates transcriptional repression alone or together with GCL (Nili *et al.*, 2001) (*Figure 6*). LAP2β was also shown to directly associate with a recently identified component of the chromatin (see below).

8.2 *HA95: a protein of the chromatin–nuclear matrix interface*

We and others have cloned a 95-kDa nuclear protein designated HA95/NAKAP95/HAP95 (Orstavik *et al.*, 2000; Seki *et al.*, 2000; Westberg *et al.*, 2000). HA95 displays partial homology to AKAP95, including the NH$_2$-terminal nuclear matrix binding domain (Akileswaran *et al.*, 2001; Yang *et al.*, 2001) but lacks a PKA-binding domain and thus is not an AKAP. Notably, in photobleaching experiments an HA95–GFP fusion protein appears as a stable protein (Martins *et al.*, 2000). HA95 co-fractionates with chromatin and the nuclear matrix (see above) (Martins *et al.*, 2000; Westberg *et al.*, 2000). HA95 associates with itself (Orstavik *et al.*, 2000) and interacts with RHA to enhance expression of a constitutive transport element involved in nuclear export of retroviral RNA (Westberg *et al.*, 2000; Yang *et al.*, 2001). *In vitro* nuclear breakdown and reassembly assays combined with antibody-blocking experiments have shown that HA95 is involved in nuclear envelope breakdown and chromatin condensation, whereas a role in nuclear membrane reassembly is unlikely (Martins *et al.*, 2000).

In vitro binding and overlay assays indicate that HA95 interacts with LAP2β via two distinct domains (Martins *et al.*, 2003). A first HA95-binding region of LAP2β lies within amino acids 137–242, whereas a second domain sufficient to bind HA95

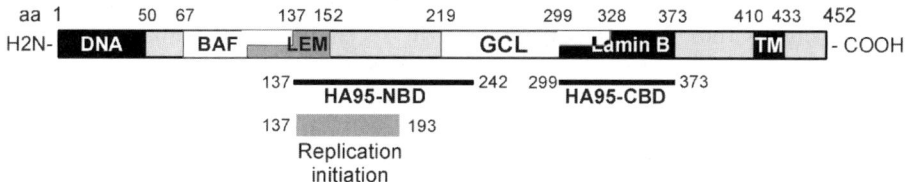

Figure 6. Known functional and binding domains of LAP2β. Identified functional regions of LAP2β (LEM and transmembrane, TM) and domains binding to indicated ligands are shown. Numbers refer to amino acid positions.

co-localizes with the lamin B-binding domain of LAP2β at residues 299–373. *In vitro* nuclear reassembly experiments argue that HA95-LAP2β interaction is not required for nuclear envelope formation (Martins *et al.*, 2003).

8.3 *Evidence for a role of the HA95–LAP2 interaction in initiation of DNA replication*

Initiation of DNA replication involves the assembly of pre-replication complexes (pre-RCs) at origins of replication in G1 phase (Bell and Dutta, 2002; Kelly and Brown, 2000). Pre-RCs include the origin recognition complex (ORC), the mini-chromosome maintenance (MCM) complex and the monomeric Cdc6 protein (Bell and Dutta, 2002). ORC recruits Cdc6 into the pre-RC in G1 and Cdc6 in turn promotes loading of MCM proteins on chromatin (Bell and Dutta, 2002). In human cells, Cdc6 levels are relatively stable during interphase (Saha *et al.*, 1998; Williams *et al.*, 1997). Nonetheless, a fraction of Cdc6 is exported out of the nucleus (Coleman *et al.*, 1996) while another remains associated with chromatin during S and G2 phases (Coverley *et al.*, 2000; Mendez and Stillman, 2000). Proteolysis of free Cdc6, not assembled into pre-RCs, has also been reported (Coverley *et al.*, 2000). Thus, after origin firing at the start of S phase, pre-RCs are dissociated, ensuring a single round of replication per cell cycle.

Recent evidence suggests that the interaction between HA95 and LAP2 is involved in the regulation of DNA replication at the G1/S phase transition. Disruption of the association of HA95 with the NH_2-terminal HA95-binding domain of LAP2β abolishes the initiation phase, but not elongation, of DNA replication in purified G1-phase nuclei incubated in an S-phase replication extract (Martins *et al.*, 2003). Moreover, inhibition of replication initiation correlates with proteasome-mediated proteolysis of the pre-RC component Cdc6 and rescue of Cdc6 degradation with proteasome inhibitors restores replication. Yet, the inhibitors do not rescue the LAP2β–HA95 interaction, therefore this association is dispensable for replication. It is likely that inhibitory LAP2 peptide causes degradation of a class of nuclear proteins, since p53 reacts similarly to Cdc6 (Martins *et al.*, 2003). This could result from a pleiotropic effect of the peptide, perhaps disturbing other proteins that interact through a similar domain.

The region of LAP2β implicated in DNA replication maps to residues 137–293 (Martins *et al.*, 2003), indicating that it may not so much be the interaction of HA95 with LAP2β *per se*, but rather the association of HA95 with intranuclear LAP2 isoforms (e.g., LAP2α) (Dechat *et al.*, 2000) that is important for the control of replication initiation. This would concur with the observation that replication initiates at many discrete sites within the nucleus, rather than specifically at the nuclear periphery. A current hypothesis is that interaction of HA95 with the NH_2-terminal domain of LAP2 proteins and with components of the pre-RC either brings components of the pre-RC to replication origins, maintains integrity of the pre-RC and/or protects pre-RC components from degradation.

9. Conclusions

AKAPs constitute an emerging class of adaptor molecules for protein kinases and phosphatases. This adaptor function appears to be restricted spatially and temporally to mediate accurate cellular responses. AKAPs can also regulate the activity of associated phosphatases in a substrate-dependent or -independent manner. Regulation of activity may also be modulated by intermolecular interactions within the AKAP complex. Importantly, AKAPs integrate signalling networks but also can perform crucial tasks during the cell cycle independently of their role in intracellular signalling.

Recent progress in our understanding of cell signalling has underlined the complexity of the intracellular signal networks. Many components of these pathways are continuously being identified, but it remains uncertain how many such pathways exit, how they interact with each other and how each is controlled (Smith and Scott, 2002). Several pathways may share common molecules, and specificity of signalling events is at least in part achieved by tethering components to discrete sites within the cytoplasm and the nucleus, preferably in the vicinity of substrates. This tethering may be cell cycle dependent, as exemplified in the G1-phase-specific association of PP1 with nuclear envelope-bound AKAP149. The growing use of techniques such as mass spectrometry, peptide arrays, genome two-hybrid screens and four-dimensional live imaging analyses are expected to allow the discovery of many more anchors for signalling networks.

Acknowledgements

Our work is supported by the Norwegian Cancer Society, the Research Council of Norway, the European Union and the Human Frontiers Science Program Organization.

References

Akileswaran, L., Taraska, J.W., Sayer, J.A., Gettemy, J.M. and Coghlan, V.M. (2001) A-kinase-anchoring protein AKAP95 is targeted to the nuclear matrix and associates with p68 RNA helicase. *J. Biol. Chem.* 276: 17448–17454.

Bell, S.P. and Dutta, A. (2002) DNA replication in eukaryotic cells. *Annu. Rev. Biochem.* 71: 333–374.

Berger, R., Theodor, L., Shoham, J., Gokkel, E., Brok-Simoni, F., Avraham, K.B., Copeland, N. G., Jenkins, N.A., Rechavi, G. and Simon, A.J. (1996) The characterization and localization of the mouse thymopoietin/lamina-associated polypeptide 2 gene and its alternatively spliced products. *Genome Res.* 6: 361–370.

Beullens, M., Van Eynde, A., Vulsteke, V., Connor, J., Shenolikar, S., Stalmans, W. and Bollen, M. (1999) Molecular determinants of nuclear protein phosphatase-1 regulation by NIPP-1. *J. Biol. Chem.* 274: 14053–14061.

Bollen, M. (2001) Combinatorial control of protein phosphatase-1. *Trends Biochem. Sci.* 26: 426–431.

Bollen, M. and Beullens, M. (2002) Signalling by protein phosphatases in the nucleus. *Trends Cell Biol.* 12: 138–145.

Bomar, J., Moreira, P., Balise, J.J. and Collas, P. (2002) Differential regulation of maternal and paternal chromosome condensation in mitotic zygotes. *J. Cell Sci.* 115: 2931–2940.

Bouniol-Baly, C., Nguyen, E., Besombes, D. and Debey, P. (1997) Dynamic organization of DNA replication in one-cell mouse embryos: relationship to transcriptional activation. *Exp. Cell Res.* 236: 201–211.

Cai, M., Huang, Y., Ghirlando, R., Wilson, K.L., Craigie, R. and Clore, G.M. (2001) Solution structure of the constant region of nuclear envelope protein LAP2 reveals two LEM-domain structures: one binds BAF and the other binds DNA. *EMBO J.* 20: 4399–4407.

Carlson, C.R., Witczak, O., Vossebein, L., Labbe, J.C., Skalhegg, B.S., Keryer, G., Herberg, F.W., Collas, P. and Tasken, K. (2001) CDK1-mediated phosphorylation of the RIIa regulatory subunit of PKA works as a molecular switch that promotes dissociation of RIIa from centrosomes at mitosis. *J. Cell Sci.* 114: 3243–3254.

Chen, Q., Lin, R.Y. and Rubin, C.S. (1997) Organelle-specific targeting of protein kinase AII (PKAII) Molecular and in situ characterization of murine A-kinase anchor proteins that recruit regulatory subunits of PKAII to the cytoplasmic surface of mitochondria. *J. Biol. Chem.* 272: 15247–15257.

Ciemerych, M.A. and Czolowska, R. (1993) Differential chromatin condensation of female and male pronuclei in mouse zygotes. *Mol. Reprod. Dev.* 34: 73–80.

Coghlan, V.M., Langeberg, L.K., Fernandez, A., Lamb, N.J. and Scott, J.D. (1994) Cloning and characterization of AKAP95, a nuclear protein that associates with the regulatory subunit of type II cAMP-dependent protein kinase. *J. Biol. Chem.* 269: 7658–7665.

Cohen, P.T. (2002) Protein phosphatase 1 – targeted in many directions. *J. Cell Sci.* 115: 241–256.

Coleman, T.R., Carpenter, P.B. and Dunphy, W.G. (1996) The *Xenopus* Cdc6 protein is essential for the initiation of a single round of DNA replication in cell-free extracts. *Cell* 87: 53–63.

Collas, P., Le Guellec, K. and Tasken, K. (1999) The A-kinase anchoring protein, AKAP95, is a multivalent protein with a key role in chromatin condensation at mitosis. *J. Cell Biol.* 147: 1167–1180.

Colledge, M. and Scott, J.D. (1999) AKAPs: from structure to function. *Trends. Cell Biol.* 9: 216–221.

Coverley, D., Pelizon, C., Trewick, S. and Laskey, R.A. (2000) Chromatin-bound Cdc6 persists in S and G2 phases in human cells, while soluble Cdc6 is destroyed in a cyclin A-cdk2 dependent process. *J. Cell Sci.* 113: 1929–1938.

Davare, M.A., Horne, M.C. and Hell, J.W. (2000) Protein phosphatase 2A is associated with class C L-type calcium channels (Cav1.2) and antagonizes channel phosphorylation by cAMP-dependent protein kinase. *J. Biol. Chem.* 275: 39710–39717.

Dechat, T., Vlcek, S. and Foisner, R. (2000) Review: Lamina-associated polypeptide 2 isoforms and related proteins in cell cycle-dependent nuclear structure dynamics. *J. Struct. Biol.* 129: 335–345.

Diviani, D., Langeberg, L.K., Doxsey, S.J. and Scott, J. D. (2000) Pericentrin anchors protein kinase A at the centrosome through a newly identified RII-binding domain. *Curr. Biol.* 10: 417–420.

Dodge, K.L., Khouangsathiene, S., Kapiloff, M.S., Mouton, R., Hill, E.V., Houslay, M.D., Langeberg, L.K. and Scott, J.D. (2001) mAKAP assembles a protein kinase A/PDE4 phosphodiesterase cAMP signalling module. *EMBO J.* 20: 1921–1930.

Donahue, R.P. (1972) Fertilization of the mouse oocyte: sequence and timing of nuclear progression to the two-cell stage. *J. Exp. Zool.* 180: 305–318.

Edwards, A.S. and Scott, J.D. (2000) A-kinase anchoring proteins: protein kinase A and beyond. *Curr. Opin. Cell Biol.* 12: 217–221.

Eide, T., Carlson, C., Tasken, K.A., Hirano, T., Tasken, K. and Collas, P. (2002) Distinct but overlapping domains of AKAP95 are implicated in chromosome condensation and condensin targeting. *EMBO Rep.* 3: 426–432.

Eide, T., Coghlan, V., Orstavik, S., *et al.* (1998) Molecular cloning, chromosomal localization, and cell cycle-dependent subcellular distribution of the A-kinase anchoring protein, AKAP95. *Exp. Cell Res.* **238**: 305–316.

Endoh, H., Maruyama, K., Masuhiro, Y., Kobayashi, Y., Goto, M., Tai, H., Yanagisawa, J., Metzger, D., Hashimoto, S. and Kato, S. (1999) Purification and identification of p68 RNA helicase acting as a transcriptional coactivator specific for the activation function 1 of human estrogen receptor alpha. *Mol. Cell Biol.* **19**: 5363–5372.

Feliciello, A., Rubin, C.S., Avvedimento, E.V. and Gottesman, M.E. (1998) Expression of a kinase anchor protein 121 is regulated by hormones in thyroid and testicular germ cells. *J. Biol. Chem.* **273**: 23361–23366.

Feschenko, M.S., Stevenson, E., Nairn, A.C. and Sweadner, K.J. (2002) A novel cAMP-stimulated pathway in protein phosphatase 2A activation. *J. Pharmacol. Exp. Ther.* **302**: 111–118.

Fill, M. and Copello, J. A. (2002) Ryanodine receptor calcium release channels. *Physiol Rev.* **82**: 893–922.

Foisner, R. and Gerace, L. (1993) Integral membrane proteins of the nuclear envelope interact with lamins and chromosomes, and binding is modulated by mitotic phosphorylation. *Cell* **73**: 1267–1279.

Furukawa, K. (1999) LAP2 binding protein 1 (L2BP1/BAF) is a candidate mediator of LAP2–chromatin interaction. *J. Cell Sci.* **112**: 2485–2492.

Furukawa, K., Panté, N., Aebi, U. and Gerace, L. (1995) Cloning of a cDNA for lamina-associated polypeptide 2 (LAP2) and identification of regions that specify targeting to the nuclear envelope. *EMBO J.* **14**: 1626–1636.

Furusawa, M., Taira, T., Iguchi-Ariga, S.M. and Ariga, H. (2002) AMY-1 interacts with S-AKAP84 and AKAP95 in the cytoplasm and the nucleus, respectively, and inhibits cAMP-dependent protein kinase activity by preventing binding of its catalytic subunit to A-kinase-anchoring protein (AKAP) complex. *J. Biol. Chem.* **277**: 50885–50892.

Furusawa, M., Ohnishi, T., Taira, T., Iguchi-Ariga, S.M. and Ariga, H. (2001) AMY-1, a c-Myc-binding protein, is localized in the mitochondria of sperm by association with S-AKAP84, an anchor protein of cAMP-dependent protein kinase. *J. Biol. Chem.* **276**: 36647–36651.

Ginsberg, M.D., Feliciello, A., Jones, J.K., Avvedimento, E.V. and Gottesman, M.E. (2003) PKA-dependent binding of mRNA to the mitochondrial AKAP121 protein. *J. Mol. Biol.* **327**: 885–897.

Hirano, T. (2000) Chromosome cohesion, condensation, and separation. *Annu. Rev. Biochem.* **69**: 115–144.

Hirano, T., Kobayashi, R. and Hirano, M. (1997) Condensins, chromosome condensation protein complexes containing XCAP- C, XCAP-E and a *Xenopus* homolog of the *Drosophila* Barren protein. *Cell* **89**: 511–521.

Houslay, M.D. and Adams, D.R. (2003) PDE4 cAMP phosphodiesterases: modular enzymes that orchestrate signalling cross-talk, desensitization and compartmentalization. *Biochem. J.* **370**: 1–18.

Huang, L.J., Durick, K., Weiner, J.A., Chun, J. and Taylor, S.S. (1997) Identification of a novel protein kinase A anchoring protein that binds both type I and type II regulatory subunits. *J. Biol. Chem.* **272**: 8057–8064.

Huang, L.J., Wang, L., Ma, Y., Durick, K., Perkins, G., Deerinck, T.J., Ellisman, M.H. and Taylor, S.S. (1999) NH_2-terminal targeting motifs direct dual specificity A-kinase-anchoring protein 1 (D-AKAP1) to either mitochondria or endoplasmic reticulum. *J. Cell Biol.* **145**: 951–959.

Jenuwein, T. and Allis, C.D. (2001) Translating the histone code. *Science* **293**: 1074–1080.

Kapiloff, M.S., Schillace, R.V., Westphal, A.M. and Scott, J.D. (1999) mAKAP: an A-kinase anchoring protein targeted to the nuclear membrane of differentiated myocytes. *J. Cell Sci.* **112**: 2725–2736.

Kapiloff, M.S., Jackson, N. and Airhart, N. (2001) mAKAP and the ryanodine receptor are part of a multi-component signalling complex on the cardiomyocyte nuclear envelope. *J. Cell Sci.* 114: 3167–3176.

Kelly, T.J. and Brown, G.W. (2000) Regulation of chromosome replication. *Annu. Rev. Biochem.* 69: 829–880.

Kimura, K. and Hirano, T. (1997) ATP-dependent positive supercoiling of DNA by 13S condensin: a biochemical implication for chromosome condensation. *Cell* 90: 625–634.

Kimura, K., Rybenkov, V.V., Crisona, N.J., Hirano, T. and Cozzarelli, N.R. (1999) 13S condensin actively reconfigures DNA by introducing global positive writhe: implications for chromosome condensation. *Cell* 98: 239–248.

Kimura, K., Cuvier, O. and Hirano, T. (2001) Chromosome condensation by a human condensin complex in *Xenopus* egg extracts. *J. Biol. Chem.* 276: 5417–5420.

Kosher, R.A., Savage, M.P. and Chan, S.C. (1979) Cyclic AMP derivatives stimulate the chondrogenic differentiation of the mesoderm subjacent to the apical ectodermal ridge of the chick limb bud. *J. Exp. Zool.* 209: 221–227.

Landsverk, H.B., Carlson, C.R., Steen, R.L., Vossebein, L., Herberg, F.W., Tasken, K. and Collas, P. (2001) Regulation of anchoring of the RIIα regulatory subunit of PKA to AKAP95 by threonine phosphorylation of RIIα: implications for chromosome dynamics at mitosis. *J. Cell Sci.* 114: 3255–3264.

Lin, F., Blake, D.L., Callebaut, I., Skerjanc, I.S., Holmer, L., McBurney, M.W., Paulin-Levasseur, M. and Worman, H.J. (2000) MAN1, an inner nuclear membrane protein that shares the LEM domain with lamina-associated polypeptide 2 and emerin. *J. Biol. Chem.* 275: 4840–4847.

Lin, R.Y., Moss, S.B. and Rubin, C.S. (1995) Characterization of S-AKAP84, a novel developmentally regulated A-kinase anchor protein of male germ cells. *J. Biol. Chem.* 270: 27804–24811.

Liu, J. and Brautigan, D.L. (2000) Glycogen synthase association with the striated muscle glycogen-targeting subunit of protein phosphatase-1. Synthase activation involves scaffolding regulated by beta-adrenergic signalling. *J. Biol. Chem.* 275: 26074–26081.

Liu, J., Ben-Shahar, T.R., Riemer, D., Treinin, M., Spann, P., Weber, K., Fire, A. and Gruenbaum, Y. (2000) Essential roles for *Caenorhabditis elegans* lamin gene in nuclear organization, cell cycle progression, and spatial organization of nuclear pore complexes. *Mol. Biol. Cell* 11: 3937–3947.

Martins, S., Eikvar, S., Furukawa, K. and Collas, P. (2003) HA95 and LAP2b mediate a novel chromatin-nuclear envelope interaction implicated in initiation of DNA replication. *J. Cell. Biol.* 160: 177–188.

Martins, S.B., Eide, T., Steen, R.L., Jahnsen, T., Skålhegg, B.S. and Collas, P. (2000) HA95 is a protein of the chromatin and nuclear matrix regulating nuclear envelope dynamics. *J. Cell Sci.* 113: 3703–3713.

Marx, S.O., Reiken, S., Hisamatsu, Y., Jayaraman, T., Burkhoff, D., Rosemblit, N. and Marks, A.R. (2000) PKA phosphorylation dissociates FKBP12.6 from the calcium release channel (ryanodine receptor): defective regulation in failing hearts. *Cell* 101: 365–376.

Mayer, W., Smith, A., Fundele, R. and Haaf, T. (2000) Spatial separation of parental genomes in preimplantation mouse embryos. *J. Cell Biol.* 148: 629–634.

McAvoy, T., Allen, P.B., Obaishi, H., Nakanishi, H., Takai, Y., Greengard, P., Nairn, A.C. and Hemmings, H.C., Jr. (1999) Regulation of neurabin I interaction with protein phosphatase 1 by phosphorylation. *Biochemistry* 38: 12943–12949.

Mendez, J. and Stillman, B. (2000) Chromatin association of human origin recognition complex, cdc6, and minichromosome maintenance proteins during the cell cycle: assembly of prereplication complexes in late mitosis. *Mol. Cell Biol.* 20: 8602–8612.

Nakajima, T., Uchida, C., Anderson, S.F., Lee, C.G., Hurwitz, J., Parvin, J.D. and Montminy, M. (1997) RNA helicase A mediates association of CBP with RNA polymerase II. *Cell* 90: 1107–1112.

Nili, E., Cojocaru, G.S., Kalma, Y., *et al.* (2001) Nuclear membrane protein LAP2β mediates transcriptional repression alone and together with its binding partner GCL (germ-cell-less) *J. Cell Sci.* 114: 3297–3307.

Orstavik, S., Eide, T., Collas, P., Han, I.O., Tasken, K., Kieff, E., Jahnsen, T. and Skålhegg, B.S. (2000) Identification, cloning and characterization of a novel nuclear protein, HA95, homologous to A-kinase anchoring protein 95. *Biol. Cell* 92: 27–37.

Pawson, T. and Scott, J.D. (1997) Signalling through scaffold, anchoring, and adaptor proteins. *Science* 278: 2075–2080.

Petersen, H.V., Serup, P., Leonard, J., Michelsen, B.K. and Madsen, O.D. (1994) Transcriptional regulation of the human insulin gene is dependent on the homeodomain protein STF1/IPF1 acting through the CT boxes. *Proc. Natl Acad. Sci. USA* 91: 10465–10469.

Polioudaki, H., Kourmouli, N., Drosou, V., Bakou, A., Theodoropoulos, P.A., Singh, P.B., Giannakouros, T. and Georgatos, S.D. (2001) Histones H3/H4 form a tight complex with the inner nuclear membrane protein LBR and heterochromatin protein 1. *EMBO Rep.* 2: 920–925.

Riabowol, K.T., Fink, J.S., Gilman, M.Z., Walsh, D.A., Goodman, R.H. and Feramisco, J.R. (1988) The catalytic subunit of cAMP-dependent protein kinase induces expression of genes containing cAMP-responsive enhancer elements. *Nature* 336: 83–86.

Saha, P., Chen, J., Thome, K.C., Lawlis, S.J., Hou, Z.H., Hendricks, M., Parvin, J.D. and Dutta, A. (1998) Human CDC6/Cdc18 associates with Orc1 and cyclin-cdk and is selectively eliminated from the nucleus at the onset of S phase. *Mol. Cell Biol.* 18: 2758–2767.

Schmiesing, J.A., Ball, A.R.J., Gregson, H.C., Alderton, J.M., Zhou, S. and Yokomori, K. (1998) Identification of two distinct human SMC protein complexes involved in mitotic chromosome dynamics. *Proc. Natl Acad. Sci USA* 95: 12906–12911.

Schmiesing, J.A., Gregson, H.C., Zhou, S. and Yokomori, K. (2000) A human condensin complex containing hCAP-C-hCAP-E and CNAP1, a homolog of *Xenopus* XCAP-D2, colocalizes with phosphorylated histone H3 during the early stage of mitotic chromosome condensation. *Mol. Cell Biol.* 20: 6996–7006.

Seki, N., Ueki, N., Yano, K., Saito, T., Masuho, Y. and Muramatsu, M. (2000) cDNA cloning of a novel human gene NAKAP95, neighbor of A-kinase anchoring protein 95 (AKAP95) on chromosome 19p13.11-p13.12 region. *J. Hum. Genet.* 45: 31–37.

Shumaker, D.K., Lee, K.K., Tanhehco, Y.C., Craigie, R. and Wilson, K.L. (2001) LAP2 binds to BAF.DNA complexes: requirement for the LEM domain and modulation by variable regions. *EMBO J.* 20: 1754–1764.

Smith, F.D. and Scott, J.D. (2002) Signalling complexes: junctions on the intracellular information super highway. *Curr. Biol.* 12: R32–R40.

Steen, R.L., Beullens, M., Landsverk, H.B., Bollen, M. and Collas, P. (2003) AKAP149 is a novel PP1 specifier required to maintain nuclear envelope integrity in G1 phase. *J. Cell. Sci.* 116: 2237–2246.

Steen, R.L., Cubizolles, F., Le Guellec, K. and Collas, P. (2000a) A-kinase anchoring protein (AKAP)95 recruits human chromosome-associated protein (hCAP)-D2/Eg7 for chromosome condensation in mitotic extract. *J. Cell Biol.* 149: 531–536.

Steen, R.L. and Collas, P. (2001) Mistargeting of B-type lamins at the end of mitosis: implications on cell survival and regulation of lamins A/C expression. *J. Cell Biol.* 153: 621–626.

Steen, R.L., Martins, S.B., Tasken, K. and Collas, P. (2000b) Recruitment of protein phosphatase 1 to the nuclear envelope by A-kinase anchoring protein AKAP149 is a prerequisite for nuclear lamina assembly. *J. Cell Biol.* 150: 1251–1262.

Taira, T., Maeda, J., Onishi, T., Kitaura, H., Yoshida, S., Kato, H., Ikeda, M., Tamai, K., Iguchi-Ariga, S.M. and Ariga, H. (1998) AMY-1, a novel C-MYC binding protein that stimulates transcription activity of C-MYC. *Genes Cells* 3: 549–565.

Tasken, K., Skalhegg, B.S., Tasken, K.A., *et al.* (1997) Structure, function, and regulation of human cAMP-dependent protein kinases. *Adv. Second Messenger Phosphoprotein Res.* 31: 191–204.

Tasken, K.A., Collas, P., Kemmner, W.A., Witczak, O., Conti, M. and Tasken, K. (2001) Phosphodiesterase 4D and protein kinase a type II constitute a signalling unit in the centrosomal area. *J. Biol. Chem.* **276**: 21999–22002.

Thompson, L.J., Bollen, M. and Fields, A.P. (1997) Identification of protein phosphatase 1 as a mitotic lamin phosphatase. *J. Biol. Chem.* **272**: 29693–29697.

Trendelenburg, G., Hummel, M., Riecken, E.O. and Hanski, C. (1996) Molecular characterization of AKAP149, a novel A kinase anchor protein with a KH domain. *Biochem. Biophys. Res. Commun.* **225**: 313–319.

Vijayaraghavan, S., Liberty, G.A., Mohan, J., Winfrey, V.P., Olson, G.E. and Carr, D.W. (1999) Isolation and molecular characterization of AKAP110, a novel, sperm-specific protein kinase A-anchoring protein. *Mol. Endocrinol.* **13**: 705–717.

Westberg, C., Yang, J.-P., Tang, H., Reddy, T.R. and Wong-Staal, F. (2000) A novel shuttle protein binds to RNA helicase A and activates the retroviral constitutive transport element. *J. Biol. Chem.* **275**: 21396–21401.

Williams, R.S., Shohet, R.V. and Stillman, B. (1997) A human protein related to yeast Cdc6p. *Proc. Natl Acad. Sci. USA* **94**: 142–147.

Worman, H.J., Yuan, J., Blobel, G. and Georgatos, S.D. (1988) A lamin B receptor in the nuclear envelope. *Proc. Natl Acad. Sci. USA* **85**: 8531–8534.

Worman, H.J., Evans, C. and Blobel, G. (1990) The lamin B receptor of the nuclear envelope inner membrane: a polytopic protein with eight potential transmembrane domains. *J. Cell Biol.* **111**: 1535–1542.

Yang, J.P., Tang, H., Reddy, T.R. and Wong-Staal, F. (2001) Mapping the functional domains of HAP95, a protein that binds RNA helicase A and activates the constitutive transport element of type D retroviruses. *J. Biol. Chem.* **276**: 30694–30700.

Ye, Q. and Worman, H.J. (1996) Interaction between and integral protein of the nuclear envelope inner membrane and human chromodomain proteins homologous to *Drosophila* HP1. *J. Biol. Chem.* **271**: 14653–14656.

Ye, Q., Callebaut, I., Pezhman, A., Courvalin, J.-C. and Worman, H.J. (1997) Domain-specific interactions of human HP1-type chromodomain proteins and inner nuclear membrane protein LBR. *J. Biol. Chem.* **272**: 14983–14989.

Yukitake, H., Furusawa, M., Taira, T., Iguchi-Ariga, S.M. and Ariga, H. (2002) AAT-1, a novel testis-specific AMY-1-binding protein, forms a quaternary complex with AMY-1, A-kinase anchor protein 84, and a regulatory subunit of cAMP-dependent protein kinase and is phosphorylated by its kinase. *J. Biol. Chem.* **277**: 45480–45492.

Zhang, Q., Carr, D.W., Lerea, K.M., Scott, J.D. and Newman, S.A. (1996) Nuclear localization of type II cAMP-dependent protein kinase during limb cartilage differentiation is associated with a novel developmentally regulated A-kinase anchoring protein. *Dev. Biol.* **176**: 51–61.

Zimmer, H.G. (1997) Catecholamine-induced cardiac hypertrophy: significance of proto-oncogene expression. *J. Mol. Med.* **75**: 849–859.

Plate 1. Schematic representation of the vertebrate nuclear pore complex and direct visualization of the cytoplasmic and nucleoplasmic faces of the vertebrate by scanning electron microscopy. The NPC has eight-fold rotational symmetry and is asymmetric about the plane of the nuclear envelope (N.E.; partially shown to the rear of the NPC). (See Chapter 6.)

Plate 2. Vertebrate nucleoporins and their shared sequence motifs. The distribution and localization of FG (red), FxFG (blue) and GLFG (green) amino acid motifs are illustrated on proportionally-sized representations of hCG1, Nup45, Nup50, Nup54, Nup58, Nup62, Nup153, Nup98, POM121, Nup214/CAN, and Nup358/RanBP2. Only the repeat motif-containing portion of Nup358/RanBP2 is included. (See Chapter 6.)

Plate 3. 'Classical' nuclear import model. (a) Import factors carry cargo to and through the NPC and are then dissociated when RanGTP binds importin β. (b) Importin β is then exported and released from Ran when it hydrolyses its GTP. (c) Importin α is exported by CAS and RanGTP and likewise is released when Ran hydrolyses its GTP. (d) GDP bound Ran is returned to the nucleus by NTF2 and RanGEF regenerates RanGTP. (See Chapter 7.)

Plate 4. Transcription centres in human cells. Patterns of transcription in human cells can be visualized by either light (left) or electron (centre) microscopy. Light microscopy shows the global distribution of the transcription centres and electron microscopy a more detailed image of the anatomy of a single transcription factory. For the first image, Hela cells were permeabilized, nascent transcripts extended in Br-UTP and cryosections (~100 nm) prepared. Br-RNA was indirectly immunolabelled with Cy3 and nucleic acids were counterstained with TOTO-3. Red and far-red images were collected using a confocal microscope. The transcription sites are bright (shown green) foci. Note the distribution of very bright nucleolar foci in the centre and many smaller sites – RNA polymerase II/III transcription centres – throughout the nucleoplasm. TOTO-stained heterochromatin is clearly visible around the nucleolus and along the nuclear periphery. Image reproduced from Pombo et al. EMBO J. 18: 2241–2253 (1999). The second image is an electron microscopy section that has been prepared in LR-white, treated using the EDTA regressive staining technique and immunolabelled using antibodies to RNA polymerase II and Biotin-RNA (labelled in vitro using Biotin-CTP). Note the relative distributions of RNA polymerase II (large, 10 nm, gold particles), Biotin-RNA (small, 5 nm, gold particles), chromatin clouds (pale areas) and RNA rich inter-chromatin channels (grey areas). The transcription centre of ~100 nm in this example corresponds to one nucleoplasmic foci in the light microscopy image. Reproduced from Current Opinion in Cell Biology, Vol 15, Jackson, The anatomy of transcription sites, pp 311–317 (2003), with permission from Elsevier. The cartoon (right) shows one interpretation of distribution of components in a transcription centre. The chromatin clouds are depicted by extended loops (dark blue) that are attached to the transcription factory through interactions within promoters and the transcription complexes (green). As transcription proceeds, a gene moves over the surface of the factory (movement is depicted by the arrowheads) and the nascent RNA is extruded into an RNA-rich region of the compartment. Processing or maturation of the mRNP takes place before the mature product leaves the transcription site. This cartoon emphasizes the three distinct zones of the transcription factory and the adjacent but spatially compartmentalized chromatin-rich regions seen by electron microscopy. (See Chapter 8.)

Plate 5. (a) Model for bipartite geminivirus movement based on studies of SqLCV. NSP binds replicated viral genomes (ssDNA) and shuttles these between the nucleus and cytoplasm. MP traps NSP–genome complexes in the cytoplasm and moves these along ER-derived tubules to and across the cells wall. In adjacent uninfected cells, NSP–genome complexes are released and NSP targets the genome to and across the nuclear pore to start a new round of replication and infection. (b) Co-expression of MP and NSP NES mutants in tobacco protoplasts. (left) NSP$^{L189A/L190A/L193A}$ (leu at positions 189, 190 and 193 changed to ala) remains in the nucleus when co-expressed with MP because the NES is not functional. Insertion of the NES from Xenopus laevis TFIIIa (NSP$^{L189A/L190A/L193A/TFIIIa-NES}$) allows the mutated NSP to exit from the nucleus and interact with MP, as evident from its relocalization to the cortical cytoplasm. (right) Table shows infectivity when these same mutations are inserted into SqLCV. (See Chapter 9.)

Plate 6. *AtNSI is a nuclear protein. Indirect immune fluorescence staining of methacrylate-embedded sections of symptomatic systemic leaves from* N. benthamiana *plants infected with PVX-AtNSI. Sections were incubated with pre-immune or AtNSI antisera, as indicated, followed by AlexaFluor 488-conjugated goat anti-rabbit antibody and imaged by confocal microscopy. Each immune fluorescence image is superimposed on the Nomarski image. Bar = 20 μm. Reproduced from McGarry et al. (2003),* The Plant Cell, *© ASPB Press, with permission. (See Chapter 9.)*

Plate 7. *(a) Comparison of lamin-B1-GFP fluorescence (green) and immunostaining with lamin A/C antibody (red) in a transfected CHO-K1 cell. Note that the GFP fluorescence in intranuclear channels is readily visible, while immunostaining in the same cell shows barely any nucleoplasmic structures (arrow). Similar observations can be made with antibodies recognizing the transfected protein. A merged image shows colocalization at the nuclear lamina and preferentially GFP signal in the nucleoplasmic areas (right). (b) Nucleus of an apoptotic cell treated with caspase 6 inhibitors. Note that while the nuclear lamina is still mainly intact (GFP-lamin C, green), the nuclear DNA is largely condensed (propidium iodide, red). (c) Apoptotic lamin B1-GFP transfected cell. Note that remnants of the nuclear lamina (green) are still present in this late apoptotic nucleus with condensed DNA (propidium iodide, red). (See Chapter 10.)*

Plate 8. *(a–d) Ethidium bromide-stained stably expressing LBR-GFP₅ tobacco leaf epidermal cells. (a) Ethidium bromide staining of chromatin; (b) LBR-GFP₅ Fluorescence localized to the nuclear rim in leaf epidermal cells (arrow). Inset: magnified nucleus, adjacent auto-fluorescent cell wall (arrow); (c) merged image of (a) and (b); (d) view of the cortex of stably transformed leaf epidermal cells showing no ER Fluorescence from LBR-GFP₅. (e) Transiently expressing sGFP₅CX tobacco leaf epidermal cell showing bright NE and ER fluorescence; (f) view of the cortex of the leaf epidermal cell shown in (e), sGFP₅CX shows intense fluorescent labelling of cortical ER. (g) Stably expressing spGFP₅-HDEL tobacco leaf epidermal cell, bright NE and ER fluorescence; (h) view of the cortex of the leaf epidermal cell shown in (g), spGFP₅-HDEL high ER fluorescence. (i) Interphase tobacco BY-2 cell stably expressing LBR-GFP₅, 14 d in suspension. (j, k) Tobacco BY-2 cell stably co-expressing LBR-GFP₅ and spYFP-HDEL, 2 d in suspension. (j) LBR-GFP₅ labelling of cortical ER. (k) The same cell as in (j), imaged for the expression of YFP, spYFP-HDEL heavily labels the cortical ER. (l) Ethidium bromide-stained metaphase tobacco BY-2 cell stably expressing LBR-GFP₅. Arrow indicates bright punctate structure. Scale bar = 10 μm. Reproduced from Irons et al. (2003). The first 238 amino acids of the human lamina B receptor are targeted to the nuclear envelope in plants. Journal of Experimental Botany, Vol 54, with permission of the Society for Experimental Biology. (See Chapter 14.)*

Plate 9. *LBR cell. A tobacco BY-2 suspension culture cell stably expressing LBR-GFP5, in interphase. 10 μm scale bar. (See Chapter 14.)*

Plate 10. *Homologies between different spectrin repeat proteins in human, mouse and fly. Spectroplakins represent a single modular unit compared to the spectrins and contain unique GAS2 domains. Generally NUANCE has fewer and more widely distributed spectrin repeats compared to other proteins of this type and contains the unique klarsicht homology domain (KLS) within the transmembrane domain (TM). (See Chapter 16.)*

Plate 11. *Proposed linkages between Spectroplakins, components of the cytoskeleton and surface adhesion proteins. Spetroplakins can link microtubules to actin stress fibres via interactions between EF hand domains and calponin homology domains respectively, or they may link microtubules to the plasma membrane either via cortical actin or surface adhesion proteins. Finally, spectroplakins anchor adherin or septate junctions again via cortical actin. (See Chapter 16.)*

Plate 12. *Speculative model illustrating how NUANCE and nesprin 1might link the nuclear lamina to the actin cytoskeleton. The model suggests that NUANCE might penetrate nuclear pores in order to interact with the lamina, while at the same time may bind to the actin cytoskeleton through its calponin homology domain. (See Chapter 16.)*

Plate 13. *In vivo dynamics of SR45. (a) Movements of GFP-SR45 speckles in a leaf pavement cell. A time-lapse sequence of images of a GFP-SR45 nucleus taken at times indicated. Arrows indicate speckles appearing and disappearing from the plane of focus. Arrowheads show what appears to be the separation and change in shape of speckles. (b) Inhibition of transcription redistributes GFP-SR45 into large speckles. (Left) The GFP-SR45 Arabidopsis seedlings were treated with actinomycin-D (50 μg/ml) or α-amanitin (50 μg/ml) in DMSO for 8 h. Controls were treated with 1% DMSO. In each treatment nuclei of two root epidermal cells are shown. (Right) Number (grey bars) and size (open bars) of speckles in the nuclei of control, actinomycin-D- and α-amanitin-treated seedlings. Over 120 nuclei per treatment in 5–7 different roots were observed. Vertical bars are standard errors. (c) Effect of heat, cold and a protein kinase inhibitor on SR45 redistribution. Heat (42°C) induced redistribution of GFP-SR45 into irregularly shaped compartments, while cold (4°C) relocalizes GFP-SR45 mostly to nucleoplasmic pool. Treatment with staurosporine, an inhibitor of protein kinase, also caused redistribution of GFP-SR45 into irregularly shaped large speckles. The GFP-SR45 seedlings were placed in an incubator set to 4°C or 42°C or treated with 10 μM staurosporine. Controls were incubated at 22°C. Cells shown are root epidermal cells (Ali et al., 2003. Nuclear localization and in vivo dynamics of a plant-specific serine/argenine-rich protein. Plant Journal Vol 38, reprinted with permission from Blackwell Publishing.). (See Chapter 17.)*

Plate 14. *CAAX modifications of lamin proteins. (a) Schematic view of the structure of human lamin proteins. B-type lamins, which include lamin B1 and lamin B2, and A-type lamins, lamin A and splice variant lamin C, share a conserved domain structure; head (yellow), rod (green) and tail (pink) domain. Mitotic phosphorylation sites (P) flank the rod domain. A nuclear localization signal (NLS) in the tail domain contributes to a chromatin-binding region with residues immediately upstream (Stierle et al., 2003). The tail domain upstream of the polyglutamate region forms an immunoglobulin (Ig)-like domain. The CAAX motif is located at the carboxyl terminus of B-type lamins and lamin A. The polyglutamate tract 30 residues upstream of the CAAX motif may contribute to CAAX endoproteolysis in B-type lamins. The larger lamin A is cleaved upstream of the CAAX motif following CAAX modification and is therefore non-farnesylated in its mature form. (b) Biochemistry of CAAX modifications of the lamin proteins. The processing steps and resulting structures of the carboxyl terminus of CAAX-modified lamin proteins are shown. (c) One of the consequences of carboxymethylation of lamin B1 is the methylation-dependent organization of the nuclear lamina. Shown here is an ultrathin cryosection of HeLa cells labelled with a polyclonal goat anti-lamin B1 antibody, a monoclonal antibody specific for endoproteolysed and methylated lamin B1 (but not unprocessed lamin B1) and the nucleic acid stain, DAPI. CAAX-processed lamin B1 is seen to occupy subdomains of the nuclear lamina compared with total lamin B1. The DNA stain shows a characteristic rim of peripheral heterochromatin underlying the lamina. Scale bar = 10 μm. (See Chapter 19.)*

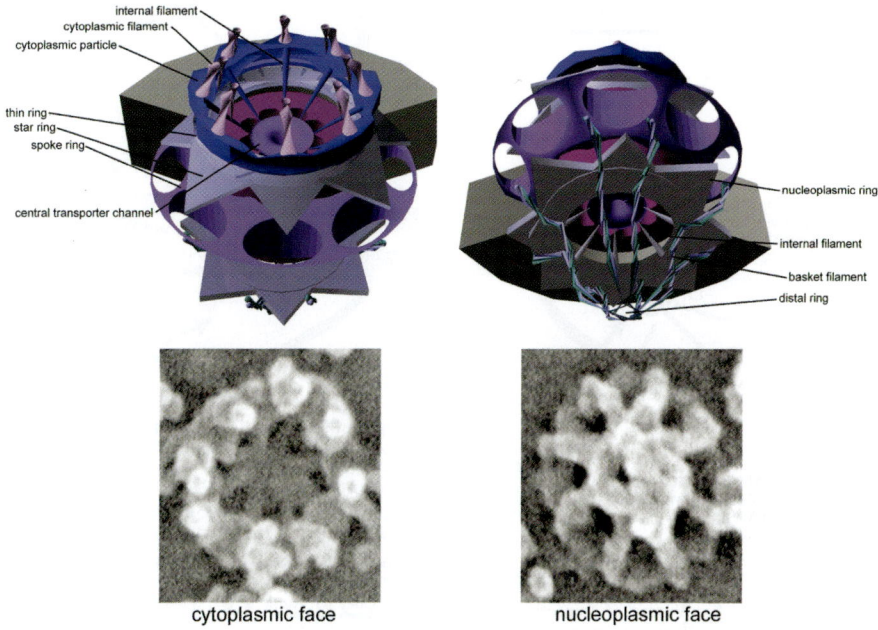

internal filament
cytoplasmic filament
cytoplasmic particle

thin ring
star ring
spoke ring

central transporter channel

nucleoplasmic ring

internal filament

basket filament

distal ring

cytoplasmic face

nucleoplasmic face

Plate 1

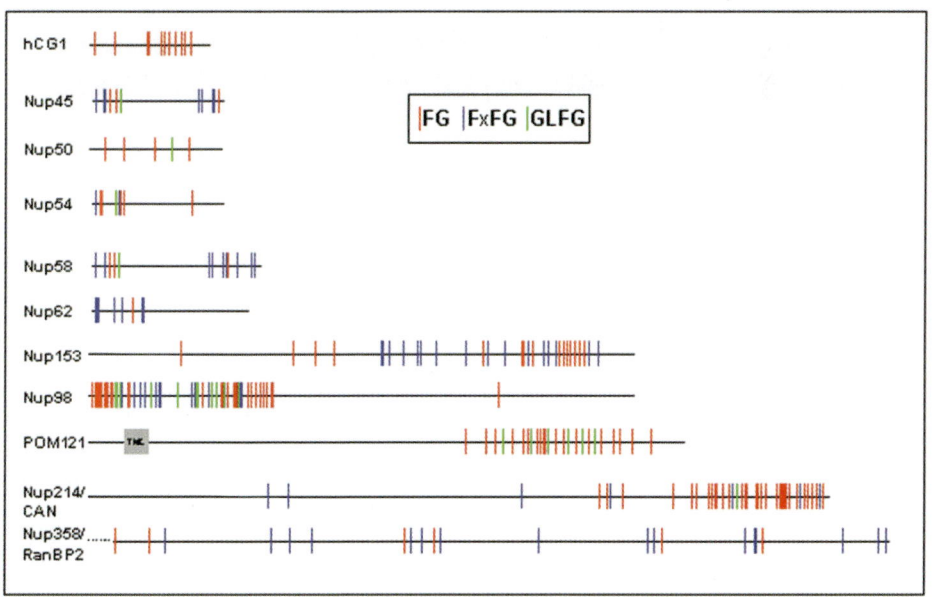

hCG1

Nup45

Nup50

Nup54

Nup58

Nup62

Nup153

Nup98

POM121

Nup214/
CAN

Nup358/
RanBP2

FG   FxFG   GLFG

TM

Plate 2

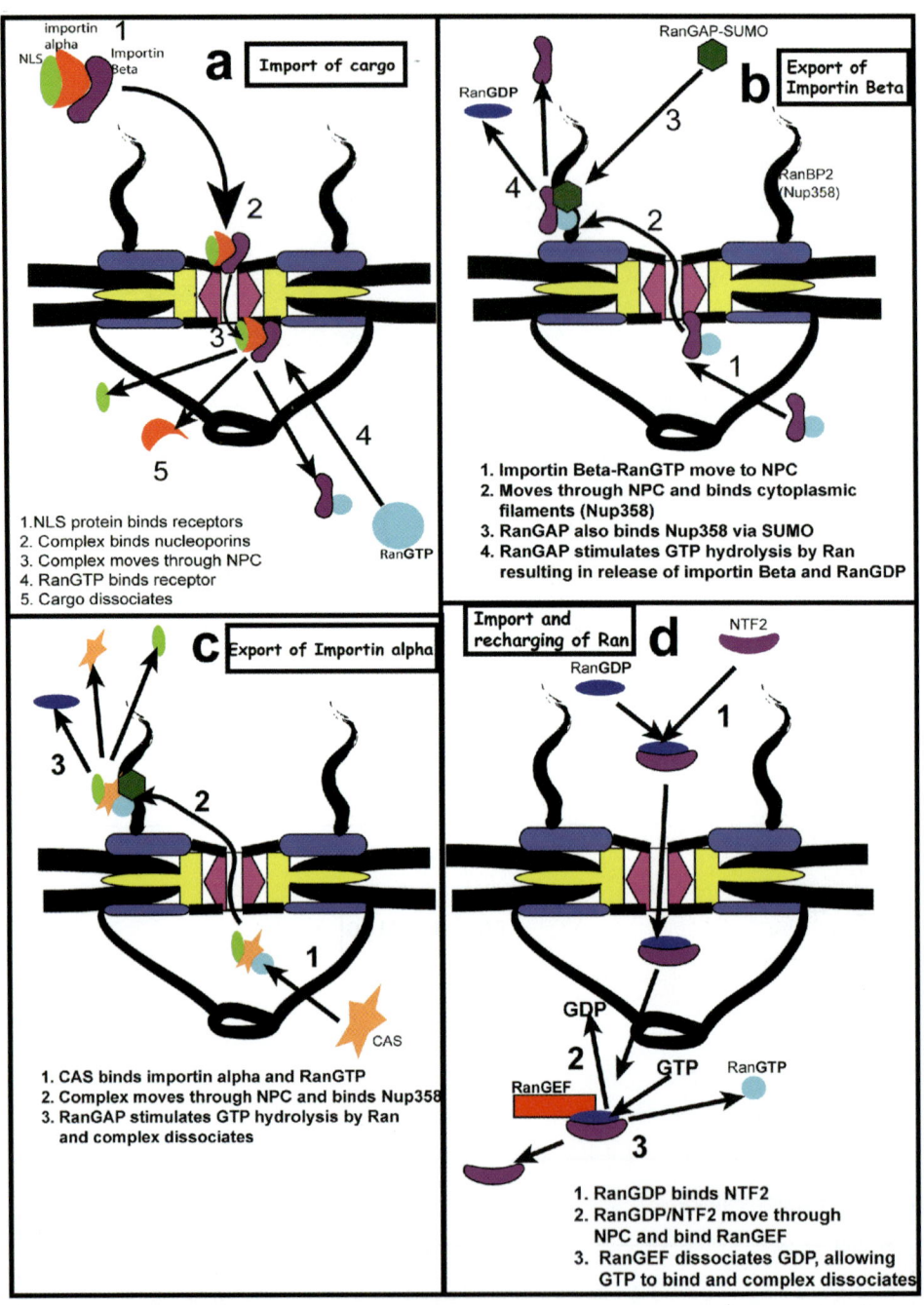

a Import of cargo

importin alpha
NLS
Importin Beta
1
2
3
5
4
RanGTP

1. NLS protein binds receptors
2. Complex binds nucleoporins
3. Complex moves through NPC
4. RanGTP binds receptor
5. Cargo dissociates

b Export of Importin Beta

RanGAP-SUMO
RanGDP
3
4
2
RanBP2 (Nup358)
1

1. Importin Beta-RanGTP move to NPC
2. Moves through NPC and binds cytoplasmic filaments (Nup358)
3. RanGAP also binds Nup358 via SUMO
4. RanGAP stimulates GTP hydrolysis by Ran resulting in release of importin Beta and RanGDP

c Export of Importin alpha

3
2
1
CAS

1. CAS binds importin alpha and RanGTP
2. Complex moves through NPC and binds Nup358
3. RanGAP stimulates GTP hydrolysis by Ran and complex dissociates

d Import and recharging of Ran

NTF2
RanGDP
1
GDP
2
GTP
RanGTP
RanGEF
3

1. RanGDP binds NTF2
2. RanGDP/NTF2 move through NPC and bind RanGEF
3. RanGEF dissociates GDP, allowing GTP to bind and complex dissociates

Plate 3

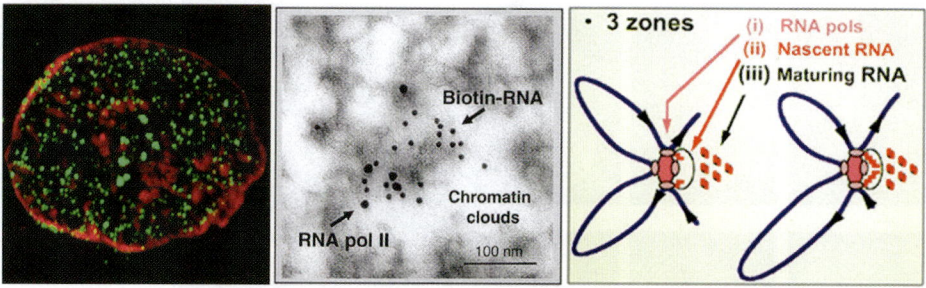

Plate 4

A

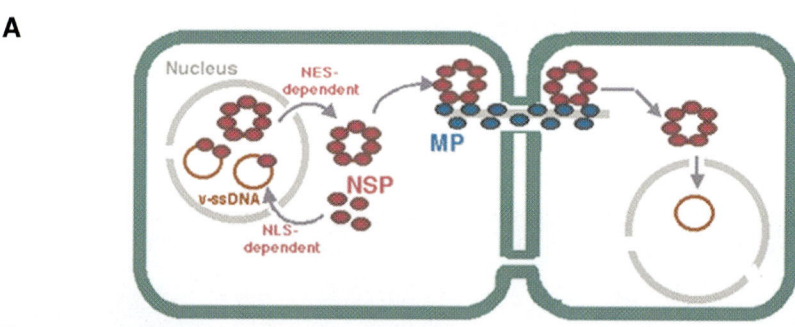

B

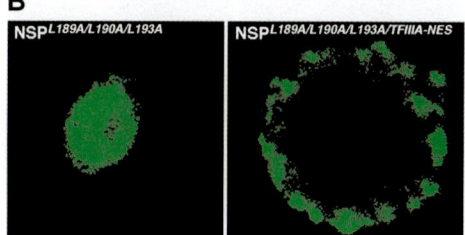

SqLCV mutant	Infectivity (%)
wild type	100
NSP$^{L189A/L190A/L193A}$	0
NSP$^{L189A/L190A/L193A/TFIIIA-NES}$	11

Plate 5

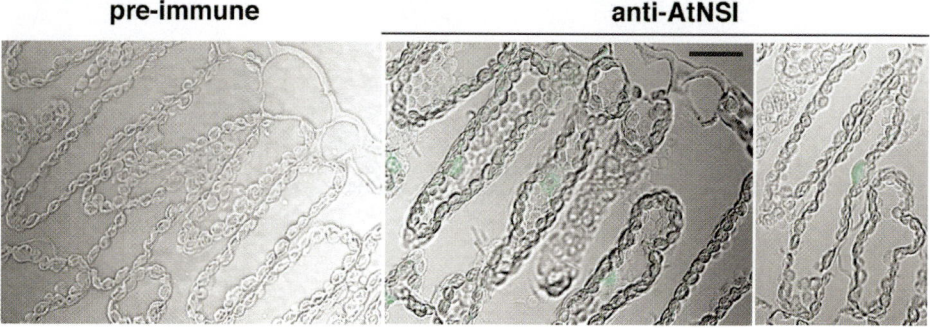

Plate 6

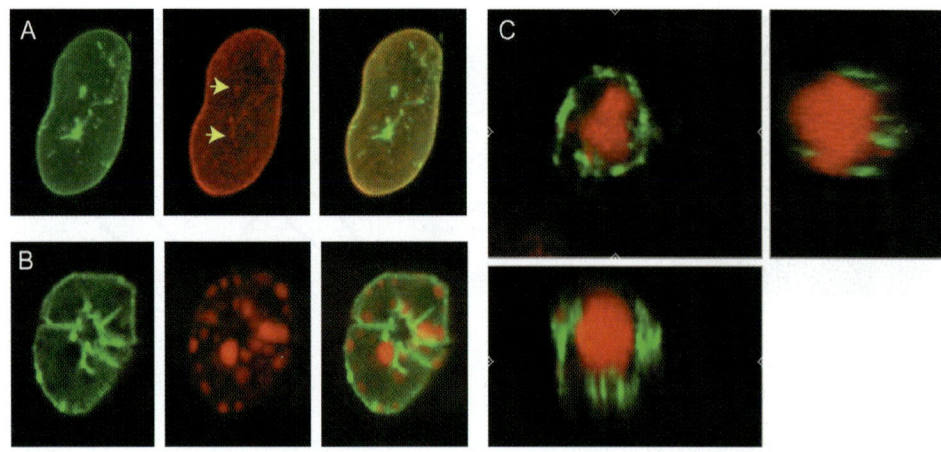

Plate 7

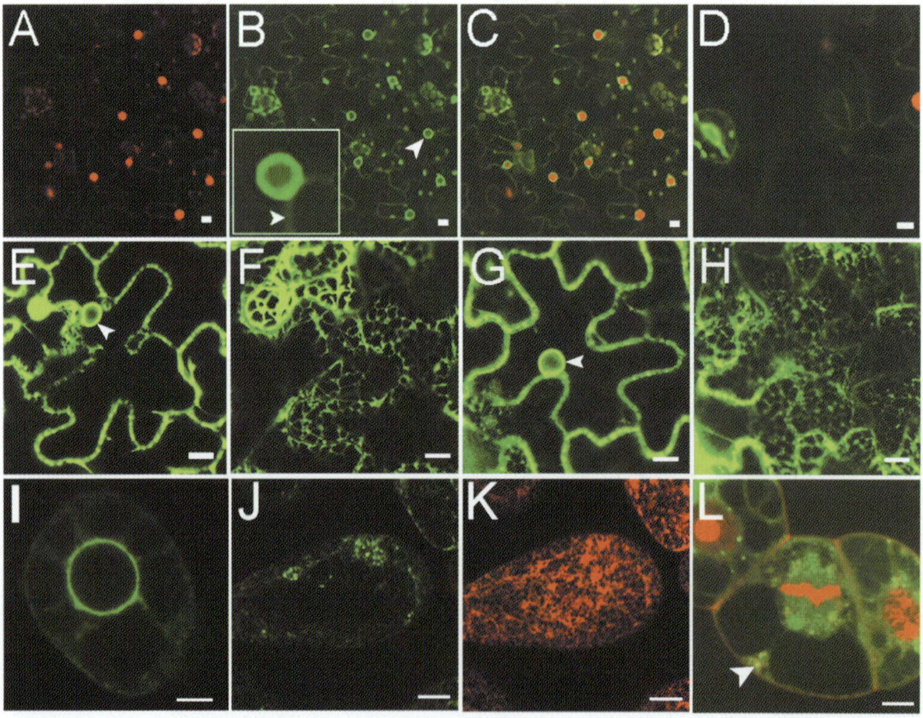

Plate 8

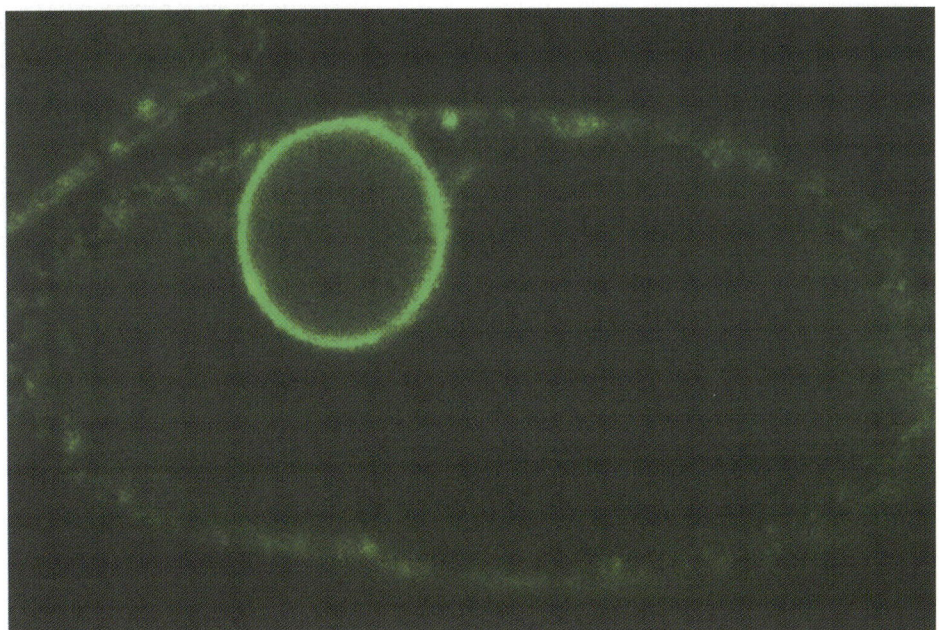

Plate 9

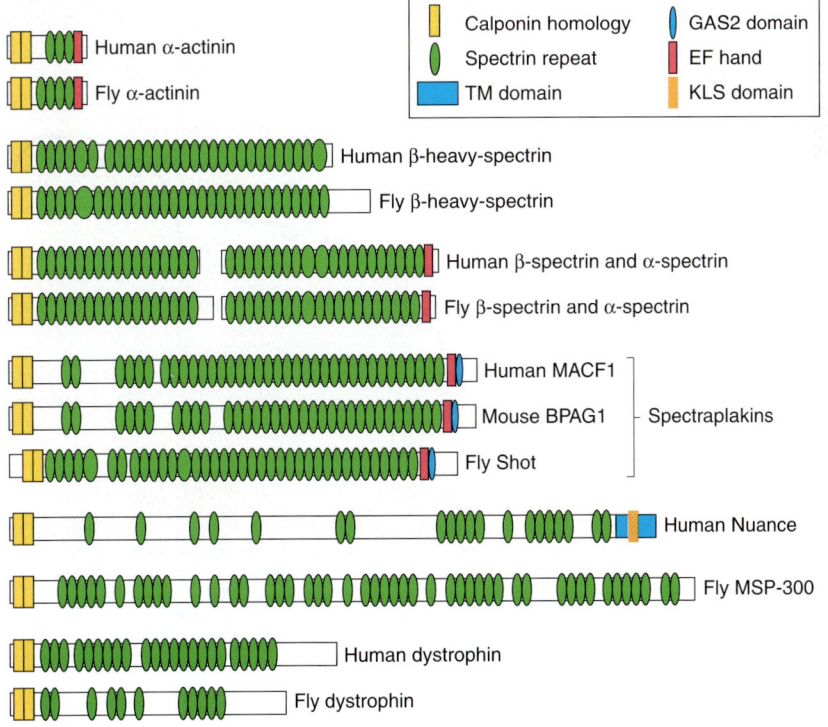

Plate 10

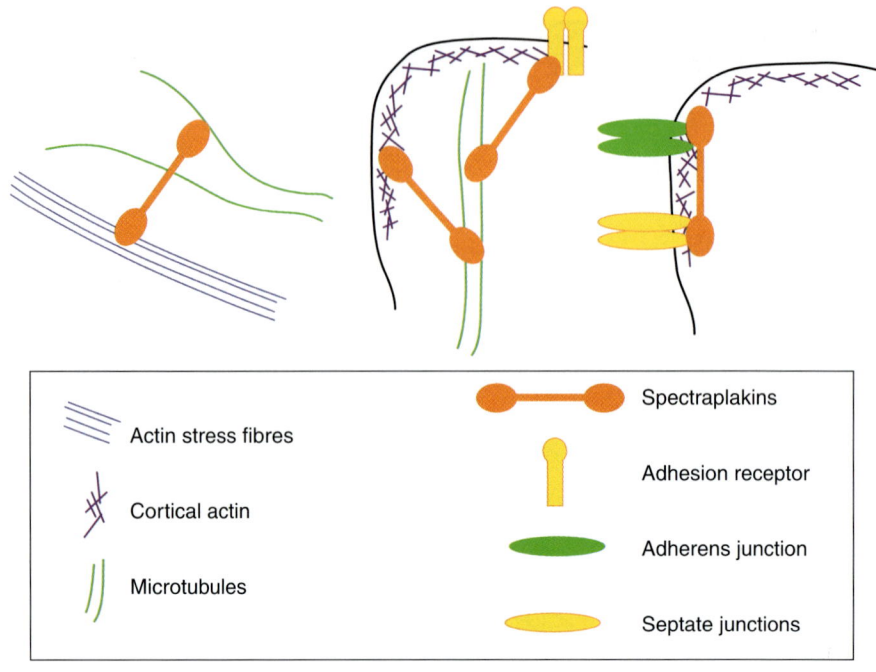

Plate 11

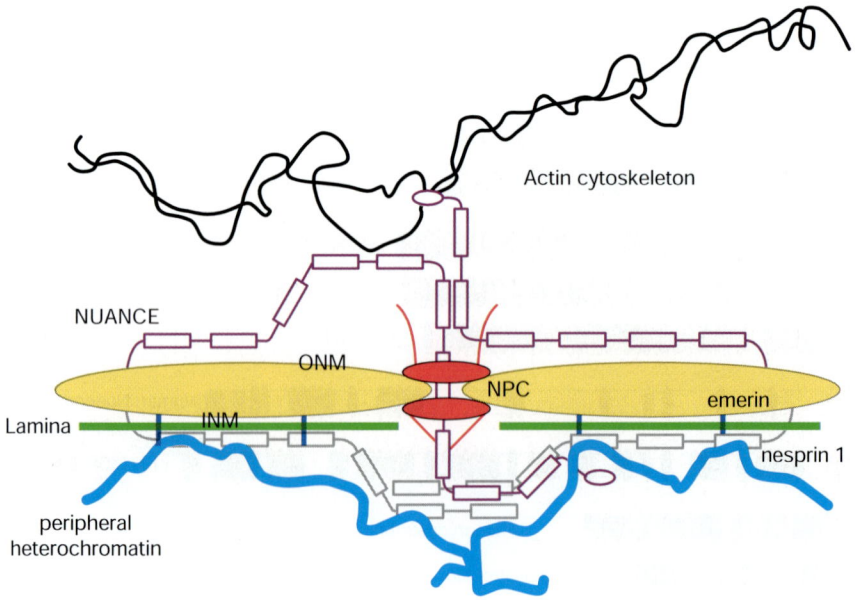

Plate 12

A

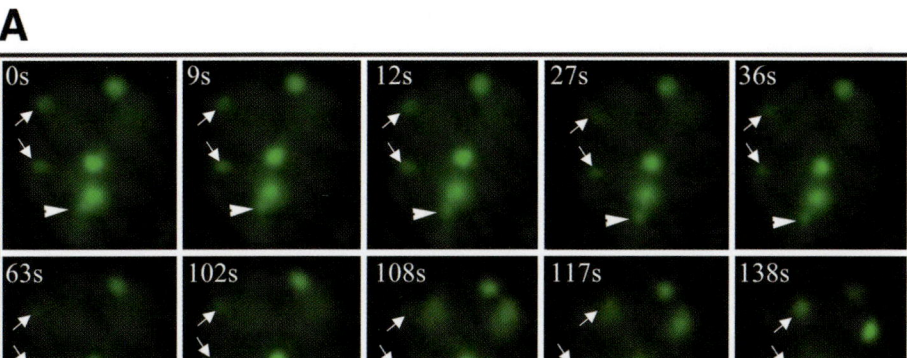

B

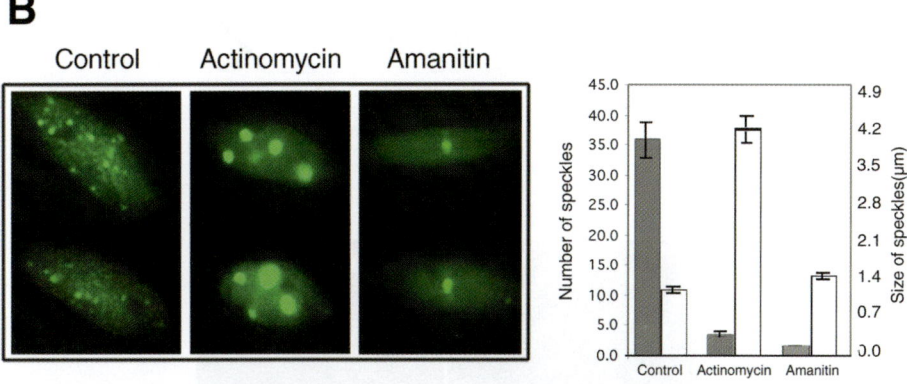

C

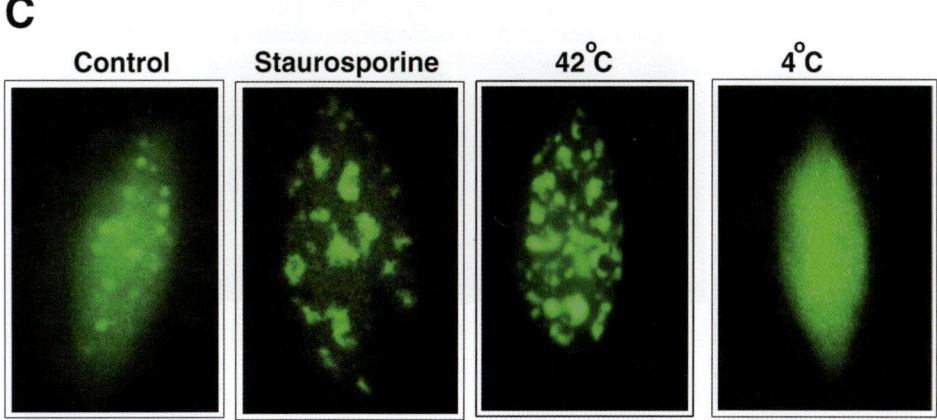

Plate 13

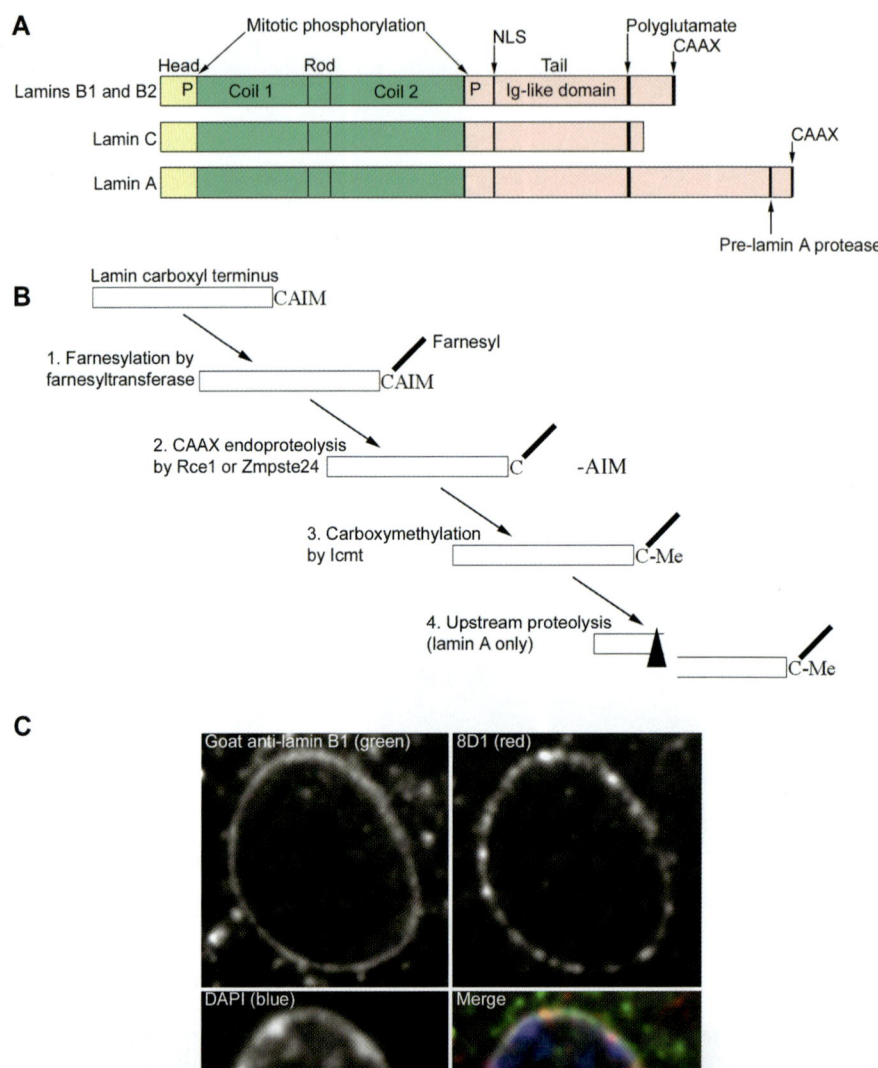

A

Head — Mitotic phosphorylation — NLS — Polyglutamate CAAX

Lamins B1 and B2 | P | Coil 1 | Rod | Coil 2 | P | Ig-like domain |

Tail

Lamin C

Lamin A — CAAX — Pre-lamin A protease

B

Lamin carboxyl terminus ___CAIM

1. Farnesylation by
farnesyltransferase ___CAIM — Farnesyl

2. CAAX endoproteolysis
by Rce1 or Zmpste24 ___C -AIM

3. Carboxymethylation
by Icmt ___C-Me

4. Upstream proteolysis
(lamin A only) ___C-Me

C

Goat anti-lamin B1 (green)   8D1 (red)

DAPI (blue)   Merge

Plate 14

16

Spectraplakins and nesprins, giant spectrin repeat proteins participating in the organization of the cytoskeleton and the nuclear envelope

Arto Määttä, Christopher J. Hutchison and
Martin D. Watson

1. Spectraplakins organize actin and microtubule cytoskeletons

Cells require an internal architecture that supports tissue integrity and allows flexibility in development and repair. To this end they possess various protein families that contribute to accurate relative positioning and dynamics of different cytoskeletal networks and organelles. One large protein family that participates in the structural organization of cells and tissues is the spectrin repeat (SR) family. The spectrin repeat is a triple helical bundle of about 106 amino acids. In this characteristic structure, three alpha-helices are separated by two loops to form a coiled coil (Pasqual *et al.*, 1997). Spectrin repeats are utilized to form flexible rods of variable length. The rod separates distinct functional domains of SR proteins. In the classical family members these comprise a calponin-homologous actin-binding domain in the N-terminus and a variable C-terminal domain that usually interacts with membrane lipids or membrane protein complexes. The classical SR family members are spectrins, α-actinins and dystrophins and their structure and functions are comprehensively covered in several recent reviews (Blake *et al.*, 2001; Brakebusch and Fässler, 2003; Broderick and Winder, 2002; Djinovic-Carugo *et al.*, 2002; Knust 2000). This chapter concentrates on the emerging new subfamilies, spectraplakins and nesprins, comprising giant cytoplasmic and nuclear proteins encoded by very large genes that are subject to complex alternative

---

The Nuclear Envelope, edited by D.E. Evans, C. Hutchison & J.A. Bryant.
© 2004 Garland Science/BIOS Scientific Publishers

splicing (Röper et al., 2002; Zhang et al., 2001, 2002). We will review the structure and predicted isoforms of these proteins and evaluate what is known about spectraplakin and nesprin functions based on cell culture studies and on phenotypes of mutated model organisms.

There are a few key characteristics that distinguish spectraplakins and nesprins from traditional SR proteins. Firstly, the new gene families encode unique C-terminal domains: a microtubule-binding GAS domain in spectraplakins (Sun et al., 2001) and a Klarsicht domain in nesprins (Zhang et al., 2001). Secondly, both spectraplakin and nesprins interact with intermediate filaments. Finally, it should be noted that the nesprins are the only known transmembrane spectrin repeat proteins and that spectraplakins are hybrid proteins of spectrin and plakin families.

1.1 *Spectraplakins harbour characteristic domains of plakins and spectrins*

Plakins are cytoskeletal linker proteins that connect intermediate filaments (IFs) to desmosomes and hemidesmosomes. In addition, plakin proteins can interconnect all three cytoskeletal networks, Ifs, actin microfilaments and microtubules to each other (Leung et al., 2002).

The cloning of the *Drosophila* plakin homologue revealed a very large protein characterized by a remarkable difference from the previously known plakins. Instead of a coiled coil rod typical for plakins the *Drosophila* homologue harboured a long rod domain comprised of 37 spectrin repeats (Gregory and Brown, 1998; Strumpf and Volk, 1998). The first full-length cDNA clones of this protein were identified in a genetic screen aimed at dissecting the molecular basis of the attachment structures of fly wing epithelial cells (Prout et al., 1997; Walsh and Brown 1998). This protein was named *kakapo* (after a flightless New Zealand parrot). Subsequently, kakapo (also known as *kopupu* or *groovin*) was recognized to be allelic to a previously known mutant phenotype *shot* (short stop) (Lee et al., 2000; Van Vactor et al., 1993). Proteins such as shot, that share characteristics of both plakin and spectrin families are called spectraplakins (Roper et al., 2002). Unlike 'traditional' spectrins, different spectraplakin isoforms can bind, in addition to actin microfilaments, IFs and microtubules.

Shot is the only spectraplakin gene in the fly genome (Roper et al., 2002). Likewise, there is one spectraplakin locus, vab-10, in *Caenorhabditis elegans* (Bosher et al., 2003) and it appears that vertebrates harbour only two genes for spectraplakins, MACF1 (microtubule actin cross-linking factor) and BPAG1 (bullous pemphigoid antigen 1). MACF1 has also been cloned and named as ACF7 (the original partial cDNA), MACF, MACF7, macrophin, trabeculin a and ABP 620 (Bernier et al., 1996; Leung et al., 1999; Okuda et al., 1999; Sun et al., 1999). BPAG1, in turn, is also referred to as dystonin and MACF2 (Brown et al., 1995; Leung et al., 2001). Despite the limited number of spectraplakin genes, there appears to exist a plethora of putative proteins. Each spectraplakin gene encodes several alternatively spliced isoforms. In fact, the large size of the genes and the number of the known and predicted combinations of the coding exons mean that the true number of the isoforms or their relative abundance is not yet known. This uncertainty is easy to appreciate when one notes that even the smallest known transcript from the BPAG1 gene encodes a 230-kDa epithelial plakin that is related to desmoplakin and plectin and does not contain a SR rod.

A prototype spectraplakin would consist of a calponin homology actin-binding domain, a plakin domain, one or two sets of plectin repeat domains, a SR rod comprising about 30 repeats, EF hands and, finally, a GAS2 domain (Röper et al., 2002). Nesprins share CH domains and the spectrin repeats with spectraplakins but have a unique C-terminus, the Klarsicht domain, and do not contain domains homologous to plakins.

1.2 Calponin-homology domain mediates interactions with actin

The N-terminal actin-binding domain (ABD) found in all spectrin family members consists of a bipartite calponin homology (CH) domain. In each case, the CH domain consists of one type 1 CH repeat immediately followed by a type 2 repeat, a configuration found in proteins where CH repeats usually form an ABD (Gimona et al., 2002). This CH1/2 tandem array seems to be required for optimal binding to actin. An isolated type 1 CH domain has ten times lower affinity than an intact tandem and a type 2 domain alone does not bind actin at all (Stradal et al., 1998; Way et al., 1992; Winder et al., 1995). Indeed, in vitro experiments have confirmed that the spectraplakin CH domains bind filamentous actin with a comparable micromolar range of dissociation constants to the ABD of dystrophin. For example, the ABD of the mouse MACF has a K_d of 0.35 μm (Karakesisoglou and Fuchs, 2000). The studies on spectraplakins have not addressed possible differences between CH1 and CH2 domains but an analysis of the closely related plectin ABD supports the view of possible independent functions for the subdomains. C-terminal truncations encompassing the whole CH2 subdomain of the ADB of plectin did not interfere with actin binding (Fontao et al., 2001). The ADB of plectin increases the rate of actin polymerization in vitro (Fontao et al., 2001; Gimona et al., 2002) and actin dynamics is impaired in plectin-deficient mouse fibroblasts (Andrä et al., 1998).

It should be noted that F-actin is not necessarily the only interacting partner for CH1/2 domains. In plectin, the binding site for β4 integrin overlaps with the ABD and consequently anchorage of plectin to hemidesmosomes, thereby preventing association with actin (Geerts et al., 1999). In addition, ABDs appear to be capable of self-association and the sequence responsible for the dimerization in plectin is largely overlapping with the CH1 subdomain (Fontao et al., 2001). Notably, the self-association of utrophin ABD involves both CH1 and CH2 domains (Keep et al., 1999) and plectin ABD was able to heterodimerise the ABD of PBAG1 but not with the corresponding domain of dystrophin (Fontao et al., 2001) indicating further divergence within the SR superfamily. Finally, the sequences preceding the ABD can influence the interaction with actin. Alternative transcription start sites result in at least four different N-termini in BPAG1 and three in MACF (summarized in Röper et al., 2002). Two of the start sites encode proteins that have different short amino acid sequences before the ABD. Interestingly, use of the third transcription start site results in an incomplete ABD that lacks the crucial CH1 residues required for actin binding and appears to constitute a microtubule-binding site (Yang et al., 1999). The nature of this N-terminal microtubule-binding site remains to be verified as it has not been studied in the context of the full-length protein (Leung et al., 1999) where only a C-terminal MT-binding domain can be demonstrated.

1.3 *N-terminal plakin domains can target proteins to cellular junctions but remain poorly characterized in spectraplakins*

The N-terminal plakin domain and an intermediate filament-binding domain harbouring plectin repeats are the characteristic features of plakin proteins (Leung *et al.*, 2002; Ruhrberg and Watt 1997). In addition to the spectraplakins, MACF and BPAG1, the plakin family comprises plectin, desmoplakin, envoplakin and periplakin (Leung *et al.*, 2002; Ruhrberg and Watt 1997).

The plakin domain is encoded by 21–23 small exons of highly conserved size (Maatta *et al.*, 2000). At the amino acid level the conservation is less striking but the plakin box is predicted to be formed by small consecutive α-helical globular domains (Green *et al.*, 1990). This said, no crystal structures for plakin N-termini are yet available. Plakin box domains target the proteins to cellular junctions. For example, the N-terminus of desmoplakin, the quintessential structural constituent of the cytoplasmic plaque of desmosomes, interacts directly with plakoglobin and plakophilins (Kowalczyk *et al.*, 1997, 1999; Smith and Fuchs, 1998). Likewise, the N-terminus of the epithelial BPAG1 variant binds BPAG2 (collagen XVII, a transmembrane component of hemidesmosomes) (Hopkinson and Jones, 2000). The subcellular distribution of the plakin domain proteins is not, however, limited to junctions. For example, the periplakin N-terminus directs the protein to the cortical actin network in addition to desmosomes (Di Colandrea *et al.*, 2000). Multiple potential subcellular localizations that are possibly cell or tissue specific seem to be the rule with the spectraplakins as well. All of the spectraplakin splice variants predicted so far carry a plakin domain (Röper *et al.*, 2002) but the localization of the proteins is widespread. Even though Shot is a component of the integrin-mediated cytoskeleton–ECM junctions in *Drosophila* epidermal cells, it is also found in the apical aspect of the same cells co-localizing with protein 4.1. superfamily member Coracle (Gregory and Brown, 1988). In mouse primary keratinocytes, adheren junctions and desmosomes start to organize rapidly after a switch to high calcium in the culture medium. In contrast, MACF displays slower kinetics and although it accumulates to cell borders does not appear to be a component of either junction (Karakesisoglou *et al.*, 2000). No direct interaction partners for the plakin domains in spectraplakins have been identified and the function of this domain in the context of spectrin repeat proteins remains to be fully characterized.

1.4 *Plectin repeats constitute a specific intermediate filament-binding domain*

The C-terminus of plakins contains variable numbers of plectin repeats. In a recent analysis of the crystal structure of desmoplakin the IF binding domain defined the plectin repeat as a 38 residues long β-hairpin that is followed by two antiparallel a-helices (Choi *et al.*, 2002). The plectin repeats fold as globular domains separated by linker sequences with 4.5 or 5 repeats per domain (Choi *et al.*, 2002; Janda *et al.*, 2001). The number of plectin repeats within the plakin and spectraplakin families varies; for example, plectin contains six such globular domains in its C-terminus. The periplakin C-terminus has no globular domains but a conserved linker domain with similarity to two plectin repeats that form the linker next to the fifth globular domain in the plectin C-terminus. This linker sequence binds intermediate filaments (DiColandrea *et al.*, 2000; Fontao *et al.*, 2003; Karashima and Watt, 2002; Kazerounian *et al.*, 2002; Nicolic *et al.*, 1996) indicating that the actual protein motif sufficient for IF interactions does not need to be folded into globular domains. It remains to be seen whether the plectin repeats in MACF- or in SR-containing isoforms of BPAG1 bind intermediate filaments. In the spectraplakins the plectin repeats are in the middle of the molecule before the spectrin repeats and no interactions between the long MACF isoforms and intermediate filaments have been demonstrated (Karakesisoglou *et al.*, 2000; Leung *et al.*, 1999). Furthermore, the conserved plectin repeats in *Drosophila* Shot remain a puzzle as the fly does not have any cytoplasmic intermediate filaments, which argues for additional functions for plectin repeats. Indeed, a non-IF binding partner, periphilin, has recently been described for the periplakin C-terminus (Kazerounian and Aho, 2003).

1.5 *A C-terminal GAS2 domain constitutes a microtubule binding site in spectraplakins*

The C-terminus of spectraplakin contains two EF hands followed by a GAS2 domain. The EF hands of α-actinin and spectrin bind calcium (Lundberg *et al.*, 1995) and it can be assumed that the corresponding domain in spectraplakins performs the same function. Spectraplakins are unique among spectrin repeat proteins in having a GAS2 domain in their C-terminus. Gas2 protein was originally identified as an up-regulated protein in growth-arrested cultured NIH-3T3 cells (Brancolini *et al.*, 1992) and in addition to Gas2 itself and spectraplakins the GAS2 domain is conserved in two other proteins. GAR17 and GAR22 also contain an actin-binding domain (Goriounov *et al.*, 2003). Full-length spectraplakins associate with microtubules in flies, *C. elegans* and in cultured mammalian cells. GAS2 domain-containing fragments of MACF1 and Shot bind to and stabilize microtubules both *in vitro* and in cultured cells (Lee and Kolodziej, 2002a; Leung *et al.*, 1999; Sun *et al.*, 2001). Conversely, when the GAS domain is deleted from a full-length isoform of Shot, the interaction with microtubules is lost. Gas and EF hand domains also interact with the microtubule plus end binding complex EB1/APC1 protein complex that is proposed to connect MTs to the cellular cortex (Subramanian *et al.*, 2003). Notably, the Gas2 protein is a substrate to caspase 3 and participates in regulating cell shape changes during apoptosis. In addition, Gas-2 binds m-calpain inhibiting calpain-dependent processing of p53 (Benetti *et al.*, 2001). It is not yet known whether spectraplakins with a GAS domain are cleaved by caspase or whether they contribute to the morphological changes during apoptosis.

1.6 *Shot and vab-10 mutations underline the importance of spectraplakins in tissue integrity and morphogenesis*

Shot alleles have been identified in many different genetic screens in *Drosophila* (Gao *et al.*, 1999; Prokop *et al.*, 1998; Prout *et al.*, 1997; Strumpf and Volk, 1998; Van Vactor *et al.*, 1993; Walsh and Brown, 1998). In all cases the recognized mutations are embryonic lethal when homozygous. Notably, Shot is the only spectraplakin in the Fly and cannot be compensated by other family members. The overriding themes of the *shot* mutations are: (i) lost cross-connections between actin and microtubule cytoskeletons; and (ii) failure of the cytoskeleton to link to plasma membrane proteins. Shot was found to be strongly expressed close to both apical and basal plasma membranes of epidermal muscle attachments cells in *Drosophila* and Shot mutations resulted in detachment of muscles from the epidermis (Gregory and Brown, 1998). Electron microscopy revealed that in mutant epidermal cells, the microtubule bundles were no longer connected to plasma membrane and the epidermal cells were consequently fragile (Prokop *et al.*, 1998). This phenotype is reminiscent of several mammalian plakin mutations where, for example, loss of plectin or the epithelial isoform of BPAG1 leads to blistering owing to severed linkage of keratin intermediate filaments and hemidesmosomes (Andrä *et al.*, 1997; Guo *et al.*, 1995). The role of Shot in the regulation of MT dynamics is further emphasized by the observation that the EB1/APC protein complex is dissociated from muscle–tendon attachment sites when Shot expression is down-regulated by RNA interference leading to elongation of MT arrays (Subramanian *et al.*, 2003).

Another prominent function of Shot is in axon extension and dentritic sprouting. Shot is required for growth cone extension of both sensory axons along neuronal substrates (Kolodziej *et al.*, 1995) and motor axons into their target fields (Van Vactor *et al.*, 1993). Moreover, the number of terminal dentritic branches of shot-deficient motoneurons is greatly reduced (Gao *et al.*, 1999; Prokop *et al.*, 1998). The ability of spectraplakins to form a bridge between F-actin and microtubules seems to be crucial for the neuronal function of shot. Lee and Kolodziej (2002b) found out that only the constructs containing both end domains of shot in the same molecule were able to rescue sensory neuron extension in shot embryos. Intriguingly, the Ca-binding EF-hand was required for the rescue in addition to the microtubule-binding GAS domain in the C-terminus indicating a possible role for Ca^{2+} ions in stabilizing cytoskeletal cross-bridges (Lee and Kolodziej, 2002b). In contrast, the actual length of the bridge appeared not so important, as the size of the SR rod could be remarkably reduced without compromising the rescue of axon growth (Lee and Kolodziej, 2002b).

A final twist in the cytoskeletal cross-linking function of Shot is the functional independence of the actin- and microtubule-binding domains in tracheal tube formation. During tracheal development the lumen of the branches is initially closed at branch tips but subsequently joins to other branches to form an open tubular network. The fusion cells at branch tips form shotgun- (E-cadherin-) and armadillo- (B-catenin-) dependent transient junctions prior to branch fusion (Uemura *et al.*, 1996). RhoA signalling localizes Shot at the E-cadherin junctions along with F-actin and *shot* mutants fail to remodel the initial contacts and make lumenal contacts between the tracheal branches (Lee and Kolodziej, 2002a). Although this activity is again dependent on interactions of Shot and cytoskeleton, there is a remarkable difference to the situation in axons. The binding sites for F-actin and microtubules act redundantly in tracheal fusion cells, since the presence of either the CH domain to bundle actin filaments or the GAS domain that can stabilize microtubules is alone sufficient for cytoskeletal organization required for tracheal fusion (Lee and Kolodziej, 2002a). Thus, in some cell types end domains of spectraplakin can have independent functions without forming bridges between the different cytoskeletal networks.

Another intriguing lesson on shot function from fly mutants is a putative role for spectraplakins as a platform or organizer for signalling molecules. In *Drosophila*, the differentiation of tendon cells commences upon activation of EGF receptor signalling by Vein, a neuregulin-like factor (Yarnitzky *et al.*, 1998. In shot mutants, Vein fails to localize at muscle–tendon junctions and the cells do not differentiate properly (Strumpf and Volk, 1998). Notably, Vein is an extracellular growth factor so it is possible that a loss of shot and the subsequent cytoskeletal disorganization lead to mis-localization of transmembrane proteins required for Vein distribution. Support of specialized membrane domains could be a common spectraplakin function since in mutant *shot* axons distribution of some membrane proteins is altered as well (Prokop *et al.*, 1998).

A recent characterization of the spectraplakin locus, vab-10 (variably abnormal 10), in the nematode worm *C. elegans* has shed further light on the role of these giant proteins in morphogenesis. Unlike *Drosophila*, the *C. elegans* genome encodes cytoplasmic intermediate filaments. Not surprisingly, the vab-10 locus gives rise to two distinct types of isoforms, vab-10A with a plectin-like C-terminus and vab-10B spectraplakin with a MACF like spectraplakin C-terminus (Bosher *et al.*, 2003). The A and B type isoforms have non-overlapping functions as supported by genetic complementation studies on different vab-10 mutations (Bosher *et al.*, 2003). The function of plectin-like vab 10A closely resembles those of vertebrate hemidesmosomal plakins, plectin and BPAG-1e. VAB-10A is associated with worm intermediate filaments in the fibrous organelle (FO) that is molecularly and functionally equivalent to hemidesmosomes. Vab-10A mutants have a reduced number of FOs and the epidermis is detached from both underlying muscle and the ECM cuticle above (Bosher *et al.*, 2003). Vab-10B mutations, in turn, lead to abnormal epidermal morphology and are required for organization of actin microfilaments (Bosher *et al.*, 2003) and can be presumed to bridge actin and microtubule cytoskeletons.

2. The identification of novel spectrin repeat proteins in the nucleus and nuclear membrane

During the past 2 years, the existence of giant spectrin repeat proteins of the nuclear envelope have been reported. Syne-1, was the first such protein to be reported as a dystrophin-like protein associated with the nuclei of skeletal muscle cells and involved in nuclear migration (Apel *et al.*, 2000). Nesprin 1 and nesprin 2 (nuclear envelope spectrin repeat) were initially identified as differentiation markers of vascular smooth muscle cells (Zhang *et al.*, 2001). NUANCE (nucleus and actin connecting element) is a giant 796-kDa protein (Zhen *et al.*, 2002). It now appears that all four proteins are alternatively spliced members of the same protein family and the list of proteins belonging to this family is set to grow. Nesprin 2 and NUANCE appear to be one and the same protein. The protein first described as Nesprin 1 is now referred to as nesprin 1α, while syne-1 is referred to as nesprin 1β. All of the proteins are characterized by the presence of multiple clustered spectrin repeats, bipartite nuclear localization signal sequences, N-terminal actin-binding domains and a conserved C-terminal single pass transmembrane spanning domain (TMD). It now appears that the nesprins are expressed ubiquitously, although levels of expression vary considerably between tissues and alternative splicing means that in different tissues variations on the exact form of protein expressed will occur.

2.1 *Nuclear envelope localization*

Nesprin 1a localizes predominantly to the nuclear envelope although it becomes relocated to the nucleoplasm and cytoplasm during skeletal muscle differentiation (Zhang *et al.*, 2001) More recently, nesprin 1a has been shown to self-associate, giving rise to the possibility that it forms oligomers. In addition, nesprin 1a binds directly to the nuclear intermediate filament proteins lamins A/C and to the inner nuclear membrane protein emerin *in vitro* (Mislow *et al.*, 2003). Thus nesprin 1α appears to be capable of being integrated into the nuclear lamina as part of an intermediate-type filament. Alternatively, it may extend beyond the lamina to form part of a nucleoskeleton–. Finally, it may link the lamina to an actin cytoskeleton, either inside or outside of the nucleus.

NUANCE is also ubiquitously expressed but localizes to the outer nuclear envelope, the nucleoplasm and the nucleolus. NUANCE associates with actin *in vivo* and *in vitro* and influences actin polymerization dynamics. In addition, the distribution of NUANCE is clearly influenced by the behaviour of the actin cytoskeleton. For example, NUANCE becomes concentrated towards the leading edge of migrating cells where it co-localizes with F-actin. In addition, when actin filaments are depolymerised with the drug Latrunculin A, the nuclear envelope became wrinkled and invaginated. NUANCE and actin both accumulated at sites of invagination. These results suggest that the size and shape of the nucleus are both influenced by the actin cytoskeleton, which in turn is linked to the nuclear envelope, probably through the lamina, by NUANCE (Zhen *et al.*, 2002).

2.2 *The transmembrane spanning domain is a unique feature of the nesprin family*

Nesprin 1α, syne-1/nesprin 1β and NUANCE are unique among spectrin repeat proteins in possessing a C-terminal TMD. Each member of the family is recruited to the NE through their C-terminal TMDs. The C-terminal domain of each protein is highly homologous to NE-targeting domains in the *Drosophila Klarsicht* protein and is therefore referred to as a *Klarsicht* domain. Moreover, *Klarsicht* and syne-1 are both involved in nuclear migration. *Klarsicht* is required for migration of nuclei in developing retinal photoreceptors (Mosely-Bishop *et al.*, 1999), while Syne-1 is involved in migration of nuclei in myotubes and anchorage of nuclei at the post-synaptic membrane (Apel *et al.*, 2000). Nuclear migration is an important process in metazoan development and is essential during fertilization, meiotic and mitotic cell division and in human disease (Mosely-Bishop *et al.*, 1999). Since nesprin 1α and NUANCE also possess *Klarsicht* domains and apparently link the actin cytoskeleton to the lamina, it is tempting to speculate that these proteins also have a role in nuclear migration. With the increasing use of RNAi in mammalian cells (Harborth *et al.*, 2001) it is now straightforward to test this hypothesis.

2.3 *Components of a nucleoskeleton?*

The identification of the nesprin family for the first time putatively allows real proteins to be ascribed to structures, which for years have remained enigmatic. The nucleoskeleton (Jackson and Cook, 1988) or nuclear matrix (He *et al.*, 1990) is a structure composed of 7–13-nm diameter filaments, revealed by resinless section electron microscopy after extraction of chromatin from the nucleus. Nucleoskeleton filaments abut with the lamina and appear to contact the cytoskeleton through nuclear pore complexes (NPCs). While it is premature to suggest that they form the nucleoskeleton this remains a possibility since forms of nesprins are both distributed throughout the nucleoplasm (Zhang *et al.*, 2001; Zhen *et al.*, 2002). In addition, NUANCE partially co-localizes with the NPCs (Zhen *et al.*, 2002) suggesting that it might contact the lamina through NPCs. In colour *Plate 10* NUANCE and nesprin 1 are represented as linkers between the cytoskeleton, the nuclear lamina and chromatin allowing direct communication between these three structures. What would be the possible functions of linking the cytoskeleton to the lamina in this way?

2.4 *Influences on cytoplasmic organization*

It has recently been shown that the lamina has a direct influence on cytoplasmic organization. Disruption of the *Drosophila* lamin Dm_0 by random insertion of P elements, leads to profound effects on cytoplasmic polarity during development. For example, a number of such mutations severely alter the outgrowth of terminal branches of trachea during embryogenesis. A subset of the same mutants also caused severe disruption of dorso-ventral polarity in eggs, ranging from mild to severe dorsalization and mis-localization of mRNA species in the egg cytoplasm (Guillenim *et al.*, 2001). These findings are consistent with the observation that RNAi knockdown of the Ce-lamin gene gives rise to sterility in *C. elegans* (Lui *et al.*, 2001). Several possible explanations have been ascribed to these observations, including the hypotheses that the lamina is directly connected to the actin and microtubule cytoskeleton (Fey *et al.*, 1984) and that these connections permit correct cytoplasmic localization of specific mRNA species. The direct linkage of the lamina to the actin cytoskeleton through NPCs, via NUANCE, provides a physical linkage that would permit directed deposition of mRNA species within the cytoplasm (colour *Plate 11*).

2.5 *Implications for disease*

There has been considerable current interest in a range of genetic diseases caused by mutations in lamins A/C and their associated proteins. The diseases include muscular dystrophies, cardiomyopathies, partial lipodystrophy and neuropathies (Hutchison, 2002). Given the range of these diseases, a number of properties of the lamina must promote the various pathophysiological phenotypes. Indeed, structural weakness within affected cells (Hutchison *et al.*, 2001) and altered gene expression (Wilson *et al.*, 2001), have both been proposed as the basis of one or more of the diseases. The nesprin family are highly homologous to the dystrophin family (Ahne and Kunkel, 1993). Given the involvement of the dystrophin family in Duchenne and Becker muscular dystrophies (Biggar *et al.*, 2002) it is now conceivable that lamin-based muscular dystrophies are promoted through direct dis-organization of the actin cytoskeleton via altered interactions between the lamins and nuclear spectrin repeat proteins.

3. Conclusions

The discovery of nuclear envelope-associated spectrin repeat proteins, related to the α-actinin superfamily, provides potential explanations for problems which are at the current forefront of cell biology. The proteins would permit direct association of the actin cytoskeleton to the nuclear lamina, via NPCs. These linkages may not only influence how nuclei are positioned in a cell, but also offer putative explanations for hitherto unexplained phenotypes resulting from mutations in nuclear lamins.

References

Ahn, A.H. and Kunkel, L.M. (1993) The structural and functional diversity of dystrophin. *Nat. Genet.* **3**: 283–291.

Andrä, K., Lassmann, H., Bittner, R., Shorny, S., Fassler, R., Propst, F. and Wiche, G. (1997) Targeted inactivation of plectin reveals essential function in maintaining the integrity of skin, muscle, and heart cytoarchitecture. *Genes Dev.* **11**: 3143–3156.

Andrä, K., Nikolic, B., Stocher, M., Drenkhahn, D. and Wiche, G. (1998) Not just scaffolding: plectin regulates actin dynamics in cultured cells. *Genes Dev.* **12**: 3442–3451.

Apel, E.D., Lewis, R.M., Grady, R.M. and Sanes, J.R. (2000) Syne-1, a dystrophin- and Klarsicht-related protein associated with synaptic nuclei at the neuromuscular junction. *J. Biol. Chem.* **275**: 31986–31995.

Benetti, R., Del Sal, G., Monte, M., Paroni, G., Brancolini, C. and Schneider, C. (2001) The death substrate Gas2 binds m-calpain and increases susceptibility to p53-dependent apoptosis. *EMBO J.* **20**: 2702–2714.

Biggar, W.D., Klamut, H.J., Demacio, P.C., Stevens, D.J. and Ray, P.N. (2002) Duchenne muscular dystrophy:current knowledge, treatment and future prospects. *Clin. Orthop.* **401**: 88–106.

Blake, D.J., Weir, A., Newey, S.E. and Davies, K.E. (2001) Function and genetics of dystrophin and dystrophin-related proteins in muscle. *Physiol. Rev.* **82**: 291–329.

Bosher, J.M., Hahn, B-S., Legouis, R., Sookhareea, S., Weimer, R.M., Gansmuller, A., Chisholm, A.D., Rose, A.M., Bessereau, J-L. and Labousse, M. (2003) The *Caenorhabditis elegans* vab-10 spectraplakin isoforms protect the epidermis against internal and external forces. *J. Cell Biol.* **161**: 757–768.

Brakebusch, C. and Fässler, R. (2003) The integrin-actin connection, and eternal love affair. *EMBO J.* **22**: 22324–22333.

Brancolini, C., Bottega, S. and Schneider, C. (1992) Gas2, a growth arrest specific protein, is a component of the microfilament network system. *J. Cell Biol.* **117**: 1251–1261.

Broderick, M.J.F. and Winder, S.J. (2002) Towards a complete atomic structure of spectrin family proteins. *J. Struct. Biol.* **137**: 184–193.

Brown, A., Bernier, G., Mathieu, M., Rossant, J. and Kothary, R. (1995) The mouse dystonia musculorum gene is a neural isoform of bullous pemphigoid antigen 1. *Nature Genet.* **10**: 301–306.

Choi, H.-J., Park-Snyder, S., Pascoe, L.T., Green, K.J. and Weis, W.I. (2002) Structures of two intermediate filament-binding fragments of desmoplakin reveal a unique repeat motif structure. *Nat. Struct. Biol.* **9**: 612–620.

DiColandrea, T., Karashima, T., Määttä, A. and Watt, F.M. (2000) Subcellular distribution of envoplakin and periplakin: insights into their role as precursors of the epidermal cornified envelope. *J. Cell Biol.* **151**: 573–585.

Djinovic-Carugo, K., Gautel, M., Ylänne, J. and Young, P. (2002) The spectrin repeat: a structural platform for cytoskeletal protein assemblies. *FEBS Lett.* **513**: 119–123.

Fey, E.G., Wan, K.M. and Penman, S. (1984) Epithelial cytoskeletal framework and nuclear matrix-intermediate filament scaffold: three dimensional organization and protein composition. *J. Cell Biol.* **98**: 1654–1665.

Fontao, L., Geerts, D., Kuikman, I., Koster, J., Kramer, D. and Sonnenberg, A. (2001) The interaction of plectin with actin: evidence for cross-linking of actin filaments by dimerization of the actin-binding domain of plectin. *J. Cell Sci.* **114**: 2065–2076.

Fontao, L., Favre, B., Riou, S., Geerts, D., Jaunin, F., Saurat, J-H., Green, K.J., Sonnenberg, A. and Borradori, L. (2003) Interaction of bullous pemphigoid antigen 1 (BP230) and desmoplakin with intermediate filaments is mediated by distinct sequences within their COOH terminus. *Mol. Biol. Cell* **14**: 1978–1992.

Gao, F.B., Brenman, J.E., Jan, L.Y. and Jan, Y.N. (1999) Genes regulating dendritic outgrowth, branching, and routing in *Drosophila*. *Genes Dev.* **13**: 2549–2561.

Geerts, D., Fontao, L., Nievers, M.G., Schaapveld, R.Q.J., Purkis, P.E., Wheeler, G.N., Lane, E.B., Leigh, I.M. and Sonnenerg, A. (1999) Binding of integrin a6β4 to plectin prevents plectin association with F-actin but does not interfere with intermediate filament binding. *J. Cell Biol.* **147**: 417–434.

Gimona, M., Djinovic-Carugo, K., Kranewitter, W.J. and Winder, S.J. (2002) Functional plasticity of CH domains. *FEBS Lett.* **513**: 98–106.

Gong, T.W., Besirli, C.G. and Lomax, M.L. (2001) MACF1 gene structure: a hybrid of plectin and dystrophin. *Mamm. Genome* **12**: 852–861.

Goriounov, D., Leung, C.L. and Liem, R.K. (2003) Protein products of human Gas2 related genes on chromosomes 17 and 22 (hGAR17 and hGAR22) associate with both microfilaments and microtubules. *J. Cell Sci.* **116**: 1045–1058.

Green, K.J., Parry, D.A., Steinert, P.M., Virata, M.L., Wagner, R.M., Angst, B.D. and Nilles, L.A. (1990) Structure of the human desmoplakins. Implications for function in the desmosomal plaque. *J. Biol. Chem.* **265**: 2603–2612.

Gregory, S.L. and Brown, N.H. (1998) kakapo, a gene required for adhesion between and within cell layers in *Drosophila*, encodes a large cytoskeletal linker protein related to plectin and dystrophin. *J. Cell Biol.* **143**: 1271–1282.

Guillemin, K., Williams, T. and Krasnow, M.A. (2001) A nuclear lamin is required for cytoplasmic organization and egg polarity in *Drosophila*. *Nat. Cell Biol.* **3**: 848–851.

Guo, L.F., Degenstein, L., Dowling, J., Yu, Q.C., Wollmann, R., Perman, B. and Fuchs, E. (1995) Gene targeting of BPAG1 abnormalities in mechanical strength and cell migration in stratified epithelia and neurologic degeneration. *Cell* **81**: 233–243.

He, D., Nickerson, J.A. and Penman, S. (1990) Core filaments of the nuclear matrix. *J. Cell Biol.* **110**: 569–580.

Hopkinson, S.B. and Jones, J.C.R. (2000) The N terminus of the transmembrane protein BP180 interacts with the N-terminal domain of BP230, thereby mediating keratin cytoskeleton anchorage to the cell surface at the site of the hemidesmosome. *Mol. Biol. Cell* **11**: 277–286.

Hutchison, C.J. (2002) Lamins: building blocks or regulators of gene expression? *Nature Rev. Mol. Cell Biol.* (in press).

Hutchison, C.J., Alvarez-Reyes, M. and Vaughan, O.A. (2001) Lamins in disease: why do ubiquitously expressed nuclear envelope proteins give rise to tissue specific disease phenotypes? *J. Cell Sci.* **114**: 9–19.

Jackson, D.A. and Cook, P.R. (1988) Visualisation of a filamentous nucleoskeleton with a 23 nm axial repeat. *EMBO J.* **7**: 3667–3677.

Janda, L., Damborsky, J., Rezniczek, G.A. and Wiche, G. (2001) Plectin repeats and modules: strategic cysteines and their presumed impact on cytolinker functions. *BioEssays* **23**: 1064–1069.

John, M.K., Mislow, J.M., Holaska, M.S., Kim, K.K., Lee, M., Segura-Totten, K.L. and McNally, E.M. (2002) Nesprin-1 self-associates and binds directly to emerin and lamin A in vitro. *FEBS Letts.* **525**: 135–140.

Karakesisoglou, I., Yang, Y. and Fuchs, E. (2000) An epidermal plakin that integrates actin and microtubule networks at cellular junctions. *J. Cell Biol.* **149**: 195–208.

Karashima, T. and Watt, F.M. (2002) Interaction of periplakin and envoplakin with intermediate filaments. *J. Cell. Sci.* **115**: 5027–5037.

Kazerounian, S. and Aho, S. (2003) Characterization of periphilin, a widespread, highly insoluble nuclear protein and potential constituent of the keratinocyte cornified envelope. *J Biol. Chem.* **278**: 36707–36717.

Kazerounian, S., Uitto, J. and Aho, S. (2002) Unique role for periplakin tail in intermediate filament association: specific binding to keratin 8 and vimentin. *Exp. Dermatol.* **11**: 5027–5037.

Keep, N.H., Winder, S.J., Moores, C.A., Walke, S., Norwood, F.L. and Kendrick-Jones, J. (1999) Crystal structure of the actin-binding region of utrophin reveals a head-to-tail dimer. *Structure Fold. Des.* **7**: 1539–1546.

Knust, E. (2000) Control of epithelial cell shape and polarity. *Curr. Opin. Genet. Devel.* **10:** 471–475.

Kolodziej, P.A., Jan, L.Y. and Jan, N.Y. (1995) Mutations that affect the length, fasciculation, or ventral orientation of specific sensory axons in the *Drosophila* embryo. *Neuron* **15:** 273–286.

Kowalczyk, A.P., Bornslaeger, E.A., Borgwardt, J.E., Palka, H.L., Dhaliwal, A.S., Corcoran, C.M., Denning, M and K.J. Green, K.J. (1997) The amino-terminal domain of desmoplakin binds to plakoglobin and clusters desmosomal cadherin-plakoglobin complexes. *J. Cell Biol.* **139:** 773–784.

Kowalczyk, A.P., Hatzfeld, M., Bornslaeger, E.A., Kopp, D.S., Borgwardt, J.E., Corcoran, C.M., Settler, A. and Green K.J. (1999) The head-domain of plakophilin-1 binds to desmoplakin and enhances its recruitment to desmosomes. Implications for cutaneous disease. *J. Biol. Chem.* **274:** 18145–18148.

Lee, S. and Kolodziej, P.A. (2002a) Short Stop provides an essential link between F-actin and microtubules during axon extension. *Development* **129:** 1195–1204.

Lee, S. and Kolodziej, P.A. (2002b) The plakin Short Stop and the RhoA GTPase are required for E-cadherin-dependent apical surface remodeling during tracheal tube fusion. *Development* **129:** 1509–1520.

Lee, S., Harris, K.L., Whitington, P.M. and Kolodziej, P.A. (2000) short stop is allelic to kakapo, and encodes rod-like cytoskeletal-associated proteins required for axon extension. *J. Neurosci.* **20:** 1096–1108.

Leung, C.L., Sun, D., Zheng, M., Knowles, D.R. and Liem, R.K. (1999) Microtubule actin cross-linking factor (MACF): a hybrid of dystonin and dystrophin that can interact with the actin and microtubule cytoskeletons. *J. Cell Biol.* **147:** 1275–1286.

Leung, C.L., Zheng, M., Prater, S.M. and Liem, R.K.H. (2001) The BPAG1 locus: alternative splicing produces multiple isoforms with distinct cytoskeletal linker domains, including predominant isoforms in neurons and muscles. *J. Cell Biol.* **154:** 691–698.

Leung, C.L., Green, K.J. and Liem, R.K. (2002) Plakins: a family of versatile cytolinker proteins. *Trends Cell Biol.* **12:** 37–45.

Lui, J., Ben-Shahar, T.R., Riemer, D., Treinin, M., Spann, P., Weber, K., Fire, A. and Gruenbaum, Y. (2000) Essential roles of *Caenorhabditis elegans* lamin gene in nuclear organisation, cell cycle progression, and spatial organization of nuclear pore complexes. *Mol. Biol. Cell.* **11:** 3937–3947.

Lundberg, S., Bjork, J., Lofvenberg, L. and Backman, L. (1995) Cloning, expression and characterization of two putative calcium-binding sites in human non-erythroid alpha-spectrin. *Eur. J. Biochem.* **230:** 658–665.

Määttä, A., Ruhrberg, C. and Watt, F.M. (2000) Structure and regulation of the envoplakin gene. *J. Biol. Chem.* **275:** 19857–19865.

Mosely-Bishop, K.L., Li, Q., Patterson, L. and Fisher, J.A. (1999) Molecular analysis of klarsicht gene and its role in nuclear migration within differentiating cells of *Drosophila* eye. *Curr. Biol.* **9:** 1211–1220.

Nikolic, B., Macnulty, E., Mir, B. and Wiche, G. (1996) Basic amino acid residue cluster within nuclear targeting sequence motif is essential for cytoplasmic plectin vimentin network junctions. *J. Cell Biol.* **134:** 1455–1467.

Okuda, T., Matsuda, S., Nakatsugawa, S., Ichigotani, Y., Iwahashi, N., Takahashi, M., Ishigaki, T. and Hamaguchi, M. (1999) Molecular cloning of macrophin, a human homologue of *Drosophila* kakapo with a close structural similarity to plectin and dystrophin. *Biochem. Biophys. Res. Comm.* **264:** 568–574.

Pasqual, J., Pful, M., Walther, D., Saraste, M. and Nilges, M. (1997) Solution structure of the spectrin repeat: left-handed antiparallel triplehelical coiled-coil. *J. Mol. Biol.* **273:** 740–751.

Prokop, A., Uhler, J., Roote, J. and Bate, M. (1998) The kakapo mutation affects terminal arborization and central dendritic sprouting of *Drosophila* motorneurons. *J. Cell Biol.* **143:** 1283–1294.

Prout, M., Damania, Z., Soong, J., Fristrom, D. and Fristrom, J.W. (1997) Autosomal mutations affecting adhesion between wing surfaces in *Drosophila melanogaster*. *Genetics* **146**: 275–285.

Röper, K., Gregory, S. and Brown, N.H. (2002) 'Spectraplakins': cytoskleletal giants with characteristics of both spectrin and plakin families. *J. Cell. Sci.* **115**: 4215–4225.

Ruhrberg, C. and Watt, F.M. (1997) The plakin family: versatile organizers of cytoskeletal architecture. *Curr. Opin. Genet. Dev.* **7**: 392–397.

Smith, E.A. and Fuchs, E. (1998) Defining the interactions between intermediate filaments and desmosomes. *J. Cell Biol.* **141**: 1229–1241.

Stradal, S.T., Kranewitter, W., Winder, S.J. and Gimona, M. (1998) CH domains revisited. *FEBS Lett.* **431**: 134–137.

Strumpf, D. and Volk, T. (1998) Kakapo, a novel cytoskeletal-associated protein is essential for the restricted localization of the neuregulin-like factor, vein, at the muscle-tendon junction site. *J. Cell Biol.* **143**: 1259–1270.

Subramanian, A., Prokop, A., Yamamoto, M., Sugimura, K., Uemura, T., Betschinger, J., Knoblich, J.A. and Volk, T. (2003) Shortstop recruits EB1/APC1 and promotes microtubule assembly at the muscle–tendon junction. *Curr. Biol.* **13**: 1086–1095.

Sun, Y., Zhang, J., Kraeft, S.-K., *et al.* (1999) Molecular cloning and characterization of human trabeculin-a, a giant protein defining a new family of actin-binding proteins. *J. Biol. Chem.* **247**: 33522–33530.

Sun, D., Leung, C.L. and Liem, R.K. (2001) Characterization of the microtubule binding domain of microtubule actin crosslinking factor (MACF): identification of a novel group of microtubule associated proteins. *J. Cell Sci.* **114**: 161–172.

Uemua, T., Oda, H., Kraut, R., Hayashi, S., Kataoka, Y., and Takeichi, M. (1996) Zygotic *Drosophila* E-cadherin expression is required for process of dynamic epithelial cell rearrangement in the *Drosophila* embryo. *Genes. Dev.* **10**: 659–671.

Van Vactor, D., Sink, H., Fambrough, D., Tsoo, R. and Goodman, C.S. (1993) Genes that control neuromuscular specificity in *Drosophila*. *Cell* **73**: 1137–1153.

Walsh, E.P. and Brown, N.H. (1998) A screen to identify *Drosophila* genes required for Integrin mediated adhesion. *Genetics* **150**: 791–805.

Way, M., Pope, B. and Weeds, A.G. (1992) Evidence for functional homology in the F-actin binding domains of gelsolin and a-actinin: implications for the requirements of severing and capping. *J. Cell Biol.* **119**: 835–842.

Wilson, K.L., Zastro, M.S. and Lee, K.K. (2001) Lamins and disease: insights into nuclear infrastructure. *Cell* **104**: 647–650.

Yang, Y., Bauer, C., Strasser, G., Wollman, R., Julien, J.P. and Fuchs, E. (1999) Integrators of the cytoskeleton that stabilize microtubules. *Cell* **98**: 229–238.

Yarnitzky, T., Min, L. and Volk, T. (1998) An interplay between two EGF-receptor ligands, vein and Spitz, is required for the formation of a subset of muscle precursors in *Drosophila*. *Mech. Dev.* **79**: 73–82.

Zhang, Q., Skepper, J., Yang, F., Davies, J.D., Hegyi, L., Roberts, R.G., Weissberg, P.L., Ellis, J.A. and Shanahan, C.M. (2001) Nesprins: a novel family of spectrin-repeat-containing proteins that localize to nuclear membrane in multiple tissues. *J. Cell Sci.* **114**: 4485–4498.

Zhang, Q., Ragnauth, C., Greener, M.J., Shanahan, C.M. and Roberts, R.G. (2002) The nesprins are giant actin-binding proteins, orthologous to *Drosophila* melanogaster muscle protein MSP-300. *Genomics* **80**: 473–481.

Zhen, Y-Y., Libotte, T., Munck, M., Noegal, A.A. and Korenbaum, E. (2002) NUANCE: a giant protein connecting the nucleus and actin cytoskeleton. *J. Cell Sci.* **115**: 3207–3222.

17

Arabidopsis U1 snRNP 70K protein and its interacting proteins: Nuclear localization and *in vivo* dynamics of a novel plant-specific serine/arginine-rich protein

Anireddy S.N. Reddy, Gul Shad Ali and Maxim Golovkin

1. Introduction

About 80% of plant nuclear genes contain one or more introns (Goodall *et al.*, 1991; Reddy, 2001). The removal of introns is an important step in the expression of most eukaryotic genes. The introns in precursor messenger RNAs (pre-mRNAs) are removed in the nucleus to produce functional mRNAs. Excision of introns takes place in a large multicomponent complex called the spliceosome by a two-step reaction involving successive transesterification reactions (Burge *et al.*, 1999). In the first step, the 5' splice site is cleaved to generate a 5' exon and a lariat intermediate. In the second transesterification reaction the 3' splice site is cleaved to generate ligated exons and release the lariat intron. Pre-mRNAs with multiple introns are often spliced in alternate patterns and produce structurally and functionally different proteins from the same gene (Brett *et al.*, 2000; Lorkovic *et al.*, 2000; Reddy, 2001). The combinatorial joining of exons by alternative splicing is an elegant way that most eukaryotes use to generate several distinct proteins from a single transcript. Recent large-scale studies involving the comparison of ESTs with corresponding gene sequences indicate that alternative splicing of pre-mRNAs accounts for a large proportion of proteomic complexity in multicellular eukaryotes. It is estimated that pre-mRNAs from about

The Nuclear Envelope, edited by D.E. Evans, C. Hutchison & J.A. Bryant.
© 2004 Garland Science/BIOS Scientific Publishers

38% of all human genes undergo alternative splicing (Brett *et al.*, 2000). Alignment of genes on human chromosome 22 with available ESTs and cDNAs indicate that about 59% of the genes are alternatively spliced to produce two or more transcripts (Consortium, 2001). In *C. elegans* about 22% of genes, for which ESTs were found, showed alternative splicing.

Exon/intron architecture varies considerably among eukaryotes. Vertebrate genes typically have large introns and small exons whereas the opposite is true in lower eukaryotes (Sterner *et al.*, 1996). In general, plant genes are shorter and have fewer introns compared to animals. Several *in vivo* pre-mRNA splicing studies in plants indicate that plant introns have unique *cis*-elements. Most plant intron-containing transcripts, with some exceptions, are either not processed or processed inaccurately in mammalian nuclear extracts (Lorkovic *et al.*, 2000; Reddy, 2001; Schuler, 1998). Furthermore, animal introns are not excised from pre-mRNA transcripts *in vivo* in plant nuclei. Also, non-intron sequences of animal or bacterial origin are sometimes cryptically spliced in plants (Schuler, 1998). These studies indicate that at least some mechanisms involved in intron recognition are likely to be different from animals and involve novel proteins that recognize the plant-specific *cis*-elements. Splicing studies with dicot introns in monocot cells and *vice versa* indicate that intron recognition signals may also be somewhat different between dicots and monocots (Schuler, 1998).

A distinguishing feature of plant introns is their compositional bias for UA- or U-rich sequences as compared to the introns from animals and yeast (reviewed in Brown and Simpson, 1998; Goodall *et al.*, 1991; Lorkovic *et al.*, 2000; Luehrsen *et al.*, 1994; Reddy, 2001). Plant introns are generally about 15% more rich in UA, primarily due to an increased number of U residues, than exons. Several reports using synthetic and natural plant introns have shown that intronic UA- or U-richness is important for recognition of 5' and 3' splice sites and for efficient splicing of introns in plant cells (reviewed in Brown and Simpson, 1998; Goodall *et al.*, 1991; Lorkovic *et al.*, 2000; Luehrsen *et al.*, 1994; Reddy, 2001; Schuler, 1998). It is likely that plants may have protein factors that recognize the U-rich region and participate in intron recognition. A few proteins that specifically interact with U-rich intron sequences have been reported (Lambermon *et al.*, 2000).

The assembly of the spliceosome involves a series of RNA–RNA, RNA–protein and protein–protein interactions. Five small nuclear (sn) RNAs (U1, U2, U4, U5 and U6) and over 300 distinct proteins are likely to be involved in splicing (Burge *et al.*, 1999; Rappsilber *et al.*, 2002; Zhou *et al.*, 2002). snRNAs and the associated proteins form small nuclear ribonucleoprotein (snRNP) particles. Each snRNP contains at least one snRNA bound to several proteins, some of which are common to all the spliceosomal snRNPs while some are unique to each snRNP (Burge *et al.*, 1999). The snRNPs bind to pre-mRNAs and along with other proteins form the spliceosome (Burge *et al.*, 1999). The snRNPs recognize splice sites and branch point sequences and aid in splicing (Burge *et al.*, 1999). Based on the conservation of RNA and protein components of the spliceosome in plants and animals, it is assumed that plants have a similar spliceosome cycle. However, *in vivo* studies indicate that plant introns have some unique *cis*-elements and are likely to require some novel proteins. So far, very little is known about the mechanisms that regulate basic and alternative splicing in plants. The lack of an *in vitro* splicing system derived from plant cells has hampered studies on the assembly and biochemical characterization of the plant spliceosome.

The U1 snRNP recognizes the 5' splice site in an ATP-independent manner to form a complex that commits the pre-mRNA to spliceosome assembly (Rosbash and Séraphin, 1991). This complex is called the early (E) complex in mammalian cells or commitment complex (CC) in yeast. In metazoans, U1 snRNP contains one U1 snRNA molecule and at least 11 proteins, including three U1snRNP-specific proteins (U1-70K, U1-A and U1-C) (Andersen and Zieve, 1991). Others (B, B', D_1, D_2, D_3, E, F, G) are present in all the major nucleoplasmic snRNPs. The highly conserved ten nucleotides at the 5' end of U1 snRNA base pair with the 5' splice site. The association of U1 snRNP with the 5' sequence is important for splicing and the binding of U1-70K to pre-mRNA appears to be dependent on the presence of U1 snRNP proteins (Rossi et al., 1996). Two U1 snRNP specific proteins (U1-A and U1-70K) of metazoans have been characterized from plants (Golovkin and Reddy, 1996; Simpson et al., 1995) and the third one (U1-C) has been found in the Arabidopsis genome database (www.Arabidopsis.org).

U1-70K is implicated in 5' splice site recognition and/or interaction with other snRNP complexes (Grabowski et al., 1991; Kuo et al., 1991). There is evidence that the U1 snRNP interacts with U2 snRNP during splicing (Black et al., 1985; Manley and Tacke, 1996). Other studies also implicate the U1 snRNP in alternative splicing (Ge and Manley, 1991; Krainer et al., 1990; Kuo et al., 1991). It has been shown that the animal U1-70K interacts with two splicing factors SC35 and ASF/SF2 that are involved in basic and alternative splicing of pre-mRNAs (Kohtz et al., 1994; Manley and Tacke, 1996; Wu and Maniatis, 1993). The interaction of SC35 and ASF/SF2 with U1-70K takes place through a specific association of their serine/arginine-rich domains (Kohtz et al., 1994; Manley and Tacke, 1996; Valcarcel and Green, 1996; Wu and Maniatis, 1993). Furthermore, over-expression of the arginine-rich region of U1-70K has been shown to inhibit pre-mRNA splicing and nucleocytoplasmic transport of mRNA (Romac and Keene, 1995; Romac et al., 1994). These studies suggest a key role for U1-70K protein in both basic and alternative splicing through complex protein–protein interactions.

2. U1-70K is essential in plants

We have isolated and characterized full-length cDNAs and genomic clones encoding U1-70K protein (Golovkin and Reddy, 1996). It is coded by a single gene, which produces two (short and long) transcripts by inclusion or exclusion of a 910 bp intron. Both U1-70K transcripts are expressed in all tissues and the level of the transcripts varies in different organs. The deduced amino acid sequence from the short transcript is similar to the animal U1-70K protein and contains an RNA recognition motif (RRM), a glycine hinge, and an arginine-rich region characteristic of the animal U1-70K protein. Plant U1-70K shares some characteristic features with animal U1-70K but differs in others (Golovkin and Reddy, 1996). The long transcript has an in-frame translational termination codon within the 910 base-included intron and produces a truncated protein containing 204 amino acids with part of the RRM containing only RNP2.

It has been shown that U1snRNP is dispensable for in vitro splicing of some animal pre-mRNAs and inactivation of U1-70K in yeast was found to be not lethal (Crispino and Sharp, 1995; Hilleren et al., 1995; Tarn and Steitz, 1994). To test if plant U1-70K is dispensable in plants we blocked the expression of *Arabidopsis thaliana* U1-70K in petals and stamens by expressing a *U1-70K* antisense transcript using the *APETALA3* (*AP3*) promoter specific to these floral organs (Golovkin and Reddy, 2003). Flowers of transgenic *Arabidopsis* plants expressing the *U1-70K* antisense transcript showed partially developed stamens and petals that are arrested at different stages of development (*Figure 1*). In some transgenic lines, flowers have rudimentary petals and stamens and are male sterile. The severity of the phenotype was correlated with the level of the antisense transcript. Molecular analysis of transgenic plants has confirmed that the observed phenotype is due to over-expression of U1-70K antisense transcript and not due to disruption of whorl-specific homeotic genes, *AP3* or *PISTILLATA*, responsible for petal and stamen development (Golovkin and Reddy, 2003). Flowers of *Arabidopsis* plants transformed with a reporter gene driven by the same promoter showed no abnormalities. These results demonstrate that expression of *U1-70K* antisense transcript in petals and stamens aborts their development and suggest that U1-70K is essential for the development of these organs.

3. *Arabidopsis* U1-70K interacts with a novel set of SR proteins

The sequence elements (*cis*-elements) in the introns, exons and at the exon/intron borders contribute to the recognition of splice sites by a large number of trans-acting factors. Serine/arginine-rich (SR) proteins, a large family of proteins, are one of the best-characterized non-snRNP proteins in the spliceosome. SR proteins play central roles in both constitutive and alternative splicing as essential splicing factors and as specific splicing regulators at multiple stages in spliceosome assembly (Caceres *et al.*, 1997; Graveley, 2000; Graveley and Maniatis, 1998; Graveley *et al.*, 1999; Manley and Tacke, 1996; Valcarcel and Green, 1996). SR proteins are involved in a network of protein–protein interactions primarily through their RS domain. These proteins, with a molecular mass ranging from 20–75 kDa, contain one or two RNA binding domains (RBDs) and an arginine/serine-rich (RS) domain with multiple RS dipeptide repeats at the C-terminus (Manley and Tacke, 1996). These proteins recruit other factors during spliceosome assembly through protein–RNA and/or protein–protein interactions involving their RS domain (Graveley, 2000; Graveley and Maniatis, 1998; Graveley *et al.*, 1999). During E- complex formation, ASF (alternative splicing factor)/SF2 (splicing factor 2), one of the SR proteins, recruits U1 snRNP to the 5' splice by interacting simultaneously with the pre-mRNA and the U1-70K protein (Kohtz *et al.*, 1994). SR proteins (e.g., SC35 and ASF/SF2) are also involved in bridging 5' and 3' splice sites by interacting concurrently with U1-70K and U2AF[35] (Stark *et al.*, 1998; Wu and Maniatis, 1993). Furthermore, SR proteins facilitate incorporation of the tri–snRNP complex (U4/U6.U5 snRNP) into the spliceosome and promote base pairing between U2 and U6 snRNA (Roscigno and Garcia-Blanco, 1995; Tarn and Steitz, 1995). The RS domain of SR proteins has been found to modulate RNA–RNA interactions directly (MacMillan *et al.*, 1997; Valcarcel and Green, 1996). In addition to their role in constitutive splicing, SR proteins have been shown to play an important role in alternative splicing by influencing 5' and 3' splice sites selection *in vitro* and *in vivo* (Graveley, 2000; Manley and Tacke, 1996).

Published reports on SR proteins coupled with *Arabidopsis* genome sequence analysis resulted in identification of at least 18 SR proteins (Golovkin and Reddy, 1998,

Figure 1. *Targeted expression of U1-70K antisense transcript in petal and stamens using the AP3 promoter. (A) Schematic diagram of U1-70K antisense construct used to transform Arabidopsis plants. AP3, promoter from AP3 gene; U1-70K antisense, U1-70K long cDNA in antisense orientation; Term, NOS3 terminator. (B) Expression of U1-70K antisense transcript results in flowers with rudimentary petals and stamens. Left panel: influorescence of a wild-type plant (top); wild type flower (bottom). Right panel: influorescence of a transgenic plant expressing U1-70K antisense transcript (top); flower from a transgenic plant lacking petals and stamens (bottom). (Golovkin and Reddy, 2003. Expression of U1 Small Nuclear Ribonucleoprotein 70k Antisense Transcript using APETALA3 promoter suppresses the development of Sepals and Petals, 1, Plant Physiology, August 2003. Vol. 132: 1884–1889, Figure 2. With permission from the American Society of Plant Biologists).*

1999; Lazar and Goodman, 2000; Lazar *et al.*, 1995; Lopato *et al.*, 1996a, 1996b, 1999a, 2002; Lorkovic *et al.*, 2000). Some plant SR proteins are homologues of metazoan SR family splicing factors whereas others are unique to plants with novel structural features (*Figure 2*). A few plant SR proteins have been shown to complement splicing-deficient S100 extract (Lopato *et al.*, 1996a, 1999a) whereas *Arabidopsis* ASF/SF2-like protein did not (Lazar *et al.*, 1995). Over-expression of one of the ASF/SF2-like proteins in transgenic plants modulated splice-site choice of several pre-mRNAs including its own pre-mRNA (Lopato *et al.*, 1999b).

Using the *Arabidopsis* full-length and C-terminal region of U1-70K in the yeast two-hybrid system we have isolated four cDNAs (*SRZ21, SRZ22, SR33 and SR45*) encoding novel serine/arginine-rich proteins that interact with U1-70K (Golovkin and Reddy, 1998, 1999). Two of these proteins (SRZ21 and SRZ22) interact with the full-

SR Proteins from Arabidopsis

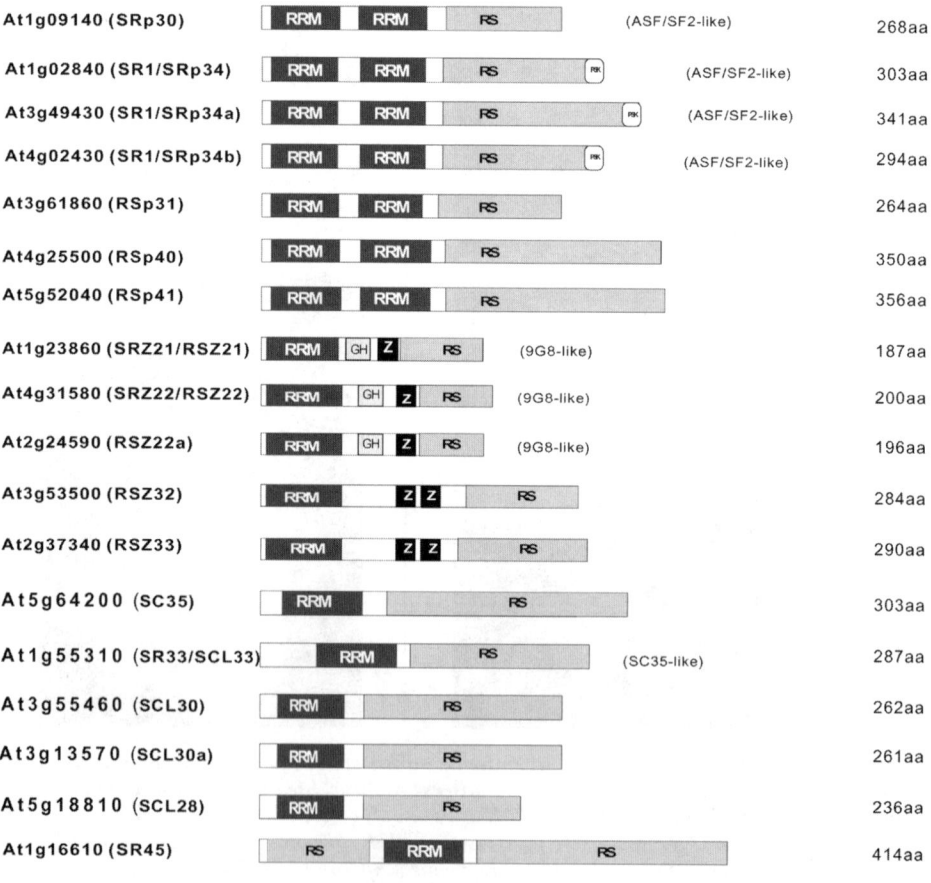

Figure 2. Schematic diagram showing various domains in Arabidopsis serine/arginine-rich proteins. RRM, RNA recognition motif; RS, arginine/serinine-rich; Z, zinc knuckle; GH; glycine hinge; PSK, a domain rich in proline, serine and lysine. Several of these proteins have alternatively spliced forms – only the longest protein is shown for each gene.

length U1-70K whereas the other two proteins (SR33 and SR45) interact with the C-terminal arginine-rich region as well as full-length U1-70K. *SRZ21* and *SRZ22* encode proteins with a molecular mass of 21 and 22 kDa, respectively. The SRZ proteins contain several well-defined modular domains characteristic of the SR family (Golovkin and Reddy, 1998). SRZ21 and SRZ22 contain a single RNA-binding domain with RNP2 and RNP1 consensus sequences, a glycine hinge, a zinc knuckle (CCHC motif) and a serine/arginine-rich region (*Figure 2*). The SRZ proteins are more similar to the human 9G8 splicing factor because it also has a CCHC motif (zinc knuckle). This motif is a feature found only in 9G8 splicing factor (Cavaloc *et al.*, 1994). However, the serine/arginine-rich region in the SRZ proteins differs from 9G8 splicing factor in lacking the consensus sequences (RRSRSXSX) (Cavaloc *et al.*, 1994). Both SRZ proteins, however, contain a glycine-rich region (glycine hinge) that is present in SF2/ASF, RSp55, SC35 and SR1 splicing factors but not in 9G8 (Manley and Tacke, 1996). These unique structural features of the SRZ proteins and the fact that the interaction occurs only with the full-length U1-70K indicate that the SRZ proteins represent a new group of serine/arginine-rich proteins.

SR33 and SR45 encode 33- and 45-kDa proteins, respectively. These proteins also have two well-defined modular domains (an RBD, and an RS domain) characteristic of proteins of the SR family. The SR33 protein contains one RBD at the N-terminus, and an arginine/serine-rich (RS) region at the C-terminus (*Figure 2*). The deduced amino acid sequence of SR33 revealed sequence similarity to animal serine/arginine-rich splicing factor, SC35. SR45 revealed some similarity between its RBD and other RNA-binding proteins whereas the RS domains did not show similarity with known SR proteins. The SR45 protein is unusual in having two distinct RS domains, one on either side of RBD (*Figure 2*) (Golovkin and Reddy, 1999). Unlike most SR proteins, SR45 has no paralogues in *Arabidopsis* or orthologues in metazoans (Golovkin and Reddy, 1999; Lorkovic *et al.*, 2000). A homologue of *Arabidopsis* SR45 is found in rice (AK070420) and maize (AY103850), which share 53–55% similarity with the AtSR45 at the amino acid level and have similar domain organization with two RS domains (*Figure 3*). In *Arabidopsis* as well as in rice a single gene codes SR45. SR33 and SR45 proteins are rich in proline content (12% in SR33 and 17% in SR45) (Golovkin and Reddy, 1999).

Among four plant U1-70K interacting proteins, SRZ22 (also called atRSZp22) has been shown to complement splicing deficient HeLa cell S100 extract (Lopato *et al.*, 1999a). In animals there is single ASF/SF2, which interacts with U1-70K. Although there are three ASF/SF2-like proteins present in *Arabidopsis* (*Figure 2*) (Lopato *et al.*, 1999a; Lorkovic *et al.*, 2000), none of these was isolated in the yeast two-hybrid screening with either full-length or C-terminal U1-70K. Unlike its animal counterpart, *Arabidopsis* ASF/SF2 does not complement S100 extract (Lazar *et al.*, 1995). These studies demonstrate that plant U1-70K interacts with a different set of SR proteins including some novel SR proteins, suggesting that early stages of spliceosome formation or splice site selection may differ from animals. The fact that *Arabidopsis* contains almost double the number of SR proteins including several plant-specific ones as compared to humans (18 in *Arabidopsis* and ten in humans) and four plant SR proteins (as opposed to two in humans) that interact with U1-70K suggests that these proteins are likely to play important roles during the splice site recognition and spliceosome assembly (Golovkin and Reddy, 1998, 1999; Lopato *et al.*, 2002; Lorkovic

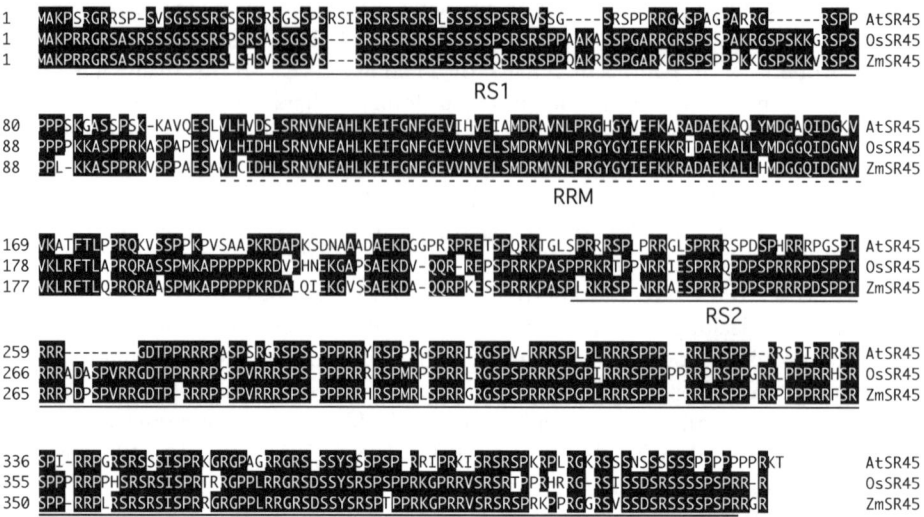

Figure 3. *Alignment of Arabidopsis SR45 (AtSR45) protein with its homologues from rice (OsSR45) and maize (ZmSR45). Protein sequences were aligned using the Clustal method. Reverse lettering shows identical amino acids. RS1, arginine/serine-rich domain one: RRM, RNA recognition motif; RS2, arginine/serine-rich domain two.*

et al., 2000). Further, the early steps of pre-mRNA splicing, especially intron recognition, are likely to differ from metazoans.

Several *Arabidopsis* SR proteins have been shown to interact with other proteins in the SR family. Interaction studies among SR proteins have shown a complex network of interactions (Golovkin and Reddy, 1999; Lopato *et al.,* 2002). The known interaction of U1-70K with plant SR proteins and interactions among SR proteins is summarized in *Figure 4*. In animals, only two SR proteins (SC35 and ASF/SF2) have been shown to interact with U1-70K. Although plants have homologues of SC35 and ASF/SF2, they are not known to interact with U1-70K. Instead, two 9G8-like SR proteins (SRZ21 and SRZ22), SR33 and SR45 have been shown to interact with U1-70K. In animals, 9G8 splicing factor is not known to interact with U1-70K. Further, the animal U1-70K has not been shown to interact with any double RS domain-containing protein.

4. A LAMMER-type protein kinase (AFC2) phosphorylates several plant SR proteins

Several recent studies suggest that phosphorylation of SR proteins is required for spliceosome assembly and splice site selection (Cao *et al.,* 1997). Furthermore, dephosphorylation of SR proteins is also necessary for the later stages of splicing, suggesting that the phosphorylation and dephosphorylation cycle of SR proteins plays a critical role in splicing (Cao *et al.,* 1997; Murray *et al.,* 1999). All SR proteins in animals are phosphoproteins and can be detected by a monoclonal antibody (mAb 104), which recognizes a phosphoepitope within the RS domain (Manley and Tacke, 1996). Several protein kinases that are capable of phosphorylating serine residues in

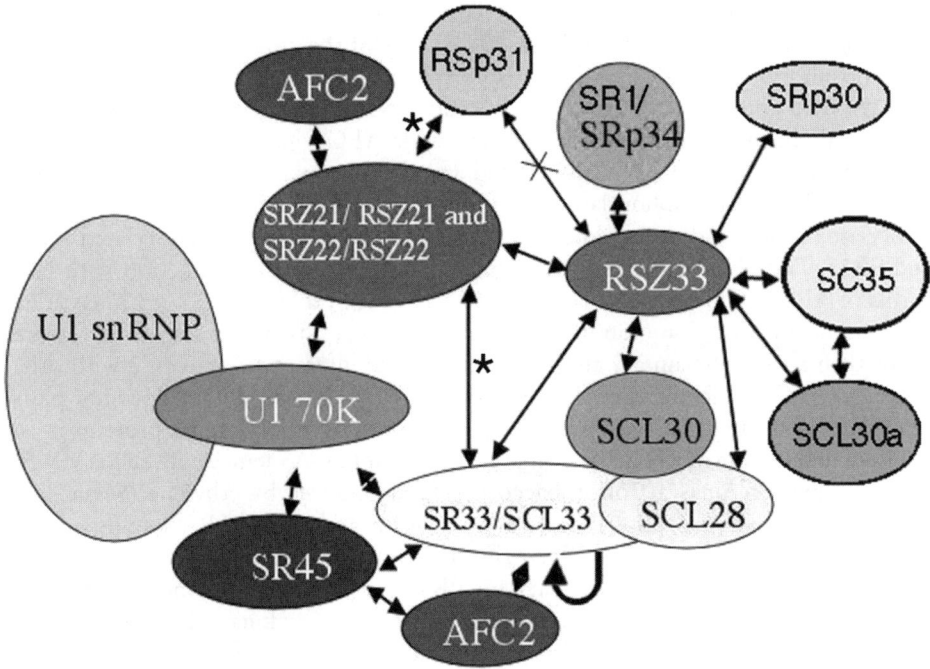

Figure 4. Diagram showing the interaction of U1-70K with plant SR proteins and the interaction between SR proteins, and SR proteins and a LAMMER-type protein kinase (AFC2). These interactions were identified using either yeast two-hybrid analysis and/or in vitro assays (Golovkin and Reddy, 1998, 1999; Lopato et al., 2002). Asterisks indicate that SR33 and RSp31 interact only with RSZ21. RSp31 and RSZ33 do not interact with each other.

the RS domain of SR proteins have been identified in animals (Colwill *et al.*, 1996a, 1996b; Duncan, *et al.*, 1998; Gui, *et al.*, 1994; Wang, *et al.*, 1998). Of these, two families of protein kinases, SRPK (SR protein-specific kinase) and Clk/Sty that differ in their substrate specificity have been extensively characterized (Colwill *et al.*, 1996a, 1996b). SRPKs are present in both the cytoplasm and the nucleus whereas Clk/Sty, a dual-specificity protein kinase that can phosphorylate serine, threonine and tyrosine residues (Colwill *et al.*, 1996a, 1996b), is present exclusively in the nucleus. The Clk/Sty kinases (also called LAMMER-type kinases) contain a unique sequence (EHLAMMERILGDLA) in subdomain X of the kinase catalytic domain. The interaction of Clk/Sty with some of its target proteins involves an RS-rich region at the N-terminus of Clk/Sty kinase. Several studies indicate that phosphorylation of SR proteins affects their mode of interaction with other proteins. Phosphorylation of ASF/SF2 in the RS domain enhances its interaction with U1-70K protein and affects its splicing activity (Xiao and Manley, 1997, 1998). Recruitment of SR proteins to sites of transcription is also mediated by phosphorylation (Misteli, 1999). SRPK and Clk/Sty can influence the distribution of SR proteins within the nucleus and an excess of SRPK can inhibit splicing. Furthermore, nuclear import of SR proteins is also regulated by phosphorylation (Yeakley *et al.*, 1999).

Three kinases that are similar to Clk/Sty kinase have been identified in the *Arabidopsis* genome (Bender and Fink, 1994). One of the kinases (AFC2) underwent autophosphorylation and heavily phosphorylated four plant SR proteins (SRZ21, SRZ22, SR33 and SR45) (Golovkin and Reddy, 1999). Co-precipitation studies have confirmed the interaction of SR proteins with AFC2 kinase and the interaction between AFC2 and SR33 is modulated by the phosphorylation status of these proteins. The non-phosphorylated form of SR33 interacts efficiently with autophosphorylated AFC2 and no binding was detected between non-phosphorylated AFC2 and SR33. The phosphorylated SR33 protein showed a weak interaction with both forms of AFC2 (Golovkin and Reddy, 1999). However, the plant Clk/sty kinase does not contain an RS-rich region in the N-terminus, suggesting that structural features other than an RS domain are also important in the interaction of AFC2 with other proteins. AFC1 which rescues the yeast double MAP kinase mutant does not phosphorylate SR proteins, suggesting that the ability to phosphorylate SR proteins is not likely to be a feature of all Clk/Sty kinases in *Arabidopsis*. A member of the LAMMER family of protein kinases from tobacco (PK12) is induced by ethylene (Sessa *et al.*, 1996). PK12 binds and phosphorylates both plant and animal SR proteins *in vitro* (Savaldi-Goldstein *et al.*, 2000). Using site-directed mutagenesis, it was demonstrated that the LAMMER motif is necessary for PK12 kinase activity but not for its interaction with the SR proteins. PK12, as expected of a Clk/Sty kinase, is localized to the nucleus. The finding that PK12 transcript and activity is induced by ethylene suggests that this hormone may influence pre-mRNA splicing in plants through phosphorylation of SR proteins. Over-expression of a tomato protein kinase related to AFC2 in *Arabidopsis* has resulted in abnormal development and changes in alternative splicing of some pre-mRNAs (Savaldi-Goldstein *et al.*, 2003). In addition to three LAMMER-type kinases there is an orthologue of SRPK in the *Arabidopsis* genome (www.Arabidopsis.org).

5. Localization and *in vivo* dynamics of SR45

5.1 *SR45 is localized to distinct domains in the nucleus*

The lack of a plant-derived *in vitro* splicing system requires that *in vivo* methods be used to identify spliceosomal proteins and interactions among them. Fluorescence resonance energy transfer (FRET) (Tsien and Tsien, 1990) and fluorescence recovery after photobleaching (FRAP) (Phair and Misteli, 2000) permit analysis of interaction of various proteins implicated in spliceosome formation and their dynamics *in vivo*. To study *in vivo* subcellular and suborganellar distribution and dynamics of SR45, we expressed a translational fusion between GFP and SR45 stably in plant cells (Ali *et al.*, 2003). The GFP–SR45 fusion protein was detected exclusively in the nucleus, indicating that the GFP–SR45 is targeted to the nucleus. Furthermore, GFP–SR45 is diffusely distributed as well as concentrated in distinct domains within the nucleus. This observation is consistent with the characteristic speckled distribution of SR splicing factors in animal cells (Misteli, 2001; Misteli *et al.*, 1997). In animals, it has been proposed that the diffused pattern of various nuclear proteins corresponds to active transcription sites and/or active sites of spliceosome assembly (Misteli and Spector, 1999; Zeng *et al.*, 1997). Therefore, the diffused distribution of SR45 is probably attributable to the presence of SR45 at the active sites of transcription as has been reported for snRNPs (Neugebauer and Roth, 1997; Wetterberg *et al.*, 2001). In all cell types of these transgenic lines, GFP–SR45 was localized to the nucleus (Ali *et al.*, 2003). Analysis of different cell types including root and shoot meristematic cells, root and hypocotyl epidermal cells, root hairs, leaf pavement cells, guard cells, mesophyll cells and trichomes revealed that SR45 was present mostly in round speckles, whereas a small amount was also present in a uniformly diffuse form throughout the nucleus. The speckled and diffused distribution was excluded from nucleoli (Ali *et al.*, 2003). The number of speckles varied greatly between and within different cell types.

5.2 *SR45 is highly dynamic in living cells*

The dynamics of SR proteins has not been studied in living plant cells. Using time-lapse microscopy, we analysed the dynamics and morphological alterations of SR45 speckles in transgenic plants. The SR45 speckles showed rapid movements detectable in real time (colour *Plate 13a*). The bigger speckles are relatively stationary, whereas smaller speckles, appearing as small dots against a diffuse background, appeared to move around a central point within a relatively small space (Ali *et al.*, 2003). Several speckles appeared and disappeared from the plane of focus indicating their movements within a 3-D space (arrows, colour *Plate 13a*). In addition, some speckles appeared to either bud off from a speckle or coalesce with another speckle (arrowheads, colour *Plate 13a*). The shape of speckles also changed over time.

5.3 *Inhibition of transcription affects intranuclear distribution of GFP-SR45*

In plants nothing is known about the effect of transcriptional activity on the spatial distribution of SR proteins. Therefore, we evaluated the effect of transcription inhibitors (actinomycin-D and α-amanitin) on the distribution of GFP-SR45. Treatment with actinomycin-D significantly changed the intranuclear distribution of GFP-SR45 by reducing the numbers of small speckles resulting in the formation of a few bigger round speckles (colour *Plate 13b*, left panel). On average the number of speckles in the actinomycin-D-treated seedlings was reduced to four per nucleus from 36 per nucleus in the untreated control (colour *Plate 13b*, right panel). In contrast, the size of speckles after actinomycin-D treatment increased four-fold the size of those in controls (colour *Plate 13b*, right panel). These data suggest that the SR45 in transcrip-tionally inactive cells is restricted to a few larger foci. Treatment of GFP-SR45 *Arabidopsis* seedlings with α-amanitin, an inhibitor of RNA polymerase II, resulted in disappearance of small speckles. Instead, only one and occasionally two bigger round speckles appeared (colour *Plate 13b*). In all these nuclei a substantial quantity of SR45 was also present in the nucleoplasmic pool, indicating that only the speckled pattern is affected by RNA polymerase II inhibition. Thus inhibition of RNA polymerase II activity restricted GFP–SR45 to one or two large speckles. Heat treatment inhibits transcription of many genes and is used routinely as one of the inhibitors of tran-scription (Boudonck *et al.*, 1999). At 42°C, drastic changes in the distribution of GFP–SR45 and the morphology of SR45-speckles were noticed (colour *Plate 13c*). Most of the nucleoplasmic distribution diminished to negligible levels and the shape of the speckles changed from normally rounded to irregularly shaped structures, which appeared coalesced all over the nucleus (colour *Plate 13c*). Exposure to cold, on the other hand, resulted in disappearance of almost all speckles and GFP–SR45 was evenly distributed in the nucleoplasm (colour *Plate 13c*). Our data indicate that the spatial dynamics of SR45 is similar to the intranuclear distribution of proteins involved in mRNA synthesis and processing including splicing factors SC35 and ASF/SF2. In control cells that are not treated with inhibitors of transcription, GFP–SR45 is present in a diffused nucleoplasmic pool as well as in many small round speckles throughout the nucleus except the nucleolus. In transcriptionally inactive cells, on the other hand, the protein relocates to a few large speckles (colour *Plate 13b*). Since SR45 interacts with U1-70K and has domains typical of splicing factors, the SR45 speckles may have a role similar to that in animals. Zeng *et al.* (1997) have demonstrated that RNA polymerase II, incorporation of bromouridine and splicing factors (detected by antiSC35 and mAB104) colocalize to numerous sites of preferred aggregation called dots and a diffuse nucleoplasmic pattern. In normal cells of GFP–SR45 plants we also observed such numerous sites in several cells. Since it has been known for some time that transcription and splicing are usually coupled (Maniatis and Reed, 2002; Neugebauer, 2002) and that several splicing factors interact with RNA polymerase II (Goldstrohm *et al.*, 2001; Maniatis and Reed, 2002), we speculate that the numerous dots that we observed are likely to be the actual sites of co-transcriptional splicing.

5.4 *Phosphorylation/dephosphorylation regulate the spatial distribution of GFP–SR45*

SR45 and several other SR proteins have been shown to be phosphorylated *in vitro* (Golovkin and Reddy, 1999; Misteli, 1999; Savaldi-Goldstein *et al.*, 2000). As the localization pattern of several SR proteins and interaction between SR proteins is known to be dependent upon their phosphorylation status (Misteli, 1999; Sacco-Bubulya and Spector, 2002), we examined the effect of protein phosphorylation and dephosphorylation on the nuclear distribution of GFP–SR45 by treating seedlings with either okadaic acid (a phosphatase inhibitor) or staurosporine (an inhibitor of protein kinases) for varying time periods. In control nuclei, a normal pattern of distribution consisting of several small speckles and diffused nucleoplasmic distribution excluding nucleoli was observed. In contrast, seedlings exposed to staurosporine showed drastically altered distribution of GFP–SR45 with several big irregularly shaped bodies, which in many cases coalesced indicating that inhibition of phosphorylation causes GFP–SR45 to move into larger irregular speckles (colour *Plate 13c*). This condensation of SR45 into large nuclear bodies was accompanied by substantial reduction in the nucleoplasmic proportion of GFP–SR45 (Ali *et al.*, 2003). The shape of these foci (mostly round) was very different from the large speckles observed in heat-shocked cells (irregular and coalesced clusters). Although both heat and actinomycin-D cause redistribution of SR45 into large structures, the pattern of speckles is distinct with each treatment, indicating that the pathways affected by these treatments may be different. The staurosporine-induced compaction of GFP–SR45 suggests that inhibition of kinase activity, hence the dephosphorylated status of SR45 leads to its aggregation. If inhibition of phosphorylation of SR45 induced the formation of large speckles, inhibition of dephosphorylation should promote disassembly of speckles. To test this hypothesis, we treated the GFP–SR45 seedlings with okadaic acid, a Ser/Thr protein phosphatase inhibitor. Okadaic acid had no effect on the localization pattern of GFP–SR45 (Ali *et al.*, 2003). However, as described later, okadaic acid did prevent the formation of big speckles in cells treated with transcription inhibitors (Ali *et al.*, 2003).

5.5 *Transcription-dependent redistribution of GFP–SR45 is inhibited by blocking phosphatase activity of the cell*

Since the inhibition of either the kinase activity or the transcriptional activity relocalized GFP–SR45 to bigger speckles (colour *Plate 13b and c*), it could be envisioned that these treatments block phosphorylation resulting in the condensation of GFP–SR45. This notion is strengthened by the fact that SR45, like other members of the SR family of splicing factors, has SR domains that could be the targets of kinases/phosphatases leading to the modulation of its interaction with other proteins and to its intranuclear distribution. If this is true then the inhibition of phosphatase activity should prevent the heat- and actinomycin-D-induced aggregation of GFP–SR45. To test this, we treated the GFP–SR45 seedlings with okadaic acid before they were heat shocked or treated with actinomycin-D. Prior treatment with okadaic acid prevented much of the redistribution of GFP–SR45 into large speckles caused by actinomycin-D treatment or irregularly shaped, large speckles induced by heat shock, suggesting that dephosphorylation leads to relocalization of SR45 into enlarged nuclear bodies (Ali *et al.*, 2003). Since cold treatment relocalized GFP–SR45 to diffused nucleoplasmic pattern (*Figure CS.15c*), we sought to know if this change in the localization was related to or affected by protein phosphorylation. The presence of okadaic acid in the medium did not affect dissipation of GFP–SR45 into the nucleoplasmic pool by cold treatment. In contrast, staurosporine not only inhibited the cold-induced dissipation of GFP–SR45 into nucleoplasm but also accentuated its speckled distribution (Ali *et al.*, 2003). Taken together these data suggest that the low- and high-temperature-induced cycling of GFP–SR45 between speckles and nucleoplasmic pool is dependent on protein phosphorylation.

We have also analysed the mobility of SR45 in the nucleus of living cells using fluorescent recovery after photobleaching (FRAP). In these studies we bleached an area in the nucleus corresponding to nucleoplasm or a speckle by a high-powered laser pulse and followed the mobility of SR45 by monitoring the recovery of fluorescence signal in the bleached area over time. Recovery of fluorescent signal was observed both in speckles and nucleoplasm. However, the movement of GFP–SR45 is much slower than GFP alone. The slow mobility of GFP–SR45 is mostly likely due to interaction of SR45 with other spliceosomal/nuclear proteins. Similar studies with fixed cells did not show the mobility of GFP or GFP-fused SR45 (Ali and Reddy, unpublished results).

6. Conclusions

Plant U1-70K interacts with at least four of the 18 SR proteins. Three of the four U1-70K-interacting proteins have novel structural features. SRZ21 and SRZ22 interact with only full-length U1-70K whereas SR33 and SR45 interact with the full-length as well as the C-terminal region of U1-70K. Furthermore, U1-70K-interacting SR proteins were found to interact with other SR proteins. AFC-2, a LAMMER-type kinase, phosphorylated all four U1-70K-interacting SR proteins. The interaction between AFC2 and SR33 is modulated by the phosphorylation status of these proteins. SR45, a plant-specific U1-70K-interacting protein, localizes to the nucleus and is distributed in both nucleoplasm and speckles of variable sizes and shapes. The cycling of SR45 between the nucleoplasm and speckles is dependent upon transcriptional and phosphorylation activities. The similarity of SR45 to the splicing factors in animal systems in its nuclear distribution in normal cells and its intranuclear relocation in response to transcription and phosphorylation status, together with previous findings that SR45 interacts with U1-70K and SR33, strongly suggest a role for this novel protein in plant pre-mRNA splicing.

Acknowledgements

We thank Dr. Irene Day for critically reading the manuscript. This work was supported by a grant from the Department of Energy, Division of Energy Biosciences (grant No. DE-FG03-01ER15199) to A.S.N.R.

References

Ali, G.S., Golovkin, M. and Reddy, A.S.N. (2003) Nuclear localization and in vivo dynamics of a plant-specific serine/arginine-rich protein. *Plant J.* **36**: 883–893.

Andersen, J. and Zieve, G.W. (1991) Assembly and intracellular transport of snRNP particles. *BioEssays.* **13**: 57–64.

Bender, J. and Fink, G. (1994) AFC1, a LAMMER kinase from *Arabidopsis thaliana*, activates STE12-dependent processes in yeast. *Proc. Natl Acad. Sci. USA* **91**: 12105–12109.

Black, D.L., Chabot, B. and Steitz, J.A. (1985) U1 as well as U1 small nuclear ribonucleoproteins are involved in pre-messenger RNA splicing. *Cell* **42**: 737–750.

Boudonck, K., Dolan, L. and Shaw, P.J. (1999) The movement of coiled bodies visualized in living plant cells by the green fluorescent protein. *Mol. Biol. Cell.* **10**: 2297–2307.

Brett, D., Hanke, J., Lehmann, G., Haase, S., Delbruck, S., Krueger, S., Reich, J. and Borka, P. (2000) EST comparison indicates 38% of human mRNAs contain possible alternative splice forms. *FEBS Lett.* **474**: 83–86.

Brown, W.W.S. and Simpson, C.G. (1998) Splice site selection in plant pre-mRNA splicing. *Annu. Rev. Plant Physiol. Plant Mol. Biol.* **49**: 77–95.

Burge, C.B., Tushl, T. and Sharp, P.A. (1999). Splicing of precursors to mRNAs by the spliceosomes. In: Gesteland, R.F., Cech, T.R. and Atkins, J.F. (eds) *The RNA World*, pp. 525–560. Cold Spring Harbor Laboratory Press: Cold Spring Harbor, New York.

Caceres, J.F., Misteli, T., Screaton, G.R., Spector, D.L. and Krainer, A.R. (1997) Role of the modular domains of SR proteins in subnuclear localization and alternative splicing specificity. *J. Cell Biol.* **138**: 225–238.

Cao, W., Jamison, S.F. and Garcia-Blanco, M.A. (1997) Both phosphorylation and dephosphorylation of ASF/SF2 are required for pre-mRNA splicing in vitro. *RNA* **3**: 1456–1467.

Cavaloc, Y., Popielarz, M., Fuchs, J.-P., Gattoni, R. and Stevenin, J. (1994) Characterization and cloning of the human splicing factor 9G8: a novel 35 kDa factor of the serine/arginine protein family. *EMBO J.* **13**: 2639–2649.

Colwill, K., Feng, L.L., Yeakley, J.M., Gish, G.D., Caceres, J.F., Pawson, T. and Fu, X.D. (1996a) SRPK1 and Clk/Sty protein kinases show distinct substrate specificities for serine/arginine-rich splicing factors. *J. Biol. Chem.* **271**: 24569–24575.

Colwill, K., Pawson, T., Andrews, B., Prasad, J., Manley, J.L., Bell, J.C. and Duncan, P.I. (1996b) The Clk/Sty protein kinase phosphorylates SR splicing factors and regulates their intranuclear distribution. *EMBO J.* **15**: 265–275.

Consortium, I.H.G.S. (2001) Initial sequencing and analysis of the human genome. *Nature* 409: 860–921.

Crispino, J.D. and Sharp, P.A. (1995) A U6 snRNA:pre-mRNA interaction can be rate-limiting for U1-independent splicing. *Genes Dev.* 9: 2314–2323.

Duncan, P.I., Stojdl, D.F., Marius, R.M., Scheit, K.H. and Bell, J.C. (1998) The Clk2 and Clk3 dual-specificity protein kinases regulate the intranuclear distribution of SR proteins and influence pre-mRNA splicing. *Exp. Cell Res.* 241: 300–308.

Ge, H. and Manley, J.L. (1991) A protein factor, ASF, controls cell-specific alternative splicing of SV40 early pre-mRNA *in vitro*. *Cell* 62: 25–34.

Goldstrohm, A.C., Greenleaf, A.L. and Garcia-Blanco, M.A. (2001) Co-transcriptional splicing of pre-messenger RNAs: considerations for the mechanism of alternative splicing. *Gene* 277: 31–47.

Golovkin, M. and Reddy, A.S. (1998) The plant U1 small nuclear ribonucleoprotein particle 70K protein interacts with two novel serine/arginine-rich proteins. *Plant Cell.* 10: 1637–1648.

Golovkin, M. and Reddy, A.S. (1999) An SC35-like protein and a novel serine/arginine-rich protein interact with *Arabidopsis* U1-70K protein. *J. Biol. Chem.* 274: 36428–36438.

Golovkin, M. and Reddy, A.S.N. (1996) Structure and expression of a plant U1 snRNP 70K gene: Alternative splicing of U1 snRNP 70K pre-mRNAs produces two different transcripts. *Plant Cell.* 8: 1421–1435.

Golovkin, M. and Reddy, A.S.N. (2003) Expression of U1snRNP 70K antisense transcript using APEATALA3 promoter suppresses the development of petals and stamens. *Plant Physiol.* 132: 1884–1891.

Goodall, G.J., Kiss, T. and Filipowicz, W. (1991) Nuclear RNA splicing and small nuclear RNAs and their genes in higher plants. *Oxford Surv. Plant Mol. Cell Biol.* 7: 255–296.

Grabowski, P.J., Nasim, F.H., Kuo, H.-C. and Burch, R. (1991) Combinatorial splicing of exon pairs by two-site binding of U1 small nuclear ribonucleoprotein particle. *Mol. Cell. Biol.* 11: 5919–5928.

Graveley, B.R. (2000) Sorting out the complexity of SR protein functions. *RNA* 6: 1197–1211.

Graveley, B.R., Hertel, K.J. and Maniatis, T. (1999) SR proteins are 'locators' of the RNA splicing machinery. *Curr. Biol.* 9: R6–R7.

Graveley, B.R. and Maniatis, T. (1998) Arginine/serine-rich domains of SR proteins can function as activators of pre-mRNA splicing. *Mol. Cell.* 1: 765–771.

Gui, J.-F., Lane, W.S. and Fu, X.-D. (1994) A serine kinase regulates intracellular localization of splicing factors in the cell cycle. *Nature* 369: 678–682.

Hilleren, P.J., Kao, H.Y. and Siliciano, P.G. (1995) The amino-terminal domain of yeast U1-70K is necessary and sufficient for function. *Mol. Cell Biol.* 15: 6341–6350.

Kohtz, J.D., Jamison, S.F., Will, C.L., Zuo, P., Lührmann, R., Garcia-Blanco, M.A. and Manley, J.L. (1994) Protein–protein interactions and 5'-splice-site recognition in mammalian mRNA precursors. *Nature* 368: 119–124.

Krainer, A.R., Conway, G.C. and Kozak, D. (1990) Purification and characterization of pre-mRNA splicing factor SF2 from HeLa Cells. *Genes Dev.* 4: 1158–1171.

Kuo, H., Nasim, F.H. and Grabowski, P.J. (1991) Control of alternative splicing by differential binding of U1 small nuclear ribonucleoprotein particle. *Science* 251: 1045–1050.

Lambermon, M.H., Simpson, G.G., Wieczorek Kirk, D.A., Hemmings-Mieszczak, M., Klahre, U. and Filipowicz, W. (2000) UBP1, a novel hnRNP-like protein that functions at multiple steps of higher plant nuclear pre-mRNA maturation. *EMBO J.* 19: 1638–1649.

Lazar, G. and Goodman, H.M. (2000) The *Arabidopsis* splicing factor SR1 is regulated by alternative splicing. *Plant Mol. Biol.* 42: 571–581.

Lazar, G., Schall, T., Maniatis, T. and Goodman, H. (1995) Identification of a plant serine-arginine-rich protein similar to the mammalian splicing factor SF2/ASF. *Proc. Natl Acad. Sci. USA* 92: 7672–7676.

Lopato, S., Mayeda, A., Krainer, A. and Barta, A. (1996a) Pre-mRNA splicing in plants: Characterization of Ser/Arg splicing factors. *Proc. Natl Acad. Sci. USA* **93**: 3074–3079.

Lopato, S., Waigmann, E. and Barta, A. (1996b) Characterization of a novel arginine/serine-rich splicing factor in *Arabidopsis*. *Plant Cell*. **8**: 2255–2264.

Lopato, S., Gattoni, R., Fabini, G., Stevenin, J. and Barta, A. (1999a) A novel family of plant splicing factors with a Zn knuckle motif: examination of RNA binding and splicing activities. *Plant Mol. Biol*. **39**: 761–773.

Lopato, S., Kalyna, M., Dorner, S., Kobayashi, R., Krainer, A.R. and Barta, A. (1999b) atSRp30, one of two SF2/ASF-like proteins from *Arabidopsis* thaliana, regulates splicing of specific plant genes. *Genes Dev*. **13**: 987–1001.

Lopato, S., Forstner, C., Kalyna, M., Hilscher, J., Langhammer, U., Indrapichate, K., Lorkovic, Z.J. and Barta, A. (2002) Network of interactions of a novel plant-specific Arg/Ser-rich protein, atRSZ33, with atSC35-like splicing factors. *J. Biol. Chem*. **277**: 39989–39998.

Lorkovic, Z.J., Wieczorek Kirk, D.A., Lambermon, M.H. and Filipowicz, W. (2000) Pre-mRNA splicing in higher plants. *Trends Plant Sci*. **5**: 160–167.

Luehrsen, K.R., Taha, S. and Walbot, V. (1994) Nuclear pre-mRNA processing in higher plants. *Nucl. Acids Res*. **47**: 149–193.

MacMillan, A.M., McCaw, P.S., Crispino, J.D. and Phillip, A.S. (1997) SC35-mediated reconstitution of splicing in U2AF-depleted nuclear extract. *Proc. Natl Acad. Sci. USA* **94**: 133–136.

Maniatis, T. and Reed, R. (2002) An extensive network of coupling among gene expression machines. *Nature* **416**: 499–506.

Manley, J.L. and Tacke, R. (1996) SR proteins and splicing control. *Genes Dev*. **10**: 1569–1579.

Misteli, T. (1999) RNA splicing: What has phosphorylation got to do with it? *Curr. Biol*. **9**: R198–R200.

Misteli, T. (2001) Protein dynamics: Implications for nuclear architecture and gene expression. *Science* **291**: 843–847.

Misteli, T., Caceres, J.F. and Spector, D.L. (1997) The dynamics of a pre-mRNA splicing factor in living cells. *Nature* **387**: 523–527.

Misteli, T. and Spector, D.L. (1999) RNA polymerase II targets pre-mRNA splicing factors to transcription sites in vivo. *Mol. Cell*. **3**: 697–705.

Murray, M.V., Kobayashi, R. and Krainer, A.R. (1999) The type 2C Ser/Thr phosphatase PP2cg is a pre-mRNA splicing factor. *Genes Dev*. **13**: 87–97.

Neugebauer, K.M. (2002) On the importance of being co-transcriptional. *J. Cell Sci*. **115**: 3865–3871.

Neugebauer, K.M. and Roth, M.B. (1997) Distribution of pre-mRNA splicing factors at sites of RNA polymerase II transcription. *Genes Dev*. **11**: 1148–1159.

Phair, R.D. and Misteli, T. (2000) High mobility of proteins in the mammalian cell nucleus. *Nature* **404**: 604–609.

Rappsilber, J., Ryder, U., Lamond, A.I. and Mann, M. (2002) Large-scale proteomic analysis of the human spliceosome. *Genome Res*. **12**: 1231–1245.

Reddy, A.S.N. (2001) Nuclear pre-mRNA processing in plants. *CRC Crit. Rev. Plant Sci*. **20**: 523–572.

Romac, J.M. and Keene, J.D. (1995) Overexpression of the arginine-rich carboxy-terminal region of U1 snRNP 70K inhibits both splicing and nucleocytoplasmic transport of mRNA. *Genes Dev*. **9**: 1400–1410.

Romac, J.M.-J., Graff, D.H. and Keene, J.D. (1994) The U1 small nuclear ribonucleoprotein (snRNP) 70K protein is transported independently of U1 snRNP particles via a nuclear localization signal in the RNA-binding domain. *Mol. Cell. Biol*. **14**: 4662–4670.

Rosbash, M. and Séraphin, B. (1991) Who's on first? The U1 snRNP-5' splice site interaction and splicing. *Trends Biochem. Sci*. **16**: 187–190.

Roscigno, R.F. and Garcia-Blanco, M.A. (1995) SR proteins escort the U4/U6.U5 tri-snRNP to the spliceosome. *RNA* **1**: 692–706.

Rossi, F., Forne, T., Antoine, E., Tazi, J., Brunel, C. and Cathala, G. (1996) Involvement of U1 small nuclear ribonucleoproteins (snRNP) in 5' splice site-U1 snRNP interaction. *J. Biol. Chem.* **271**: 23985–23991.

Sacco-Bubulya, P. and Spector, D.L. (2002) Disassembly of interchromatin granule clusters alters the coordination of transcription and pre-mRNA splicing. *J. Cell Biol.* **156**: 425–436.

Savaldi-Goldstein, S., Aviv, D., Davydov, O. and Fluhr, R. (2003) Alternative splicing modulation by a LAMMER kinase impinges on developmental and transcriptome expression. *Plant Cell.* **15**: 926–938.

Savaldi-Goldstein, S., Sessa, G. and Fluhr, R. (2000) The ethylene-inducible PK12 kinase mediates the phosphorylation of SR splicing factors. *Plant J.* **21**: 91–96.

Schuler, M.A. (1998) Plant pre-mRNA splicing. In: Bailey-Serres, J. and Gallie, D.R. (eds) *A Look Beyond Transcription: Mechanisms Determining mRNA Stability and Translation in Plants*, pp. 1–19. American Society of Plant Physiologists Baltimore, MD, USA.

Sessa, G., Raz, V., Savaldi, S. and Fluhr, R. (1996) PK12, a plant dual-specificity protein kinase of the LAMMER family, is regulated by the hormone ethylene. *Plant Cell.* **8**: 2223–2234.

Simpson, G.G., Clark, G.P., Rothnie, H.M., Boelens, W., van Venrooij, W. and Brown, J.W.S. (1995) Molecular characterization of spliceosomal proteins U1A and U2B" from higher plants. *EMBO J.* **14**: 4540–4550.

Stark, J.S., Bazett-Jones, D.P., Herfort, M. and Roth, M.B. (1998) SR proteins are sufficient for exon bridging across an intron. *Proc. Natl Acad. Sci. USA* **95**: 2163–2168.

Sterner, D.A., Carlo, T. and Berget, S.M. (1996) Architectural limits on split genes. *Proc. Natl Acad. Sci. USA* **93**: 15081–15085.

Tarn, W.-Y. and Steitz, J.A. (1994) SR proteins can compensate for the loss of U1 snRNP functions in vitro. *Genes Dev.* **8**: 2704–2717.

Tarn, W.Y. and Steitz, J.A. (1995) Modulation of 5' splice site choice in pre-messenger RNA by two distinct steps. *Proc. Natl Acad. Sci. USA* **92**: 2504–2508.

Tsien, R.W. and Tsien, R.Y. (1990) Calcium channels, stores, and oscillations. *Ann. Rev. Cell Biol.* **6**: 715–760.

Valcarcel, J. and Green, M.R. (1996) The SR protein family: pleiotropic functions in pre-mRNA splicing. *TIBS* **21**: 296–301.

Wang, H.-Y., Lin, W., Dyck, J.A., Yeakley, J.M., Songyang, Z., Cantley, L.C. and Fu, X.-D. (1998) SRPK2: A differentially expressed SR protein-specific kinase involved in mediating the interaction and localization of pre-mRNA splicing factors in mammalian cells. *J. Cell. Biol.* **140**: 737–750.

Wetterberg, I., Zhao, J., Masich, S., Wieslander, L. and Skoglund, U. (2001) In situ transcription and splicing in the Balbiani ring 3 gene. *EMBO J.* **20**: 2564–2574.

Wu, J.Y. and Maniatis, T. (1993) Specific interactions between proteins implicated in splice site selection and regulated alternative splicing. *Cell* **75**: 1061–1070.

Xiao, S.-H. and Manley, J. (1998) Phosphorylation-dephosphorylation differentially affects activities of splicing factor ASF/SF2. *EMBO J.* **17**: 6359–6367.

Xiao, S.-H. and Manley, J.L. (1997) Phosphorylation of the ASF/SF2 RS domain affects both protein-protein and protein-RNA interactions and is necessary for splicing. *Genes Dev.* **11**: 334–344.

Yeakley, J.M., Tronchere, H., Olesen, J., Dyck, J.A., Wang, H.Y. and Fu, X.D. (1999) Phosphorylation regulates in vivo interaction and molecular targeting of serine/arginine-rich pre-mRNA splicing factors. *J. Cell Biol.* **145**: 447–455.

Zeng, C., Kim, E., Warren, S.L. and Berget, S.M. (1997) Dynamic relocation of transcription and splicing factors dependent upon transcriptional activity. *EMBO J.* **16**: 1401.

Zhou, Z., Licklider, L.J., Gygi, S.P. and Reed, R. (2002) Comprehensive proteomic analysis of the human spliceosome. *Nature* **419**: 182–185.

18

Calcium/calmodulin-binding transcription activators in plants and animals

Fawzi Taleb and Hillel Fromm

1. Calcium: a versatile biological signal

Specific cell membrane receptor proteins interact with primary messengers (hormones, growth factors, external physical and chemical stumili) to trigger a sequence of biochemical events which work as a channel of information transfer. Following the interaction of primary messengers with their receptors, systems of 'second messengers' transduce extracellular signals into a biochemical and biophysical language understood by the cell machinery and transfer the signal from the membrane to target signalling proteins. This results in the cells performing the biological effect triggered by the signal (for example: DNA synthesis and cell division by growth factors) or, if the signal is aggressive, in inducing them to reply appropriately by a defence and adaptive mechanism. It is fascinating to consider that the cell contains thousands of signalling proteins and only about ten second messengers. The calcium ion (Ca^{2+}) is a second messenger playing a key role in responses of the cell to environmental and hormonal signals as well as in some intrinsic developmental processes. Additional information may be present in transient calcium signals as a result of single channel opening (termed 'blips') or the opening of synchronized clusters of channels (termed 'puffs') spatially restricted within the cell. Oscillation frequency, amplitude, duration and number can also be involved in the regulation of given cellular responses (Adler, 2002).

In addition to calcium channels of the plasma and nuclear membranes, many are found on the endoplasmic reticulum (ER) in animals and on the tonoplast (vacuolar membrane) and ER in plants. There is evidence that channels interact with calmodulin (CaM), cNMPs and other secondary messengers and it is likely that they play an important role in co-ordinating cross talk between these different regulatory messengers. Recent literature reported a number of intriguing candidate genes encoding Ca^{2+} channels. A unique gene in *Arabidopsis*, *TPC1* (At4 g03560), encodes a

The Nuclear Envelope, edited by D.E. Evans, C. Hutchison & J.A. Bryant.
© 2004 Garland Science/BIOS Scientific Publishers

channel that includes two EF hands. TPC1 expression enhances Ca^{2+} uptake in a yeast Ca^{2+} channel mutant (Furuichi *et al.*, 2001). In addition, other Ca^{2+} traffic pathways regulated by different agents exist in plants and animals.

Cytolosic Ca^{2+} influx is regulated by internal (the principal intracellular Ca^{2+} store is the vacuole in plants and the ER in animals) or external Ca^{2+} stores according to the type of the signal and the kind of the targeted cell. For example, in plants, cold induces extracellular Ca^{2+} uptake, wind induces internal Ca^{2+} stores release, whereas ABA induces Ca^{2+} release from both internal and external stores. It is noteworthy that nuclear and cytolosic Ca^{2+} homeostasis and regulation are different. Van der Luit *et al.* (1999) demonstrated distinct nuclear and cytosolic Ca^{2+} augmentation in tobacco plants in response to mechanical (wind) and cold shock, respectively.

It is well established that a high concentration of free cytosolic Ca^{2+} is toxic. Therefore the intracellular concentration of Ca^{2+} is tightly regulated by calcium-transporting systems. High affinity Ca^{2+}-pumps (Ca^{2+}-ATPases, especially in plants) and Ca^{2+} exchangers (Na^+/Ca^{2+} exchangers, particularly in animals) play a crucial role in raising and restoring the cytolosic Ca^{2+} levels in response to various stimuli. Several Ca^{2+}-ATPases (ECA, ACA) have been characterized. Two classes of Ca^{2+}-ATPase have been reported. In mammals type IIA are targeted to the endomembranes whereas type IIB are targeted exclusively to the plasma membrane. Only type IIB members are Ca^{2+}/calmodulin (CaM) regulated. In plants CaM-regulated ATPases are found in endomembranes including the ER and tonoplast as well as in the plasma membrane (Snedden and Fromm, 2001). It is noteworthy that type IIB Ca^{2+}-ATPases have an N-terminal CaM-binding domain in plants and a C-terminal CaM domain in mammals (Carafoli and Brini, 2000). H^+ antiporters (calcium exchanger) have been identified from *Arabidopsis* (*CAX* 1 and *CAX* 2) and mung bean. The increased antiport activity of the *CAX* 1 gene product causes severe depletion of free cytosolic Ca^{2+} levels due to pumping of Ca^{2+} from cytosol to vacuole (Reddy, 2001). These systems have a high affinity for Ca^{2+} and keep the cytosolic free Ca^{2+} concentration as low as 100–200 nM. Maintenance of low cytosolic Ca^{2+} is achieved by the ATP or proton motive force-driven removal of Ca^{2+} by pumps or transporters. This is a constant energy-consuming process as the concentration outside the cell could be 10 000-fold higher. Calcium versatility, amplitude and spatio-temporal patterning are maintained perfectly orchestrated with the other aspects of cell physiology with which they interact. Transporters become activated when the level of cytosolic calcium rises to roughly 1000 nM. Failure to maintain calcium activity within spatial and temporal boundaries can result in cell death through both necrosis and apoptosis (Berridge *et al*, 2000; Demaurex and Distelhorst, 2003).

2. Calmodulin as a calcium sensor

Calcium signals are transduced by sensors, predominantly proteins. The Ca^{2+}-binding protein CaM is one of the best-characterized Ca^{2+} sensors. It is a small soluble and acidic protein (148 amino acids, 17 kDa, pI ~ 4), with two globular domains, each with a pair of EF hands, linked to a central helix. Because of the presence of four EF-hand (helix-loop-helix) motifs, each of which is about 30 residues in length, only a tiny segment of the CaM molecule does not participate in Ca^{2+} binding. The binding of Ca^{2+} to CaM, between an aspartate (D) and glutamate (E) residue 12 units apart, is followed by a conformational change that exposes the hydrophobic pockets of CaM at each globular end permitting interaction with target proteins. The role of CaM as a calcium signal transducer within cells is well established. There is also evidence of extracellular CaM; however, its role is not clear. This ubiquitous Ca^{2+} receptor has the ability to modulate transient cytosolic or nuclear stoichiometric calcium fluctuations into a cascade of reiterative physiological, biochemical and genetic responses (membrane rearrangement, cyclic-nucleotide formation/breakdown, protein phosphorylation and dephosphorylation, and gene transcription).

CaM seems to exist abundantly in all eukaryotic cells (1–10 μM) and it is well conserved throughout evolution so that the CaM of mammals and eukaryotic microbes differ in only a few amino acids, leaving them functionally similar. However, it is noteworthy that unlike animals, plants use a large family of CaM isoforms and CaM-like proteins. This permits CaM genes to be differentially expressed in response to different stimuli. A single point mutation in a CaM isoform could alter its function. Several interesting studies have shown that CaM isoforms have different Ca^{2+} or protein target sensitivity. Furthermore there are differences in tissue and cell type expression, stimuli response or expression during development. This provides the cell with a complex and delicate regulatory control system taking into account responsiveness to Ca^{2+}, stoichiometry and target affinity (reviewed by Snedden and Fromm, 2001). Analysis of the human genome predicts the presence of common calcium-binding motifs similar to CaM (83 proteins). Another 73 proteins have been found to contain a C2 domain, a conserved calcium-binding motif originally described in protein kinase C (Klee and Means, 2002). It is interesting to note that Anandalakshmi *et al.* (2000) reported a plant CaM-like protein that suppresses post-transcriptional gene silencing (PTGS), an ancient eukaryotic regulatory mechanism in which a particular RNA sequence is targeted and destroyed.

A wide variety of ion channels are Ca^{2+}/CaM-regulated. The list includes voltage-gated Ca^{2+} channels (VGCCs), various Ca^{2+}- or ligand-gated channels, Trp family channels, and even the Ca^{2+}-induced Ca^{2+} release channels from organelles (Siami and Kung, 2002). Ehlers and Augustine (1999(Q1)) proposed three models for CaM-dependent regulation of VGCCs. In each model there is a tethering site, where CaM is constitutively bound, and a CaM-effector site through which Ca^{2+}-associated CaM modulates channel gating. Once the channel is activated, tethered Ca^{2+}-CaM binds to the effector site, which inactivates or facilitates the channel. The CaM-binding IQ-like motif on the cytoplasmic tail of the channel may act as either the tethering or the effector site or as both.

On the other hand, this Ca^{2+}-binding protein interacts with and regulates various proteins including CaM-binding transcription activators, CaM-dependent protein kinases (CaMKs), calcineurin, NF-AT and AP-1, which are all involved in transcriptional regulation. More fascinating is the diversity of these binding partners. The principal differences in CaM interaction with various peptides are associated with the N-terminal domain of CaM (Kurokawa *et al.*, 2001). It is clear that calcium dynamics and various CaM-binding proteins affect the three-dimensional structure of CaM. These phenomena occur not only in CaM but also in other proteins (for example prions). Among the large number of interactions, there are at least three different cases (Hoeflich and Ikura, 2002):

- Auto-inhibitory domain (AID) displacement used for example by CaM-dependent serine/threonine protein kinases (such as CaM kinase I/II/IV) and the phosphatase calcineurin
- Active site remodelling used for example by anthrax adenyl cyclase
- CaM-induced dimerization of membrane proteins used as a system for channel activation
- In addition to CaM, Other Ca^{2+} sensors are active and contribute to a new layer of Ca^{2+} signal fine-tuning processes. They can be divided into three major groups:
- CaM-like and other EF-hand containing Ca^{2+}-binding proteins
- Ca^{2+}-regulated protein kinases with EF hands found only in plants and protozoan
- Ca^{2+}-binding proteins without EF-hand motifs.

3. Regulation of Ca^{2+}/CaM flux by the nuclear envelope

The boundary between the nucleus and the cytoplasm in eukaryotic cells is defined by the nuclear envelope (NE). It has two typical unit membrane structures enclosing a flattened sac or lumen and connected at circular apertures: nuclear pore complexes (NPCs). The NPC is a complex and dynamic transport machine with dual permeability characteristics. For many years the area of nuclear/cytoplasmic calcium fluxes and their control remained controversial. NPCs could be seen as large 'gate channels' for the free and passive osmotic diffusion of ions and small macromolecules. Molecules of up to 20 kDa can pass through freely and they are able to transport macromolecules with diameters of up to circa 39 nm. If this were the case, nuclear Ca^{2+} signatures could be seen as mere cytosolic Ca^{2+} waves generated in the nucleus from passive transmission through the NPC. However, patch clamp studies show that the NE between the cytoplasm and nucleoplasm plays an active and essential role in the regulation of Ca^{2+} signals inside the nucleus. It contains proteins that regulate and respond to changes in nucleosolic Ca^{2+} concentration. For example, patch clamping of the NE shows ionic conductances that close during the translocation of macromolecules through the NPC. Nuclei from plant cells are also able to generate their own calcium signals independently of changes in calcium ion concentration in the cytosol (Pauly et al., 2000). Moreover, the NE of animals contains also numerous specific calcium-transporting systems: a calcium pump (Lanini et al., 1992), an InsP 3-sensitive calcium channel and a cADPr-ryanodine modulated channel (Gerasimenko et al., 1995) and the receptor InsP 4 (Koppler et al., 1996.). Evidence for a NE plant Ca^{2+}-pump has been obtained using light and electron microscope immunocytochemistry and antibodies raised to a plant homologue of the mammalian sarcoplasmic/ER (ER/SR) Ca^{2+} pump (SERCA) This work suggests that a SERCA pump is present at the NE of plant as well as animal cells (Downie et al., 1998).

Ca^{2+} signals might also originate from sources within the nucleus itself. In this case, two models have been proposed: either the nuclear Ca^{2+} derives from the Ca^{2+} liberated in the immediate perinuclear vicinity of the NE, or it is liberated from the nuclear reservoir (via IP3R on the inner membrane, eventually via ryanodine receptors, RyR) located in the lumen of the NE. In all cases, the calcium signal propagates across the nucleoplasm by simple diffusion. It is dissipated by storage into the ER or the nuclear Ca^{2+} pool. This pool results from the continuity of the ER with the lumen of the NE or through the nuclear Ca^{2+}-ATPase IP_4 receptor that is located on the outer membrane of the NE. Nuclear IP$_3$R or cADP ribose receptors could be involved in the capacitative influx of Ca^{2+} (Petersen *et al.*, 1998; Rogue and Anant, 1999).

There is evidence that the nucleus contains more intricate calcium signalling machinery that can release calcium locally in discrete regions of the nuclear interior. Echevarría *et al.* (2003) identified a nucleoplasmic reticulum in SKHep1 epithelial cells. It is a fine, branching intranuclear network that was continuous with the NE and the ER. They showed that these structures are a nuclear calcium-storing network. The nucleoplasmic reticulum can give rise to localized calcium gradients in the nuclear interior, and calcium-dependent events can be regulated differentially in the nucleus, just as they are in the cytosol.

Compartmentalization of the cell creates a physical separation between transcription and translation and gives rise to a need for inter-compartmental transport of macromolecules. Transcription-regulatory proteins have to undergo nucleocytoplasmic transport in order to fulfil their function. Controlling the access of transcriptional regulators to their target genes, at the level of transport complex formation is a very efficient way to adapt gene expression to the cell environment. Molecules above 40 kDa are imported into the nucleus by an active signal-mediated process. It involves nuclear localization sequences (NLS) recognized by specific importins or nuclear export sequences (NES) recognized by specific exportins, binding of the protein to the NPC, its translocation through the central channel of the pore and its release in the nucleoplasm or the cytoplasm (Chapters 7, 10). Importins interact with the NPC through their high affinity to FG-rich nucleoporin sequences (Stewart *et al.*, 2001). After crossing the nuclear pore, the importin cargo complex binds to nucleoplasmic GTP-Ran. This will trigger the dissociation of the import substrate/receptor interaction and release of the cargo into the nucleus (Bayliss *et al.*, 2002; Siebrasse and Peters, 2002). CaM enters the nucleus apparently in an ATP-independent way. Data suggest that CaM diffuses freely through nuclear pores and that CaM-binding proteins in the nucleus act as a sink for Ca^{2+}-CaM, resulting in accumulation of CaM in the nucleus on elevation of intracellular free Ca^{2+} (Liao *et al.*, 1999). On the other hand, a different mode of nuclear import for proteins with classical NLSs depending on cytoplasmic calcium and CaM has been reported. This transport process is not affected by GTP and is inhibited by CaM antagonists. In addition, it requires ATP. It was proposed that CaM-dependent nuclear transport of cNLS substrates is activated in response to external stimuli. The increase of intracellular calcium following cell stimulation inhibits GTP-dependent nuclear transport; the calcium spark then acts through CaM to stimulate the novel GTP-independent mode of import (Stochaj and Rother, 1999). Therefore, this pathway may function only under special stress conditions.

4. CaM-binding proteins

The 115-Mb *Arabidopsis* genome is estimated to contain 25 498 genes that are derived from about 11 000 gene families (Bennetzen, 2001). Based on screening of expression libraries with labelled CaM or sequence similarity alignments, Reddy *et al.* (2002) identified a total of 27 CaM-binding protein (CBP) families in *Arabidopsis*. Among these, 16 families were represented by 2–20 members (a table representing all the 27 families and their respective members can be viewed online at: http://www.arabidopsis.org/info/genefamily/CBP.html). Their analyses showed that plants have a unique set of

CBPs, and several CBPs have a large number of paralogues. Including all paralogues, they reported about 100 CBPs. Some CBPs are common to plants and animals, whereas others are unique to one kingdom. The primary sequence of the CaM-binding domain (CBD) in different CBPs is not conserved (Rhoads and Friedberg, 1997). CBP sequences share minimal structural characteristics. They all have the potential to fold into a basic, amphiphilic alpha helix. They display large hydrophobic residues in conserved positions either 1-5-10 or 1-8-14, which point to one face of a presumed α helical conformation. These structural characteristics have been used to predict several dozen potential CaM-binding sites from DNA and protein databases, as well as to engineer synthetic CaM-binding peptides (Vetter and Leclerc, 2003). Recent bioinformatics efforts have divided more than 300 CBPs in the CaM target database into five classes (1-14 motif, 1-10 motif, 1-16 motif, IQ motif, other CBPs). Valuable software related to CaM-binding site search and analysis in new putative amino-acid sequences is also available (http://calcium.uhnres.utoronto.ca/ctdb/flash.htm).

Using these resources, O'Day (2003) proposed a classification of CBPs into three categories:

- Ca^{2+}-dependent grouped into two related motifs, called 1-8-14 and 1-5-10 based on the position of conserved hydrophobic residues
- Ca^{2+}-independent
- Ca^{2+}-inhibited.

However this classification, like the previous ones (see the CaM target database http://calcium.uhnres.utoronto.ca/ctdb/flash.htm) has many exceptions and misfits.

Although the presence of a modulator in second messenger pathways is common, it is not a canonical rule. Sometimes Ca^{2+} can bind directly to its functional target. An example is the DREAM protein (downstream regulatory element antagonist modulator). It is a multifunctional Ca^{2+}-binding transcription factor. This protein variously named calsenilin, DREAM, and KChIP3 (potassium channel interacting protein-3) is expressed in the mammalian brain and belongs to the recoverin sub-branch of the EF-hand superfamily. It possesses four EF-hand Ca^{2+}-binding motifs (but binds only three Ca^{2+}, EF_1 being inactive and EF_2 binding Mg^{2+} as well). At low Ca^{2+} levels DREAM is a Ca^{2+} sensor that binds, in a tetrameric form, to the DRE silencer element of the prodynorphin gene and represses its transcription. Transcription occurs upon stimulation, when Ca^{2+} levels rise. Ca^{2+}-bound DREAM forms dimers that dissociate from the DRE site and may regulate cytosolic target proteins such as presenilin-2 or voltage-gated potassium channels (An et al., 2000; Ikura et al, 2002). The eukaryotic parasite Entamoeba histolytica contains a 22-kDa EF-hand Ca^{2+}-binding protein, named URE3-BP, that regulates the expression of virulence factor genes (Gilchrist et al., 2001). It binds specifically to the upstream regulatory element-3 (URE3) of the lectin heavy subunit gene and might regulate its expression in a DREAM-like way, by calcium-dependent phosphorylation of the URE3-BP protein. In conclusion, DREAM-like proteins form a new family of transcription activators able to interact directly with Ca^{2+} without a CaM mediation. Moreover the Janus kinase-signal transducer and activator of transcription (JAK-STAT) pathway transmits extracellular signals directly to target gene promoters in the nucleus, providing a mechanism for transcriptional regulation without second messengers. Evolutionarily conserved in eukaryotic organisms from slime moulds to humans, JAK-STAT signalling appears to be an early mechanism of intercellular communication that has co-evolved with a myriad of cellular signalling events (Aaronson and Horvath, 2002).

5. Regulation of transcription

Snedden and Fromm (2001) discussed possible alternative roles of CaM in regulating transcription in plants. In general transcription factor function can be regulated in three possible ways.

5.1 Transcription factor changes its activity upon ligand binding (from activator to repressor or the reverse pathway)

For instance, the S100B protein, a member of the S100 subfamily of the EF-hand superfamily, interacts with and regulates, in a Ca^{2+}-dependent manner, the tumour suppressor protein, p53 which functions as a transcriptional activator or repressor of numerous genes. The S100B protein interacts with the C-terminal portion of p53 (but does not require the extreme C-terminal end of p53 to inhibit its activity) and inhibits both p53 tetramerization and phosphorylation (Lin et al., 2001).

A second interesting example is the Id subfamily of helix-loop-helix (HLH) proteins. They interact with other proteins involved in regulating cellular proliferation and differentiation. Roberts et al. (2001) reported interactions between Id proteins and members of the Pax-2/-5/-8 subfamily (regulating several developmental processes) of paired domain transcription factors, resulting in the disruption of DNA-bound complexes containing Pax-2, Pax-5 and Pax-8.

5.2 Ligand binding to transcription factors triggers their movement to their targets

CaM binding to its targets may facilitate nucleocytoplasmic transport, different from the Ran-GTPase translocation mechanism. For example, two members of the NF-kappa B/Rel family of transcription factors, c-Rel and RelA, interact directly with Ca^{2+}-loaded CaM. CaM binds c-Rel and RelA after their release from I kappa B and can inhibit nuclear import of c-Rel while letting RelA translocate to the nucleus and act on its target genes. CaM can therefore differentially regulate the activation of NF-kappa B/Rel proteins following stimulation (Antonsson et al., 2003). The function of NF-kappa B is primarily regulated by I-kappa B family members, which ensure cytoplasmic localization of the transcription factor in the resting state. Upon stimulus-induced I-kappa B degradation, the NF-kappa B complexes move to the nucleus and activate NF-kappa B-dependent transcription. Over the years, a second regulatory mechanism, independent of I-kappa B, has become generally accepted. Changes in NF-kappa B transcriptional activity have been assigned to phosphorylation of the p65 subunit by a large variety of kinases in response to different stimuli (Vermeuden et al., 2002).

5.3 Phosphorylation can directly modulate the activity of transcription factors on at least five levels of regulation

In addition to a direct regulation of a given transcription factor, phosphorylation and dephosphorylation can also control transactivation by affecting interacting proteins, the general transcription machinery, or chromatin architecture (Holmberg *et al.*, 2002). In animals, Ca^{2+}/CaM-dependent protein kinases (CaMK) react in a cascade manner. In plants, calcium-dependent protein kinases or CaM-like domain protein kinases (CDPKs) are responsible for numerous Ca^{2+}-stimulated protein kinase activities. The *Arabidopsis* genome is predicted to encode 34 different CDPKs. Four distinct domains typify CDPK family members: a non-conserved N-terminal domain, a protein kinase domain, an auto-inhibitory domain, and a CaM-like domain. The latter contains Ca^{2+}-binding EF hands allowing the protein to function as a Ca^{2+} sensor. The number and position of EF hands in a specific CDPK is related to variations in the allosteric properties of Ca^{2+} binding and the activation threshold. Under low Ca^{2+} concentrations, the auto-inhibitory domain is bound by the kinase domain, keeping the protein activity low. Upon binding Ca^{2+} via the EF hand motifs, CDPKs undergo conformational changes that release the auto-inhibitory domain from the catalytic site thus activating the protein. In addition to Ca^{2+}, reversible phosphorylation also may regulate CDPK kinase activity. Although the role of autophosphorylation in the activities of CDPKs is unclear, it is possible that other protein kinases modulate CDPK activation (Cheng *et al.*, 2002).

The effect of phosphorylation can be classified into four major groups: stability of transcription factors; cellular localization of transcription factors; protein–protein interactions and oligomerization; DNA binding and transcription factor activity. Cross talk between the different classes and specific cases are common. These groups will be considered in more detail below.

Stability of the transcription factor
Phosphorylation of p53 protects against its degradation. N-terminal phosphorylations are important for stabilizing p53 and are crucial for acetylation of C-terminal sites, which in combination lead to the full p53-mediated response to genotoxic stresses (Appella and Anderson, 2001).

Cellular localization of a transcription factor
The Ca^{2+} activation of calcineurin, also known as protein phosphatase 2B, is mediated by two Ca^{2+}-binding proteins, the integral subunit calcineurin B(CnB) and CaM. NF-AT1 is a transcription complex consisting of nuclear (NF-ATn) and cytoplasmic components (NF-ATc). NF-ATc undergoes cytoplasmic-to-nuclear translocation upon dephosphorylation by calcineurin. Only the non-phosphorylated NF-ATc moves into the nucleus and binds DNA in association with partner proteins such as AP-1 (Fos/Jun) transcription factors promoted, phosphorylated and activated by protein kinase C (PKC)/Ras (Macian *et al.*, 2001).

Protein–protein interactions and oligomerization (binding of mediators)
The transcription factor CREB (cyclic AMP response element-binding protein) plays a role in glucose homeostasis, growth-factor-dependent cell survival, and is believed to be implicated in learning and memory. CREB is phosphorylated in response to various signals. Phosphorylation of CREB at Ser133 promotes recruitment of the transcriptional co-activator CRBP (CREB-binding protein) and its paralogue p300. CRBP, in turn, mediates transcriptional activation through its association with RNA polymerase II (Pol II) complexes and through intrinsic histone acetyltransferase activity. Target gene activation is

terminated by the serine/threonine phosphatase PP-1-mediated dephosphorylation of CREB (Mayr and Montminy, 2001). Phosphorylation of serine 142 in CREB by CaM Kinase II (CaMKII) leads to dissociation of the CREB dimer without impeding DNA binding capacity. CaMKII-modified CREB binds to DNA efficiently as a monomer; however, monomeric CREB is unable to recruit the CRBP even when phosphorylated at serine 133. Thus, CaMKII confers a dominant inhibitory effect on transcription by preventing dimerization of CREB (Wu and McMurray, 2001).

A second example in this category is the regulation of CDPKs by 14-3-3 isoforms upon phosphorylation. Putative 14-3-3 consensus binding sites, R-S/T-X-S-X-P, were found in the AtCPKs, AtCPK24 and AtCPK28 (*Arabidopsis* CDPKs). In the presence of Ca^{2+}, CDPK autophosphorylation may be induced. 14-3-3 proteins may regulate the activities of these enzymes by binding specific phosphorylated residues (Camoni *et al.*, 1998).

DNA-binding activity

The binding of transcription factors to promoter sites can be regulated directly or indirectly by protein phosphorylation. Phosphorylation within or near these domains introduces a negative charge that may be incompatible for efficient DNA binding. DNA-binding activity of transcription factors may also be regulated indirectly by phosphorylation at residues remote from the DNA-binding domain. The phosphorylation-dependent oligomerization of the STAT or SMAD families, which facilitates their nuclear import, is also essential for DNA binding (Whitmarsh and Davis, 2000). In addition, Yin *et al.* (1995) reported that the DNA binding of purified bacterially expressed dCREB2 is inhibited by phosphorylation *in vitro* by calcium/CaM-dependent protein kinase II.

Activity of a transcription factor

A good example of this type of activity modulation was reported by Ikura *et al.* (2002) in vertebrates. When the Ca^{2+} concentration increases within the cell, CaM binds and activates its target kinases including CaM kinase IV (CaMKIV) and CaM kinase kinase (CaMKK). CaMKK phosphorylates and fully activates CaM-KIV, resulting in transcriptional activation through phosphorylation of transcriptional factors. CaMKK also phosphorylates protein kinase B (PKB), which leads to the inhibition of apoptosis.

Another example is NtCDPK2 (*Nicotiana tabacum* CDPK). It is apparently activated through direct phosphorylation by an upstream protein kinase.

Dephosphorylation is as important as phosphorylation in controlling signalling pathways. A soluble phosphoserine phosphatase from winged bean shoots dephosphorylates an autophosphorylated winged bean CDPK1 (WbCDPK1) *in vitro* (Ganguly and Singh, 1999). It is thought that this action releases the inhibitory effect of autophosphorylation. These findings indicate an intricate and dynamic interplay between protein kinases and phosphatases in regulating some CDPK activities.

6. CaM-regulated transcription factors

By screening cDNA expression libraries with labelled CaM, Bouché *et al.* (2002) identified a group of six *Arabidopsis* genes encoding nuclear CaM-binding proteins. This family of proteins was characterized and designated CaM-binding transcription activators (CAMTAs). Independently, Yang and Poovaiah (2002) reported the six AtSR genes (*Arabidopsis thaliana* signal-responsive genes) encoding for six CBPs. These genes are the same as those coding for CAMTAs (see *Table 1*).

CAMTAs are widely distributed among multicellular organisms (including human CAMTAs designated HsCAMTA) and contain a conserved domain organization. Present predominantly in the nucleus, they are composed of a transcription activation domain and two types of DNA-binding domains designated the CG-1 domain and the transcription factor immunoglobulin domain (TIG), ankyrin repeats, and a varying number of IQ CaM-binding motifs (see *Figure 1*).

It is possible that the 'IQ motifs' (IQxxxRGxxxR), which appear in tandem repeats could bind multiple CaM molecules in a Ca^{2+}-independent manner. These motifs are used by many CaM-dependent proteins (calcineurin, IP_3 receptors, plasma membrane Ca^{2+} pump, CaM-binding transcription activators or CAMTA), but are far from being a *sine qua non* criterion to identify CaM-interacting proteins (Hoeflich and Ikura, 2002). A putative NLS was detected in the CG-1 domain of all CAMTAs identified so far, this signal directs CAMTAs to the nucleoplasm. Rhee *et al.* (2000) have designed an assay to detect active NLSs and NESs. A modified bacterial protein (mLexA) was fused to the yeast Gal4p (Gal4AD) activation domain with or without the SV40 large T-antigen NLS. In the import assay, NLS functional proteins fused to mLexA-Gal4AD enter the cell nucleus and activate the reporter gene expression. In the export assay, NES functional proteins fused to mLexA-SV40 NLS-Gal4AD exit into the cytoplasm, decreasing the reporter gene expression. This approach could be used as an alternative way to localize CaM-binding proteins.

6.1 *DNA-binding specificity of CAMTAs*

The control of transcription is the result of an intricate network of cross talk between various channels, CaM isoforms, CBPs and other regulatory elements. During the last two decades, several transcription factors with a sequence-specific DNA binding site have been identified. Sequence-specific DNA-binding (SSDB) factors play central roles in transcription, DNA replication, recombination and repair. Wang *et al.* (1998) described a simple procedure for high-throughput identification of SSDB activities without prior knowledge of their target genes or binding sequences. Using common gel-retardation assays and PCR, oligonucleotide molecule factors from a population of completely random oligonucleotides that specifically bind to cellular SSDB, sequences are selected, amplified and cloned. Oligonucleotide DNA libraries enriched in DNA

Table 1. CAMTA genes, their AtSR equivalents and their respective gene accession numbers

Gene name by Bouché et al., 2002	Gene name by Yang and Poovaiah, 2002	Gene accession number
CAMTA.1	AtSR2	At5g09410
CAMTA.2	AtSR4	At5g64220
CAMTA.3	AtSR1	At2g22300
CAMTA.4	AtSR5	At1g67310
CAMTA.5	AtSR6	At4g16150
CAMTA.6	AtSR3	At3g16940

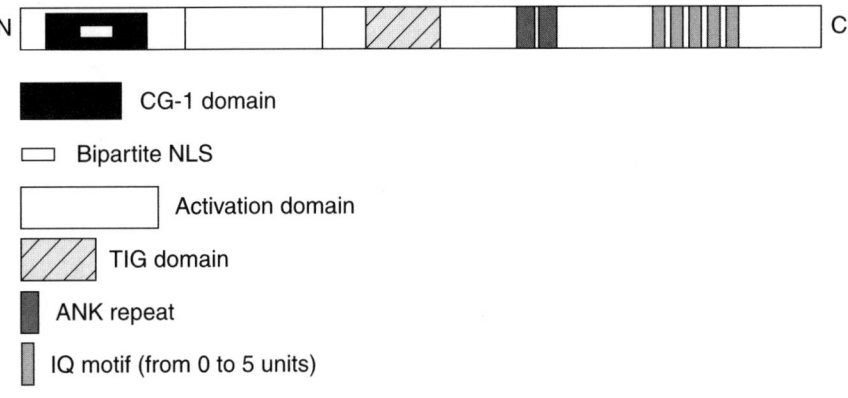

Figure 1. Predicted structure of a CAMTA protein.

molecules containing specific sequences recognized by SSDB factors are rapidly screened to identify a large number of SSDB activities, including those that are differentially regulated by developmental and environmental signals. With identified oligonucleotides as probes, the corresponding SSDB factors can be isolated and analysed with respect to their structures, regulation and functions. Da Costa e Silva (1994) reported that parsley CG-1 bound to a DNA fragment in which CGCG was crucial for DNA binding. Yang and Poovaiah (2002) used a pool of 30 random sequences of oligonucleotides in order to test whether the nuclear protein AtSR1 (or CAMTA 3, At2g22300) had a specific interacting DNA element. Gel-retarded oligonucleotide DNA molecules showed a common 6-bp element ACGCGG (or CCGCGT) in 50% of the cases. All of them shared the sequence CGCG in the middle of the (A/C/G)CGCG(C/G/T) consensus CGCG-box. Similar experiments, using genes with CGCG cis-acting elements (EIN3-330-295, CaM2-221-199, phyA-162-104) revealed that AtSR1 had specific binding activity for these DNA fragments. We compared all the probe sequences used above and a sequence from the DREB1C promoter and we found a broader consensus – AAA(C)(Xn)(A/C/G) CGCG(C/G/T) with $n < 18$ (see *Figure 2*).

Yang and Poovaiah (2002) reported the presence of the CGCG-box in the promoters of a large number of genes from various multicellular organisms and showed that *cis*-acting elements ACGCGG/CCGCGT were present in the promoter regions of about 130 genes (more than two copies) in the *Arabidopsis* genome. From the TAIR database and using the patmatch program (http://www.arabidopsis.org/cgi-bin/patmatch/nph-patmatch.pl), we extracted and classified genes containing AAACGCGG in the 1-kb region upstream from their UTRs. In 298 hit sequences, we obtained 174 genes with known or putative function. Fifty-two genes coding for proteins related to transcriptional regulation and cellular signalling are presented to illustrate this (see *Table 2*).

Our results complement those reported by Yang and Poovaiah (2002). A full investigation of the results from the broader consensus: AAA(C)(Xn, $n <$ 18)(A/C/G)CGCG(C/G/T) against the TAIR database should be attempted.

```
DREB1C    GTTTCTTATCCACGTGGCATTCACAGAGACAGAAACTC                    CGCGT
EIN3                          CTCCGCGTAAGAAACAAT    TAATTACATACCGCGT
Parsley (inverted)               GATCTTCCCCAAACAACCTCTAATT       GCGC
CAM2                           GATCCGCGTAGAAAC                   CGCGT
Probe1                     GTCGCAGTTAAGTAAAA                     CGCGG
PhyA        TCACGATTCGCGTGAGAAGAACTAACCCTGAAA                     CGCGG
(Complemented and inverted)
                                       AAA (C)      Xn    (A/C/G)CGCG(C/G/T)
```

Figure 2. Nucleic acid alignment of several sequences containing the CGCG-box. DREB1C: gene related to drought response in Arabidopsis. EIN3: gene related to ethylene and ABA signalling in Arabidopsis. Parsley: sequence reported by da Costa e Silva (1994). CAM2: calmodulin 2 gene. PhyA: phytochrome A gene. Probe 1: sequence reported by Yang and Poovaiah (2002).

A preliminary search, using patmatch indicated that there are 7895 hits in 5801 gene promoters (1 kb upstream from the UTR). In general there is a ratio of four to two hits by sequence although more than ten hits by sequence and a unique hit by promoter are also noticed. From the literature, 12 DNA-binding sequences containing the CGCG motif fitting in the frame proposed by Yang and Poovaiah (2002) ((A/C/G)CGCG(C/G/T)) were selected for further study (see *Table 3*).

In conclusion, the data may indicate a core motif CGCG generating different regulatory elements in accordance with the combinatorial control of genes. The CGCG motif might generate a secondary structure playing a role in the deceleration or site recognition of various transcription factors.

6.2 *The physiological role of CAMTA under stress conditions*

Plant productivity is greatly affected by environmental stresses such as drought, salt loading and freezing. The *cis*-acting promoter element, the dehydration response element DRE (TA<u>CCGAC</u>AT), plays an important role in regulating gene expression in response to these stresses. The transcription factor DREB specifically interacts with the DRE and induces expression of stress tolerance genes (for example rd29A, kin1, cor6.6/kin2, cor47/rd17, cor15a, and erd10 in *Arabidopsis*). On the other hand, sequence analysis of the dehydration responsive element binding proteins and genes (DREB2B, DREB2A, DREB1C, DREB1B, DREB1A) suggested that some of them might be triggered, directly or in a cascade manner by CAMTAs. (see *Figure 3*).

We were able to identify putative CaM-binding sites in DREB1C, DREB1B, DREB1A, DREB2A and DREB2B using the CaM target database (http://calcium.uhnres.utoronto.ca/ctdb/flash.htm) (*Figure 4* shows the alignments with the CaM binding site framed). An N-terminal (between position 50 to 100) 19 amino acid sequence is identified in each DREB protein. It is noteworthy that this putative CaM-binding region is conserved between the DREB2 elements (drought- and salt-induced) and separately, conserved between DREB1 elements (cold-induced). It is important to note that DREB1C has two putative CaM-binding sites. One seems to be derived from the DREB2 consensus and the second one is proper to the DREB1 family. DREB1C might be the evolutionary link between DREB1 and DREB2 protein families and it may play a leading role in DREB regulation pathways.

Table 2. *Putative genes under the CGCG-box (CAAACGCGG) control in the Arabidopsis genome*

Transcription related elements (21)

AT1G66400	calmodulin-related protein, putative, similar to touch-induced
AT1G79650	DNA repair protein RAD23 –related
AT1G16190	DNA repair protein RAD23, putative
AT2G45750	dehydration-induced protein family,
AT5G66400	dehydrin RAB18-related protein
AT1G58520	ERD4 protein-related
AT3G61190	BON1-associated protein 1 (BAP1),
AT2G40000	nematode-resistance protein-related
AT2G34930	disease resistance protein family
AT5G42410	auxin-induced (indole-3-acetic acid induced) protein family
AT5G35735	auxin-induced protein family
AT5G13200	GRAM domain-containing protein, similar to ABA-responsive
AT1G10910	membrane-associated salt-inducible protein
AT5G57620	myb family transcription factor
AT3G10113	myb family transcription factor
AT2G37260	WRKY family transcription factor
AT2G30250	WRKY family transcription factor
AT5G22880	histone H2B, putative
AT5G09250	transcriptional co-activator (KIWI)-related
AT3G61260	DNA-binding protein-related
AT3G25990	DNA-binding protein, GT-1-related

Kinases and phosphatases (17)

AT4G31770	calcineurin-like phosphoesterase family
T1G07010	calcineurin-like phosphoesterase family
AT5G43380	serine/threonine protein phosphatase type on (TOPP7)
AT5G58720	PRL1 associated protein-related
AT3G19980	protein phosphatase,
AT1G18890	calcium-dependent protein kinase (CDPK1),
AT5G66880	serine/threonine protein kinase, putative
AT5G57035	protein kinase family
AT5G12090	protein kinase family
AT4G26890	protein kinase family
AT4G01190	phosphatidylinositol kinase-related
AT3G50500	protein kinase
AT3G07960	phosphatidylinositol-4-phosphate 5-kinase-related
AT2G40850	phosphatidylinositol 3- and 4-kinase family
AT1G74325	protein kinase-related
AT1G07870	protein kinase family
AT4G28880	casein kinase I, putative

Protein–protein interacting elements (14)

AT5G61270	bHLH protein family
AT3G19860	bHLH protein family
AT5G48180	Kelch repeats protein family
AT1G16250	Kelch repeat containing F-box protein family
AT4G35450	ankyrin repeat-containing protein 2
AT3G18190	chaperonin, putative
AT3G22700	F-box protein family
AT1G71290	F-box protein-related
AT1G53790	F-box protein family
AT4G21180	DnaJ protein family
AT1G18700	DnaJ protein family
AT1G73710	pentatricopeptide (PPR) repeat-containing protein
AT4G19440	pentatricopeptide (PPR) repeat-containing protein
AT4G21170	pentatricopeptide (PPR) repeat-containing protein

Table 3. The presence of the CGCG box in different DNA binding sequences

DNA binding sequences	Name	Reference
GGA<u>CGCG</u>TGGC	Synthetic element (hex-3) related to response to ABA and to desiccation	Busk and Pages (1998)
GTACGTG<u>GCGC</u>	ABA-responsive element of rice (O.s.) rab16 and alpha-amylase genes	Busk and Pages M (1998)
<u>GC</u>CGCGTGGC	'Motif III' found in the promoter of rice (O.s.) rab16B gene	Ono *et al.* (1996)
GATC<u>CGCG</u>XnACCAATCS (n=14)	'Type III element'; Oct-containing composite element found in tobacco (N.t.) histone gene promoter	Taoka *et al.* (1999)
TCA<u>CGCG</u>GATC	Oct-containing composite element Type II found in tobacco (N.t.) histone gene promoter	Taoka *et al.* (1999)
CCACGTCANCGATC<u>CGCG</u>	Oct-containing composite element Type I found in tobacco (N.t.) histone gene promoter	Taoka *et al.* (1999)
<u>CGCG</u>GATC	'Octamer motif' found in promoter of wheat (T.a.) histone genes H3 and H4	Taoka *et al.* (1999)
<u>CGCG</u>GCAT	Octamer motif found in histone-gene-specific consensus sequences	Chaubet *et al.* (1986)
CATGGG<u>CGCG</u>G	RE1 (putative repressor element) responsible for Pfr-directed repression of oat (A.s.) phyA3 phytochrome gene	Ngai *et al.* (1997)
A<u>CGCG</u>TGTCCTC	Coupling element 3 of ABRC3 in barley HVA1 gene	Busk and Pages (1998)
A<u>CGCG</u>CCTCCTC	Coupling element 3 in maize (Z.m.) rab28 gene promoter	Busk and Pages M (1998)
AA<u>CGCG</u>TGTC	Coupling Element 3 found in the promoter of the rice (O.s.) Osem gene	Hobo *et al.* (1999)

*GGCAAA*GCTTAGCTGTTTCTTATCCACGTGGCATTCACAGAGACAG<u>AAAC</u>T CGCGT TCGACCCCACAAA

TATCCAAATATCTTCCGGCCAATATAAACAGCAAGCTCTC ACTCCAAC *GAATTCCG*

Figure 3. *A possible CAMTA binding sequence in DREB1C promoter. The AAAC motif next to the GCCG box is underlined.*

Under stress condition CaM might trigger the CAMTAs in a Ca²⁺-dependent manner. CAMTAs may induce transcription of DREB genes. One of the possibilities is that CaM might have an inhibitory effect on DREB proteins, acting in a feedback manner. Hence one might suggest that CaM binding to DREB might interfere with the DRE recognition by DREB proteins. A second possibility is that CaM binding sites in DREB proteins may be recruited to the nucleus in an ATP-dependent/ CaM-binding mechanism. Predictions on PSORTII (http://psort.ims.u-tokyo.ac.jp/form2.html) show that with the exception of DREB1A, all DREBs are mainly nuclear. A third possibility is that Ca²⁺/CaM binds to DREB proteins in order to trigger a new cascade in a completely new signalling pathway. A fourth possibility is

```
DREB2B ....1  MAVYEQTGTE  QPKKRKSRAR  AGGLTVADRL  KKWKEYNEIV  EASAVKEGEK
       .....  0000000000  0000022222  2222222222  2222200000  0000000000
DREB2A ....1  MAVYDQSGDR  NRTQIDTSRK  RKSRSRGDGT  TVAERLKRWK  EYNETVEEVS
       .....  0000000000  0000000000  0000000000  0000000000  0000000000
DREB1C ....1  MNSFSAFSEM  FGSDYESPVS  SGGDYSPKLA  TSCPKKPAGR  KKFRETR‾HPI‾
       .....  0000000000  0000000000  0000000000  0000000000  0000004999
DREB1B ....1  MNSFSAFSEM  FGSDYEPQGG  DYCPTLATSC  PKKPAGRKKF  RETRHPIYRG
       .....  0000000000  0000000000  0000000000  0000000000  0000000000
DREB1A ....1  MNSFSAFSEM  FGSDYESSVS  SGGDYIPTLA  SSCPKKPAGR  KKFRETRHPI
       .....  0000000000  0000000000  0000000000  0000000000  0000000000

DREB2B ...51  PKRKVPAKGS  KKGCMKGKGG  PDNSHC‾SFRG  VRQRIWGKWV  AEIRE‾PKIGT
       .....  0000000000  0000000000  0000247999  9999999999  9999742000
DREB2A ...51  TKKRKVPAKG  SKKGCMKGKG  GPENSR‾CSFR  GVRQRIWGKW  VAEIRE‾PNRG
       .....  0000000000  0000000000  0122245788  9999999999  9877754211
DREB1C ...51  ‾YRGVRQRNSG  KWVCEL‾REPN  ‾KKTRIWLGTF  QTAEMAARA‾H  DVAAIALRGR
       .....  9999999999  9999994004  9999999999  9999999994  0000000000
DREB1B ...51  VRQRNSGKWV  SEVREPN‾KKT  RIWLGTFQTA  EMAARA‾HDVA  ALALRGRSAC
       .....  0000000000  0000004999  9999999999  9999994000  0000000000
DREB1A ...51  YRGVRRRNSG  KWVCEVREPN  ‾KKTRIWLGTF  QTAEMAARA‾H  DVAALALRGR
       .....  0000000000  0000000004  9999999999  9999999994  0000000000
```

Figure 4. Amino-acid alignment showing putative CaM binding domains in DREB proteins. Scores range from 0 to 9. A score less than 5 is considered irrelevant. A 19 amino acid sequence is framed in each DREB protein. It is important to note that DREB1C has two putative calmodulin binding sites.

that CaM binding increases DREB stability. This function of CaM has already been reported in literature where the binding of CaM directly affects the stability and therefore the steady-state level of oestrogen receptors (Li *et al.*, 2001).

7. Future perspectives

The transcription machinery in plants is far from being fully described. However, a growing number of plant transcriptional activators have been identified. In order to study the role of transcription factors in plants over-expression and antisense technology have been used. Because of the lethal or strong pleiotropic effects in transgenic plants and the limitations of transgenic inactivation by an antisense approach, an alternative technique, widely used over the past decade, has been developed. It employs the polymerase chain reaction to screen large populations of plants containing T-DNA or transposon insertions and thus it allows the identification and characterization of mutants in specific genes (Schwechheimer *et al.*, 1998). We are using this technique in order to isolate *CAMTA* knockouts. Different mutants could be crossed to generate double mutants. Comparatives studies under stress conditions between the knockouts and the wild type would give valuable insights on the pathways controlled by these different transcription factors and the eventual cross talk between them. In addition, we are investigating, *in vivo*, the CGCG binding affinity of CAMTA. We constructed a set of synthetic and natural sequences with CGCG blocks in a GUS/pCambia1380 plasmid, next to a minimal promoter (-64 nucleotides upstream of the transcription initiation site of the cauliflower mosaic virus 35S gene promoter). Using the comparisons between the GUS reporter activity in the wild type and the mutant transgenic plants, after transactivation experiments (bombardment with a full CAMTA gene driven by the full 35S CaMV promoter) will allow us to gather valuable information and new insights about the nature and the identity of the DNA binding site of this family of transcription activators. Finally, we are using microarrays to compare the transcriptome of wild-type and CAMTA knockout mutants in order to identify CAMTA target genes.

Analysis of CAMTAs in other organisms is also of great interest. Following the characterization of plant and human CAMTAs as transcription activators (Bouché et al., 2002) a recent paper (Katoh and Katoh, 2003) characterized the human *CAMTA1* gene as a candidate suppressor of neuroblastoma. However, the downstream gene targets of CAMTAs are unknown in any organism. Also unknown are the mechanisms of CAMTA function and regulation, and possible interacting proteins. The spatial and temporal dynamics of CAMTA distribution in the nucleus, and regulation of movement from the cytosol to the nucleus are likely to be resolved by fluorescent microscopy techniques and protein tagging. Clearly, as CAMTAs similar in structure operate in all multicellular organisms (Bouché et al., 2002), a concerted effort to resolve their function may involve studies in various model systems.

References

Aaronson, D.S. and Horvath, C.M. (2002) A road map for those who don't know JAK-STAT. *Science.* **31**: 1653–1655.

Adler, E. (2002) Keeping count of calcium. *Science* **298**: 707.

An, W.F., Bowlby, M.R., Betty, M., *et al.* (2000) Modulation of A-type potassium channels by a family of calcium sensors. *Nature* **403**: 553–556.

Anandalakshmi, R., Marathe, R., Ge, X., Herr, J.M. Jr., Mau, C., Mallory, A., Pruss, G., Bowman, L. and Vance, V.B. (2000) A CaM-related protein that suppresses posttranscriptional gene silencing in plants. *Science* **290**: 142–144.

Antonsson, A., Hughes, K., Edin, S. and Grundstrom, T. (2003) Regulation of c-Rel nuclear localization by binding of Ca^{2+}/CaM. *Mol. Cell Biol.* **23**:1418–1427.

Appella, E. and Anderson, C.W. (2001) Post-translational modifications and activation of p53 by genotoxic stresses. *Eur. J. Biochem.* **268**: 2764–2772.

Bayliss, R., Littlewood, T., Strawn, L.A., Wente, S.R. and Stewart, M. (2002) GLFG and FxFG nucleoporins bind to overlapping sites on importin. *J. Biol. Chem.* **277**: 50597–50606.

Bennetzen, J.L. (2001) *Arabidopsis* arrives. *Nat. Genet.* **27**: 3–5.

Berridge, M.J., Lipp, P. and Bootman, M.D. (2000) The versatility and universality of calcium signalling. *Nat. Rev. Mol. Cell Biol.* **1**: 11–21.

Bouché, N., Scharlat, A., Snedden, W., Bouchez, D. and Fromm, H. (2002) A novel family of CaM-binding transcription activators in multicellular organisms. *J. Biol. Chem.* **277**: 21851–21861.

Bunney, T.D., Shaw, P.J., Watkins, P.A.C., Taylor, J.P., Beven, A.F., Wells, B., Calder, G.M. and Drobak, B.K. (2000) ATP-dependent regulation of nuclear Ca2+ levels in plant cells. *FEBS Lett.* **476**: 145–149.

Busk, P.K. and Pages, M. (1998) Regulation of abscisic acid-induced transcription. *Plant Mol. Biol.* **37**: 425–435.

Camoni, L., Harper, J.F. and Palmgren, M.G. (1998) 14-3-3 proteins activate a plant calcium-dependent protein kinase(CDPK). *FEBS Lett.* **430**: 381–384.

Carafoli, E. and Brini, M. (2000) Calcium pumps: structural basis for and mechanism of calcium transmembrane transport. *Curr. Opin. Chem. Biol.* **4**: 152–161.

Chaubet, N., Philipps, G., Chaboute, M-E., Ehling, M. and Gigot, C. (1986) Genomic organization of the corn histone H3 and H4 genes. *Plant Mol. Biol.* **6**: 253–263.

Cheng, S-H., Willmann, M.R., Chen, H-C. and Sheen, J. (2002) Calcium signalling through protein kinases. The *Arabidopsis* calcium-dependent protein kinase gene family. *Plant Physiol.* **129**: 469–485.

Da Costa e Silva, O. (1994) CG-1, a parsley light-induced DNA-binding protein. *Plant Mol. Biol.* **25**: 921–924.

Demaurex, N. and Distelhorst, C. (2003) Apoptosis: the calcium connection. *Science* **300**: 65–67.

Downie, L., Priddle, J., Hawes, C. and Evans, D.E. (1998) A calcium pump at the higher plant NE? *FEBS Lett.* **429**: 44–48.

Echevarría, W., Leite, M.F., Guerra, M.T., Zipfel, W.R. and Nathanson, M.H. (2003) Regulation of calcium signals in the nucleus by a nucleoplasmic reticulum. *Nature Cell Biol.* **5**: 440–446.

Furuichi, T., Cunningham, K.W. and Muto, S. (2001) A putative two-pore channel AtTPC1 mediates Ca 2 flux in *Arabidopsis* leaf cells. *Plant Cell Physiol.* **42**: 900–905.

Ganguly S. and Singh M. (1999) Purification and characterization of a protein phosphatase from winged bean. *Phyto-chemistry* **52**: 239–246.

Gerasimenko, O.V., Gerasimenko, J.V., Tepikin, A.V. and Petersen, O.H. (1995) ATP-dependent accumulation and inositol trisphosphate-mediated or cyclic ADP-ribose-mediated release of Ca^{2+} from the NE. *Cell* **80**: 439–444.

Gilchrist, C.A., Holm, C.F., Hughes, M.A., Schaenman, J.M., Mann, B.J. and Petri, W.A. (2001) Identification and characterization of an *Entamoeba histolytica* up-stream regulatory element 3 sequence-specific DNA-binding protein containing EF-hand motifs. *J. Biol. Chem.* **276**: 11838–11843.

Hobo, T., Kowyama, Y. and Hattori, T. (1999) A bZIP factor, TRAB1, interacts with VP1 and mediates abscisic acid-induced transcription. *Proc. Natl Acad. Sci. USA* **96**: 15348–15353.

Hoeflich, K.P. and Ikura, M. (2002) CaM in action: diversity in target recognition and activation mechanisms. *Cell* **108**: 739–742.

Holmberg, C.I., Tran, S.E.F., Eriksson, J.E. and Sistonen, L. (2002) Multisite phosphorylation provides sophisticated regulation of transcription factors. *Trends Biochem. Sc.* **27**: 619–627.

Ikura, M., Osawa, M. and Ames, J.B. (2002) The role of calcium-binding proteins in the control of transcription: structure to function. *BioEssays* **24**: 625–636.

Katoh, M. and Katoh, M. (2003) Identification and characterization of *FLJ10737* and *CAMTA1* genes on the commonly deleted region of neuroblastoma at human chromosome 1p36.31-p36.23. *Int. J. Oncol.* **23**: 1219–1224.

Klee, C.B. and Means, A.R. (2002) Keeping up with calcium. *EMBO J.* **3**: 823–827.

Koppler, P., Mersel, M., Humbert, J.P., Vignon, J., Vincendon, G. and Malviya, A.N. (1996) High affinity inositol 1,3,4,5-tetrakisphosphate receptor from rat liver nuclei:purification, characterization, and amino-terminal sequence. *Biochemistry* **35**: 5481–5487.

Kurokawa, H., Osawa, M., Kurihara, H., Katayama, N., Tokumitsu, H., Swindells, M.B., Kainosho, M. and Ikura, M. (2001) Target-induced conformational adaptation of CaM revealed by the crystal structure of a complex with nematode Ca^{2+} CaM-dependent kinase peptide. *J. Mol. Biol.* **312**: 59–68.

Lanini, L., Bachs, O. and Carafoli, E. (1992) The calcium pump of the liver nuclear membrane is identical to that of ER. *J. Biol. Chem.* **267**: 11548–11552.

Lin, J., Blake, M., Tang, C., Zimmer, D., Rustandi, R.R., Weber, D.J. and Carrier, F. (2001) Inhibition of p53 transcriptional activity by the S100B calcium-binding protein. *J. Biol. Chem.* **276**: 35037–35041.

Li, Z., Joyal, J.L. and Sacks, D.B. (2001) CaM enhances the stability of the estrogen receptor. *J. Biol. Chem.* **20**: 17354–17360.

Liao, B., Paschal, B.M. and Luby-Phelps, K. (1999) Mechanism of Ca 21-dependent nuclear accumulation of CaM. *Proc. Natl Acad. Sci. USA* **96**: 6217–6222.

Luan, S., Kudla, J., Rodriguez-Concepcion, M., Yalovsky, S., and Gruissem, W. (2002) CaMs and calcineurin B-like proteins: Calcium sensors for specific signal response coupling in plants. *Plant Cell.* 14: 389–400.

Mayr, B. and Montminy, M. (2001) Transcriptional regulation by the phosphorylation-dependant factor CREB. *Nat. Rev. Mol. Cell Biol.* 2: 599–609.

Macian, F., Lopez-RodròCguez, C. and Rao, A. (2001) Partners in transcription: NFAT and AP-1. *Oncogene* 20: 2476–2489.

Ngai, N., Tsai, F.Y. and Coruzzi, G. (1997) Light-induced transcriptional repression of the pea AS1 gene: identification of cis-elements and transfactors. *Plant J.* 12: 1021–1034.

O'Day, D.H. (2003) CaMBOT: profiling and characterizing CaM-binding proteins. *Cell. Signal.* 15: 347–354.

Ono, A., Izawa, T., Chua, N-H. and Shimamoto, K. (1996) The rab16B promoter of rice contains two distinct abscisic acid-responsive elements. *Plant Physiol.* 112: 483–491.

Pauly, N., Knight, M.R., Thuleau, P., Van der luit, A.H., Moreau, M., Trewavas, A.J., Ranjeva, R. and Mezars, C. (2000) Control of free calcium in plant cell nuclei. *Nature* 405: 754.

Petersen, O.H., Gerasimenko, O.V., Gerasimenko, J.V., Mogami, H. and Tepikin, A.V. (1998) The calcium store in the NE. *Cell Calcium* 23: 87–90.

Reddy, A.S.N. (2001) Calcium: silver bullet in signalling. *Plant Sci.* 160: 381–404.

Reddy, V.S., Ali, G.S. and Reddy, A.S.N. (2002) Genes encoding CaM-binding proteins in the *Arabidopsis* genome. *J. Biol. Chem.* 277: 9840–9852.

Rhee, Y., Gurel, F., Gafni, Y., Dingwall, C. and Citovsky, V. (2000) A genetic system for detection of protein nuclear import and export. *Nature Biotechnol.* 18: 433–437.

Rhoads, A.R. and Friedberg, F. (1997) Sequence motifs for CaM recognition. *FASEB J.* 11: 331–340.

Roberts, E.C., Deed, R.W., Inoue, T., Norton, J.D. and Sharrocks, A.D. (2001) Id helix-loop-helix proteins antagonize Pax transcription factor activity by inhibiting DNA binding. *Mol. Cell. Biol.* 21: 524–533.

Rogue, P.J. and Anant, N.M. (1999) Calcium signals in the cell nucleus. *EMBO J.* 18: 5147–5152.

Saimi, Y. and Kung, C. (2002) CaM as an ion channel subunit. *Annu. Rev. Physiol.* 64: 289–311.

Schwechheimer, C., Zourelidou, M. and Bevan, M.W. (1998) Plant transcription factor studies. *Annu. Rev. Plant Physiol. Plant Mol. Biol.* 49: 127–150.

Siebrasse, J.P. and Peters, R. (2002) Rapid translocation of NTF2 through the nuclear pore of isolated nuclei and Nes. *EMBO Rep.* 3: 887–892.

Snedden, W.A and Fromm, H. (2001) CaM as a versatile calcium signal transducer in plants. *New Physiol.* 151: 35–66.

Stewart, M., Baker, R.P., Bayliss, R., Clayton, L., Grant, R.P., Littlewood T. and Matsuura, Y. (2001) Molecular mechanism of translocation through NPCs during nuclear protein import. *FEBS Lett.* 498: 145–149.

Stochaj, U. and Rother, K.L. (1999) Nucleocytoplasmic trafficking of proteins: with or without Ran? *BioEssays* 21: 579–589.

Taoka, K., Kaya, H., Nakayama, T., Araki, T., Meshi, T. and Iwabuchi, M. (1999) Identification of three kinds of mutually related composite elements conferring S phase-specific transcriptional activation. *Plant J.* 18: 611–623.

Van der Luit, A.H., Olivari, C., Haley, A., Knight, M.R. and Trewavas, A.J. (1999) Distinct calcium signalling pathways regulate CaM gene expression in tobacco. *Plant Physiol.* 121: 705–714.

Vermeuden, L., De Wilde, G., Notebaert, S., Berghe, W.V. and Haegeman, G. (2002) Regulation of the transcriptional activity of the nuclear factor kappa B p65 unit. *Biochem. Pharmacol.* 64: 963–970.

Vetter, S.W. and Leclerc, E. (2003) Novel aspects of CaM target recognition and activation. *Eur. J. Biochem.* 270: 404–414.

Wang, Z., Yang, P., Fan, B. and Chen, Z. (1998) An oligo selection procedure for identification of sequence-specific DNA-binding activities associated with the plant defence response. *Plant J.* **16:** 515–520.

Whitmarsh, A.J. and Davis, R.J. (2000) Regulation of transcription factor function by phosphorylation. *Cell. Mol. Life Sci.* **57:** 1172–1183.

Wu, X. and McMurray, C.T. (2001) CaM kinase II attenuation of gene transcription by preventing cAMP response element-binding protein (CREB) dimerization and binding of the CREB-binding protein. *J. Biol. Chem.* **276:** 1735–1741.

Yang, T. and Poovaiah, B.W. (2002) A CaM-binding/CGCG box DNA-binding protein family involved in multiple signalling pathways in plants. *J. Biol. Chem.* **277:** 45049–45058.

Yin, J.C., Del Vecchio, M., Zhou H. and Tully T. (1995) CREB as a memory modulator: induced expression of a dCREB2 activator isoform enhances long-term memory in Drosophila. *Cell* **81:** 107–115.

Xiong, L., Schumaker, K.S. and Zhu, J-K. (2002) Cell signalling during cold, drought, and salt stress. *Plant Cell* **14:** S165–S183.

19

CAAX-dependent modifications of the lamin proteins in the organization of the nuclear periphery

C.P. Maske and D.J. Vaux

1. Introduction

The important biological functions of the nuclear envelope (NE) and its associated structures include the compartmentalization of the nucleoplasm and cytoplasm, provision of a regulated, bi-directional nucleocytoplasmic transport mechanism and finally the organization of the nuclear periphery (Goldman *et al.*, 2002). There is no doubt that minor roles for the nuclear envelope may also exist related to specific functions of individual NE proteins, for example the steroid reductase activity of the lamin B receptor (Silve *et al.*, 1998). In order to achieve heterogeneity of interactions at the nuclear periphery a certain degree of organization is required within the nuclear envelope superstructure, which may relate to the organization of the underlying chromatin or sites of chromatin attachment to the nuclear envelope. Transcriptionally poor heterochromatin is known preferentially to occupy peripheral nuclear sites (Cremer and Cremer 2001). Moreover, chromosomes partition between the nuclear periphery and nuclear interior dependent on the gene content and level of transcriptional activity (Boyle *et al.*, 2001). The positioning of chromatin domains is transmitted to daughter nuclei, suggesting that the epigenetic factors that influence gene activity and therefore differentiation are stereotypical between mother and daughter cells for a given cell type (Gerlich *et al.*, 2003). This organization may be driven by the peripheral chromatin, dictated by domains within NE components, or, perhaps more likely, result from a dynamic interaction of the two.

The many ways by which higher-order organization may be introduced into nuclear envelope structures include differential expression patterns of proteins in all structures of the nuclear envelope, modifications to underlying chromatin, e.g. in the form of methylation or acetylation (Strahl and Allis, 2000), and post-translational modifications

The Nuclear Envelope, edited by D.E. Evans, C. Hutchison & J.A. Bryant.
© 2004 Garland Science/BIOS Scientific Publishers

of the nuclear envelope and nuclear lamina proteins (Stuurman *et al.,* 1998). While expression patterns of proteins may govern long-term differentiation states (e.g. lamin A/C is expressed only in terminally differentiated cells), post-translational modifications may further diversify the range of interactions and provide reversible mechanisms for nuclear organization. Nuclear envelope proteins are modified by phosphorylation at mitosis, a mechanism that rapidly alters affinities between these proteins and allows the depolymerisation and breakdown of the nuclear envelope during open mitosis in higher eukaryotes (Collas and Courvalin, 2000). There is however evidence which suggests that interphase modifications of nuclear envelope proteins may be important in the maintenance of the nuclear periphery. These modifications are likely to include phosphorylation of multiple components of the nuclear envelope (Kill and Hutchison, 1995) and also prenylation of the nuclear lamins (see colour *Plate 14a*).

In this chapter we explore the nature and significance of prenylation and prenyl-dependent modifications of the nuclear lamins in the organization of the nuclear periphery. We further assess the role of normal or defective processing in the pathogenesis of the nuclear laminopathies and the effects of novel cancer chemotherapeutics that target the enzymes necessary for such modifications in the stability of the nuclear lamina.

2. The biochemistry of lamin CAAX processing

Prenylation involves the addition of lipid moieties to proteins, usually with consequent changes to the localization of the protein in the cell and therefore ability of the protein to function in its correct cellular context. This is elegantly illustrated in the ras GTPase signalling pathway, amongst other proteins (see *Table 1*) (Downward, 2003). The consensus sequence for prenylation in eukaryotic cells, from yeast to mammals, is the CAAX motif (Maurer-Stroh *et al.,* 2003). More specifically, proteins terminating in a CAAX motif (cysteine, aliphatic, aliphatic, any of several amino acids) at the carboxyl terminus are substrates for the prenyltransferase enzymes. In general, the CAAX modifications include the addition of a prenyl moiety to the side chain of the cysteine residue by covalent thioether linkage catalysed by the prenyltransferase enzymes, followed by endoproteolytic cleavage of the last three amino acids of the protein (i.e. -AAX is removed), followed by carboxymethylation of the farnesylated cysteine residue forming the novel carboxyl-terminus (Colour *Plate 14b*). Each of these steps will be discussed in more detail, but the final outcome of the CAAX-dependent modifications is the generation of a hydrophobic carboxyl-terminus from a more hydrophilic one. Not only does the prenyl moiety provide hydrophobicity, but the neutralization of the carboxylate anion by carboxymethylation removes the negative charge associated with a free carboxyl-terminus. Generating membrane-associated proteins from soluble proteins is a common theme in cellular processes and is central to many signal transduction pathways (see *Table 1*).

CAAX motifs are recognized by one of two soluble prenyltransferases; farnesyltransferase and geranylgeranyltransferase I (Sebti and Hamilton, 2001). The difference between the two enzymes relates to the length of lipid chain that is used in the prenyl transfer reaction, where farnesyltransferase adds a 15-carbon moiety and geranylgeranyltransferase adds a 20-carbon moiety, and in the specificity of the enzymes for different substrates. Both enzymes are heterodimers of an alpha and beta subunit. In fact, both

Table 1. Putative human CAAX modified proteins. A search of Swissprot (September 2003) with the motif C(A,V,I,L,M,S,P) (A,V,I,L,M,S,P) (A,V,I,L,M,S,P)X* produced a list of proteins from which human proteins were selected. Type II membrane proteins and those with signal sequences were excluded, as well as a large group of GTP binding proteins (ras, rho, rab and gbp families). The resulting list contains 26 proteins

Swissprot name	Length	Upstream seq	CAAX seq	Gene	Protein location	Comments
AS17_HUMAN	231	VKEIK	CLSS	asb17	Cytoplasmic	ankyrin repeat and socs-box protein
CD4_HUMAN	458	RFQKT	CSPI	cd4	Type I membrane protein	cytoplasmic C terminus (2 palmitoylation sites)
CNRC_HUMAN	858	KKSKT	CLML	pde6c	Cytoplasmic	Cone cGMP 3,5 phosphodiesterase
CU80_HUMAN	424	HFHTV	CLLV	c21orf80	Cytoplasmic	Large brain protein on Chr 21
CXB1_HUMAN	283	EKSDR	CSAC	cx32	4 TM membrane protein	Connexin 32 liver gap junction protein
CXX1_HUMAN	209	PPETS	CVLA	cxx1	Cytoplasmic	Cerebral CAAX box protein
DRNL_HUMAN	302	LSPQL	CPAA	dnas1/1	Nuclear	muscle DNAse 1 like protein
DS10_HUMAN	87	QAPVP	CMPV	dscr10	Cytoplasmic	Down syndrome critical region protein 10
GCM1_HUMAN	436	MLLNL	CPLR	gcm1	Cytoplasmic/nuclear	chorion specific transcription factor
GRK7_HUMAN	553	SKSGV	CLLL	gprk7	Cytoplasmic membrane	GPCR kinase 7
I5P1_HUMAN	412	HVHKC	CVVQ	inpp5a	Cytoplasmic membrane	Type 1 IP3 5 phosphatase
IP3T_HUMAN	2671	VDVQN	CISR	iprp3	6 TM membrane protein	Type 3 IP3 receptor
KPB1_HUMAN	1223	LPHSI	CAMQ	phka1	Cytoplasmic	Phosphorylase b kinase alpha regulatory chain
KPBB_HUMAN	1092	NNDDP	CLIS	phkb	Cytoplasmic	Phosphorylase b kinase beta regulatory chain
LMO6_HUMAN	615	ARDKN	CIVA	lmo6	Cytoplasmic	triple lim domain protein 6
MTA3_HUMAN	594	DELTC	CVSD	mta3	Cytoplasmic/nuclear	metastasis associated zinc finger nuclear protein
NADE_HUMAN	111	HHDEF	CLMP	ngfrap1	Cytoplasmic	p75ntr associated cell death executor
PALM_HUMAN	387	HRCKC	CSIM	palm	Cytoplasmic vesicle	paralemmin vesicle associated coiled coil protein
P12R_HUMAN	386	EASVA	CSLC	ptgir	7 TM membrane protein	prostacyclin GPCR (2 palmitoylation sites)
PPGA_HUMAN	528	ERRPC	CLLM	ppp1r16a	Cytoplasmic membrane	protein phosphatase 1 regulatory subunit 16a
PXF_HUMAN	299	ASGEQ	CLIM	pex19	Outer surface peroxisomes	33kDa farnesylated peroxisomal protein
RK_HUMAN	563	SKSGM	CLVS	rhok	Cytoplasmic membrane	Rhodopsin kinase
SLUG_HUMAN	268	EESGC	CVAH	snai2	Cytoplasmic/nuclear	snai2 neural crest transcription factor
SX18_HUMAN	384	VYYSA	CISG	sox18	Cytoplasmic/nuclear	Transcription factor sox-18
TTC3_HUMAN	2025	QELPS	CSSR	ttc3	Cytoplasmic/nuclear	Tetratricopeptide repeat protein 3
Z147_HUMAN	630	ATLSI	CSPK	trim25	Cytoplasmic/nuclear	Zn-finger protein 147

enzymes share the same alpha subunit (the product of a single gene) but differ in their beta subunits. This difference allows discrimination of potential CAAX-containing substrates, where farneysltransferase will recognize proteins with methionine, isoleucine or serine at the -X position and geranylgeranyltransferase I will recognize proteins terminating in leucine. This specificity, although not absolute, is responsible for the development of agents that act to inhibit the ras-signalling pathway on tumour cells, namely the farnesyltransferase inhibitors. There is one further prenyltransferase enzyme, geranylgeranyltransferase II, which recognizes proteins terminating in cysteine-cysteine or cysteine-X-cysteine (where X is any amino acid) and prenylates rab GTPase proteins with a geranylgeranyl moiety.

The lamin proteins, with the exception of splice-variant lamin C, are farnesylated (see *Figure CS.16a*). Functional farnesyltransferase exists in both the nuclear and cytoplasmic compartments, although evidence from *Xenopus* oocytes suggests that prenylation of the lamins may take place in the cytoplasm prior to nuclear import (Firmbach-Kraft and Stick, 1995). There are two possible proteases that may cleave the terminal three amino acids of farnesylated CAAX proteins, namely Rce1 or Ste24p (called Zmpste24 in mammals). These membrane proteins of the endoplasmic reticulum are conserved throughout the eukaryotic world (Otto *et al.*, 1999). They also exhibit a degree of specificity, with most prenylated proteins being substrates for Rce1 (Ras converting enzyme 1) (Boyartchuk *et al.*, 1997). Ste24p has one substrate in yeast, namely the mating pheromone a factor (Tam *et al.*, 1998). While mammalian substrates for Rce1 overlap those in yeast and include the ras proteins and subunits of heterotrimeric G proteins, a mammalian substrate for Zmpste24 was not known until recently. A mouse knockout model of the gene *Zmpste24* revealed that cells contained a lamin A processing defect at the carboxyl terminus along with consequent phenotypic changes (Pendas *et al.*, 2002). Subsequent studies have confirmed that Zmpste24 is almost certainly involved in the processing of the carboxyl terminus of lamin A and is probably the lamin A CAAX endoprotease (Bergo *et al.*, 2002). Like a factor, lamin A undergoes further proteolytic cleavage 15 residues upstream of the farnesylcysteine residue. It is not known whether Zmpste24 is responsible for this second upstream cleavage event (which it is in a factor), because previous biochemical studies have indicated that the prelamin A protease, as opposed to the lamin A CAAX endoprotease, is a serine protease and not a zinc metalloprotease like Zmpste24 (Kilic *et al.*, 1999).

Still more recently it has been revealed that the lamin B1 CAAX endoprotease, and probably the endoprotease for lamin B2, is the enzyme Rce1 (Maske *et al.*, 2003). The evolutionary significance of different CAAX endoproteases for two types of lamin protein is not clear, but processing of lamin proteins may be different from the other prenylated proteins. In the first instance, endoproteolytic cleavage of farnesylated lamin B1 requires at least 40 residues of the carboxyl-terminus to be present (Maske *et al.*, 2003), suggesting that Rce1 requires more than just a farnesylcysteine and a CAAX motif to recognize lamin B1. Previous studies using ras sequences as substrates have indicated that the protease requires little more than the farnesylated CAAX motif for recognition of the substrate (Choy *et al.*, 1999; Dolence *et al.*, 2000). The difference may be due to secondary or tertiary structure differences between the tails of the ras proteins and B-type lamins, to post-translational modifications present in the tails of lamin proteins or to secondary recognition motifs, for example the conserved polyglutamate tract approximately 30 residues upstream of the CAAX motif in B-type lamins (*Figure CS.16a*).

The final step in the triad of sequential reactions involved in CAAX processing is carboxymethylation. This step is catalysed by isoprenylcysteine carboxyl methyltransferase (Icmt), an integral membrane protein of the endoplasmic reticulum. There is a single carboxyl methyltransferase activity in eukaryotic cells (Bergo *et al.*, 2001) and all CAAX proteins, including the nuclear lamins, are therefore substrates. Only in lamin A is there a further proteolytic cleavage following the triad of CAAX modifications, which removes the carboxyl-terminus containing the modifications.

3. Biological consequences of CAAX processing of the nuclear lamins

The lamin proteins polymerise to form the filamentous polymer of the nuclear lamina, which provides much of the structural integrity of the nucleus (Stuurman *et al.*, 1998) and forms a scaffold for the attachment of nuclear pore complexes (Daigle *et al.*, 2001) and peripheral chromatin (Belmont *et al.*, 1993; Paddy *et al.*, 1990; Schirmer *et al.*, 2001; see Chapter 2).

Intermediate filament proteins self assemble into stable fibrils without needing a membrane association to organize them. However, the delivery of lamin B1 to the nuclear periphery is clearly more efficient in the presence of farnesylation. In cells in which farnesylation has been inhibited, lamin B1 may still gain access to the nuclear periphery, and presumably remains associated by interactions between the intermediate filament rod domain and the nuclear lamina or inner nuclear membrane proteins, but significant amounts of protein accumulate in the nucleoplasm (Maske *et al.*, 2003). Nucleoplasmic foci of lamin B1 (i.e. non-membrane associated) have been described in the nucleus, suggesting that at least a proportion of the lamin entering the nucleus may remain in a non-farnesylated state (Maske *et al.*, 2003; Moir *et al.*, 1994). Nucleoplasmic lamin B1 has been shown to be highly immobile in certain studies (Moir *et al.*, 2000b), however more recently the non-farnesylated, nucleoplasmic pool of lamin B1 was shown to be relatively more soluble than peripheral lamin B1 (Maske *et al.*, 2003). The role of intranuclear B-type lamins is controversial; they may be a component of a nuclear skeleton (Moir *et al.*, 2000b) and have also been localized to sites of DNA replication (Moir *et al.*, 1994, 2000a). It is clear, however, that proper organization of the nuclear lamina (Ellis *et al.*, 1997; Spann *et al.*, 1997) and increase in nuclear size with cell cycle progression (Yang *et al.*, 1997) are crucial for intranuclear processes such as replication to proceed normally. The quantitative farnesylation of lamins, determining the delivery of lamin to the nuclear periphery or retention in nucleoplasm, is likely to be important in all of these processes.

Since farnesylation is enough in itself to deliver lamins efficiently to the nuclear periphery, the role of the subsequent CAAX processing steps remains enigmatic. It is evident from work on ras proteins that a charged carboxyl terminus decreases the membrane affinity of the ras proteins where carboxyl methylation has been inhibited (Bergo *et al.*, 2000). In the case of lamin B1, where intermediate filament assembly in the rod domain of the protein results in an already stable association with the nuclear envelope, a more subtle role for carboxyl methylation may be present. A recent study has shown that mammalian cells may possess a carboxymethyl-lamin receptor, that is a receptor in the nuclear envelope that specifically interacts with the methylated form of lamin B1, recognizing the structural differences that the methyl group brings to the carboxyl-terminus of the protein (Maske *et al.*, 2003). In this study it was shown that the last 40 residues of lamin B1, including the CAAX motif, were sufficient to cause retention of a reporter in the nuclear envelope. Furthermore, this nuclear envelope localization was dependent on carboxymethylation of the expressed protein, suggesting that a protein–protein interaction required this modification. Numerous inner nuclear membrane proteins have been reported and characterized (Dreger *et al.*, 2001; Schirmer *et al.*, 2003). Future studies should reveal the identity of the putative carboxymethyl-lamin receptor and whether such methylation-dependent interactions extend to other lamin proteins, including prelamin A.

The biological consequences of such a methylation-dependent interaction may include higher organization of the nuclear lamina. Using an antibody specific for the CAAX-processed form of lamin B1, subdomains of the nuclear lamina could be discerned in cells treated with a farnesyltransferase inhibitor (allowing depletion of the nascent pool of lamin B1) (Maske *et al.*, 2003). Such subdomains indicate in the first instance that lamins are not uniformly distributed around the nuclear lamina, in agreement with previous studies (Belmont *et al.*, 1993; Paddy *et al.*, 1990). However, a novel concept introduced by this experiment is that lamins may be organized within the nuclear lamina polymer by CAAX-dependent modifications (Colour *Plate 14c*). Foci of mature methylated lamin B1 are likely to arise from interactions with a putative carboxymethyl-lamin receptor. The relationship between structural components of the NE and gene expression has been highlighted in yeast, in which a boundary activity (the ability of sequences and interactions to inhibit the spread of transcriptional activity or silencing to neighbouring genes) was found to be dependent on the Nup2p receptor of the nuclear pore complex (Ishii *et al.*, 2002). In mammals, which unlike yeast contain lamin proteins, similar functional roles may be the domain of the nuclear lamina, in addition to the nuclear pore complex. A methylation-dependent organization of the lamina, if it is to have significance as a means of differentially regulating events at the nuclear periphery, demands that pools of modified and unmodified lamin should exist. It is likely that, using lamin B1 as an example, lamin proteins exist in differentially CAAX-modified states (Maske *et al.*, 2003). These states of lamin B1 include a pool of non-farnesylated lamin B1 that can be depleted following protein synthesis inhibition, suggesting that there is direct precursor relationship between this and further modified pools. Furthermore, pools of methylated and non-methylated lamin B1 exist in the nuclear periphery that are stable through the cell cycle. Therefore, a pool of farnesylated, endoproteolysed but non-methylated lamin B1 is part of the nuclear lamina but is prevented from carboxymethylation, possibly by additional post-translational modifications in the upstream sequence. The mechanisms underlying the regulation of CAAX processing may include phosphorylation of residues immediately upstream of the CAAX motif, although evidence for this has not yet emerged.

The CAAX-processed state of mature A-type lamins differs from the B-type lamins. Lamin C is a truncated splice variant of lamin A and lacks the carboxyl-terminal CAAX motif. Lamin C is located at the nuclear periphery like other lamins, in addition to intranuclear foci, and is likely to associate with the nuclear envelope through interactions with inner nuclear membrane proteins and integration into the lamina polymer. Lamin A is farnesylated, is localized at the nuclear periphery where it undergoes CAAX endoproteolysis by Zmpste24 and subsequent methylation, and is then endoproteolysed upstream of the CAAX motif. In the mature state lamin A therefore does not contain the CAAX modifications of B-type lamins. The differences in CAAX modification state of A- and B-type lamins, and the ability of B-type lamins to interact in a methylation-dependent fashion, may encompass one of the crucial biological differences between these proteins.

4. Defects of lamin CAAX processing: drugs and diseases

The importance of the CAAX modifications of lamins for nuclear homeostasis has emerged from studying mouse models and human genetic diseases. Both CAAX endo-proteases and the carboxyl methyltransferase enzymes have been deleted in mouse knockout models. Homozygous deletion of the *Rce1* gene and *Icmt* results in embryonic lethality in mid-gestation (Bergo *et al.*, 2001; Kim *et al.*, 1999). In the case of Icmt deficiency this may be the result of agenesis of the liver (Lin *et al.*, 2002). The multiple substrates of these enzymes make elucidation of the molecular pathology difficult; the ras proteins, G-protein subunits and B-type lamins are all affected by the defective processing.

The phenotype involved with deletion of the *Zmpste24* gene is less severe. The homozygous mice survive uterine life but develop characteristic phenotypic changes later, which include spontaneous bone fractures, muscle weakness and growth retardation (Bergo *et al.*, 2002; Pendas *et al.*, 2002). These phenotypic features are shared in part by mice in which the lamin A gene has been deleted (Sullivan *et al.*, 1999), and in certain mutations of human lamin A/C (Burke and Stewart, 2002). Recently human mutations in *Zmpste24* have been documented as a cause of mandibuloacral dysplasia, another phenotype common to lamin A/C mutations (Agarwal *et al.*, 2003). Lamin A is currently the only known substrate for mammalian Zmpste24 and the pathology associated with *Zmpste24* deletion or mutation is therefore assumed to be wholly the result of defective processing of lamin A. The processing defect of lamin A in these mice would therefore involve a lack of CAAX endoproteolysis following normal farnesylation. Furthermore, methylation is lost from the carboxyl-terminus and this results in loss of the upstream protease activity, which is dependent on correct CAAX processing (Bergo *et al.*, 2002).

The processing defect in lamin A resulting from the Zmpste24 deletion in mice or inactivating mutations of Zmpste24 in humans results in two important consequences; firstly lamin A is constitutively farnesylated and therefore almost exclusively occupies a peripheral nuclear localization, and secondly the protein will contain 18 residues at the carboxyl-terminus that would not normally be present in mature lamin A. A soluble nuclear protein, NARF, which binds specifically to pre-lamin A but not protein that has been proteolysed upstream of the CAAX motif, has been described (Barton and Worman, 1999). Therefore the presence of constitutively farnesylated lamin A at the nuclear periphery may alter the localization of NARF in addition to depleting the nucleoplasm of lamin A pools. Lamin A foci have been implicated in transcription (Spann *et al.*, 2002) and it is possible that depletion of nucleoplasmic lamin A may have an effect on the organization or processivity of transcriptional sites.

A human mutation in the lamin A protein has recently been described in patients suffering from Hutchinson–Gilford progeria syndrome (Eriksson *et al.*, 2003; see Chapter 3). A single base change results in the activation of a cryptic splice site and the deletion of 50 residues near the carboxyl terminus of lamin A. The deletion includes the upstream protease site but leaves the CAAX motif intact. The phenotype of these patients is severe, despite the fact that at least a proportion of the lamin A transcripts are spliced normally and produce wild-type protein. It is expected that the lamin A harbouring the 50-residue deletion would be farnesylated, endoproteolysed and methylated in the mature form, but will not be cleaved by upstream proteolysis. This mutant is subtly different from the Zmpste24 deletion in which farnesylation of lamin A is not followed by CAAX endoproteolysis or methylation. Carboxymethylated lamin A without the upstream cleavage event is more reminiscent of B-type lamin maturation. Is it possible that such a form of lamin A may bind to a carboxymethyl-lamin receptor in the nuclear envelope? Should this be the case, disruption of methylation-dependent interactions of the B-type lamins may contribute to the pathogenesis of this disease. Since the phenotype of the disease includes features of premature ageing, it is conceivable that the molecular changes mooted may result in a final common pathway of precocious changes to gene activity at the nuclear periphery.

The molecular pathogenesis of laminopathies requires more experimental data before such hypotheses can be confirmed or refuted. There is, however, a more tangible clinical example in which CAAX processing is specifically inhibited for therapeutic effect. Activating mutations in the ras proteins have been implicated in a variety of human cancers, including solid tumours and leukaemias (Downward, 2003). Since ras requires full CAAX processing to localize to the plasma membrane (trafficking takes place through the Golgi apparatus) (Choy et al., 1999), inhibitors of any of the enzymes involved in ras processing may be effective anti-tumour agents. Therefore, farnesyltransferase inhibitors have been developed to inactivate ras by preventing plasma membrane association (Sebti and Hamilton, 2001). Numerous drugs have progressed to clinical trials, although the efficacy of these agents is limited by cross-prenylation of proteins (by geranylgeranyltransferase) and toxicity. However, despite such problems it is clear that the anti-proliferative effect of these agents is not limited to their ability to inhibit ras signalling. Extra-ras targets of the FTI drugs that may contribute include the Rho proteins and their ability to inhibit the actin cytoskeleton (Prendergast et al., 1994). Farnesylation of the lamin proteins is clearly inhibited (Dalton et al., 1995), and recent evidence suggests that FTI treatment may have profound effects in the nuclear envelope (Maske et al., 2003). Both pre-lamin A- and B-type lamins accumulate in the nucleoplasm in sites excluded from nucleoli (Maske et al., 2003; Sinensky et al., 1994). At the nuclear periphery the total lamin B1 appears to be normal (Dalton et al., 1995; Maske et al., 2003), but examination with reagents that specifically label the mature, processed lamin B1 show a quantitative decrease in the amount of mature protein. This is reflected in a loss of structural integrity of the lamina as seen in differential extraction experiments of purified nuclear envelopes from treated cells. Mutations in the lamin A protein are thought to alter the structural integrity of the nuclear lamina; this phenomenon may be produced iatrogenically by administration of FTI drugs. Moreover, these data illustrate that the intermediate filament-type interactions produced by the rod domains of the lamin proteins confer only a part of the structural integrity of the lamina. Maturation of the lamina depends on CAAX processing and multiple interactions with the nuclear envelope proteins. Mutations in the lamin-A-binding protein emerin are a testament to the importance of each one of these interactions.

The potential importance of methylation on the ability of lamin B1 to organize the nuclear periphery into functional domains has been discussed previously in this chapter. What, then, may be the consequences of disrupting methylation of lamin B1? A recent discovery involving one of the oldest and most effective chemotherapeutic agents, methotrexate, has highlighted this emerging issue. Cells treated with this anti-folate agent increase the levels of a metabolite, S-adenosylhomocysteine, due to inhibition of a folate-dependent enzyme step in conversion of homocysteine to methionine. The result of an increase in S-adenosylhomocysteine is the inhibition of Icmt, the enzyme involved in carboxymethylation of CAAX substrates. The anti-proliferative effect of methotrexate is therefore at least in part due to the inhibition of ras signalling (Winter-Vann *et al.*, 2003). It is further possible that inhibition of lamin methylation would have the predicted effect of pre-lamin A accumulation at the nuclear envelope, and lack of methylation-dependent organization of the B-type lamins. The consequences of this are unknown: whether this may contribute to the anti-proliferative effect, or result in changes in gene expression and predispose to dedifferentiation/resistance has yet to be determined. Specific inhibitors of Rce1 and Icmt are being sought, due to the growth-inhibitory effect associated with loss of enzyme acitivity (Downward, 2003). The development of such agents, including the farnesyltransferase inhibitors, should take into account the effects on the lamin proteins and the organization of the nuclear lamina.

5. Conclusions

The CAAX-dependent modifications of lamin proteins affect their localization within the nucleus and are further involved in maturation of the lamina and determining protein–protein interactions. These modifications are disrupted in the context of inherited and acquired mutations in the lamin proteins or in the enzymes involved in post-translational modification, and in drug therapy. Indeed, drugs that affect these enzymes include the very new, in the form of the farnesyltransferase inhibitors, and the very old, as evidenced by methotrexate.

A more detailed analysis of the mechanisms underlying nuclear organization, including the role of CAAX modifications in health and disease states will allow a greater understanding of anti-tumour therapy in both successful and unsuccessful cancer treatment. Inherited mutations in lamin and enzyme genes may provide the model systems needed for further elucidation of these mechanisms. Furthermore, it is clear that reagents that can identify specific post-translational modifications will promote a higher-resolution understanding of the molecular events of the nuclear lamina and envelope. It is therefore likely that individual site-specific phospho-lamin antibodies will provide the next great insight into the role of interphase phosphorylation of lamins and its role in regulating nuclear events.

References

Agarwal, A.K., Fryns, J.P., Auchus, R.J. and Garg, A. (2003) Zinc metalloproteinase, ZMPSTE24, is mutated in mandibuloacral dysplasia. *Hum. Mol. Genet.* 12: 1995–2001.
Barton, R.M. and Worman, H.J. (1999) Prenylated prelamin A interacts with Narf, a novel nuclear protein. *J. Biol. Chem.* 274: 30008–30018.

Belmont, A.S., Zhai, Y. and Thilenius, A. (1993) Lamin B distribution and association with peripheral chromatin revealed by optical sectioning and electron microscopy tomography. *J. Cell Biol.* 123: 1671–1685.

Bergo, M.O., Leung, G.K., Ambroziak, P., Otto, J.C., Casey, P.J. and Young, S.G. (2000) Targeted inactivation of the isoprenylcysteine carboxyl methyltransferase gene causes mislocalization of K-Ras in mammalian cells. *J. Biol. Chem.* 275: 17605–17610.

Bergo, M.O., Leung, G.K., Ambroziak, P., Otto, J.C., Casey, P.J., Gomes, A.Q., Seabra, M.C. and Young, S.G. (2001) Isoprenylcysteine carboxyl methyltransferase deficiency in mice. *J. Biol. Chem.* 276: 5841–5845.

Bergo, M.O., Gavino, B., Ross, J., *et al.* (2002) Zmpste24 deficiency in mice causes spontaneous bone fractures, muscle weakness, and a prelamin A processing defect. *Proc. Natl Acad. Sci. USA* 99: 13049–13054.

Boyartchuk, V.L., Ashby, M.N. and Rine, J. (1997) Modulation of Ras and a-factor function by carboxyl-terminal proteolysis. *Science* 275: 1796–1800.

Boyle, S., Gilchrist, S., Bridger, J.M., Mahy, N.L., Ellis, J.A. and Bickmore, W.A. (2001) The spatial organization of human chromosomes within the nuclei of normal and emerin-mutant cells. *Hum. Mol. Genet.* 10: 211–219.

Burke, B. and Stewart, C.L. (2002) Life at the edge: the nuclear envelope and human disease. *Nat. Rev. Mol. Cell. Biol.* 3: 575–585.

Choy, E., Chiu, V.K., Silletti, J., Feoktistov, M., Morimoto, T., Michaelson, D., Ivanov, I.E.,and Philips, M.R. (1999) Endomembrane trafficking of ras: the CAAX motif targets proteins to the ER and Golgi. *Cell* 98: 69–80.

Collas, I. and Courvalin, J.C. (2000) Sorting nuclear membrane proteins at mitosis. *Trends Cell. Biol.* 10: 5–8.

Cremer, T. and Cremer, C. (2001) Chromosome territories, nuclear architecture and gene regulation in mammalian cells. *Nat. Rev. Genet.* 2: 292–301.

Daigle, N., Beaudouin, J., Hartnell, L., Imreh, G., Hallberg, E., Lippincott-Schwartz, J. and Ellenberg, J. (2001) Nuclear pore complexes form immobile networks and have a very low turnover in live mammalian cells. *J. Cell Biol.* 154: 71–84.

Dalton, M.B., Fantle, K.S., Bechtold, H.A., DeMaio, L., Evans, R.M., Krystosek, A. and Sinensky, M. (1995) The farnesyl protein transferase inhibitor BZA-5B blocks farnesylation of nuclear lamins and p21ras but does not affect their function or localization. *Cancer Res.* 55: 3295–3304.

Dolence, E.K., Dolence, J.M. and Poulter, C.D. (2000) Solid-phase synthesis of a farnesylated CaaX peptide library: inhibitors of the Ras CaaX endoprotease. *J. Comb. Chem.* 2: 522–536.

Downward, J. (2003) Targeting RAS signalling pathways in cancer therapy. *Nat. Rev. Cancer* 3: 11–22.

Dreger, M., Bengtsson, L., Schoneberg, T., Otto, H. and Hucho, F. (2001) Nuclear envelope proteomics: Novel integral membrane proteins of the inner nuclear membrane. *Proc. Natl Acad. Sci. USA* 98: 11943–11948.

Ellis, D.J., Jenkins, H., Whitfield, W.G. and Hutchison, C.J. (1997) GST-lamin fusion proteins act as dominant negative mutants in Xenopus egg extract and reveal the function of the lamina in DNA replication. *J. Cell Sci.* 110(Pt 20): 2507–2518.

Eriksson, M., Brown, W.T., Gordon, L.B., *et al.* (2003) Recurrent de novo point mutations in lamin A cause Hutchinson-Gilford progeria syndrome. *Nature* 423: 293–298.

Firmbach-Kraft, I. and Stick, R. (1995) Analysis of nuclear lamin isoprenylation in Xenopus oocytes: isoprenylation of lamin B3 precedes its uptake into the nucleus. *J. Cell Biol.* 129: 17–24.

Gerlich, D., Beaudouin, J., Kalbfuss, B., Daigle, N., Eils, R. and Ellenberg, J. (2003) Global chromosome positions are transmitted through mitosis in mammalian cells. *Cell* 112: 751–764.

Goldman, R.D., Gruenbaum, Y., Moir, R.D., Shumaker, D.K. and Spann, T.P. (2002) Nuclear lamins: building blocks of nuclear architecture. *Genes Dev.* 16: 533–547.

Ishii, K., Arib, G., Lin, C., Van Houwe, G. and Laemmli, U.K. (2002) Chromatin boundaries in budding yeast: the nuclear pore connection. *Cell* 109: 551–562.

Kilic, F., Johnson, D.A. and Sinensky, M. (1999) Subcellular localization and partial purification of prelamin A endoprotease: an enzyme which catalyzes the conversion of farnesylated prelamin A to mature lamin A. *FEBS Lett.* 450: 61–65.

Kill, I.R. and Hutchison, C.J. (1995) S-phase phosphorylation of lamin B2. *FEBS Lett.* 377: 26–30.

Kim, E., Ambroziak, P., Otto, J.C., Taylor, B., Ashby, M., Shannon, K., Casey, P.J. and Young, S.G. (1999) Disruption of the mouse Rce1 gene results in defective Ras processing and mislocalization of Ras within cells. *J. Biol. Chem.* 274: 8383–8390.

Lin, X., Jung, J., Kang, D., Xu, B., Zaret, K.S. and Zoghbi, H. (2002) Prenylcysteine carboxylmethyltransferase is essential for the earliest stages of liver development in mice. *Gastroenterology* 123: 345–351.

Maske, C.P., Hollinshead, M.S., Higbee, N.C., Bergo, M.O., Young, S.G. and Vaux, D.J. (2003) A carboxyl-terminal interaction of lamin B1 is dependent on the CAAX endoprotease Rce1 and carboxymethylation. *J. Cell Biol.* 162: 1223–1232.

Maurer-Stroh, S., Washietl, S. and Eisenhaber, F. (2003) Protein prenyltransferases. *Genome Biol.* 4: 212.

Moir, R.D., Montag-Lowy, M. and Goldman, R.D. (1994) Dynamic properties of nuclear lamins: lamin B is associated with sites of DNA replication. *J. Cell Biol.* 125: 1201–1212.

Moir, R.D., Spann, T.P., Herrmann, H. and Goldman, R.D. (2000a) Disruption of nuclear lamin organization blocks the elongation phase of DNA replication. *J. Cell Biol.* 149: 1179–1192.

Moir, R.D., Yoon, M., Khuon, S. and Goldman, R.D. (2000b) Nuclear lamins A and B1: different pathways of assembly during nuclear envelope formation in living cells. *J. Cell Biol.* 151: 1155–1168.

Otto, J.C., Kim, E., Young, S.G. and Casey, P.J. (1999) Cloning and characterization of a mammalian prenyl protein-specific protease. *J. Biol. Chem.* 274: 8379–8382.

Paddy, M.R., Belmont, A.S., Saumweber, H., Agard, D.A. and Sedat, J.W. (1990) Interphase nuclear envelope lamins form a discontinuous network that interacts with only a fraction of the chromatin in the nuclear periphery. *Cell* 62: 89–106.

Pendas, A.M., Zhou, Z., Cadinanos, J., et al. (2002) Defective prelamin A processing and muscular and adipocyte alterations in Zmpste24 metalloproteinase-deficient mice. *Nat. Genet.* 31: 94–99.

Prendergast, G.C., Davide, J.P., deSolms, S.J., Giuliani, E.A., Graham, S.L., Gibbs, J.B., Oliff, A. and Kohl, N.E. (1994) Farnesyltransferase inhibition causes morphological reversion of ras-transformed cells by a complex mechanism that involves regulation of the actin cytoskeleton. *Mol. Cell Biol.* 14: 4193–4202.

Schirmer, E.C., Florens, L., Guan, T., Yates, J.R., 3rd and Gerace, L. (2003) Nuclear membrane proteins with potential disease links found by subtractive proteomics. *Science* 301: 1380–1382.

Schirmer, E.C., Guan, T. and Gerace, L. (2001) Involvement of the lamin rod domain in heterotypic lamin interactions important for nuclear organization. *J. Cell Biol.* 153: 479–489.

Sebti, S. and Hamilton, A.D. (2001) *Farnesyltransferase Inhibitors in Cancer Therapy*. Humana Press, Totowa, New Jersey.

Silve, S., Dupuy, P.H., Ferrara, P. and Loison, G. (1998) Human lamin B receptor exhibits sterol C14-reductase activity in Saccharomyces cerevisiae. *Biochim. Biophys. Acta* 1392: 233–244.

Sinensky, M., Fantle, K., Trujillo, M., McLain, T., Kupfer, A. and Dalton, M. (1994) The processing pathway of prelamin A. *J. Cell Sci.* 107(Pt 1): 61–67.

Spann, T.P., Moir, R.D., Goldman, A.E., Stick, R. and Goldman, R.D. (1997) Disruption of nuclear lamin organization alters the distribution of replication factors and inhibits DNA synthesis. *J. Cell Biol.* 136: 1201–1212.

Spann, T.P., Goldman, A.E., Wang, C., Huang, S. and Goldman, R.D. (2002) Alteration of nuclear lamin organization inhibits RNA polymerase II-dependent transcription. *J. Cell Biol.* 156: 603–608.

Stierle, V., Couprie, J., Ostlund, C., Krimm, I., Zinn-Justin, S., Hossenlopp, P., Worman, H.J., Courvalin, J.C. and Duband-Goulet, I. (2003) The carboxyl-terminal region common to lamins A and C contains a DNA binding domain. *Biochemistry* **42**: 4819–4828.

Strahl, B.D. and Allis, C.D. (2000) The language of covalent histone modifications. *Nature* **403**: 41–45.

Stuurman, N., Heins, S. and Aebi, U. (1998) Nuclear lamins: their structure, assembly, and interactions. *J. Struct. Biol.* **122**: 42–66.

Sullivan, T., Escalante-Alcalde, D., Bhatt, H., Anver, M., Bhat, N., Nagashima, K., Stewart, C.L. and Burke, B. (1999) Loss of A-type lamin expression compromises nuclear envelope integrity leading to muscular dystrophy. *J. Cell Biol.* **147**: 913–920.

Tam, A., Nouvet, F.J., Fujimura-Kamada, K., Slunt, H., Sisodia, S.S. and Michaelis, S. (1998) Dual roles for Ste24p in yeast a-factor maturation: NH2-terminal proteolysis and COOH-terminal CAAX processing. *J. Cell Biol.* **142**: 635–649.

Winter-Vann, A.M., Kamen, B.A., Bergo, M.O., Young, S.G., Melnyk, S., James, S.J. and Casey, P.J. (2003) Targeting Ras signaling through inhibition of carboxyl methylation: an unexpected property of methotrexate. *Proc. Natl Acad. Sci. USA* **100**: 6529–6534.

Yang, L., Guan, T. and Gerace, L. (1997) Lamin-binding fragment of LAP2 inhibits increase in nuclear volume during the cell cycle and progression into S phase. *J. Cell Biol.* **139**: 1077–1087.

20

All in the family: evidence for four new LEM-domain proteins Lem2 (NET-25), Lem3, Lem4 and Lem5 in the human genome

Kenneth K. Lee and Katherine L. Wilson

1. Summary

LEM-domain proteins share a folded structure, the 'LEM-domain', which binds a conserved chromatin protein named BAF. Most LEM-domain proteins are found at the nuclear membrane, but some are nucleoplasmic. All characterized members of this family bind nuclear lamin filaments. We summarize the 'founding' LEM-domain proteins LAP2, emerin and MAN1 ('SANE' or 'XMAN' in *Xenopus*) and their emerging roles in gene regulation and nuclear assembly. These roles are placed in the context of human diseases ('laminopathies') caused by mutations in either emerin or A-type lamins. Other LEM-domain proteins might modify the phenotype or severity of human laminopathy, or cause new laminopathies. We summarize evidence that the human genome encodes at least four additional LEM-domain proteins, designated Lem2 (NET-25), Lem3, Lem4 and Lem5. Early adaptation of a consistent nomenclature, such as the "Lem" names proposed here, will facilitate rapid progress in this field. Further investigation of 'founder' and novel members of this family will be important to understand nuclear structure, and presents new opportunities to understand human disease.

2. Introduction to LEM-domain proteins, their functions in the nucleus and links to 'laminopathy' disease

The nuclear envelope (NE) consists of two membranes (inner and outer), nuclear pore complexes and an underlying network of filaments formed by lamin proteins (Cohen

The Nuclear Envelope, edited by D.E. Evans, C. Hutchison & J.A. Bryant.
© 2004 Garland Science/BIOS Scientific Publishers

et al., 2001). Molecules must pass through pore complexes to enter or exit the nucleus; passage is regulated by import and export signals. The architecture of the NE depends on stable lamin filaments, which are reviewed elsewhere (Goldman *et al.*, 2002; Gruenbaum *et al.*, 2003; Hutchison, 2002). Humans have three lamin genes; two encode essential B-type lamins, and one encodes non-essential A-type lamins (Goldman *et al.*, 2002). Lamins determine the shape and mechanical resilience of the nucleus, in part through binding to various proteins anchored at the nuclear inner membrane (IM) (Burke and Stewart, 2002; Foisner, 2003; Zastrow *et al.*, 2004). Many of these IM proteins in turn also bind to chromatin. In differentiated cells, the 'peripheral' chromatin near the IM is often transcriptionally silenced. During mitosis all aspects of nuclear architecture are lost when the nucleus disassembles. After mitosis, the nuclear envelope reforms around daughter chromosomes and then mediates import and nuclear 'growth' to reach full size (Gant and Wilson, 1997). Nuclear membrane proteins are involved from the earliest stage of assembly, starting when membranes attach to the surface of daughter chromosomes (Wilson and Newport, 1988).

Nuclear membrane proteins are a new frontier in cell biology, with the potential to reveal the functional boundaries of the 'chromatin world'. Until recently, few IM proteins were known in metazoans. The first IM proteins to be discovered, lamin B receptor (LBR), lamina associated polypeptide-1 (LAP1) and LAP2, were thought to have structural roles because they interacted with lamins and chromatin (Foisner and Gerace, 1993; Gerace and Blobel, 1982; Östlund and Worman, 2003; Worman *et al.*, 1990). Then came more IM proteins and the discovery that inherited tissue-specific disorders were caused by defects in NE proteins. The X-linked recessive form of Emery–Dreifuss muscular dystrophy (EDMD) was linked to the loss of emerin, a previously unknown lamin-binding membrane protein related to LAP2β (Bione *et al.*, 1994; Burke and Stewart, 2002; Manilal *et al.*, 1996; Nagano *et al.*, 1996; Bengtsson and Wilson, 2004). Then the more rare autosomal dominant form of EDMD was linked to missense mutations in A-type lamins (Bonne *et al.*, 1999). In rapid succession, a stunning variety of tissue-specific diseases affecting muscle, heart, fat, neurons or bone were likewise linked to A-type lamins (reviewed by Östlund and Worman, 2003). One recent 'laminopathy', Hutchinson–Gilford progeria syndrome, mimics premature ageing and causes death typically in the mid-teens (De Sandre-Giovannoli *et al.*, 2003; Eriksson *et al.*, 2003). A second form of progeria, atypical Werner syndrome, is also linked to A-type lamins (Chen *et al.*, 2003).

How do we account for these diseases, when the functional repertoires of nuclear membrane proteins and the lamina network are unknown? To find molecular mechanisms and help patients, we must understand the functions of each protein implicated in disease, and integrate this knowledge into the three-dimensional network that constitutes nuclear structure (Wilson *et al.*, 2001). Characterized nuclear membrane proteins now include LBR, LAP1 (three isoforms), LAP2 (over six isoforms), otefin, MAN1, nurim, ring finger binding protein (RFBP), LUMA, AKAP149, UNC-84 and nesprin-1α (reviewed by Cohen *et al.*, 2001; Foisner, 2003; Gruenbaum *et al.*, 2003; Zhang *et al.*, 2002). Nuclear membrane proteins are diverse, and there is growing evidence that each one has multiple roles. Many characterized IM proteins bind directly to A- or B-type lamins (or both), as well as a chromatin partner (Foisner, 2003; Gruenbaum *et al.*, 2003). Most IM proteins have a single transmembrane domain.

However, there are now examples of IM proteins with two (MAN1), four (nurim, LUMA) or eight (LBR, RFBP) transmembrane domains. The functions of most IM proteins are still unknown. The magnitude of this task and the depth of our ignorance were revealed by a proteomic study of liver tissue, which identified more than 60 predicted new IM proteins (Schirmer *et al.*, 2003). Other tissues, particularly those affected by laminopathy, are likely to express additional novel IM proteins.

Much interest has focused on emerin (Bengtsson and Wilson, 2004). As noted above, recessive mutations in emerin or dominant mutations in lamins A/C can cause the same disease, Emery–Dreifuss muscular dystrophy. Thus, emerin can provide unique insight into the mechanism of EDMD. Besides emerin, we know of no other nuclear membrane protein that can independently give rise to *LMNA*-linked laminopathy. However, the phenotype and severity of laminopathies can vary even within the same family, suggesting that many other genes influence (or depend on) lamina function (Bonne *et al.*, 2003; Morris, 2001). Interestingly emerin belongs to the 'LEM-domain' family of nuclear proteins, other members of which might modulate the severity or phenotype of laminopathy. The first recognized members of this family (LAP2, Emerin, MAN1) share a conserved ~40-residue motif, termed the 'LEM' domain (Lin *et al.*, 2000). Other LEM-domain proteins include otefin and two alterna-tively-spliced isoforms of a protein named Bocksbeutel in *Drosophila* (Goldberg *et al.*, 1998; Wagner *et al.*, 2004) and Lem3, which is conserved from *Caenorhabditis elegans* to humans (Lee *et al.*, 2000) but has not yet been characterized or localized. The only known biological function of the LEM domain is to interact with BAF (barrier-to-autointegration factor), a highly conserved chromatin protein that binds DNA and is essential for cell viability (Lee and Craigie, 1998; Zheng *et al.*, 2000; reviewed by Segura-Totten and Wilson, 2004). Direct binding to BAF, mediated by the LEM domain, has been shown *in vitro* for LAP2 (Furukawa, 1999; Shumaker *et al.*, 2001), emerin (Holaska *et al.*, 2003; Lee *et al.*, 2001) and MAN1 (Liu *et al.*, 2003). In living cells, GFP-BAF binds directly but transiently to emerin at the inner nuclear membrane (Shimi *et al.*, 2004; reviewed by Bengtsson and Wilson, 2004).

It is unclear how BAF fits the disease picture. BAF is small (89 residues), forms dimers and binds double-stranded DNA *in vitro* (Lee and Craigie, 1998). When mixed with short (21 bp) DNA molecules, BAF dimers form discrete nucleoprotein complexes consisting of six BAF dimers and at least two molecules of DNA (Zheng *et al.*, 2000). BAF can co-complex with emerin and lamin A *in vitro* (Holaska *et al.*, 2003), and in living cells is required to recruit and stabilize emerin during nuclear assembly (Haraguchi *et al.*, 2001). BAF is highly mobile during interphase, and its interactions with emerin during interphase appear too transient to be 'structural' in any conventional sense (Shimi *et al.*, 2004). Moreover, BAF binds directly to home-odomain proteins such as Crx, and represses Crx-dependent gene expression in retinal tissue (Wang *et al.*, 2002). Unlike emerin, which is dispensable at the cellular level, BAF is essential in embryonic cells (Furukawa *et al.*, 2003; Zheng *et al.*, 2000). Further studies are clearly needed to understand if BAF plays a role in disease.

The LEM domains of LAP2 and emerin are ~40% identical and adapt similar struc-tural folds, as shown by NMR (Cai *et al.*, 2001; Laguri *et al.*, 2001; Wolfe *et al.*, 2001). This fold is almost identical to a fold previously seen in a bacterial RNA-binding protein (transcription termination factor, rho) and a DNA-binding phage protein (T4 endonuclease VII), despite no recognizable amino acid sequence homology (Cai *et al.*,

2001). Interestingly, LAP2 also has a 'LEM-like' domain, which adapts the same fold but has different surface-exposed residues that confer binding to DNA, rather than BAF (Cai *et al.*, 2001). Site-directed mutagenesis of LAP2 and emerin was used to show that LEM-domain residues are essential for binding to BAF (Lee *et al.*, 2001; Shumaker *et al.*, 2001). A complementary analysis of 25 BAF missense mutants identified residues essential for binding to emerin, and mapped these residues to a 'valley' on the BAF dimer surface (Cai *et al.*, 1998; Segura-Totten *et al.*, 2002; Umland *et al.*, 2000).

The function of emerin appears to overlap with both LAP2β and MAN1, at several levels. All three LEM proteins interact with lamins and BAF, and two (emerin and LAP2β) also interact with a transcription repressor named 'germ cell-less' *in vitro* (Holaska *et al.*, 2003; Liu *et al.*, 2003; Nili *et al.*, 2001). Furthermore emerin and MAN1 have overlapping essential functions in early *C. elegans* embryos, since the combined loss of emerin plus 90% reduction of MAN1 is lethal at the 100-cell stage (Liu *et al.*, 2003). One very exciting new development is that MAN1 regulates signalling. Two independent laboratories working with *Xenopus* have found that probable orthologues of MAN1 (see below) known as 'XMAN' or 'SANE' bind directly to Smad1, and antagonize bone morphogenetic protein (BMP)-dependent signalling during embryogenesis (Osada *et al.*, 2003; Raju *et al.*, 2003). Furthermore MAN1 ('SANE') also inhibits BMP-mediated osteoblast differentiation in cell culture (Raju *et al.*, 2003). The amino acid sequences of *Xenopus* SANE (Raju *et al.*, 2003) and XMAN (Osada *et al.*, 2003) are nearly identical to each other; they probably originate from different MAN1 alleles in *Xenopus*, which is partially tetraploid.

In addition to binding Smad1, MAN1/SANE co-immunoprecipitated with BMP type-I receptors when both proteins were over-expressed in *Xenopus* oocytes (Raju *et al.*, 2003). BMP receptors are located at the cell surface, not the nuclear IM, thus the physiological relevance of receptor binding to SANE is unknown. Raju *et al.* (2003) speculated that *Xenopus* SANE has only one transmembrane span and that its N-terminal domain was exposed outside the cell, but did not localize endogenous SANE. However, we find that both *Xenopus* proteins align well against full-length human MAN1, with remarkable conservation in the C-terminal domain and both transmembrane domains, suggesting to us that SANE and XMAN1 are probably *Xenopus* alleles of human MAN1 (authors' unpublished data). Nonetheless functional criteria will be needed to determine if SANE and XMAN1 are orthologous to MAN1 *versus* Lem2.

Over 50% of patients with clinically diagnosed EDMD have normal emerin and normal A-type lamins (Bonne *et al.*, 2003), suggesting that other LEM-domain proteins might be involved in EDMD or other forms of laminopathy. In particular the Smad1-inhibitory functions of XMAN1/SANE and its roles in bone formation might be specifically relevant to bone defects seen in two laminopathies, mandibuloacral dysplasia (Agarwal *et al.*, 2003; Novelli *et al.*, 2002) and Hutchinson–Gilford progeria syndrome (Bergo *et al.*, 2002; Eriksson *et al.*, 2003).

3. Evidence for more LEM-domain proteins in the human genome

In *C. elegans* the LEM-domain family appears to consist of three proteins: emerin, MAN1 and an unusual third protein named Lem3 (Lee *et al.*, 2000). The Lem3 protein is predicted to be soluble (no transmembrane domain) and its LEM domain is located near the C- rather than N-terminus (Lee *et al.*, 2000). Thus Lem3 could either function in the nucleoplasm (like LAP2α in humans; Dechat *et al.*, 1998, 2000), or the cyto-plasm. ESTs homologous to Lem3 are present in the human database, suggesting Lem3 is conserved in humans (*Table 1*).

A *Drosophila* protein named otefin has a divergent LEM-domain (Goldberg *et al.*, 1998), which has not yet been tested for binding to BAF. Another nuclear membrane protein, nesprin-1α, was reported to have a 'broken LEM motif' (Mislow *et al.*, 2002a). However when tested, no binding to BAF was detected *in vitro* (Mislow *et al.*, 2002b).

To identify novel LEM-domain proteins, we searched the human database using a relatively strict consensus sequence, based on residues conserved in both the human and *C. elegans* orthologues of emerin, MAN1 and Lem3 (Lee *et al.*, 2000, 2001). This search yielded three additional LEM-domain genes, here designated *LEM2, LEM4* and *LEM5* (*Figure 1; Table 1*). The structure and aligned amino acid sequences of numerous LEM-domains can be viewed on the web (http://pfam.wustl.edu/cgi-bin/getdesc?acc=PF03020).

LEM2 is located on chromosome 6 (NM_181336) at position 6p21, and encodes a predicted 503-residue 56-kDa polypeptide with an N-terminal LEM domain and two predicted transmembrane domains (*Figure 1*). The Lem2 protein is ~40% similar to human MAN1, which also has two transmembrane domains (*Figure 1*). We view human Lem2 as the true orthologue of *C. elegans* MAN1 (encoded by *lem-2*), since their lengths are similar and both lack the C-terminal tail extension seen in human MAN1(authors, unpublished data). Human Lem2

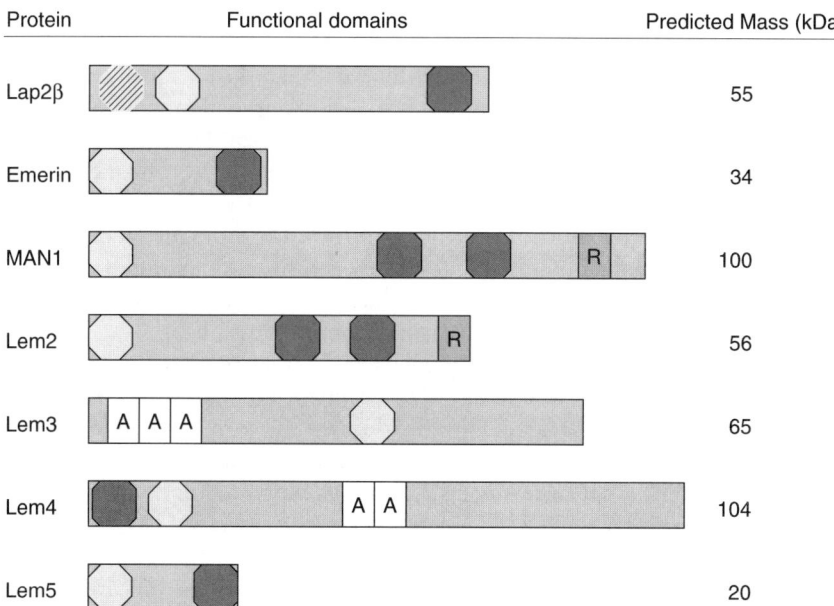

Figure 1. *Schematic diagrams of human LEM-domain proteins. Known or predicted transmembrane domain(s)are depicted in black. LEM domains are depicted as yellow hexagons; the DNA-binding 'LEM-like' domain in LAP2 is striped. MAN1 and Lem2 are highly related to each other, with two predicted transmembrane domains, and a C-terminal RNA-Recognition Motif (RRM; here denoted "R"; Birney et al., 1993). Lem3 and Lem4 have three and two ankyrin repeats ("A"), respectively. See Table 1 for gene accession numbers.*

Table 1. Proposed nomenclature for human LEM-domain proteins.

Protein	Chromosome	Accession #	Other names	Comments*
LAP2	12q22	NP_003267	Thymopoietin	At least 6 splicing isoforms
Emerin	Xq28	NP_000108	*STA*, EDMD	Loss causes EDMD
MAN1	12q14	NP_055134	XMAN?, SANE?	RRM domain
Lem2	6p21.31	NP_851853	LEMD2, dJ482C21.1	RRM domain
Lem3	19p13.12	NP_689576	FLJ39369	Ankyrin repeats; no TM
Lem4	12q24.33	AAH43157		Ankyrin repeats; N-term TM
Lem5	1q32.1	XP_371338	LOC93273	C-terminal TM

*TM, transmembrane domain; EDMD; Emery-Dreifuss muscular dystrophy; RRM, RNA-recognition motif (Birney *et al.*, 1993).

is expressed in many human tissues, according to the EST database (http://www.ncbi.nlm.nih.gov/UniGene/clust.cgi?ORG=Hs&CID=444845). There is also strong evidence that Lem2 is conserved in mammals. Rat peptides homologous to human Lem2 were identified by proteomic analysis of the rat liver nuclear membrane, and annotated as nuclear envelope transmembrane protein 25 (NET-25) by Schirmer *et al.* (2003). The human database currently annotates *LEM2* as 'LMD2'.

LEM4 is located on chromosome 12 (AAH43157) at position 12q24.33, and encodes a predicted 938-residue polypeptide. Interestingly *LEM4* is located near the genes encoding LAP2 and MAN1 (positions 12q22 and 12q14, respectively; see Table 1). Lem4 is the only LEM-domain protein with an N-terminal transmembrane domain. Lem4 also has two predicted ankyrin repeats, in common with Lem3, which has three ankyrin repeats near its N-terminus (Figure 1). Ankyrin repeats are 33-residue motifs, frequently clustered, which fold cooperatively and mediate protein-protein interactions. We speculate that Lem3 and Lem4 might each interact with a common partner via this motif. Mice encode a *LEM4* orthologue (accession number NP_082198.1) but so far no orthologues have been found in *Xenopus* or *Drosophila*. Lem4 is expressed ubiquitously at similar levels in a variety of mouse and human tissues (J. Gotzmann and R. Foisner, personal communication).

LEM5 is located on chromosome 1 (XP_371338) at position 1q32.1. This gene encodes a predicted 181-residue 20-kDa polypeptide with an N-terminal LEM domain and C-terminal transmembrane domain (*Figure 1*). Outside its LEM-domain, Lem5 does not resemble any other LEM protein and its subcellular location is unknown. The EST database suggests Lem5 is expressed in many human tissues (http://www.ncbi.nlm.nih.gov/UniGene/clust.cgi?ORG=Hs&CID=104830&MAX EST=27). Based on structural predictions made by SWISS-prot software, the LEM-domain of Lem5 folds like a *bona fide* LEM domain, whereas the LEM domains of Lem3 and Lem2 (surprisingly) cannot. These predictions obviously need to be tested by determining if Lem3 and Lem2 can bind BAF. It would be interesting to find LEM-domain protein(s) that bind a ligand other than BAF, since this would expand the range of functions of the LEM-domain protein family. Lem5 was not detected in the proteomic study of rat liver nuclear membrane proteins reported by Schirmer *et al.* (2003). While negative results are uninformative, we suspect that Lem5 is either not located at the nuclear envelope, or not expressed in liver cells. Supporting the latter possibility, we did not find ESTs for Lem5 in either heart or liver.

4. Other new LEM-domain proteins?

Schirmer *et al.* (2003) identified NET-66 as a nuclear envelope protein with limited similarity to MAN1. We find that NET-66 aligns with a region of human MAN1 *outside* the LEM domain of MAN1, but we found no actual LEM domain in NET-66. Only a few partial ESTs are available for NET-66 in the database, thus more information is needed. Note that our search that yielded Lem2, Lem3, Lem4 and Lem5 was based on highly conserved residues in the LEM domain. Future, less-stringent searches might reveal additional *bona fide* LEM-domain proteins.

There is no defined roadmap for studying novel LEM-domain proteins, aside from the ever-useful touchstones of subcellular localization, binding partners and null phenotype(s). 'Founder' LEM-domain proteins have multiple roles (e.g., nuclear architecture and gene regulation) and function at multiple times (e.g., embryonic development and adult somatic cells; Foisner, 2003; Schoft et al., 2003) possibly in specific tissues (Berger *et al.*, 1996; Harris *et al.*, 1994; Lang *et al.*, 1999; Osada *et al.*, 2003; Raju *et al.*, 2003) or specific stages of the cell cycle (Furukawa *et al.*, 2003; Martins *et al.*, 2003; Yang *et al.*, 1997). For example, loss of BAF is lethal in *Drosophila* embryos due to an apparent mitotic arrest (Furukawa *et al.*, 2003). Do cells arrest because BAF is required for nuclear assembly, or because BAF regulates genes required for mitosis, or both? The same uncertainty applies to cells with lethal reductions of both emerin and MAN1 (Liu *et al.*, 2003). Novel members of the LEM-domain family might have similarly complicated roles. To understand this intriguing group of nuclear proteins will require the combined efforts of many laboratories and the use of many complementary systems and approaches. The identification of new LEM-domain family members is just the beginning. *Bon apetit*!

Acknowledgements

We are indebted to Josef Gotzmann, Roland Foisner and Malini Mansharamani for discussions and comments on this manuscript. K.K.L. is a Damon Runyon Cancer Research Fellow. This work was supported by a grant from the Muscular Dystrophy Association USA (K.L.W.) and the 2003 Scott B. Deutschman Memorial Research Award of the American Heart Association (to K.L.W.).

References

Agarwal, A.K., Fryns, J.-P., Auchus, R.J. and Garg, A. (2003) Zinc metalloproteinase, *ZMPSTE24*, is mutated in mandibuloacral dysplasia. *Hu. Mol. Genet.* **12**: 1995–2001.

Bengtsson, L. and Wilson, K.L. (2004) Multiple and surprising new functions for emerin, a nuclear membrane protein. *Curr. Opin. Cell Biol.* **16**: 73–79.

Berger, R., Theodor, L., Shoham, J., Gokkel, E., Brok-Simoni, F., Avraham, K.V., Copeland, N.G., Jenkins, N.A., Rechavi, G. and Simon, A.J. (1996) The characterization and localization of the mouse thymopoietin/lamina-associated polypeptide 2 gene and its alternatively spliced products. *Genome Res.* **6**: 361–370.

Bergo, M.O., Gavino, B., Ross, J., *et al*. (2002) Zmpste24 deficiency in mice causes spontaneous bone fractures, muscle weakness, and a prelamin A processing defect. *Proc. Natl Acad. Sci. USA* 99: 13049–13054.

Bione, S., Maestrini, E., Rivella, S., Mancini, M., Regis, S., Romeo, G. and Toniolo, D. (1994) Identification of a novel X-linked gene responsible for Emery-Dreifuss muscular dystrophy. *Nat. Genet*. 8: 323–327.

Birney, E., Kumar, S. and Krainer, A.R. (1993) Analysis of the RNA-recognition motif and RS and RGG domains: conservation in metazoan pre-mRNA splicing factors. *Nucleic Acids Res*. 21: 5803–5816.

Bonne, G., Di Barletta, M.R., Varnous, S., *et al*. (1999) Mutations in the gene encoding lamin A/C cause autosomal dominant Emery-Dreifuss muscular dystrophy. *Nat. Genet*. 21: 285–288.

Bonne, G., Yaou, R.B., Beroud, C., *et al*. (2003) 108th ENMC International Workshop, 3rd Workshop of the MYO-CLUSTER project: EUROMEN, 7th International Emery-Dreifuss Muscular Dystrophy (EDMD) Workshop, 13–15 September 2002, Naarden, The Netherlands. *Neuromuscul Disord*. 13: 508–515.

Burke, B. and Stewart, C.L. (2002) Life at the edge: the nuclear envelope and human disease. *Nat. Rev. Mol. Cell. Biol*. 3: 575–585.

Cai, M., Huang, Y., Zheng, R., Wei, S.Q., Ghirlando, R., Lee, M.S., Craigie, R., Gronenborn, A.M. and Clore, G.M. (1998) Solution structure of the cellular factor BAF responsible for protecting retroviral DNA from autointegration. *Nat. Struc. Biol*. 5: 903–909.

Cai, M., Huang, Y., Ghirlando, R., Wilson, K.L., Craigie, R. and Clore, G.M. (2001) Solution structure of the constant region of nuclear envelope protein LAP2 reveals two LEM-domain structures: one binds BAF and the other binds DNA. *EMBO J*. 20: 4399–4407.

Chen, L., Lee, L., Kudlow, B.A., *et al*. (2003) *LMNA* mutations in atypical Werner's syndrome. *Lancet* 362: 440–445.

Cohen, M., Lee, K.K., Wilson, K.L. and Gruenbaum, Y. (2001) Transcriptional repression, apoptosis, human disease and the functional evolution of the nuclear lamina. *Trends Biochem. Sci*. 26: 41–47.

De Sandre-Giovannoli, A., Bernard, R., Cau, P., *et al*. (2003) Lamin A truncation in Hutchinson-Gilford progeria. *Science* 300: 2055.

Dechat, T., Gotzmann, J., Stockinger, A., Harris, C.A., Talle, M.A., Siekierka, J.J. and Foisner, R. (1998) Detergent-salt resistance of LAP2α in interphase nuclei and phosphory-lation-dependent association with chromosomes early in nuclear assembly implies functions in nuclear structure dynamics. *EMBO J*. 17: 4887–4902.

Dechat, T., Korbei, B., Vaughan, O.A., Vlcek, S., Hutchison, C.J. and Foisner, R. (2000) Lamina-associated polypeptide 2α binds intranuclear A-type lamins. *J. Cell Sci*. 113: 3473–3484.

Eriksson, M., Brown, W.T., Gordon, L.B., *et al*. (2003) Recurrent de novo point mutations in lamin A cause Hutchinson-Gilford progeria syndrome. *Nature* 423: 293–298.

Foisner, R. (2003) Cell cycle dynamics of the nuclear envelope. *ScientificWorldJournal* 3: 1–20.

Foisner, R. and Gerace, L. (1993) Integral membrane proteins of the nuclear envelope interact with lamins and chromosomes, and binding is modulated by mitotic phosphorylation. *Cell* 73: 1267–1279.

Furukawa, K. (1999) LAP2 binding protein 1 (L2BP1/BAF) is a candidate mediator of LAP2-chromatin interaction. *J. Cell Sci*. 112: 2485–2492.

Furukawa, K., Sugiyama, S., Osouda, S., Goto, H., Inagaki, M., Horigome, T., Omata, S., McConnell, M., Fisher, P.A. and Nishida, Y. (2003) Barrier-to-autointegration factor plays crucial roles in cell cycle progression and nuclear organization in *Drosophila*. *J. Cell Sci*. 116: 3811–3823.

Gant, T.M. and Wilson, K.L. (1997) Nuclear assembly. *Annu. Rev. Cell Dev. Biol*. 13: 669–695.

Gerace, L. and Blobel, G. (1982) Nuclear lamina and the structural organization of the nuclear envelope. *Cold Spring Harb. Symp. Quant. Biol*. 46: 967–978.

Goldberg, M., Lu, H., Stuurman, N., Ashery-Padan, R., Weiss, A.M., Yu, J., Bhattacharyya, D., Fisher, P.A., Gruenbaum, Y. and Wolfner, M.F. (1998) Interactions among *Drosophila* nuclear envelope proteins lamin, otefin, and YA. *Mol. Cell. Biol.* **18**: 4315–4323.

Goldman, R.D., Gruenbaum, Y., Moir, R.D., Shumaker, D.K. and Spann, T.P. (2002) Nuclear lamins: building blocks of nuclear architecture. *Genes Dev.* **16**: 533–547.

Gruenbaum, Y., Goldman, R.D., Meyuhas, R., Mills, E., Margalit, A., Fridkin, A., Dayani, Y., Prokocimer, M. and Enosh, A. (2003) The nuclear lamina and its functions in the nucleus. *Int. Rev. Cytol.* **226**: 1–62.

Haraguchi, T., Koujin, T., Segura-Totten, M., Lee, K.K., Matsuoka, Y., Yoneda, Y., Wilson, K.L. and Hiraoka, Y. (2001) BAF is required for emerin assembly into the reforming nuclear envelope. *J. Cell Sci.* **114**: 4575–4585.

Harris, C.A., Andryuk, P.J., Cline, S., Chan, H.J., Natarajan, A., Siekierka, J.J. and Goldstein, G. (1994) Three distinct human thymopoietins are derived from alternatively spliced mRNAs. *Proc. Natl Acad. Sci. USA* **91**: 6283–6287.

Holaska, J.M., Lee, K.K, Kowalski, A.K. and Wilson, K.L. (2003) Transcriptional repressor germ cell-less (GCL) and barrier to autointegration factor (BAF) compete for binding to emerin *in vitro*. *J. Biol. Chem.* **278**: 6969–6975.

Hutchison, C.J. (2002) Lamins: building blocks or regulators of gene expression? *Nat. Rev. Mol. Cell Biol.* **3**: 848–858.

Laguri, C., Gilquin, B., Wolff, N., Romi-Lebrun, R., Courchay, K., Callebaut, I., Worman, H.J. and Zinn-Justin, S. (2001) Structural characterization of the LEM motif common to three human inner nuclear membrane proteins. *Structure (Camb.)* **9**: 503–511.

Lang, C., Paulin-Levasseur, M., Gajewski, A., Alsheimer, M., Benavente, R., and Krohne, G. (1999) Molecular characterization and developmentally regulated expression of *Xenopus* lamina-associated polypeptide 2 (XLAP2). *J. Cell Sci.* **112**: 749–759.

Lee, K.K., Gruenbaum, Y., Spann, P., Liu, J. and Wilson, K.L. (2000) *C. elegans* nuclear envelope proteins emerin, MAN1, lamin, and nucleoporins reveal unique timing of nuclear envelope breakdown during mitosis. *Mol. Biol. Cell* **11**: 3089–3099.

Lee, K.K., Haraguchi, T., Lee, R.S., Koujin, T., Hiraoka, Y. and Wilson, K.L. (2001) Distinct functional domains in emerin bind lamin A and DNA-bridging protein BAF. *J. Cell Sci.* **114**: 4567–4573.

Lee, M.S. and Craigie, R. (1998) A previously unidentified host protein protects retroviral DNA from autointegration. *Proc. Natl Acad. Sci. USA* **95**: 1528–1533.

Lin, F., Blake, D.L., Callebaut, I., Skerjanc, I.S., Holmer, L., McBurney, M.W., Paulin-Levasseur, M. and Worman, H.J. (2000) MAN1, an inner nuclear membrane protein that shares the LEM domain with lamina-associated polypeptide 2 and emerin. *J. Biol. Chem.* **275**: 4840–4847.

Liu, J., Ben-Shahar, T.R., Riemer, D., Treinin, M., Spann, P., Weber, K., Fire, A. and Gruenbaum, Y. (2000) Essential roles for *Caenorhabditis elegans* lamin gene in nuclear organization, cell cycle progression, and spatial organization of nuclear pore complexes. *Mol. Biol. Cell* **11**: 3937–3947.

Liu, J., Lee, K.K., Segura-Totten, M., Neufeld, E., Wilson, K.L. and Gruenbaum, Y. (2003) MAN1 and emerin have overlapping function(s) essential for chromosome segregation and cell division in *Caenorhabditis elegans*. *Proc. Natl Acad. Sci. USA* **100**: 4598–4603.

Manilal, S., Nguyen, T.M., Sewry, C.A. and Morris, G.E. (1996) The Emery-Dreifuss muscular dystrophy protein, emerin, is a nuclear membrane protein. *Hum. Mol. Genet.* **5**: 801–808.

Martins, S., Eikvar, S, Furukawa, K. and Collas, P. (2003) HA95 and LAP2β mediate a novel chromatin-nuclear envelope interaction implicated in initiation of DNA replication. *J. Cell Biol.* **160**: 177–188.

Mislow, J.M., Kim, M.S., Davis, D.B. and McNally, E.M. (2002a) Myne-1 associates with lamin-A/C. *J. Cell Sci.* **115**: 61–70.

Mislow, J.M., Holaska, J.M., Kim, M.S, Lee, K.K., Segura-Totten, M., Wilson, K.L. and McNally, E.M. (2002b) Nesprin-1α self-associates and binds directly to emerin and lamin A in vitro. *FEBS Lett.* **525**: 135–140.

Morris, G.E. (2001) The role of the nuclear envelope in Emery-Dreifuss muscular dystrophy. *Trends Mol. Med.* **7**: 572–577.

Nagano, A., Koga, R., Ogawa, M., Kurano, Y., Kawada, J., Okada, R., Hayashi, Y.K., Tsukahara, T. and Arahata, K. (1996) Emerin deficiency at the nuclear membrane in patients with Emery-Dreifuss muscular dystrophy. *Nat. Genet.* **12**: 254–259.

Nili, E., Cojocaru, G.S., Kalma, Y., *et al.* (2001) Nuclear membrane protein LAP2β mediates transcriptional repression alone and together with its binding partner GCL (germ-cell-less). *J. Cell Sci.* **114**: 3297–3307.

Novelli, G., Muchi, A., Sangiuolo, F., *et al.* (2002) Mandibuloacral dysplasia is caused by a mutation in LMNA-encoding lamin A/C. *Am. J. Hum. Genet.* **71**: 426–431.

Osada, S., Ohmori, S.Y. and Taira, M. (2003) XMAN1, an inner nuclear membrane protein, antagonizes BMP signaling by interacting with Smad1 in *Xenopus* embryos. *Development* **130**: 1783–1794.

Östlund, C. and Worman, H.J. (2003) Nuclear envelope proteins and neuromuscular diseases. *Muscle Nerve* **27**: 393–406.

Raju, G.P., Dimova, N., Klein, P.S. and Huang, H.C. (2003) SANE, a novel LEM domain protein, regulates bone morphogenetic protein signaling through interaction with Smad1. *J. Biol. Chem.* **278**: 428–437.

Schirmer, E.C., Florens, L., Guan, T., Yates, J.R. and Gerace, L. (2003) Nuclear membrane proteins with potential disease links found by subtractive proteomics. *Science* **301**: 1380–1382.

Schoft, V.K., Beauvais, A.J., Lang, C., Gajewski, A., Prufert, K., Winkler, C., Akimenko, M.A., Paulin-Levasseur, M. and Krohne G. (2003) The lamina-associated polypeptide 2 (LAP2) isoforms beta, gamma and omega of zebrafish: developmental expression and behavior during the cell cycle. *J. Cell Sci.* **116**: 2505–2517.

Segura-Totten, M., Kowalski, A.K., Craigie, R. and Wilson, K.L. (2002) Barrier-to-autointegration factor: major roles in chromatin decondensation and nuclear assembly. *J. Cell Biol.* **158**: 475–485.

Segura-Totten, M. and Wilson, K.L. (2004) BAF: roles in chromatin, nuclear structure and retrovirus integration. *Trends Cell Biol.* **147**: 31–41.

Shimi, T., Koujin, T., Segura-Totten, M., Wilson, K.L., Haraguchi, T. and Hiraoka, Y. (2004) Dynamic interaction between BAF and emerin revealed by FRAP, FLIP and FRET analyses in living HeLa cells. *J. Struct. Biol.* (in press)

Shumaker, D.K., Lee, K.K., Tanhehco, Y.C., Craigie, R. and Wilson, K.L. (2001) LAP2 binds to BAF-DNA complexes: requirement for the LEM domain and modulation by variable regions. *EMBO J.* **20**: 1754–1764.

Umland, T.C., Wei, S.-Q., Craigie, R. and Davies, D.R. (2000) Structural basis of DNA bridging by barrier-to-autointegration factor. *Biochemistry* **39**: 9130–9138.

Vlcek, S., Just, H., Dechat, T. and Foisner, R. (1999) Functional diversity of LAP2α and LAP2β in postmitotic chromosome association is caused by an α-specific nuclear targeting domain. *EMBO J.* **18**: 6370–6384.

Vlcek, S., Korbei, B. and Foisner, R. (2002) Distinct functions of the unique C terminus of LAP2α in cell proliferation and nuclear assembly. *J. Biol. Chem.* **277**: 18898–18907.

Wagner, N., Schmitt, J. and Krohne, G. (2004) Two novel LEM-domain proteins are splice products of the annotated *Drosophila melanogaster* gene CG9424 (Bocksbeutel). *Eur. J. Cell Biol.* **82**: 605–616.

Wang, X., Xu, S., Rivolta, C., *et al.* (2002) Barrier to autointegration factor interacts with the cone-rod homeobox and represses its transactivation function. *J. Biol. Chem.* **277**: 43288–43300.

Wilson, K.L. and Newport, J.W. (1988) A trypsin-sensitive receptor on membrane vesicles is required for nuclear envelope formation in vitro. *J. Cell Biol.* **107**: 57–68.

Wilson, K.L, Zastrow, M. and Lee, KK. (2001) Lamins and disease: insights into nuclear infrastructure. *Cell* 104: 647–650.

Wolfe, N., Gilquin, B., Courchay, K., Callebaut, I., Worman, H.J. and Zinn-Justin, S. (2001) Structural analysis of emerin, an inner nuclear membrane protein mutated in X-linked Emery-Dreifuss muscular dystrophy. *FEBS Lett.* 501: 171–176.

Worman, H.J., Evans, C.D. and Blobel, G. (1990) The lamin B receptor of the nuclear envelope inner membrane: a polytopic protein with eight potential transmembrane domains. *J. Cell Biol.* 111: 1535–1542.

Yang, L., Guan. T. and Gerace, L. (1997) Lamin-binding fragment of LAP2 inhibits increase in nuclear volume during the cell cycle and progression into S phase. *J. Cell Biol.* 139: 1077–1087.

Zastrow, M.S., Vlcek, S. and Wilson, K.L. (2004) Proteins that bind A-type lamins: integrating isolated clues. *J. Cell Sci.* 117: 979-987.

Zhang, Q., Ragnauth, C., Greener, M.J., Shanahan, C.M. and Roberts, R.G. (2002) The nesprins are giant actin-binding proteins, orthologous to *Drosophila melanogaster* muscle protein MSP-300. *Genomics* 80: 473–481.

Zheng, R., Ghirlando, R., Lee, M.S., Mizuuchi, K., Krause. M. and Craigie, R. (2000) Barrier-to-autointegration factor (BAF) bridges DNA in a discrete, higher-order nucleoprotein complex. *Proc. Natl Acad. Sci. USA* 97: 8997–9002.

Index